James D. Ford • Lea Berrang-
Editors

Climate Change Adaptation in Developed Nations

From Theory to Practice

 Springer

Editors
Dr. James D. Ford
Department of Geography
McGill University
805 Sherbrooke Street West
Montréal, QC H3A 2K6
Canada
james.ford@mcgill.ca

Dr. Lea Berrang-Ford
Department of Geography
McGill University
805 Sherbrooke Street West
Montréal, QC H3A 2K6
Canada
lea.berrangford@mcgill.ca

Editorial Assistants
Michelle Maillet
Department of Geography
McGill University
Montreal, QC
Canada
michelle.maillet@mail.mcgill.ca

Carolyn Poutiainen
Department of Geography
McGill University
Montreal, QC
Canada
carolyn.poutiainen@mail.mcgill.ca

ISSN 1574-0919
ISBN 978-94-007-0566-1   e-ISBN 978-94-007-0567-8
DOI 10.1007/978-94-007-0567-8
Springer Dordrecht Heidelberg London New York

Library of Congress Control Number: 2011931286

Cover illustration: Cover image courtesy of Martin Flamand http://www.martinflamand.com/

Cover design: SPi Publisher Services

Printed on acid-free paper

Springer is part of Springer Science+Business Media (www.springer.com)

Climate Change Adaptation in Developed Nations

ADVANCES IN GLOBAL CHANGE RESEARCH

VOLUME 42

For further volumes:
http://www.springer.com/series/5588

Acknowledgments

A number of people contributed to the development of this volume and deserve recognition. Firstly, we kindly thank the chapter authors for their contributions, whose work on a wide range of scientific and regional case studies form the substantive basis of this book. More than 60 anonymous reviewers provided thoughtful and detailed comments that have contributed to the quality, clarity, and scientific rigor of the information herein. Intellectual guidance and support during the early conceptualization of this volume were provided by Barry Smit and Tristan Pearce. The managing editor for this volume, Margaret Deignan, has been highly supportive and efficient during book development and was a pleasure to work with. Contributing funds were provided by the Department of Geography at McGill University toward an editorial research assistant, and are greatly appreciated. Finally, several research assistants dedicated enormous effort and time to copyediting, author communications, and general book administration: we thank Carolyn Poutiainen, Michelle Maillet, Tanya Smith, and William Vanderbilt.

Contents

Contributors

Ann Albihn Section of Environment and Biosecurity, National Veterinary Institute, Uppsala, Sweden, ann.albihn@sva.se

Sylvia Allan MWH NZ Ltd, Wellington, New Zealand, sylvia.allan@ihug.co.nz

Yvonne Andersson Department of Epidemiology, Swedish Institute for Infectious Disease Control, Stockholm, Sweden, yvonne.andersson@smi.se

Karin André Centre for Climate Science and Policy Research, Linköping University, Norrköping, Sweden, karin.andr@tema.liu.se

Lea Berrang-Ford Department of Geography, McGill University, 805 Sherbrooke Street West, Montréal, QC H3A 2K6, Canada, lea.berrangford@mcgill.ca

Peter Berry Climate Change and Health Office, Health Canada, Ottawa, ON, Canada, peter_berry@hc-sc.gc.ca

Emily Boyd School of Earth and Environment, University of Leeds, Leeds, LS2 9JT UK, e.boyd@leeds.ac.uk

Michelle Boyle Institute for Resources, Environment and Sustainability, University of British Columbia, AERL, 422–2202 Main Mall, Vancouver, BC V6T 1Z4, Canada, mboyle@ires.ubc.ca

Michael Brklacich Department of Geography and Environmental Studies, Carleton University, Ottawa, ON, Canada, michael_brklacich@carleton.ca

Joseph Bump Department of Forest, Rangeland and Watershed Stewardship, College of Natural Resources, Colorado State University, Campus Delivery 1472, Fort Collins, CO 80523–1472, USA, joseph.bump@colostate.edu

Ian Burton Emeritus, Adaptation and Impacts Research Group (AIRG), Meteorological Service of Canada, Environment Canada, Downsview, ON, Canada Emeritus, Institute for Environmental Studies, University of Toronto, Toronto, ON, Canada, Ian.Burton@ec.gc.ca

Kaila-Lea Clarke Climate Change and Health Office, Health Canada, Ottawa, ON, Canada, kaila-lea_clarke@hc-sc.gc.ca

Stewart J. Cohen Adaptation & Impacts Research Division (AIRD), Environment Canada, Department of Forest Resources Management, University of British Columbia, 4617-2424 Main Mall, Vancouver, BC V6T 1Z4, Canada, scohen@forestry.ubc.ca

Christina L. Cook Institute of Resources, Environment and Sustainability, University of British Columbia, Vancouver, BC, Canada, clcook@interchange.ubc.ca

Caroline E. Cowan Natural England, Hercules House, Hercules Road, London SE1 7DU, UK, caroline.cowan@naturalengland.org.uk

Rob C. de Loë Water Policy and Governance Group, Department of Environment and Resource Studies, University of Waterloo, 200 University Avenue West, Waterloo, ON N2L 3G1, Canada, rdeloe@uwaterloo.ca

Thea Dickinson Burton Dickinson Consulting, 600 Kingston Road, Suite 204, Toronto, ON M4E 1R1, Canada, thea.dickinson@rogers.com

Stephen Dovers The Fenner School of Environment and Society, The Australian National University, Canberra, ACT, Australia, stephen.dovers@anu.edu.au

Hadi Dowlatabadi Institute of Resources, Environment and Sustainability and Liu Institute for Global Issues, University of British Columbia, AERL, 422–2202 Main Mall, Vancouver, BC, V6T 1Z4 Canada, hadi.d@ubc.ca

Kristie L. Ebi ClimAdapt LLC, 424 Tyndall Street, Los Altos, CA 94022, USA Department of Global Ecology, Stanford University, Stanford, CA, USA, krisebi@essllc.org; krisebi@stanford.edu

Éva Erdélyi Corvinus University of Budapest, Budapest, Hungary, eva.erdelyi@uni-corvinus.hu

Manon Fleury Environmental Issues Division, Centre for Food-borne, Environmental & Zoonotic Infectious Diseases, Public Health Agency of Canada, Guelph, ON, Canada, manon_d_fleury@phac-aspc.gc.ca

James D. Ford Department of Geography, McGill University, 805 Sherbrooke Street West, Montréal, QC H3A 2K6, Canada, james.ford@mcgill.ca

Patricia Gallaugher Department of Biological Sciences, Simon Fraser University, Burnaby, BC, Canada, gallaugher@sfu.ca

Megan Gawith UK Climate Impacts Programme, Environmental Change Institute, Oxford, UK, megan.gawith@ukcip.org.uk

Monique Helfrich School of Public Policy, George Mason University, 1600 N Oak St, Unit 1830, Arlington, VA 22209, USA, mhelfric@gmu.edu

Hayley Hesseln Department of Bioresource Policy, Business, and Economics, College of Agriculture and Bioresources, University of Saskatchewan, Saskatoon, SK, Canada, h.hesseln@usask.ca

Yasushi Honda Graduate School of Comprehensive Human Sciences, University of Tsukuba, Tsukuba, Ibaraki, Japan, honda@taiiku.tsukuba.ac.jp

Kelly Hopping Department of Forest, Rangeland and Watershed Stewardship, College of Natural Resources, Colorado State University, Campus Delivery 1472, Fort Collins, CO 80523–1472, USA, kelly.hopping@colostate.edu

David Hutton United Nations Relief and Works Agency (UNRWA), West Bank Field Office, Jerusalem 97200, Israel, d.hutton@unrwa.org

Stéphane Isoard European Environment Agency, Kongens Nytorv 6, Copenhagen DK-1050, Denmark, stephane.isoard@eea.europa.eu

Mark Johnston Saskatchewan Research Council, Saskatoon, SK, Canada, johnston@src.sk.ca

Kay Johnstone UK Climate Impacts Programme, Environmental Change Institute, Oxford, UK, kay.johnstone@ukcip.org.uk

Noni Keys Sustainability Research Centre, University of the Sunshine Coast, Macoochydore, DC 4558 Australia, noni.keys@gmail.com

Julia A. Klein Department of Forest, Rangeland and Watershed Stewardship, College of Natural Resources, Colorado State University, Campus Delivery 1472, Fort Collins, CO 80523–1472, USA, jklein@warnercnr.colostate.edu

Richard J.T. Klein Stockholm Environment Institute, Stockholm, Sweden, richard.klein@sei.se

Paul Kovacs Institute for Catastrophic Loss Reduction, Toronto, ON, Canada, pkovacs@pacicc.ca

Howard Larsen Ministry for the Environment, Wellington, New Zealand, howard.larsen@mfe.govt.nz

Geraldine Li The Fenner School of Environment and Society, The Australian National University, Canberra, ACT, Australia, geraldine.li@anu.edu.au

Elisabet Lindgren Division of Global Health / IHCAR, Department of Public Health, Karolinska Institute, SE-171 77 Stockholm, Sweden, elisabet.lindgren@ki.se

Kate Lonsdale UK Climate Impacts Programme, Environmental Change Institute, Oxford, UK, kate.lonsdale@ukcip.org.uk

Nicholas A. Macgregor Natural England, Hercules House, Hercules Road, London SE1 7DU, UK, nicholas.macgregor@naturalengland.org.uk

Michelle Maillet Department of Geography, McGill University, Montreal, QC, Canada, michelle.maillet@mail.mcgill.ca

Robert A. McLeman Department of Geography, University of Ottawa, Simard Hall 015, 60 University, Ottawa, ON, K1N 6N5, Canada, rmcleman@uottawa.ca

Gerry Metcalf UK Climate Impacts Programme, Environmental Change Institute, Oxford, UK, Gerry.metcalf@ukcip.org.uk

Susanne C. Moser University of California-Santa Cruz and Susanne Moser Research & Consulting, 134 Shelter Lagoon Drive, Santa Cruz, CA 95060, USA, promundi@susannemoser.com

Martin Mulligan Globalism Research Centre, RMIT University, GPO Box 2476, Melbourne, VIC 3001, Australia, martin.mulligan@rmit.edu.au

Yaso Nadarajah Globalism Research Centre, RMIT University, GPO Box 2476, Melbourne, VIC 3001, Australia, yaso.nadarajah@rmit.edu.au

Harry Nelson Faculty of Forestry, Forest Sciences Centre, University of British Columbia, Vancouver, BC, Canada, hnelson@forestry.ubc.ca

Laurie Newton UK Climate Impacts Programme, Environmental Change Institute, Oxford, UK, laurie.newton@ukcip.org.uk

Yonten Nyima Department of Geography, University of Colorado, Boulder, CO, USA, yy2161@gmail.com

Aynslie Ogden Forest Management Branch, Yukon Department of Energy Mines and Resources, Whitehorse, YT, Canada, aeogden@gov.yk.ca

Nicholas Hume Ogden Zoonoses Division, Centre for Food-borne, Environmental & Zoonotic Infectious Diseases, Public Health Agency of Canada, Jeanne Mance Building, 200 Eglantine, Tunney's Pasture, AL 1906B, Ottawa, ON K1A 0K9, Canada, nicholas_ogden@phac-aspc.gc.ca

Masaji Ono Association of International Research Initiatives for Environmental Studies, Tokyo, Japan, ono@airies.or.jp

Mark Pajot MES Candidate, York University, Toronto, ON, Canada, pajotm@peelregion.ca

Sylvie Parey Électricité de France (EDF) R&D, Chatou CEDEX, France, sylvie.parey@edf.fr

Jennifer Penney Clean Air Partnership, Toronto, ON, Canada, jpenney@cleanairpartnership.org

Carolyn Poutiainen Department of Geography, McGill University, Montreal, QC, Canada, carolyn.poutiainen@mail.mcgill.ca

Vivek Prasad Department of Environmental Science and Public Policy, George Mason University, Fairfax, VA, USA, vprasad1@gmu.edu

John M. Reilly Center for Environmental Policy Research, Joint Program on Global Change, Massachusetts Institute of Technology Sloan School of Management, Cambridge, MA, USA, jreilly@mit.edu

Andy Reisinger Climate Change Research Institute, Victoria University, Wellington, New Zealand, andy.reisinger@vuw.ac.nz

Benno Rothstein University of Applied Forest Sciences, Rottenburg am Neckar, Baden-Württemberg, Germany, rothstein@hs-rottenburg.de

Renate Sander-Regier Department of Geography, University of Ottawa, Ottawa, ON, Canada, rsand071@uottawa.ca

Jan C. Semenza Head of Future Threats and Determinants Section, Scientific Advice Unit, European Centre for Disease Prevention and Control (ECDC), Tomtebodavägen 11A, Stockholm S-171 83, Sweden, jan.semenza@ecdc.europa.eu

Louise Simonsson Centre for Climate Science and Policy Research, Linköping University, Norrköping, Sweden, louise.simonsson@tema.liu.se

Jodi-Anne Michelle Smith Global Cities Institute, RMIT University, GPO Box 2476, Melbourne, VIC 3001, Australia, jodi-anne.smith@rmit.edu.au

Timothy Frederick Smith Sustainability Research Centre, University of the Sunshine Coast, Macoochydore, DC 4558 Australia, Tim.Smith@usc.edu.au

Paul Sockett Communicable Disease Control Division, Primary Health Care and Public Health Directorate, First Nations and Inuit Health Branch, Health Canada, Ottawa, ON, Canada, paul_sockett@hc-sc.gc.ca

Erik Sparling CSA Standards, 155 Queen Street, Suite 1300, Ottawa, ON, K1P 6L1, Canada, erik.sparling@csa.ca

Paul Steenhof CSA Standards, 155 Queen Street, Suite 1300, Ottawa, ON, K1P 6L1, Canada, paul.steenhof@csa.ca

Roger Street UK Climate Impacts Programme, Environmental Change Institute, Oxford, UK, roger.street@ukcip.org.uk

Iwan Supit Wageningen University and Research Centre, Wageningen, The Netherlands, iwan.supit@wur.nl

Åsa Gerger Swartling Stockholm Environment Institute, Stockholm, Sweden, asa.swartling@sei.se

Dana C. Thomsen Sustainability Research Centre, University of the Sunshine Coast, Macoochydore, DC 4558 Australia, dthomsen@usc.edu.au

Laird Van Damme KBM Forestry Consultants, Thunder Bay, ON, Canada, vandamme@kbm.on.ca

Kelly Vodden Department of Geography, Memorial University of Newfoundland, St. John's, NL, Canada, kvodden@mun.ca

Oskar Wallgren Stockholm Environment Institute, Stockholm, Sweden, oskar.wallgren@sei.se

Saskia E. Werners Wageningen University and Research Centre, Wageningen, The Netherlands, werners@mungo.nl

Thomas J. Wilbanks Climate Change Science Institute, Oak Ridge National Laboratory, Oak Ridge, TN 37831-6103, USA, wilbankstj@ornl.gov

Tim Williamson Natural Resources Canada/Canadian Forest Service, Edmonton, AB, Canada, twilliam@nrcan.gc.ca

Elizabeth Willmott Willmott Sherman L.L.C., Seattle, WA, USA, elizabethwillmott1977@gmail.com; Mlle.willmott@gmail.com

Johanna Wolf Labrador Institute of Memorial University, 490, Stn. B, Happy Valley-Goose Bay, NL A0P 1E0 Canada, jwolf@mun.ca

Maureen Woodrow Telfer School of Management, University of Ottawa, Ottawa, ON, Canada, woodrow@telfer.uottawa.ca

David Wratt National Institute of Water and Atmospheric Research (NIWA), Wellington, New Zealand, d.wratt@niwa.co.nz

Emily Yeh Department of Geography, University of Colorado, Boulder, CO, USA, emily.yeh@colorado.edu

List of Figures

Part I
Introduction and Overview

Chapter 1
Introduction

James D. Ford and Lea Berrang-Ford

Introduction

It is widely accepted that the climate is changing, with implications for human systems already documented (Fussel 2009; Smith et al. 2009; Stott et al. 2010). Climate models indicate continued and accelerated climate change in the future (Solomon et al. 2007). Research is only beginning to examine the potential implications of climate change for human systems and indicates significant vulnerabilities (Hulme 2008). Society will not be static as the climate changes, however, undergoing social, cultural, economic, and political changes that will affect how human systems experience climate change and determine adaptive capacity to respond. Some of these developments will moderate vulnerability: poverty, for instance, is a major determinant of climate vulnerability the world-over, and decreasing poverty rates with economic development offers considerable opportunity to reduce sensitivity to climatic risks and enhance adaptability. Aging populations, population growth in high-risk locations (e.g., coastal zones), increasing inequality, and weakening of social networks are trends that are likely to exacerbate vulnerability.

In light of the risks posed by climate change, climate policy has become a key area of debate and research. Reducing greenhouse gas emissions (i.e., mitigation) is central to efforts to tackle climate change. Mitigation will not be enough, however. Even with the most aggressive targets, historic emissions mean that some degree of climate warming is inevitable over the coming decades and will probably surpass the 2°C threshold held by many as indicative of "dangerous" interference with the climate system (Ramanathan and Feng 2008; Parry et al. 2009b; Smith et al. 2009). In this context, Adger and Barnett (2009) have called for a new realism on climate

J.D. Ford (✉) • L. Berrang-Ford
Department of Geography, McGill University, 805 Sherbrooke Street West, Montréal,
QC H3A 2K6, Canada
e-mail: james.ford@mcgill.ca; lea.berrangford@mcgill.ca

J.D. Ford and L. Berrang-Ford (eds.), *Climate Change Adaptation in Developed Nations:*   3
From Theory to Practice, Advances in Global Change Research 42,
DOI 10.1007/978-94-007-0567-8_1, © Springer Science+Business Media B.V. 2011

change, one that recognizes the significant likelihood that means warming will be 4°C or greater. What is clear is that adaptation is unavoidable if we are to reduce the risks of significant damage (Ford and Smit 2004; Smit and Wandel 2006; Pielke et al. 2007; Parry et al. 2008, 2009b; Adger and Barnett 2009; Costello et al. 2009; Smith et al. 2009).

The good news is that opportunities for adaptation to climate change are available, feasible, and in many cases can be mainstreamed into existing policy priorities and programming (Stern 2006; Garnaut 2008; Lemmen et al. 2008; Seguin 2008; Costello et al. 2009; Karl et al. 2009). Indeed, the challenge of adaptation to policymakers and managers is not necessarily new, as humans have lived with climatic variability for a long time and developed management decisions to cope with this variability (Glantz 1988; Burton et al. 2002; Ford and Smit 2004; Smit and Wandel 2006; Dovers 2009; Ford et al. 2010b). It is through changes in the magnitude and frequency of existing climatic variability that climate change will be experienced. As Dovers (2009) argues, we already have developed capacities and understanding in many sectors to provide a basis for addressing even significantly enhanced variability. Furthermore, the costs of adaptation, while daunting – scoping studies have suggested between $9 billion per year to greater than $300 billion per year (Parry et al. 2009a) – are only a fraction of global GDP.

Despite these opportunities, the bad news is that formidable challenges to climate change adaptation exist. First, given the scale of projected impacts and experience of climate change already, the window of opportunity for adaptation is narrow (Adger and Barnett 2009; Parry et al. 2009b). Second, social, environmental, institutional, and economic stresses are likely to further exacerbate impacts and constrain adaptive responses for vulnerable people and regions. Third, despite changes in weather extremes and increasing awareness (although not universal) of climate-change risks, adaptation activities are still poorly embedded in planning systems (Moser and Luers 2008; Tribbia and Moser 2008; Repetto 2009; Berrang-Ford et al. 2010; Ford et al. 2010c). Finally, as Adger and Barnett (2009) caution, maladaptation abounds where adaptations being undertaken today are not sustainable in the long term.

These challenges in part stem from the lack of attention given to adaptation in policy discussions on climate change at local, national, and international levels. For too long adaptation has been the poor cousin of mitigation. They also stem from the nature of adaptation research. As Barnett (2010) argues, adaptation has largely been investigated as a scientific and technical problem with little research that has sought to inform decision-makers. Thus there is a well-developed literature assessing vulnerability, identifying generic opportunities for adaptation, and examining adaptation processes, but fewer studies outlining specific adaptation entry points, developing adaptation plans, evaluating progress of adaptation plans, or profiling examples of best practice (Gagnon-Lebrun and Agrawala 2007; Barnett 2010; Ford et al. 2010b). This edited book is in response to this deficit and was conceived with the aim of profiling cases from different sectors and regions in developed nations where *specific* adaptation measures have been identified, implemented, and evaluated; the focus is "from theory to practice." The contributions provide practical advice and guidance that can help guide adaptation planning in multiple contexts.

In this introductory chapter we begin by explaining why the book focuses on developed nations and then document observed and projected climate changes affecting the developed world. We then present the book outline, review contributions of specific chapters, and finish by outlining future research directions raised by the authors.

Why Developed Nations?

Definition and Characteristics of "Developed Nations"

For the purposes of this book we define "developed nations" broadly as the 41 Parties identified under Annex I to the United Nations Framework Convention on Climate Change (UNFCCC) (Table 1.1, Fig. 1.1). This includes 28 of the 30 member states of the Organization for Economic Cooperation and Development (OECD), excluding South Korea and Mexico, and includes several Economies in Transition that are not part of the OECD. Annex 1 nations are industrialized in character, have high levels of economic development, low prevalence of poverty, aging and in some cases declining populations, well-developed institutional capacity, and have the majority of the population living in urban areas with livelihoods not directly

Table 1.1 List of Annex I parties to the UNFCCC

Australia	Liechtenstein
Austria	Lithuania
Belarus	Luxembourg
Belgium	Monaco
Bulgaria	Netherlands
Canada	New Zealand
Croatia	Norway
Czech Republic	Poland
Denmark	Portugal
Estonia	Romania
European Community	Russian Federation
Finland	Slovakia
France	Slovenia
Germany	Spain
Greece	Sweden
Hungary	Switzerland
Iceland	Turkey
Ireland	Ukraine
Italy	United Kingdom
Japan	United States of America
Lativa	

Estimated total population: 1,774,753,827

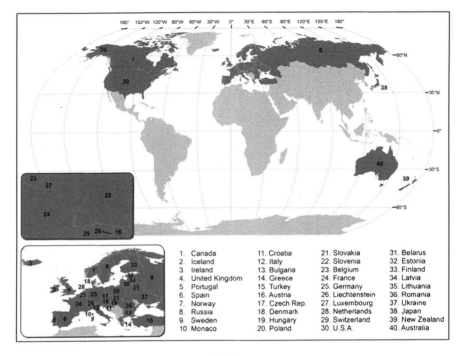

Fig. 1.1 Map of Annex I parties to the UNFCCC

linked to climate-sensitive natural resources. They have a combined population of approximately 1.7 billion people, or 25% of the global population. We also note that we include one chapter profiling a case study from Tibet given important lessons for Indigenous populations in developed nations.

Observed and Projected Changes in Developed Countries

Developed nations in the northern hemisphere have experienced some of the most pronounced changes in climatic conditions globally over the past century. In North America, the greatest warming has occurred in Arctic regions: between 1948 and 2005, for instance, winter temperatures in the Yukon and Mackenzie regions of Canada have warmed by 4.5°C and 4.3°C, respectively, with an annual warming of 2.2°C and 2.0°C (Prowse et al. 2009). The Arctic as a whole is warming at about twice the global average (Solomon et al. 2007). In Europe, the annual mean temperature has increased 0.91°C from 1901 to 2005, with a rate of warming of 0.41°C per decade from 1979 to 2005 (Solomon et al. 2007). In Japan, temperature increased by 1.0°C in the twentieth century, with estimates of 2–3°C increases for some large cities (Ichikawa 2004). In Australia, annual mean surface air temperature has increased by 0.9°C since 1910 and 0.16°C per decade since 1950 (CSIRO and Australian Bureau of Meteorology 2007). The average maximum temperature in Australia increased by 0.6°C, while the minimum temperature increased by 1.2°C (Nicholls and Collins 2006; Solomon et al. 2007). Changing precipitation regimes have also been noted, showing greater regional variability than temperature.

Climate models project that warming will be greater than the global average for most of North America and Europe and close to the global average for more southern latitude developed countries like Australia. The greatest warming is projected to occur in Arctic regions of developed nations: over the western Canadian Arctic, for instance, median temperature changes of +6°C are projected by 2080, ranging from +3.5°C to +12.5°C depending on climate model and scenario used. Either way, Arctic regions are projected to experience the greatest changes in climate globally (ACIA 2005; Solomon et al. 2007; Timlin and Walsh 2007; Barber et al. 2008). The Intergovernmental Panel on Climate Change (IPCC) estimates that Europe will warm less than North America. Average temperatures in northern and central Europe are projected to increase a few degrees by 2050 and 3–5°C by 2090. Northern and central Europe will experience greater yearly maximum temperatures than other regions (Räisänen et al. 2004; Solomon et al. 2007; Kjellström et al. 2007). Japan is projected to experience warming in most areas, increased heat waves, increased annual precipitation, and more intense rainfall on extreme precipitation days, including those associated with typhoons (Solomon et al. 2007). Average annual precipitation increases are expected over most of northern Europe, the Arctic, Canada, northeastern United States and northern Pacific, while decreases will be experienced in most of the Mediterranean and the US Southwest. In general, there will be reduced rainfall over continental interiors during summer months due to increases in evaporation, with increased drought frequency projected for much of Australia (Solomon et al. 2007).

Climate Change Adaptation in Developed Nations: Reasons for Concern

Developed nations are generally assumed to have limited vulnerability to climate change (Gagnon-Lebrun and Agrawala 2007; Solomon et al. 2007; Costello et al. 2009). On the whole, the assumption holds: individuals, communities, and governments in developed nations have access to significant resources and engage in a range of actions to manage and control climate-sensitive outcomes in a range of sectors including health, agriculture, infrastructure, business, education, and industry. Awareness of climate change is also generally high, research capacity well developed, and vulnerability and impacts assessments completed. This assumption, however, does not adequately consider the persistence of within-country socioeconomic inequities and their implications for vulnerable populations, and overlooks the fact that many observed and projected changes are considerably greater in temperate latitudes (particularly Arctic nations). Moreover, significant vulnerabilities in developed nations have been highlighted by a number of climate-related disasters including the European heat wave, Australian drought, Mountain Pine Beetle infestation in the forests of Western Canada, and Hurricane Katrina (Hulme 2003; Lagadec 2004; Parkins and MacKendrick 2007; Ebi and Burton 2008; Ford et al. 2010a). In the United States, Repetto (2009) has argued that public and private bodies charged with adaptation have failed to deal with current risks and prepare for emerging future risks. O'Brien et al. (2004; 2006) voice concern over what they see as complacency in Norway and Europe more generally that climate change will have limited impacts, identifying significant vulnerabilities at a local level. What these and other studies demonstrate is that adaptive capacity in developed nations will not necessarily translate into adaptation; planned adaptation will be necessary.

Key concerns expressed in the literature regarding the ability of developed nations to adapt to climate change include:

- *Information deficit*: Adaptation requires recognition of the necessity to adapt, knowledge about available options, the capacity to assess them, and the ability to implement most suitable ones. Vulnerability assessment is an important first step for providing the necessary information for adaptation. Most developed nations have initiated impacts assessments, including the development of climate change scenarios, and have assessed impacts at a national level (Gagnon-Lebrun and Agrawala 2007). Some nations have also examined vulnerabilities, conducted sectoral assessments, and identified generic adaptation options. Notable progress has been made by the United Kingdom, Canada, Australia, and the Netherlands, yet comprehensive approaches to implementing adaptation and mainstreaming are lacking (Gagnon-Lebrun and Agrawala 2007). In a European context, O'Brien et al. (2006) further critique the scope of understanding, with studies disproportionately focusing on direct impacts on biophysical systems and economic sectors, with indirect effects and differential vulnerabilities often ignored. For specific sectors (e.g., industry) and at local and regional levels, the deficit between what we need to know to facilitate adaptation and what we know (the "adaptation deficit") is particularly large, especially for marginal populations (e.g., Indigenous peoples) (Ford et al. 2010a, c).
- *Economic resources*: It is widely recognized that access to economic resources directly affects vulnerability to climate change through its implications for institutional capacity and the ability of households to prevent, prepare for, avoid, and recover from climate-related hazards (IPCC 2001, 2007). Developed nations are expected to be less vulnerable to climate change on account of economic well-being. National level indicators, however, hide significant disparities, perhaps the most glaring of which are among Indigenous populations in developed nations who have been referred to as the "fourth world" on account of high rates of poverty and experience of housing, water, and food insecurity (O'Neill 1986; Ford et al. 2010a). Differential vulnerabilities exist at multiple levels and

for multiple populations. However, very little has been written on vulnerable populations in the developed world compared to the developing nations (O'Brien et al. 2006).

- *Institutional capacity*: High income nations are generally believed to have well-developed institutional capacity underpinning the ability to identify, recognize, evaluate, anticipate, and respond to climate-related risks including those associated with climate change (IPCC 2007). Institutions are generally well developed, funded, and staffed by a professional and highly educated workforce, with accountability ensuring proactive identification of future risks, planning for future burdens, and underpinning institutional learning. However, recent experience – from the 2003 European heat wave, to SARS, to Hurricane Katrina – has challenged the notion that institutions will insulate developed nations from climate-change impacts. In these instances, vertical institutional weaknesses, including flow of information between decision-makers at different scales, and horizontal challenges, including conflicting and unspecified jurisdiction, power politics, and institutional defensive routines, overwhelmed institutional capacity with dramatic consequences (Kates et al. 2006; Kovats and Ebi 2006; Boettke et al. 2007).
- *Technological capacity*: Availability and accessibility of technology at various levels will affect vulnerability to climate change. Technological capacity will play an important role in buffering developed nations against the effects of climate change and provides a strong basis for adaptive planning. Surveillance and early warning systems, for instance, play a crucial role in decreasing sensitivity to climate risks and increasing adaptive capacity, helping to anticipate and respond to risks that will become more prevalent with climate change. However, access to technology is often uneven in developed nations and technology can also create new vulnerabilities.
- *Political challenges*: Climate change will result in the emergence of risks which cross borders, extend over multiple spatial-temporal scales, and span jurisdictions of several government departments. Addressing these risks will require the creation of new governance structures, including increased participation of vulnerable peoples in decision-making, increased accountability, and financial commitments, and will entail potentially unpopular decisions by national governments (Costello et al. 2009). Political challenges exist to achieving these goals, evident by the lack of coordinated and comprehensive commitments on climate change adaptation at national and international levels, and slow progress in policy development (Pielke et al. 2007; Jinnah et al. 2009). Political will to meaningfully address climate change is also often lacking, with adaptation investments in some cases having limited short-term political benefit. Hurricane Katrina provides a case in point, where despite repeated warnings about the risks posed by flooding, decades of underinvestment in flood defenses and hazard planning and development in high-risk areas were allowed to continue. This reflected the political culture favoring low taxes and economic growth at the expense of safety, and the unwillingness of the public to see tax money spent on what were perceived as less pressing problems (Kates et al. 2006). The parallels for climate change adaptation are legion.

- *Societal trends*: Developed nations (as well as some emerging economies, e.g., China) face societal challenges associated with aging populations. For many climate-change risks, age will create unique sensitivities and challenges to adaptation. Epidemiological analysis of mortality during heat waves, for example, has highlighted that the elderly are at significantly higher risk, and in the majority of developed nations the magnitude and frequency of occurrence of heat waves is projected to increase (McGeehin and Mirabelli 2001; Brown and Walker 2008; Michelozzi et al. 2009). Research has also highlighted the greater emotional toll that climate hazards place on the elderly (Marion et al. 1999; Kovats 2004; Hajat et al. 2005). Older people and retirees are also attracted to warmer climates in general (Duncombe et al. 2003; Howard 2008). As a result, some esthetically pleasing coastal areas that are highly vulnerable to extreme weather events may become increasingly populated by older age groups. There are also unique economic challenges to aging populations. For example, a shrinking workforce will have to support a growing dependent population, and this will place increasing pressure on government expenditure (Leibfritz et al. 1995; Bryant et al. 2004; Guest 2006). In such a circumstance, funding allocated to climate-change adaptations may be constrained.

With these challenges in mind, examining successful initiatives and learning from failed ones offer important insights on the adaptation process and how we can integrate adaptation planning in policy programming at multiple levels. Herein, this book documents examples of adaptation in practice from across developed nations. The practical focus reflects the urgency of the adaptation challenge. It also reflects the maturation of adaptation science: while theoretical and methodological debates continue, and vulnerability assessments are ongoing, in a number of sectors and regions we are beginning to know enough to begin adaptation planning.

Book Organization

This book is organized into seven themed parts:

- Part I: Introduction and overview
- Part II: Adaptation in the public health sector
- Part III: Adaptation in the industrial sector
- Part IV: Adaptation in the urban environment
- Part V: Adaptation in the agricultural sector
- Part VI: Adaptation in rural and resource-dependent communities
- Part VII: Future directions

Chapters in Part I review progress on adaptation in developed countries at a national level. Parts II–VI have between three and seven chapters, with each theme beginning with a summary chapter outlining the key contributions of the chapters, evaluation of progress toward the development of adaptation interventions, and identification of research and policy needs and challenges. Summary chapter authors were recruited by the editors according to their experience and expertise in the theme area, and all have been IPCC lead authors. Chapter authors were recruited to each theme using an open call for proposals in spring 2008 posted on numerous climate change listservs. Contributions were sought from the scientific community and policy stakeholders, defined to include individuals involved in the development and/or implementation of adaptation strategies. The participation of stakeholders as well as specialists aimed to document the practical experience of adaptation, challenges, and opportunities from an "on the ground" perspective. Eighty-eight proposals were received and evaluated by a review committee and screened according to their focus on *practical adaptation initiatives*, which was defined to include studies profiling the development of adaptation plans, identifying practical interventions for adaptation, evaluating existing interventions, or making significant methodological contributions to help identify adaptations. Forty full chapters were requested and were subject to double blind peer review, with 34 accepted for publication. Reflecting the practical focus of the book and appeal to a general readership, chapters were limited to 3,500 words and authors were asked to focus on only practical adaptation, avoiding lengthy discussion of methodological or theoretical considerations.

Contributions of the Book

Part I: Introduction and Overview

Chapters in the introductory section review the experience of adaptation at a national/regional level in selected developed nations/regions. Wolf begins by challenging the notion that developed nations are immune from the effects of climate change by drawing upon research from multiple developed nations. Examining the psychological, social, and cultural aspects of adaptation, she highlights how perceived adaptive capacity among vulnerable groups, sectors, and regions is likely to constrain adaptive action: if vulnerabilities are not perceived, action is unlikely to be taken. Moser then charts the evolution of adaptation as a response to climate change in the United States. Reviewing how adaptation is being addressed at federal, state, and local levels, along with the non-governmental sector, she profiles numerous examples of policy action, reflective of the increasing importance of adaptation. The chapter finishes with a note of caution; however, with widespread public engagement and mobilization on adaptation lacking and investment in research capacity still limited. Smith and colleagues argue that Australia is well advanced in anticipatory approaches to adaptation planning at all levels of government, within industry, and increasingly within communities, reflected in a significant increase in adaptation funding in recent years. Nevertheless, the authors question the extent to which such funds are being invested in long-term adaptations with implications for the sustainability of current policies. Boyd and colleagues chart the evolution of adaptation in the United Kingdom, a leader in many areas of climate policy. The authors demonstrate how adaptation policy in the United Kingdom is building adaptive capacity across institutions and networks, with adaptation actions already underway. Isoard highlights the important role of the European Union (EU) in promoting adaptation among member states, including integrating adaptation into EU sectoral policies, fostering research, coordinating and enhancing information sharing on adaptation, and providing methodological guidance for adaptation policy development. This is reflected in the recently developed EU Adaptation White Paper that provides a two-phase strategy focusing on improving understanding of opportunities for adaptation leading to the development of an EU strategy from 2013 onward.

Part II: Adaptation in the Public Health Sector

Climate change is expected to increase health risks in most developed nations, increasing the importance of adaptation as a means to prevent, reduce, and manage negative health outcomes. Semenza begins by identifying three key policy entry points for health adaptation in urban areas: interventions that advance social capital to enhance community capacity; physical improvement of the built environment to reduce exposure to climate-related health effects; and social services developments that integrate multiple sectors through emergency plans for risk reduction. A key feature of these response options is that they are based on the concept of lateral public health where interventions are grounded in community-based participation in the decision-making process. Lindgren and colleagues then report on key health findings of the Swedish Government's Commission on Climate and Vulnerability. Focusing on water-related health risks (e.g., water-borne diseases), they profile adaptations recommended by the Commission to various government departments, including information campaigns, education of public health staff, enhanced surveillance, identification of risks sources (e.g., landfills that may leach pollutants), and upgrading of waste treatment facilities. Since the Commission report, the Swedish government has increased adaptation support and a number of local adaptation projects have been developed. Honda and colleagues examine how the health risks of heat waves are managed in Japan, as climate change will increase the frequency and magnitude of heat waves. A number of initiatives are described, including the publication of a heatstroke manual for health workers, awareness-raising posters, TV messages and web advertisement, and the delivery of warning messages to cell phones and email. Such initiatives offer promise for adapting to climate change, although the authors note the need to broaden the focus of interventions from heatstroke to heat-related health risks in general. Ogden and colleagues identify adaptation challenges posed by the effects of climate change on food-borne, water-borne, and vector-borne diseases in Canada. They argue that while Canada is relatively well prepared for the health risks of climate change, a number of constrains exist, including: limited understanding of the climate sensitivity and climate impacts of a number of diseases; lack of planning for long-term health implications of climate change; limited ability to predict and detect current, emerging, or new health events; and preparedness at municipal levels. Berry and colleagues complement Ogden et al. by reporting on the results of a survey of Canadians on how the health risks of climate change are perceived. While many respondents reported being concerned about climate change, knowledge about impacts was generally limited, with few reports taking proactive action to prepare for potential emergencies. Hutton develops this analysis at a regional level, reporting on a case study from urban and rural municipalities in the Canadian province of Manitoba to examine motivations affecting people's willingness to prepare for climate change. Similar to Berry et al., participants in the study were generally aware of some of the risks of climate change but showed little concern or willingness to adjust activities and prepare. A key responsibility for adapting to the potential health impacts of climate change falls to individuals and households, and both studies support the need for adaptation interventions to target vulnerable groups, engage individuals and households, and link climate change adaptation activities to everyday issues.

Part III: Adaptation in the Industrial Sector

Despite the sensitivity of industry to climate change, research has historically neglected this sector. Addressing this deficit, Rothstein and Parey begin by iden-tifying adaptation options to minimize the risks of climate change to the energy industry, focusing on two companies: Energie Baden-Württemberg AG (Germany) and Electricité de France. Both companies have significant sensitivities to climate (supply disruption, generation disruption, climate-induced demand fluctuation) and utilize a number of strategies to manage these risks, which offer potential for climate change adaptation. Over the short term, for instance, load management allows electricity production to be adjusted to reflect generating capacity at a specific time and will underpin short-term adaptation. Over the long term, however, the authors note a need for enhanced research into impacts and adaptation. Climate-related events regularly affect built infrastructure and have resulted in the development of codes, standards, and related instruments (CSRI) to reduce risks. Steenhof and Sparling argue that CSRI must accurately reflect recent and current climate conditions, as well as the likelihood and magnitude of future change, if adaptation is to take place with regards infrastructural investments. Drawing upon the Canadian experience, however, they highlight how CSRI often act as a barrier to adaptation: codes are often based on out-of-date climate data; and tools for integrating climate change into CSRI are limited, with a lack of consensus among practitioners about the risks posed by climate change. Cook and Dowlatabadi then examined the extent to which insurance companies will facilitate anticipatory actions that reduce climate change impacts on built property. Drawing upon past experience to natural disasters, they highlight limited reliance on predictive risk modeling in the industry, with new initiatives evolving reactively in response to extreme events and legal judgment of liabilities. In this context they argue that anticipatory adaptation prompted by insurers is unlikely, necessitating proactive risk mitigation implemented by government. Johnston and colleagues finish this section by drawing upon interviews with stakeholders in the Canadian forestry sector to examine potential adaptation options. The study indicates that forest managers generally feel that they have the technical capability to adapt. However, the authors argue that policy barriers and lack of resources are likely to constrain adaptation, requiring forest management to become more flexible and forward-looking, focusing on expected future outcomes under potentially different conditions. In particular, it is argued that forest tenure arrangements need to be reassessed to provide forest managers with more flexibility to allow more timely adaptation.

Part IV: Adaptation in the Urban Environment

Urban areas, with their economic functions, large and concentrated populations, and locations in often vulnerable areas, will face unique adaptation needs. In Part IV, chapters identify important lessons from local experience of adaptation planning. Reisinger and colleagues begin by examining how local governments in New Zealand are integrating adaptation into their planning. Examining barriers to practical action and experience in overcoming them, a number of recommendations for moving adaptation forward are offered, including: separating discussion of adaptation from mitigation to avoid alienating certain groups; presenting climate change projections and their uncertainties within a relevant local context; valuing and accessing local information on adaptation through community workshops; informing practicing engineers of climate-change risks; and developing legal requirements for planning to take climate change into consideration. Wilmott and Penney describe the experiences of King County, Washington, in developing and implementing a climate change adaptation plan. One of the first Counties in North America to undertake this, the experience offers a number of lessons, including the importance of strong and clear leadership from the County, collaborations between decision-makers and scientists, commitment of human resources at different government levels, and the incorporation of climate change into existing concerns alongside long-term planning. Moving to Sweden, Simonssen and colleagues examined perceptions of climate-change risks and adaptive capacity among practitioners and experts within municipal and regional administration in Stockholm and Gothenburg. While they document significant adaptive capacity, a number of barriers to adaptation are identified, including: limited coordination within organizations and between actors; conflicting interests; lack of will and opportunity to prioritize adaptation measures; and absence of legislation and division of responsibilities for adaptation. In particular, they demonstrate a need for coordination on adaptation issues within and between municipalities, and across multiple levels of government. Finishing in Australia, Li and Dovers summarize lessons for assessment methodologies and practical adaptive responses for small- to medium-sized urban areas, drawing on integrated assessment of five settlements. They document a number of challenges to adaptation planning, including time, data, and training requirements for vulnerability assessment, and connecting climate change planning to local decision-making cycles.

Part V: Adaptation in the Agricultural Sector

There is a long history of adaptation research in the agricultural sector, and Part V begins with Macgregor and Cowan describing a project initiated by the UK Department of Environment, Food, and Rural Affairs to develop adaptation options for the agricultural sector. Developing a framework to guide adaptation decision-making, they identify a suite of policy options and describe opportunities and challenges to implementation. They argue that despite the considerable potential for autonomous adaptation, sustainable adaptation will require the government to intervene to influence the practices of farmers and land managers to encourage and reinforce forward-looking behavior. Moving to Hungary, Werners and colleagues use Modern Portfolio Theory (MPT) to evaluate diversification of agricultural land use as an adaptation to climate risk in the Tisza River Basin. Specifically, they argue that use of MPT allows benefits from agricultural diversification to be quantified, with the current mix between (1) intensive agriculture protected by flood levees and (2) water retention areas with extensive cattle breeding and orchards found to be optimal within current climatic conditions. Diversification strategies are identified for a changing climate, but ultimately there is an upper limit beyond which risks cannot be "diversified away." Helfrich and Prasad finish this section by examining and comparing adaptation in the wheat sector in Canada and Australia. Reviewing major research efforts and current practices to encourage adaptation in the two countries, they highlight the importance of government leadership working with farming stakeholders, and the importance of networks of stakeholders and scientists to produce adaptation research. Drawing upon the Australian example, they highlight the problem of fragmented approaches to adaptation and research.

Part VI: Adaptation in Rural and Resource-Dependent Communities

While the number of people living in resource-dependent communities in developed nations is small, they face significant vulnerabilities. De Loë begins this section by asking the question "how can water managers at the local level, who already face significant challenges, adapt to climate change?" Drawing upon experience from drinking water supply in rural Ontario, he argues that successful adaptation needs to be mainstreamed into ongoing efforts to protect drinking water sources. Specifically, opportunities for climate change to be integrated into planning exist under Ontario's Clean Water Act, which provides the regulatory framework within which municipalities develop source water protection plans. Challenges to achieving this are large, however, with climate change only marginally integrated into planning to implement the Act. Smith and colleagues describe a scenario thinking workshop conducted with residents and stakeholders in the Hamilton region of Australia, a prosperous farming district highly sensitive to climate risks. During the workshop, participants created their own climate change scenarios in which they predicted impacts and explored adaptation options. In this way, the scenario thinking exercise sought to create opportunities for social learning through dialogue, create networks of interested stakeholders, and identify priority areas for future action, offering a successful first step toward initiating adaptation planning. McLeman and colleagues draw upon case studies with resource-dependent communities in Canada to examine the forces that influence the ability of decision-making structures to respond to climate change. They document significant adaptive capacity at a local level but also identify a myriad of challenges typical of small remote communities: inadequate resources including over-reliance on volunteers in decision-making; regulations that fail to take into account local circumstances; and lack of locally relevant climate change information. Addressing these non-local determinants requires building linkages across government and providing communities the resources to adapt, and will ultimately depend on how the problem is perceived at non-local levels. Boyle and Dowlatabadi examine the potential for adaptation planning in Canada's Arctic territory of Nunavut. They argue that the prerequisites for conventional long-term strategic planning are absent in the territory, constrained by limited financial resources, short planning horizons, information and knowledge on climate change and vulnerability determinants, and limited history and practice of formal planning in the territory. These challenges are compounded by a myriad of social and economic problems facing Nunavut. In this context, they argue for sustained support for effective institutional capacity building and emphasize that efforts to mainstream adaptation are needed. Klein and colleagues complete this section by examining how rangeland management policies in the Tibetan Plateau affect the vulnerability and adaptability of Tibetan pastoralists. While not a developed country case study, this chapter nevertheless offers important insights for resource-dependent, Indigenous communities in developed nations. Specifically, they highlight how current rangeland policies based on restricting access to grazing lands, promoted as an adaptation by the Chinese state, are actually increasing sensitivity and decreasing the capacity of the ecological and social system to adapt to climate change. Alternatively, they argue the importance of supporting strategies of resource-dependent communities for living in uncertain and variable environments, which will also enhance adaptive capacity to climate change.

Conclusion and Future Research Directions

As the recent Conference of the Parties meetings in Copenhagen and Cancun demonstrated, adaptation is now firmly on the climate policy agenda. Politicians and stakeholders at multiple levels are finally recognizing that adaptation will be unavoidable and that planning for climate change impacts today can bring immediate benefits alongside reducing future climate vulnerability. Chapters in this book profile examples from developed nations where adaptations are being developed and implemented. In a field that has been dominated by studies focusing on adaptation in a conceptual sense, the practical focus of these chapters represents a new direction for adaptation science, providing transferable lessons to multiple contexts. Much remains to be done however, as Ian Burton cautions in the final chapter. Practical adaptation studies are still limited; for many sectors and regions it remains unclear if adaptation is taking place, and interventions remain constrained by limited understanding of vulnerability. Few studies have formally evaluated the success of adaptation interventions. This reflects the recent focus on adaptation in climate policy and limited criteria from which to judge effectiveness, although research is beginning to address the later (e.g., De Bruin et al. 2009). Resources committed to existing adaptation and policy frameworks also differ significantly between nations. Biesbroek et al. (2010) review these differences at a European level but comparable studies have not been completed for other developed nations. And finally, few studies have examined adaptation in light of worst case scenarios, both in terms of what adaptation would be needed and sustainability of current interventions. What is clear, however, is that adaptation is an essential component of climate policy and this book provides a starting point from which to begin moving from theory to practice.

References

Adger WN, Barnett J (2009) Four reasons for concern about adaptation to climate change. Environ Plan A 41:2800–2805

Arctic Climate Impact Assessemt [ACIA], Arctic Monitoring and Assessment Programme [AMAP], International Arctic Science Committee [IASC] (2005) Arctic climate impact assessment. Cambridge University Press, Cambridge

Barber DG, Lukovich JV, Keogak J et al (2008) The changing climate of the Arctic. Arct 61 (Suppl 1):7–26

Barnett J (2010) Adapting to climate change: three key challenges for research and policy: an editorial essay. Wiley Interdiscip Rev Clim Chang 1(3):314–317

Berrang-Ford L, Ford JD, and Paterson J (2010). Are we adapting to climate change? Glob Environ Chang 21(2):25–33

Biesbroek GR, Swart RJ, Carter TR et al (2010) Europe adapts to climate change: comparing national adaptation strategies. Glob Environ Chang 20(3):440–450

Boettke P, Chamlee-Wright E, Gordon P et al (2007) The political, economic, and social aspects of Katrina. South Econ J 74(2):363–376

Brown S, Walker G (2008) Understanding heat wave vulnerability in nursing and residential homes. Build Res Inf 36(4):363–372

Bryant J, Teasdale A, Tobias M et al (2004) Population ageing and government health expenditures in New Zealand, 1951–2051 – Working paper 04/14. New Zealand Treasury, Wellington

Burton I, Huq S, Lim B et al (2002) From impacts assessment to adaptation priorities: the shaping of adaptation policy. Clim Policy 2(2):145–159

Costello A, Abbas M, Allen A et al (2009) Managing the health effects of climate change. Lancet 373(9676):1693–1733

CSIRO Australian Bureau of Meteorology (2007) Climate change in Australia: technical report 2007. CSIRO Marine and Atmospheric Research Division

De Bruin KC, Dellink RB, Tol RSJ (2009) AD-DICE: an implementation of adaptation in the DICE model. Clim Chang 95(1–2):63–81

Dovers S (2009) Normalizing adaptation. Glob Environ Chang 19(1):4–6

Duncombe W, Robbins M, Wolf DA (2003) Place characteristics and residential location choice among the retirement-age population. J Gerontol Ser B 58(4):S244–S252

Ebi KL, Burton I (2008) Identifying practical adaptation options: an approach to address climate change-related health risks. Environ Sci Policy 11(4):359–369

Ford JD, Smit B (2004) A framework for assessing the vulnerability of communities in the Canadian arctic to risks associated with climate change. Arctic 57(4):389–400

Ford JD, Berrang Ford L, King M, Furgal C (2010a) Vulnerability of Aboriginal health systems in Canada to climate change. Glob Environ Chang 20(4):668–680

Ford JD, Keskitalo ECH, Smith T et al (2010b) Case study and analogue methodologies in climate change vulnerability research. Wiley Interdiscip Rev Clim Change 1(3):374–392

Ford JD, Pearce T, Prno J et al (2010c) Perceptions of climate change risks in primary resource use industries: a survey of the Canadian mining sector. Reg Environ Chang 10(1):65–81

Fussel HM (2009) An updated assessment of the risks from climate change based on research published since the IPCC Fourth Assessment Report. Clim Chang 97(3–4):469–482

Gagnon-Lebrun F, Agrawala S (2007) Implementing adaptation in developed countries: an analysis of progress and trends. Climate Policy 7(5):392–408

Garnaut R (2008) The Garnaut climate change review. Cambridge University Press, Cambridge

Glantz M (1988) Societal responses to climate change: forecasting by analogy. Westview, Boulder

Guest R (2006) Population ageing, fiscal pressure and tax smoothing: a CGE application to Australia. Fisc Stud 27(2):183–203

Hajat S, Ebi KL, Kovats RS et al (2005) The human health consequences of flooding in Europe: a review. In: Kirch W, Menne B, Bertollini R (eds) Extreme weather events and public health responses. Springer, Berlin

Howard RW (2008) Western retirees in Thailand: motives, experiences, wellbeing, assimilation and future needs. Ageing Soci 28:145–163

Hulme M (2003) Abrupt climate change: can society cope? Philos Trans Royal Soci A 361(1810):2001–2019

Hulme M (2008) Geographical work at the boundaries of climate change. Trans Inst Br Geogr 33(1):5–11

Ichikawa A (ed) (2004) Global warming – the research challenges: a report of Japan's global warming initiative. Springer, Dordrecht

IPCC (2001) Climate change 2001: synthesis report – contribution of working groups I, II and III to the third assessment report of the Intergovernmental Panel on Climate Change. Cambridge University Press, Cambridge

IPCC (2007) Climate change 2007: synthesis report – contribution of working group I, II and III to the fourth assessment report of the Intergovernmental Pannel on Climate Change [IPCC]. IPCC, Geneva

Jinnah S, Bushey D, Munoz M et al (2009) Tripping points: barriers and bargaining chips on the road to Copenhagen. Environ Res Lett 4(3):1–6

Karl TR, Melillo JM, Peterson TC (2009) Global climate change impacts in the United States: U.S. global change research program. Cambridge University Press, Cambridge

Kates RW, Colten CE, Laska S et al (2006) Reconstruction of New Orleans after hurricane Katrina: a research perspective. Proc Natl Acad Sci USA 103(4):14653–14660

Kjellström E, Bärring L, Jacob D et al (2007) Variability in daily maximum and minimum temperatures: recent and future changes over Europe. Clim Chang 81(Suppl 1):249–265

Kovats SR (2004) Will climate chante really affect our health? Results from a European assessment. J Br Menopause Soci 10(4):139–144

Kovats SR, Ebi KL (2006) Heatwaves and public health in Europe. Eur J Public Health 16(6): 592–599

Lagadec P (2004) Understanding the French 2003 heat wave experience: beyond the heat, a multi-layered challenge. J Conting Crisis Manag 12(4):160–169

Leibfritz W, Roseveare D, Fore D et al (1995) Ageing populations, pension systems and government budgets – Economics Department working papers no.156. Organization for Economic Co-operation and Development, Paris

Lemmen D, Warren F, Lacroix J et al (eds) (2008) From impacts to adaptation: Canada in a changing climate 2007. Government of Canada, Ottawa

Marion SA, Agbayewa MO, Wiggins S (1999) The effect of season and weather on suicide rates in the elderly in British Columbia. Can J Public Health 90(6):418–422

McGeehin MA, Mirabelli M (2001) The potential impacts of climate variability and change on temperature-related morbidity and mortality in the United States. Environ Health Perspect 109(Suppl 2):185–189

Michelozzi P, Accetta G, De Sario M et al (2009) High temperature and hospitalizations for cardiovascular and respiratory causes in 12 European cities. Am J Respir Crit Care Med 179:383–389

Moser SC, Luers AL (2008) Managing climate risks in California: the need to engage resource managers for successful adaptation to change. Clim Chang 87(Suppl 1):S309–S322

Nicholls N, Collins D (2006) Observed climate change in Australia over the past century. Energy Environ 17(1):1–12

O'Brien K, Sygna L, Haugen JE (2004) Vulnerable or resilient? A multi-scale assessment of climate impacts and vulnerability in Norway. Clim Chang 64(1–2):193–225

O'Brien K, Eriksen S, Sygna L et al (2006) Questioning complacency: climate change impacts, vulnerability, and adaptation in Norway. AMBIO J Hum Environ 35(2):50–56

O'Neill JD (1986) The politics of health in the fourth world: a northern Canadian example. Hum Organ 45(2):119–128

Parkins JR, MacKendrick MA (2007) Assessing community vulnerability: a study of the mountain pine beetle outbreak in British Columbia, Canada. Glob Environ Chang 17(3–4):460–471

Parry M, Palutikof J, Hanson C et al (2008) Squaring up to reality. Nat Rep Clim Chang. doi:10.1038/climate.2008.50

Parry M, Arnell M, Berry P et al (2009a) Assessing the costs of adaptation to climate change: a review of the UNFCCC and other recent estimates. International Institute for Environment and Development, London

Parry M, Lowe J, Hanson C (2009b) Overshoot, adapt and recover. Nature 458:1102–1103

Pielke RA, Prins G, Rayner S et al (2007) Climate change 2007: lifting the taboo on adaptation. Nature 445:597–598

Prowse TD, Furgal C, Bonsai BR et al (2009) Climatic conditions in northern Canada: past and future. AMBIO J Hum Environ 38(5):257–265

Räisänen J, Hansson U, Ullerstig A et al (2004) European climate in the late 21st century: regional simulations with two driving global models and two forcing scenarios. Clim Dyn 22(1):13–31

Ramanathan V, Feng Y (2008) On avoiding dangerous anthropogenic interference with the climate system: formidable challenges ahead. Proc Natl Acad Sci USA 105(38):14245–14250

Repetto R (2009) The climate crisis and the adaptation myth – working paper no.13. Yale School of Forestry and Environmental Studies, New Haven

Seguin J (2008) Human health in a changing climate: a Canadian assessment of vulnerabilities and adaptive capacity. Health Canada, Government of Canada, Ottawa

Smit B, Wandel J (2006) Adaptation, adaptive capacity and vulnerability. Glob Environ Chang 16(3):282–292

Smith JB, Schneider SH, Oppenheimer M et al (2009) Assessing dangerous climate change through an update of the Intergovernmental Panel on Climate Change (IPCC) "reasons for concern". Proc Natl Acad Sci USA 106(11):4133–4137

Solomon S, Qin D, Manning M et al (eds) (2007) Climate change 2007: the physical science basis – contribution of working group I to the fourth assessment report of the Intergovernmental Panel on Climate Change. Cambridge University Press, Cambridge

Stern N (2006) Stern review: the economics of climate change. Cambridge University Press, Cambridge

Stott PA, Gillett NP, Hegerl GC et al (2010) Detection and attribution of climate change: a regional perspective. Wiley Interdiscip Rev Clim Change 1(2):192–211

Timlin MS, Walsh JE (2007) Historical and projected distributions of daily temperature and pressure in the Arctic. Arctic 60(4):389–400

Tribbia J, Moser SC (2008) More than information: what coastal managers need to plan for climate change. Environ Sci Policy 11(4):315–328

Chapter 2
Climate Change Adaptation as a Social Process

Johanna Wolf

Abstract Research on the impacts of climate change suggests that developed countries are not immune to the effects of a changing climate. The assumption that because of their high adaptive capacity, developed countries will adapt effectively is beginning to be dispelled by empirical evidence. While advancements in projections have facilitated a move from the study of impacts to concrete adaptation strategies, research that focuses on the social process of adaptation has been relatively neglected. Yet, when viewed as a social process, the psychological, social, and cultural aspects of adaptation are exposed, which brings into focus the effects that values and power dimensions have on actual adjustments and their outcomes for adaptation. This chapter demonstrates that some of the critical barriers to adaptation in developed countries arise from perceptions and values. It argues that these barriers are hindering adaptation now and will continue to do so unless the intricacies of the social processes underpinning adaptation are taken into explicit consideration in research and policy. Drawing on recent theoretical and empirical studies, the examples highlighted here show that narratives of immunity to the impacts of climate change in developed countries, confidence in technology, and perceived lack of immediacy about climate change impacts have resulted in an unwarranted complacency about adaptation. Effective adaptation strategies should be informed by a deeper understanding of the social process of adaptation and need to address a wide range of barriers.

Keywords Climate change • Adaptation • Vulnerability • Risk perception • Values • Adaptive capacity • Developed countries • Adaptation barriers • Social process • Vulnerable groups

J. Wolf
Labrador Institute of Memorial University, 490, Stn. B, Happy Valley-Goose Bay, NL A0P 1E0 Canada
e-mail: jwolf@mun.ca

J.D. Ford and L. Berrang-Ford (eds.), *Climate Change Adaptation in Developed Nations: From Theory to Practice*, Advances in Global Change Research 42, DOI 10.1007/978-94-007-0567-8_2, © Springer Science+Business Media B.V. 2011

Introduction

The European heat wave of 2003 and its effects are a stark reminder that there are groups in the population of developed countries that are vulnerable to the effects of climate change (Lagadec 2004; Kovats and Koppe 2005; Kinney et al. 2008). This event, no longer an isolated one, raises questions about other groups vulnerable to the impacts of climate change occurring now and in the future, and how best to adapt to these changes. Vulnerable groups exist in all developed countries, including, for example, the black population in New Orleans, the elderly in France, indigenous peoples in northern Canada, and Sami in Scandinavia. These groups will be disproportionately affected by the impacts of climate change and need to adjust both in response to and preparation for climatic events.

Adaptation – that is, adjustments made to practices, processes, and structures in order to take into account changing climatic conditions (McCarthy et al. 2001) – has been gaining momentum as a response to climate change during at least the past decade. An increasing body of empirical evidence is refining the theoretical approaches to adaptation research (e.g., Eakin et al. 2009). Much of this empirical evidence stems from developing countries and points to underlying vulnerabilities as key issues that adaptation practice needs to address. A smaller but growing body of evidence from developed countries suggests that vulnerability and adaptation need to be explicitly addressed in developed countries as well (Bryant et al. 2000; O'Brien et al. 2006).

Vulnerability is a key concept in adaptation research and practice. It highlights that individuals, communities, and regions are differentially at risk from the effects of climate change. In developing countries, using a vulnerability approach is intuitive and follows directly from social vulnerability, natural hazards, food security, and development research (cf., e.g., Wisner et al. 2004). Insights gleaned from this approach are relevant for research in developed countries (see e.g., Wescoat 1991). Yet, in many developed countries, the notion of being vulnerable is seemingly at odds with the presumed high adaptive capacity that is associated with wealth, democratic institutions, and (post-) industrial development. However, some authors have recently argued that the often presumed high adaptive capacity in developed countries must be examined more critically (O'Brien et al. 2006; Pielke et al. 2007; Moser and Luers 2008). Recent research has pointed out that some groups may not be adapting effectively in part because of perceptions of impacts (Wolf et al. 2009), and that there are significant regional and social differences in the adaptive capacity of developed countries (O'Brien et al. 2006). In sum, the literature on vulnerability in the context of climate change in developed countries is beginning to solidify evidence for a link between social and cultural characteristics of local places and people on one hand, and whether and how they adapt to change on the other.

The purpose of this chapter is to examine evidence on adaptation in developed countries in order to explore the social process of adaptation and identify barriers emerging from this process. It argues that adaptation is underpinned by societal perceptions, values, and decision-making structures, and that these need to be considered explicitly in adaptation research and practice. The following section conceptualizes the process of adaptation, including its individual and social characteristics, and highlights empirical examples. The final section concludes with reflections on the potential long-term implications of this research for adaptation in developed countries.

Individual and Social Dimensions of Adaptation

Empirical evidence suggests that adaptation is highly context-specific (e.g., Risbey et al. 1999; Eriksen et al. 2005), and that socioeconomic characteristics, social networks, local knowledge, and non-climatic pressures all play key roles in shaping adaptation measures. Despite these findings, there is still a tendency to apply climate change impacts research and projections directly, so as to identify concrete adaptation options and strategies without considering how underlying social and cultural conditions affect the adaptation process (O'Brien et al. 2006). Adaptation strategies developed in this fashion fail to take account of their implications for equity, legitimacy, and social acceptability. Research on the social dimensions of adaptation highlights that aspects of local context, such as social capital, cultural norms, and ongoing economic and demographic change in a location, are important in the context of adaptation because they determine how societies interact with climate change and variability (Adger 1999, 2003; O'Brien et al. 2006; Moser and Tribbia 2008; Wolf et al. 2009). The following sections address four specific dimensions of local context: perception of vulnerability and impacts; cognitive and behavioral aspects; social and institutional context; and values.

Perception of Impacts and Vulnerability

The consensus of perception research about climate change in developed countries suggests that climate change is considered a remote risk by laypeople, removed from direct personal experience in space and time (e.g., Bord et al. 1998; Kirby 2004; Lowe et al. 2006; Stamm et al. 2000). Climate change often has a lower cognitive presence than other events and issues that are more immediately experienced by individuals (e.g., Germany, Höhle 2002). Many laypeople in developed countries, like the United Kingdom and the United States, feel there will be little or no effects of climate change on them personally (Bord et al. 1998; Kirby 2004). Even in areas that could be considered vulnerable to the effects of climate change, individuals have difficulties relating impacts of climate change to their local surroundings and their everyday life (e.g., United Kingdom, Bickerstaff et al. 2004). Perceptions of local changes are heightened among indigenous people (e.g., Canadian Arctic, Berkes and Jolly 2001; Riedlinger and Berkes 2001) and remote resource-dependent communities (Sydneysmith 2007). Climate change as a remote-risk renders the problem a vague and distant notion, and this explains, in part, why still little proactive adaptation is taking place by the public in developed countries.

Part of understanding the individual and social dimensions of adaptation depends on how vulnerability is conceptualized. The term vulnerability is often defined as a function of exposure, sensitivity, and adaptive capacity (McCarthy et al. 2001). This conceptualization, however, says little about why the effects of climate change are socially differentiated. In the context of understanding the process of adaptation, vulnerability can be usefully defined as resulting from those "characteristics of a person or group and their situation that influence their capacity to anticipate, cope with, resist, and recover" from an impact (Wisner et al. 2004). Attempting to outline adaptation options without considering the underlying socio-cultural conditions means that the characteristics mentioned by Wisner et al. are neglected and the mechanisms that produce differentiated impacts are obscured.

Research conducted on islands on the Canadian west coast indicates that perceptions of vulnerability to the effects of climate change are substantially shaped by knowledge and perceptions of pre-existing local underlying vulnerabilities (Wolf 2006). In the case of Salt Spring Island, for example, local residents are well aware that freshwater shortages are recurring issues during drought years. When presented with information about other likely impacts of climate change that are already occurring, however, these impacts fail to register with residents. In the case of this island, the local media adamantly report on freshwater issues in a context of local development pressures, while remaining silent on other climate-related impacts (e.g., landslides and flooding) and portraying the effects of climate change as relevant to developing countries such as Bangladesh. This portrayal has a direct impact on residents' perception that local impacts are absent. Combined with local narratives about resilience and self-sufficiency, it is at least plausible that this absence of local impacts portrayed in media prevents information about impacts from being taken up due to biased assimilation (Rachlinski 2000) until they have been experienced directly. At that time, however, it may be too late for anticipatory adaptation.

In sum, these results raise profound questions about how the public perceives and responds to (governmental) measures to adapt to the impacts of climate change. Further research has yet to elaborate just how significant the obstacles posed by this mismatch between ongoing and perceived impacts may be.

Cognitive and Behavioral Aspects

Risk perception and perceived adaptive capacity have been found to be important to individual-level proactive adaptation. In a study conducted in Germany, a model that included socio-cognitive aspects explained more variance than a standard socioeconomic model (Grothmann and Patt 2005). Reasons for this are that income measures alone do not account for how adaptation decisions are made. Whether proactive adaptation occurs depends on perceived ability to adapt and perceived risk from impacts.

Perceived individual agency plays a role in whether or not people act on an issue. A related cognitive mechanism is self-efficacy, which affects the perceived ability to act (Bandura 1977) in the face of risk. Both high and low self-efficacy can lead to counterproductive outcomes for adaptation. This is demonstrated by a study of elderly people's perceptions of heat wave risk and vulnerability (Wolf et al. 2009) which suggests that proactive adaptation may be prevented by individuals either insisting on their independence and abilities to cope (characteristics which are aligned with high self-efficacy), or believing that nothing can be done to prevent the health effects of heat effectively (a characteristic aligned with low self-efficacy).

Accepting that one is at risk of exposure to an impact of climate change is not simply a function of knowledge or information. Rather, direct experience, for example with flooding, has been shown to be a key factor in individuals accepting that flooding is a personal risk (Keller et al. 2006). But not even direct experience necessarily facilitates acceptance of risk and behavioral response to it. In a study of residents in an area prone to flooding in China, the vast majority of respondents believed that flooding is inevitable and cannot be prevented (Wong and Zhao 2001). As a result, residents have become accustomed to flooding and return to business as usual soon after the flood waters have receded. Further, no clear causal link is perceived between flooding and climate change among residents in flood-prone areas in the United Kingdom (Whitmarsh 2008). This suggests that residents do not perceive any likelihood of flooding reoccurring in a context of climate change. Hence, when individuals feel the controllability of an event is low, this perception often leads to inaction rather than preparatory action.

The way in which dangerous climate change is framed provides useful insight into the different approaches to conceptualizing danger, and the potential implications of this for adaptation. External definitions of dangerous climate change rely on indicators of physical or social vulnerability (Dessai et al. 2004). Internal definitions, however, consider perceived and experienced consequences associated with climate change and their effects on life and livelihood (Dessai et al. 2004), and are ultimately determined by what is valued. In the absence of a comprehensive understanding of which impacts are viewed as dangerous by different communities and cultures, and leaving aside non-climatic stimuli for adaptation, little insight exists into which aspects of climate change and variability actually motivate adaptation.

A study conducted with elderly people in the United Kingdom suggests that members of this age group, identified by epidemiological research as vulnerable (e.g., Mc Farlane 1978; McMichael et al. 1996), do not perceive themselves as at risk from the effects of heat waves (Wolf et al. 2009). Social contacts may be unaware who is vulnerable to the effects of heat, which medical conditions and medications exacerbate risk, and what can be done to reduce heat effects (Wolf et al. 2009). These findings point to a less than straightforward relationship between social networks and health outcomes. The underlying social and individual characteristics in their cultural context are important factors that shape whether, and if so how, adaptation takes place.

The Role of the Social and Institutional Context

The social and institutional contexts of adaptation have been demonstrated to be important in shaping adaptive responses. Among institutional factors, governance in the form of institutional structures and mechanisms, including decision processes, is influential in shaping the process of adaptation. Literature on the institutional dimensions of adaptation (Adger 2001; McBeath 2003; Næss et al. 2005) suggests that institutional barriers are ubiquitous in developed countries such as Norway, the United Kingdom, and the United States. Weak incentives for proactive flood management, institutional overlaps, and poor coordination between various scales of governance prevent efforts to adapt. The evidence overall suggests that governance for effective adaptation, even in highly developed nations, remains elusive (Moser 2009).

Some argue that "social capital offers ways of understanding the role of fundamental social attributes that contribute toward building capacity for social collectives and individuals to respond to climate change" (Pelling and High 2005, p. 317). Empirical evidence, too, suggests that the ability of societies to adapt is determined, in part, by the ability to act collectively (Adger 2003). In developing countries, such as Vietnam, vulnerability is often enhanced by institutional dynamics such as the breakdown of collective action to protect from extreme events (Adger 1999). But the effects of social networks do not always contribute to positive outcomes. Investigating water use in Australia, Miller and Buys (2008) find that different aspects of social capital can have different implications, with some aspects having negative consequences on the community as a whole. Strong bonding ties can contribute to vulnerability of a population rather than reduce it, as suggested by a recent study of elderly people's responses to heat wave risk in the United Kingdom (Wolf et al. 2010).

Technology-intensive societies in the global north have created local and global risks (Beck 1996). The use of technology to extract energy from fossil fuels, for example, contributes to anthropogenic climate change and the associated risks. In Australia, Bulkeley (2001) suggests that complex linkages between government and industry are sustained by policy responses to climate change that negate the challenges climate change poses. Technological development, in part driven by a social demand, can therefore produce paradox outcomes, on one hand generating risks such as climate change, and on the other hand identifying technology as a way to manage the risks. A perspective of "risk society" (Beck 1996) has not informed adaptation research and policy to date, yet such an approach could help foster an understanding of the structural barriers to adaptation.

There are important interactions between individuals' perceptions of climate variability and impacts and the institutions that govern affected resources, as shown by several case studies. First, in a study in British Columbia, Canada, the public was found to be ill-informed and disconnected from water management due to a combination of negative factors, namely: multiple layers of governance of water resources; overlapping mandates at the local level; opaque mandates and regulatory uncertainty at the provincial level; and poor public communication about the impacts of water management (Wolf 2006). The public has questioned the efficacy of efforts by local governments that have the mandate for water management. The discontent has shaped informal networks to research the issue and attempt to engage in managing the resource (Wolf 2006). Second, the importance of institutional factors for climate change adaptation is highlighted in the agricultural sector in Norway. There is little evidence of a direct positive relationship between high adaptive capacity and actual adaptation taking place in this sector (O'Brien et al. 2006). Rather, the study suggests that in many areas in Norway, adaptation is unlikely to take place without significant institutional and financial support (O'Brien et al. 2006). Third, the results of a study conducted in California that examined coastal managers' information needs and barriers to adaptation suggest that while coastal managers are aware of the impacts of climate change on the coast, they have to date done little to prepare for impacts (Moser and Tribbia 2008). Constrained staff time and financial resources and a lack of official mandates inhibit managers' ability to address climate change concerns.

Similar studies in forestry and agriculture settings (Romsdahl 2009) highlight similar issues. This suggests that the outcomes are not in fact sectoral, but rather pervasive findings emerging from the underlying social structures that sectors have in common. Engaging resource managers in issues of climate change is necessary but insufficient to assess underlying vulnerabilities and prepare for impacts (Moser and Luers 2008; Moser and Tribbia 2008; Tribbia and Moser 2008).

The sum of this evidence underscores that unless institutional and decision structures are deliberately included in efforts to adapt, and the values that found the decisions are made explicit, the barriers arising from governance mechanisms are unlikely to be addressed.

Values and the Goals of Adaptation

Underlying any adaptation strategy and any decision about governance for adaptation is a set of values that shape what is perceived to be worth the effort of adaptation. In this context, values are defined as standards that guide decisions, choices, and attitudes (Rokeach 2000). These values also define what the goals of adaptation should be, for example, preserving the status quo of a particular ecosystem or improving quality of life and well-being. The values and goals of adaptation also determine what is perceived as a limit to adaptation and to what adjustments are deemed possible (Adger et al. 2009). While this may seem obvious, there is surprisingly little research about what these values are, indeed whose values they are and what they imply for adaptation outcomes. In fact, how values shape adaptation has, until recently, not been considered in any detail. Adaptations that address concerns of one set of values may not address those of a different and competing set of values. In a Norwegian context, O'Brien (2009) argues that the various worldviews and value systems of Norwegians could arguably lead to different adaptations to snow cover changes as a result of climate change. Some of these adaptations do not meet the same goals and hence fail to qualify as successful for some people while seeming adequate to others.

Dynamic and competing interests and goals also shape adaptation in coastal settings. Diverse sets of coastal stakeholders favor different adaptation strategies (Few et al. 2007). Recognizing the diverse interests of stakeholders has prompted some research to cater explicitly to these differences and make any trade-offs between alternate adaptation strategies and goals explicit (Tompkins et al. 2008). Unlike in coastal settings, where adaptive management has been and continues to be vital, other decision spaces on adaptation remain relatively untouched by the recognition that values are key to the process and outcomes of adaptation.

A study of flood management in Norway suggested that large-scale flood events were perceived to be outside the responsibility of municipalities and that damages should be covered by the national government (Næss et al. 2005). When local political and economic interests coincided with national political will, flood prevention measures were implemented at the expense of other local priorities. Despite opposition from non-governmental actors, decision-makers resorted to technical measures, ignoring local interests in protected areas and fishing. The authors identified that differences in culture and perception between the local and the national level can slow the process of social learning and thus hinder adaptation (Næss et al. 2005).

On the question what "dangerous" climate change might mean, some scientists believe that is not the task of the scientific community to provide a definition (Schneider and Lane 2006). Instead, it is argued, this question involves value judgments about the various impacts and about what impacts, if any, are acceptable (Schneider and Azar 2001; Mastrandrea and Schneider 2004). When conceptualizing adaptation as a social process, however, it cannot be denied that values are inherently embedded in how climate risk is perceived and in decision-making about how to adapt. For this reason there is an urgent need for a more explicit analysis of the role of values for adaptation.

The evidence outlined here suggests at least four key factors that shape the social process of adaptation. First, perceptions, both those of government officials and the public, come to bear on how vulnerability is conceived, which in turn has direct consequences for adaptation regardless of what the adaptive capacity might be. Second, institutional structures at every level operate in a socio-cultural context which influences how and in response to what decisions are made. High adaptive capacity does not necessarily translate into actual effective adaptation measures. Third, social networks, including informal associations, can play an important role in shaping adaptation; but the presence of and involvement in such networks does not necessarily lead to positive outcomes. Fourth, underlying all three factors are complex and competing values and goals that must be considered explicitly if adaptation research and policy are to be successful.

Conclusion

This chapter argues that climate change adaptation is an inherently social process, underpinned by socio-cultural characteristics of the society or group that is adapting. A key finding of adaptation and vulnerability research is that there are vulnerable groups within developed countries that are already and will continue to be disproportionately affected by the changing climate. The research examined here demonstrates that perceptions of climate change impacts and vulnerability to the impacts, perceived adaptive capacity and individual agency, social networks, formal institutions, and their mandates, and values and goals all play key roles in shaping whether and how adaptation takes place.

The impacts of climate change materialize in part due to underlying non-climatic vulnerabilities and this is as relevant in developed as in developing countries. If the role of government is to protect the most vulnerable from the impacts, an explicit consideration of the social and cultural context that embeds adaptation is necessary. This is particularly the case given the lack of perceived risk among some vulnerable groups, such as the elderly, but applies also to vulnerable communities that struggle with access to resources and loss of cultural identity, such as in the Canadian North. As part of the underlying vulnerability, economic globalization will continue to have profound effects in developed countries. Environmental changes, including the impacts of climate change and economic globalization interact to shape unprecedented pressures, not only for marginal communities (Leichenko and O'Brien 2008). The effects of climate change therefore cannot be isolated from the larger social, cultural, and economic context of a locality and must be understood as only one factor among many that produce unfavorable outcomes for those who are vulnerable.

There is an urgent need to resolve questions of responsibility for adaptation. Considering present mandates of municipal and national governments, and the conflicts potentially arising from differences between scales of governance, it seems prudent to assess whether and how such mandates can be expanded. While however national governments may be perceived as having responsibility, adaptation is ultimately implemented at scales closer to the local level. It is at the local scale that efforts carried out are scrutinized by actors with interests that compete with those in power, and while deliberation and democratic process are expected, neither necessarily works in favor of those most vulnerable.

While the discussion above focuses on adaptation, there are similar barriers to public engagement on climate change in general (Lorenzoni et al. 2007), including on mitigation. Therefore, in the context of developed countries with significant mitigation responsibilities, there are important lessons to be learned for and from mitigation as a social process.

Fundamentally, faith in the efficacy of the high adaptive capacity of developed countries seems misplaced if social and cultural factors, such as perceptions, goals, and values, either prevent adaptation from occurring or produce short-sighted and potentially maladapted strategies that ignore underlying vulnerabilities, reproduce existing power structures, and are unacceptable to the public.

Acknowledgments This chapter has benefitted from discussions with and comments from Karen O'Brien and Irene Lorenzoni.

References

Adger WN (1999) Social vulnerability to climate change and extremes in coastal Vietnam. World Dev 27(2):249–269

Adger WN (2001) Scales of governance and environmental justice for adaptation and mitigation of climate change. J Int Dev 13(7):921–931

Adger WN (2003) Social capital, collective action, and adaptation to climate change. Econ Geogr 79(4):387–404

Adger WN, Dessai S, Goulden M et al (2009) Are there social limits to adaptation to climate change? Clim Change 93(3–4):335–354

Bandura A (1977) Self-efficacy: toward a unifying theory of behavioral change. Psychol Rev 84(2):191–215

Beck U (1996) World risk society as cosmopolitan society? Ecological questions in a framework of manufactured uncertainties. Theory Cult Soc 13(4):1–32

Berkes F, Jolly D (2001) Adapting to climate change: social-ecological resilience in a Canadian Western Arctic community. Conserv Ecol 5(2):18–39

Bickerstaff K, Simmons P, Pidgeon N (2004) Public perceptions of risk, science and governance: main findings of a qualitative study of five risk cases. Centre for Environmental Risk, University of East Anglia, Norwich

Bord RJ, Fisher A, O'Connor RE (1998) Public perceptions of global warming: United States and international perspectives. Clim Res 11(1):75–84

Bryant CR, Smit B, Brklacich M et al (2000) Adaptation in Canadian agriculture to climatic variability and change. Clim Change 45(1):181–201

Bulkeley H (2001) Governing climate change: the politics of risk society? Trans Inst Br Geogr 26(4):430–447

Dessai S, Adger WN, Hulme M et al (2004) Defining and experiencing dangerous climate change. Clim Change 64(1–2):11–25

Eakin H, Tompkins EL, Nelson DR et al (2009) Hidden costs and disparate uncertainties: trade-offs involved in approaches to climate policy. In: Adger WN, Lorenzoni I, O'Brien K (eds) Adapting to climate change: governance, values and limits. Cambridge University Press, Cambridge

Eriksen SH, Brown K, Kelly PM (2005) The dynamics of vulnerability: locating coping strategies in Kenya and Tanzania. Geogr J 171(4):287–305

Few R, Brown K, Tompkins EL (2007) Climate change and coastal management decisions: insights from Christchurch Bay, UK. Coast Manag 35(2–3):255–270

Grothmann T, Patt A (2005) Adaptive capacity and human cognition: the process of individual adaptation to climate change. Glob Environ Change Part A 15(3):199–213

Höhle E (2002) Chapter 5: global climate change as perceived by the public (Joint working report). Centre of Technology Assessment in Baden-Württemberg and University of Stuttgart, Sociology of Technologies and Environment, Stuttgart

Keller C, Siegrist M, Gutscher H (2006) The role of the affect and availability heuristics in risk communication. Risk Anal 26(3):631–639

Kinney PL, O'Neill MS, Bell ML et al (2008) Approaches for estimating effects of climate change on heat-related deaths: challenges and opportunities. Environ Sci Policy 11(1):87–96

Kirby A (2004) Britons unsure of climate costs – polling results. BBC News Online. http://news. bbc.co.uk/2/hi/science/nature/3934363.stm. Cited 20 Jan 2008

Kovats RS, Koppe C (2005) Heat waves: past and future impacts on health. In: Ebi KL, Smith JB, Burton I (eds) Integration of public health with adaptation to climate change: lessons learned and new directions. Taylor & Francis, London, pp 136–160

Lagadec P (2004) Understanding the French 2003 heat wave experience: beyond the heat, a multi-layered challenge. J Conting Crisis Manag 12(4):160–169

Leichenko R, O'Brien K (2008) Environmental change and globalization: double exposures. Oxford University Press, Oxford

Lorenzoni I, Nicholson-Cole S, Whitmarsh L (2007) Barriers perceived to engaging with climate change among the UK public and their policy implications. Glob Environ Change 17(3–4): 445–459

Lowe T, Brown K, Dessai S et al (2006) Does tomorrow ever come? Disaster narrative and public perception of climate change. Public Underst Sci 15(4):435–457

Mastrandrea M, Schneider SH (2004) Probabilistic integrated assessment of 'dangerous' climate change. Science 304(5670):571–575

Mc Farlane A (1978) Daily mortality and environment in English conurbations: II. Deaths during summer hot spells in Greater London. Environ Res 15(3):332–341

McBeath J (2003) Institutional responses to climate change: the case of the Alaskan transportation system. Mitig Adapt Strateg Glob Change 8(1):3–28

McCarthy JJ, Canziani OF, Leary NA et al (eds) (2001) Climate change 2001: impacts, adaptation and vulnerability – contribution of working group II to the third assessment report of the Intergovernmental Panel on Climate Change. Cambridge University Press, Cambridge

McMichael AJ, Haines A, Sloof R et al (eds) (1996) Climate change and human health. World Health Organization, Geneva

Miller E, Buys L (2008) The impact of social capital on residential water-affecting behaviors in a drought-prone Australian community. Soc Nat Resour 21(3):244–257

Moser S (2009) Whether our levers are long enough and the fulcrum strong? – Exploring the soft underbelly of adaptation decisions and actions. In: Adger WN, Lorenzoni I, O'Brien K (eds) Adapting to climate change: governance, values and limits. Cambridge University Press, Cambridge

Moser S, Luers AL (2008) Managing climate risks in California: the need to engage resource managers for successful adaptation to change. Clim Change 87(1):307–322

Moser S, Tribbia J (2008) Vulnerability to inundation and climate change impacts in California: coastal managers' attitudes and perceptions. Mar Technol Soc 40(4):35–44

Næss LO, Bang G, Eriksen S et al (2005) Institutional adaptation to climate change: flood responses at the municipal level in Norway. Glob Environ Change Part A 15(2):125–138

O'Brien KL (2009) Climate change and values: do changing values define the limits to successful adaptation? In: Adger WN, Lorenzoni I, O'Brien K (eds) Adapting to climate change: governance, values and limits. Cambridge University Press, Cambridge

O'Brien KL, Eriksen S, Sygna L et al (2006) Questioning complacency: climate change impacts, vulnerability and adaptation in Norway. AMBIO J Hum Environ 35(2):50–56

Pelling M, High C (2005) Understanding adaptation: what can social capital offer assessments of adaptive capacity? Glob Environ Change Part A 15(4):308–319

Pielke RA Jr, Prins G, Rayner S et al (2007) Lifting the taboo on adaptation. Nature 445:597–598

Rachlinski JJ (2000) The psychology of global climate change. Univ Ill Law Rev 1:299–319

Riedlinger D, Berkes F (2001) Contributions of traditional knowledge to understanding climate change in the Canadian Arctic. Polar Rec 37(203):315–328

Risbey JS, Kandlikar M, Dowlatabadi H et al (1999) Scale, context, and decision making in agricultural adaptation to climate variability and change. Mitig Adapt Strateg Glob Change 4:137–165

Rokeach M (2000) Understanding human values: individual and societal. Free Press, New York

Romsdahl R (2009) Addressing institutional challenges in adaptation planning for climate change impacts in the U.S. Northern Great Plains: a case study of North Dakota. Interdiscip Environ Rev XI(1&2):35–56

Schneider SH, Azar C (2001) Are uncertainties in climate and energy systems a justification for stronger near-term mitigation policies? Paper presented at the Pew Center on global climate change's workshop on the timing of climate change policies, Washington, DC, 11–12 Oct 2001

Schneider SH, Lane J (2006) Dangers and thresholds in climate change and the implications for justice. In: Adger WN, Paavola J, Huq S et al (eds) Fairness in adaptation to climate change. MIT, Cambridge, MA

Stamm KR, Clark F, Eblacas PR (2000) Mass communication and public understanding of environmental problems: the case of global warming. Public Underst Sci 9(3):219–237

Sydneysmith R (2007) The co-management of climate change in coastal communities of British Columbia: social capital, trust and capacity. University of British Columbia, Vancouver

Tompkins EL, Few R, Brown K (2008) Scenario-based stakeholder engagement: incorporating stakeholders preferences into coastal planning for climate change. J Environ Manage 88(4):1580–1592

Tribbia J, Moser S (2008) More than information: what coastal managers need to prepare for climate change. Environ Sci Policy 11(4):315–328

Wescoat J (1991) Managing the Indus River basin in light of climate change: four conceptual approaches. Glob Environ Change 1(5):381–395

Whitmarsh L (2008) Are flood victims more concerned about climate change than other people? The role of direct experience in risk perception and behavioural response. J Risk Res 11(3):351–374

Wisner B, Cannon T, Blaikie PM et al (2004) At risk, 2nd edn. Routledge, London

Wolf J (2006) Climate change and citizenship: a case study of responses in Canadian coastal communities. Unpublished PhD thesis, University of East Anglia, Norwich

Wolf J, Lorenzoni I, Few R et al (2009) Conceptual and practical barriers to adaptation: an interdisciplinary analysis of vulnerability and adaptation to heat waves in the UK. In: Adger WN, Lorenzoni I, O'Brien K (eds) Adapting to climate change: governance, values and limits. Cambridge University Press, Cambridge

Wolf J, Adger WN, Lorenzoni I et al (2010) Social capital, individual responses to heat waves and climate change adaptation: an empirical study of two UK cities. Glob Environ Change 20(1):44–52

Wong KK, Zhao XB (2001) Living with flood: victim's perceptions in Beijiang, Guangdong, China. Area 33(2):190–201

Chapter 3
Entering the Period of Consequences: The Explosive US Awakening to the Need for Adaptation

Susanne C. Moser

Abstract Since the early years of the twenty-first century, the United States has been awakening rapidly to the fact that climate change is underway and that adaptation to the unavoidable impacts of climate change is needed and must be begun now. This chapter provides an historical overview of the public, political, and scientific concern with adaptation in America. It begins by describing the shift from the early concerns with climate change and adaptation to the more recent awakening to the need for a comprehensive approach to managing the risks from climate change, as reflected in the news media. This shift is evident from the recent debates and drafting of adaptation-related bills in Congress; to the rapidly expanding activities at the state and local government levels; to the increasing engagement of nongovernmental organizations, professional associations, scientists, and consultants. This policy rush is not underlain, however, by widespread public engagement and mobilization, nor does it rest on a solid research foundation. To help the United States prepare adequately for the impacts of climate change, funding for vulnerability and adaptation research must be significantly increased. This will facilitate establishing adequate decision support mechanisms; effective communication and public involvement; and building the necessary capacity in science, the consulting world, and in government agencies.

Keywords Adaptation • Federal adaptation policy • State adaptation planning • Local adaptation planning • Media coverage • Public debate • Science policy • Adaptation research • Vulnerability • Barriers to adaptation • Nongovernmental organizations • Actor network • Scale • Cross-scale coordination • United States

S.C. Moser (✉)
University of California-Santa Cruz and Susanne Moser Research & Consulting, 134 Shelter Lagoon Drive, Santa Cruz, CA 95060, USA
e-mail: promundi@susannemoser.com

J.D. Ford and L. Berrang-Ford (eds.), *Climate Change Adaptation in Developed Nations: From Theory to Practice*, Advances in Global Change Research 42, DOI 10.1007/978-94-007-0567-8_3, © Springer Science+Business Media B.V. 2011

Introduction

Owing to past neglect, in the face of the plainest warnings,
we have now entered upon a time of great danger . . .
The era of procrastination, of half-measures, of soothing and
baffling expedients, of delays, is coming to a close.
In its place we are entering a period of consequences . . .
We cannot avoid this period, we are in it now . . .

Winston S. Churchill
November 12, 1936

On December 19, 2008, an article in *The San Francisco Chronicle* called for "ideas on living in a warming world" and reported on a competition asking designers to "climate proof" the Bay Area (King 2008). Twenty years earlier, a news article in *The Boston Globe* reported on an international conference which urged coastal populations to prepare for the impacts of climate change (Dumanoski 1988). Between these two geographic and temporal bookends lies a country that has been largely untroubled by the question of adaptation to the impacts of anthropogenic climate change.

At least, this was the case until recently. Since the early years of the twenty-first century, the United States has been rapidly awakening to the fact that climate change is underway and that even if stringent greenhouse gas mitigation were implemented, adaptation to the unavoidable impacts of climate change is needed now. This sudden awakening has led to a flurry of activities at the federal, state, and local levels and in the nongovernmental community.

This chapter provides an historical overview of the public, political, and scientific concern with adaptation in the United States. It begins by describing the shift from the early concerns with climate change and adaptation, to the more recent explosive awakening to the need for a comprehensive approach to managing climate risks. This trajectory of change is reflected in media attention paid to adaptation over the past 20 years. The concurrent political context has shaped a national science policy that has tended to relegate vulnerability and adaptation-related research to the "back burner" of federal research and agency support.

Turning to the more recent history, the chapter overviews some of the adaptation efforts initiated in the last 3–5 years at the federal, state, and local levels, by government and nongovernmental entities, and in different climate-sensitive sectors (for a more detailed accounting see Moser 2009). It concludes with a discussion of some plausible implications for the state of US preparedness to deal with climate change impacts, opportunities, and barriers to adaptation. These implications suggest important future research directions and decision support needs, and the need to engage the American public in a broader dialogue on comprehensive climate risk management and resilience in the face of rapid change.

"In the Face of the Plainest Warnings": Adaptation in the Media

How have concern and public debate about adaptation changed over time in US media? News coverage in various media markets provides an interesting indicator of how the United States is shifting from "business-as-usual" coping with climate variability to an explicit concern with adaptation to climate change.

Figure 3.1 illustrates the "explosive" awakening to the need for adaptation: 67% of all articles appeared between 2006 and 2008; a nearly fourfold increase in the number of articles occurred from 2006 to 2007 alone. Yet, this compilation may underestimate the true news coverage on adaptation over time (as might be revealed by using other, but related search terms). Moreover, web-based information experienced a large rise in relative importance. There is no apparent reason to assume, however, that additional news sources should follow a completely different pattern of frequency; rather, technological and market trends suggest that they may magnify the "explosive" pattern even more (Pew Research Center for the People and the Press 2008; Croteau and Hoynes 2006; Alexander et al. 2004; Compaine and Gomery 2000).

Maybe more interesting than these patterns is the dearth of adaptation coverage in the 1980s–1990s. For example, the first US-wide climate change impacts assessment that discussed (and could have raised public interest in) adaptation was a 1989 assessment led by the Environmental Protection Agency (EPA) (Smith and Tirpak 1989); yet news coverage was virtually absent. The first assessment by the Intergovernmental Panel on Climate Change (IPCC) was released a year later – albeit with impacts and vulnerability/adaptation still underdeveloped – but did not generate any mention of the need for adaptation in US newspapers. The second IPCC assessment fared little better. From 1997 to 2001, coverage of

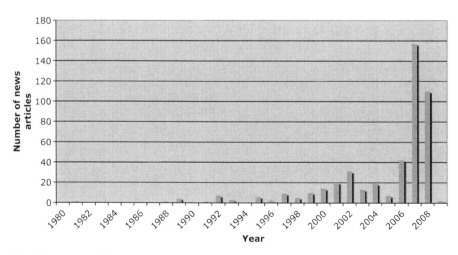

Fig. 3.1 Number of articles in US print media on adaptation to climate change (Source: Moser 2009)

adaptation increased moderately, possibly reflecting the explicit interest in impacts and adaptation in the *First US National Assessment of the Potential Consequences of Climate Variability and Change*. In 2001, the IPCC released its third assessment, contributing to the steady increase in coverage in those years. Since then, adaptation coverage has remained higher than historical levels, due to Al Gore's movie *An Inconvenient Truth*, mounting evidence of climate change impacts and a growing recognition of the need for a comprehensive approach to managing climate risks. The release of the fourth IPCC assessment and a range of extreme events produced the spike in 2007.

The extensive adaptation coverage since 2007 reflects several important developments. First, scientific insights on vulnerabilities, impacts, and adaptation have matured over the past decade (Parry et al. 2007). Second, there is "unequivocal" evidence that climate change is occurring and that the warming observed over the last half of the twentieth century is, with over 90% confidence, human-induced (Solomon et al. 2007). These developments have palpably shifted the political debate in the United States, as reflected in reporters' recent shifts from "balanced" to "weight-of-evidence" reporting (Boykoff and Boykoff 2004; Boykoff 2007a, b). Reporters have also sought a fresh angle on climate change reporting through a greater focus on local aspects, impacts, and responses. Moreover, nongovernmental and media groups have made deliberate efforts around the release of the IPCC reports to educate reporters about impacts and adaptation issues.

In summary, trends in the news media suggests that public attention in the United States is being redirected increasingly from questions of whether or not climate change is occurring and human-caused to ones about its local impacts and how to deal with them. This shift is remarkable in light of the challenging political context in which adaptation has been debated over the past two decades.

The "Era of Baffling Expedients": Adaptation in Public Debate

In 2007, at the height of news coverage on adaptation, a commentary in *Nature* suggested that there had been a "taboo" on talking about adaptation to climate change until that time (Pielke et al. 2007). Immediately following the publication of that article, the conservative newspaper *The Washington Times* published a commentary by Henry I. Miller, a fellow at Stanford University's Hoover Institution, who argued,

> Mr. Pielke and his colleagues criticize the political obsession with the idea that climate risks can be reduced by cutting emissions, because it distracts attention from other, more cost-effective approaches. However, for many activists, emissions reduction has become an article of faith: Al Gore dismissed adaptation as a kind of laziness, an arrogant faith in our ability to react in time to save our skins (Miller 2007).

Former Vice President Al Gore indeed once rejected adaptation as a distraction from the need to reduce the causes of climate change. In a 1989 interview, Gore stated,

The advice to adapt rather than prevent these [climate] changes is unwise counsel. The changes are occurring so rapidly that adaptation will not be possible without major disruptions in our civilization. The advice to adapt is an obstacle to the correct political response, which is prevention (quoted in McMasters 1989).

This brief caricature reflects a long-standing framing of climate change solutions in US public debate in which mitigation and adaptation are pitched against each other. In the absence of a solid scientific foundation on impacts and adaptation, and substantial political resistance to mitigating greenhouse gas emissions (across Democratic and Republican administrations), environmental nongovernmental organizations (NGOs) and other emission reduction advocates have avoided talking about adaptation, fearing that it would distract from the need for mitigation or be perceived as defeatist (e.g., Burton 1994). By contrast, "Climate contrarians either deny the reality of climate change or promote a view that suggests Americans have the capacity to adapt should climate change materialize, and that therefore there is no need to take action now to prepare or to mitigate" (Moser and Luers 2008, p. S31).

These entrenched positions have begun to be loosened up since the IPCC's 2007 assessment. Increasingly, the necessity of adaptation has been accepted, along with rational economic arguments to adapt now rather than delay until disaster strikes. Occasionally, questions emerged whether communities and sectors have the capacity to adapt without significant suffering or economic losses. The most constructive new frames view mitigation and adaptation as complementary, potentially synergistic approaches, and pay careful attention to trade-offs and mutual support, without one taking away from the necessity or urgency of the other.

"Owing to Past Neglect": The State of Adaptation Research

As the United States begins to develop adaptation strategies, the need for relevant science on climate change impacts, vulnerabilities, and adaptation options and constraints is growing rapidly. The supply of spatially and temporally relevant, high-resolution climate change projections, however, is still constrained by the capabilities of current climate models (Bader et al. 2008). The studies that do exist have produced important information on the climatic hazards that regions of the United States may face, yet probabilistic information quantifying the risks is not yet available.

By far the greatest lack of scientific information needed to adequately inform regional and local adaptation planning efforts lies in the ecological and human dimensions of global change (Moser 2010). Little is known about the on-the-ground degree of vulnerability of habitats, sectors, and specific communities to climatic hazards; and only spotty and preliminary research exists assessing the economic costs and benefits, feasibility, and collateral consequences of different adaptation options (NRC 2010). A long-standing research tradition in hazards, risks, and disaster studies has demonstrated that even so-called "natural" disasters are never just a function of the physical characteristics of an extreme event, but always the product of the interaction between that event and the exposed human and environmental systems (White 1974; Cannon 1994; Cutter 1996; Abramovitz and Starke 2001; Kumagai et al. 2006). Thus, taking the specific characteristics of the human or environmental system into account would yield a rather different, and spatially and socially differentiated, picture of who and what is vulnerable to climate change (e.g., Adger et al. 2007; Turner et al. 2003). Moreover, insights into the deeper causes and differences in vulnerability would better inform preparedness, planning, and adaptation interventions.

This lack of attention to vulnerability and adaptation research is rooted deeply in federal science policy and an historical imbalance in research funding for global change science. Annual budget allocations to the Climate Change Science Program (CCSP) and its predecessor have varied considerably over time but have been in steep decline in absolute terms from 2003 to 2008 (AAAS 2009). The largest portion of the CCSP budget every year has gone to NASA and less than half of the total budget to the 12 remaining CCSP agencies. While most agencies conduct some climate impacts research, only a handful – the National Science Foundation (NSF), National Oceanic and Atmospheric Administration (NOAA), Environmental Protection Agency (EPA), and to a lesser extent the United States Department of Agriculture (USDA), and National Institutes of Health (NIH) – conduct or sponsor research on social, economic, or behavioral science relevant to climate change. According to the National Research Council's (NRC) Committee on the Human Dimensions of Global Change, probably only a fraction of 1% of annual federal climate-related research funds goes to increasing the scientific understanding of vulnerability and adaptation (personal communication with Tom Wilbanks, 19 Dec 2008).

Not surprisingly, the NRC's Committee on Strategic Advice to the CCSP concluded in its 2007 evaluation of the Program that

> Progress in human dimensions research has lagged progress in natural climate science, and the two fields have not yet been integrated in a way that would allow the potential societal impacts of climate change and management responses to be addressed (NRC 2007, p. 5).

Lack of financial support is exacerbated by decades of inadequate collection, management, integration, and accessibility of relevant data; organizational barriers in the federal government that hinder the integration of social science knowledge; and institutional, cultural, and human resource/capacity hurdles undermining the effective collaboration of physical and social scientists on climate change (Stern and Wilbanks 2008).

Consequently, US states, communities, and industries are now attempting to develop adaptation strategies, yet lack the scientific foundation – particularly regarding on-the-ground vulnerabilities and the feasibility of various adaptation options – on which rational, scientifically informed policy could be made.

The Close of the "Era of Half-Measures": Emerging Adaptation Efforts Across the United States

In 2004, the Pew Center for Global Climate Change released a report on the role of adaptation in the United States (Easterling et al. 2004). Its authors pointed to a few adaptive changes in industry and some proactive measures taken in long-term transportation and water infrastructure planning. Notably absent were any mentions of state-level adaptation planning efforts or changes at the federal level. Since then, significant policy developments have occurred at all levels of governance.

Federal Policy Developments and Planning for Climate Impacts

The development of federal climate change policy became an active area of Congressional activity after the 2006 elections. Democratic leadership in key committees has opened the door to hearings and discussion of various climate-related bills. The 110th and 111th Congresses (January 3, 2007 to January 3, 2011) have produced a variety of bills in both the House and the Senate that address adaptation.

One of the bills already enacted into law authorized the America's Climate Choices suite of studies, initiated in late 2008 by the NRC (http://www.americasclimatechoices.org). It represents a Congressional request for (1) scientific input into the development of mitigation and adaptation policies and (2) guidance on research priorities, including how to build climate-related decision support capability. This NRC effort builds on other recent studies and advisory activities and will be highly influential over the coming years (NRC 2003, 2004, 2007, 2009a, b).

Outside the legislative branch of federal government, federal agencies have sponsored limited adaptation-specific research and assessments or made changes in their own operations. For example, individual agency researchers have investigated climate impacts and adaptation options, and some of the CCSP's "Synthesis and Assessment Products" have begun to address vulnerability, adaptive capacity, adaptation options, and constraints (e.g., Anderson et al. 2009; Fagre et al. 2008; Julius et al. 2008; Wilbanks et al. 2007; Gamble et al. 2008; Savonis et al. 2008). Several others address issues that will affect the United States' ability to adapt, such as extreme and abrupt climate and environmental changes and the ability to use climate change-related information in decision making (e.g., Karl et al. 2008; Clark et al. 2008; Backlund et al. 2008; CCSP 2008; Morgan et al. 2009; Beller-Simms et al. 2008).

In 2007, the Government Accountability Office highlighted the lack of guidance on managing the effects of climate change for federal land and resource managers, and noted the growing vulnerability of federal and private insurers to the impacts of weather-related disasters (GAO 2007a, b). In response, federal agencies have significantly increased their internal review and strategic planning efforts (GAO 2009), though actual management or rule changes affecting operations on the ground are still very limited (Anderson et al. 2009).

Some federal agencies have quietly adopted policies that require consideration of climate change in long-term decisions (e.g., in the Department of Interior through a 2001 Secretarial Order, Anderson et al. 2009). The National Park Service has been collaborating with the US Geological Survey to assess the physical exposure to sea level rise in its coastal parks (Pendleton et al. 2004). The Fish and Wildlife Service is also incorporating climate change impacts in its Comprehensive Conservation Plans, but the agency does not have a comprehensive strategy for addressing climate change impacts at this time.

In other agencies, the threat of climate change impacts has been recognized, even included in official guidance documents, but has not led to an appreciable change in actual decision-making (e.g., the US Army Corps of Engineer's consideration of sea level rise in coastal protection projects, Knuuti 2002; Anderson et al. 2009). The impact of more recent guidance to consider sea level rise in its projects has yet to be seen (USACE 2009). Some federal agencies face statutory hurdles to integrating climate change. For example, the Federal Emergency Management Agency at present cannot require the use of data reflecting future conditions for purposes such as floodplain management or insurance ratings unless statutory and regulatory changes are made to the National Flood Insurance Program (Anderson et al. 2009; Association of State Floodplain Managers 2007).

In summary, important movement toward adaptation planning is now observable at the federal level, though most of these efforts are still in the early stages. Lack of federal leadership, funding, understanding, intra- and inter-agency coordination and mandates, as well as political opposition, competing priorities, and legal obstacles have hindered greater engagement on adaptation (GAO 2009; Stern and Wilbanks 2008), but there is considerable more activity at this time than suggested in Repetto's (2008) assessment.

State Planning for Climate Change Impacts

Ironically, some of the earliest evidence of US states beginning to address and plan for the impacts of anthropogenic climate change comes from states which received federal financial and/or technical assistance to assess impacts and vulnerabilities (Moser 2005). In other instances, states' interest in adaptation is rooted in the *First US National Assessment of the Potential Consequences of Climate Variability and Change* (National Assessment Synthesis Team 2000). In yet other instances, state and regional efforts evolved out of existing concerns with climate variability (e.g., in the Pacific Northwest, Mote et al. 2003; Snover et al. 2003). Several states initiated their adaptation efforts while or after developing mitigation plans. State and regional adaptation planning now underway is not a federally guided or coordinated effort, and the bulk of the necessary resources, leadership, and staffing comes from the state level.

There is considerable regional variation in the level of effort and recognition of the need for adaptation. Adaptation planning has tended to appear on the political agenda first in some of the northeastern and western coastal states. Seven states – Arizona, Colorado, Utah, Arkansas, North Carolina, South Carolina, and Vermont – currently recommend creating plans for adaptation in their climate action plans (Pew Center for Global Climate Change 2008). More recently, Connecticut, Minnesota, and Illinois have joined this group of states or have begun more extensive adaptation planning (Dinse et al. 2009; personal communication with Don Wuebbles, 8 April 2009; and http://www.ctclimatechange.com). Eight states have launched more comprehensive assessment and planning efforts – Alaska, California, Maryland, Oregon, Florida, Washington, Massachusetts, and New Hampshire (Pew Center on Global Climate Change 2008). More than a dozen states are currently revising their state climate action plans or are charged by executive orders to develop such plans (e.g., Virginia, Kansas, Michigan, and Wisconsin). States with comprehensive efforts underway initiated those no more than 2 or 3 years ago, adding further weight to the notion that the United States is in a period of sudden awakening to the need for adaptation.

This observation notwithstanding, it would be a mistake to overlook the many earlier, often sector-specific policies that preceded these more comprehensive efforts, or which exist outside official state climate action plans. Such dispersed efforts can be found across the United States, though neither a systematic nor comprehensive compilation currently exists (but see NRC 2010). Maine, for example, is currently in the process of developing a state-wide adaptation policy, but the state was the first in the United States to revise its coastal law to include sea-level rise driven by anthropogenic climate change (Moser 2006, 2005). Several other coastal states have begun incorporating climate change and sea-level rise into their coastal management plans or have established Task Forces to develop policies and rules (e.g., in North Carolina, New York, Rhode Island, Maryland, and Delaware, Pew Center on Global Climate Change 2008; CSO 2007, 2008; Rubinoff et al. 2008; NOAA Coastal Services Center 2009). Most of these early efforts, however, have not fundamentally changed coastal hazards management approaches.

A similar picture emerges for the water sector. Many states experiencing drought in recent years have developed drought plans, though none currently include adaptation to climate change. New Mexico's 2006 Drought Plan acknowledges climate change, but does not explicitly account for it in its planning, monitoring, or preparedness efforts (New Mexico Drought Task 2006). In Arizona, water managers from local, state, tribal, and federal institutions came together in 2008 to discuss adaptation strategies and, specifically, ways to improve managers' access to useful decision support (Garfin et al. 2008). California's Department of Water Resources, concerned with climate change for several years, already included it in its 2005 *State Water Plan Update* (California DWR 2005, 2006). The Agency's 2008 White Paper on adaptation proposes bold general strategies, but the suggested changes are still in the planning stages and implementation has yet to pass challenging state budget hurdles and stakeholder concerns (California DWR 2008).

Several other examples could be cited from agriculture, forestry, winter tourism, public health, wildlife management, conservation, and ecosystem restoration (e.g., California Department of Fish and Game 2005; Galbraith 2008; Moser and Luers 2008). Many of these early state efforts share an initial focus on assessing the potential or most critical impacts of climate change. Only a small number of state efforts to date, however, involve vulnerability assessments informed by social science (e.g., Maryland; California is doing so at the time of this writing). These observations clearly reflect the common lack of scientific information on local or system-wide vulnerabilities, as well as the lack of expertise among staff and many consultants. Most state strategies recognize that adaptation involves costs and that these costs, while not trivial, are likely to be lower than "doing nothing." Recognizing that the past is no longer an adequate guide to the future, existing state plans hesitate to make bold policy changes, but instead call for more monitoring, improved scientific understanding, new planning guidelines, and more flexible procedures. They also recommend improving the preparedness for already apparent climate risks while limiting future liabilities. Most initial state adaptation strategies thus consist of best practices and no-regrets options, efforts to avoid future harm, win–win strategies that appear robust in the face of a range of climate futures, and ways to improve preparedness. Engagement of the wider public so far is very limited.

Local Adaptation Efforts

Local adaptation initiatives much resemble those at the state level and for some of the bigger metropolitan areas, such as New York City or Chicago; they are hardly less complex or challenging. These efforts can be grouped into three categories: (1) those initiated and supported by the Center for Clean Air Policy (CCAP) with funding, since 2008, through the Rockefeller Foundation's Climate Change Resilience Initiative; (2) those initiated and facilitated by ICLEI–Local Governments for Sustainability; and (3) independent efforts. Several of these independent efforts are closely related to various regional initiatives, through scientific support, staffing, and the underlying political mobilization process. For example, these independent efforts are linked to: regional assessments previously conducted as part of the National Assessment (e.g., New York City); regionally based, NOAA-sponsored Regional Integrated Sciences and Assessment (RISA) centers (e.g., King County, Washington); or other regional assessments (e.g., Chicago).

Communities of all sizes have begun assessing their physical vulnerabilities and developing an initial set of strategies to reduce them. These pioneers appear to be committed to early action because of their perceived or already experienced vulnerability to climate variability and change; their previous engagement (and often leadership) in climate mitigation policy; and the presence of supportive or facilitative organizations and experts.

Most of these initial adaptation plans focus only on a few high-risk areas. New York City, not surprisingly, focuses primarily on protecting its critical infrastructure from sea-level rise and storm surges, but also on establishing an ongoing adaptation process that involves neighborhood and environmental justice groups (Rosenzweig et al. 2007; City of New York 2007). Boston, San Francisco, Los Angeles, and Seattle share similar infrastructure concerns. Chicago's adaptation plan focus primarily on reducing vulnerability to extreme heat events, stormwater management, reducing the impacts from extreme events on buildings, infrastructure and equipment, and on protecting urban ecology (Parzen 2008). None of these large metropolitan areas is seriously considering (partial) relocation, though several Alaskan villages are actively relocating inland as coastal erosion threatens to destroy their homes.

Some cities have only commissioned impacts assessment to date, but have not substantially changed local policies (e.g., San Diego, California; Aspen, Colorado). Others, by contrast, astonish by the level of detail and how long they have been thinking about adaptation (e.g., King County, Washington; Keene, New Hampshire; Miami-Dade County, Florida). Boston, for example, while only recently engaging in a comprehensive adaptation planning effort, built its Deer Island sewage treatment plant back in 1993, already taking climate change-driven sea-level rise into account. San Francisco Bay's coastal management agency – the Bay Conservation and Development Commission – has been assessing sea-level rise risks for two decades, but because of its very limited regulatory power has focused most of its efforts on generating scientific understanding, raising awareness to the inevitable challenges associated with sea-level rise, educating local governments about vulnerability assessments, and providing discussion forums (http://www.bcdc.ca.gov/).

In addition to these locally based adaptation initiatives, several organizations representing local interests are beginning to be interested in adaptation. For example, the Conference of Mayors established its Climate Protection Center in 2007 (http://usmayors.org/climateprotection/about.htm) – a resource centre primarily supporting local greenhouse gas emissions reductions, but increasingly also interested in adaptation. The National Association of City and County Health Officials (NACCHO) is particularly outspoken on urging governments at all levels to prepare for the public health impacts of climate change (http://www.naccho.org/topics/environmental/climatechange/).

In most of these local efforts, the initial emphasis typically is on city operations, public lands, or city-controlled processes, bringing staff and political leaders on board with adaptation. Engagement of citizens is recognized but remains limited to date. In an ongoing, iterative process, where all involved still have to learn about adaptation, this may be appropriate at the outset, and it remains to be seen how governments will proceed to fully and meaningfully engage their communities in shaping future adaptation strategies (Lowe et al. 2009).

Non-state Actors and the Emerging Actor Network

While citizen involvement has been limited to date, non-state actors have helped mobilize and assist governmental adaptation efforts. NGOs, scientists, and consultants are principally involved, with professional organizations also playing important networking and information disseminating roles.

The active engagement from some of the advocacy groups in the last few years is surprising at first, given the stance many of them held historically toward adaptation as a kind of "capitulation." Several of these groups have strong conservation interests, however, and are themselves confronted with managing climate-change impacts (e.g., The Nature Conservancy, World Wildlife Fund). Some work with communities in assessing risks and developing strategies for ecosystem management (e.g., National Wildlife Federation). Others – through their commitment to representing the state of the science or leadership on regional impacts assessments – have brought attention to impacts and the need for adaptation (e.g., Union of Concerned Scientists).

Some NGOs serve important clearinghouse functions (e.g., the Pew Center for Global Climate Change or the virtual Adaptation Network, http://adaptationnetwork. org) or have emerged as active partners in adaptation, such as the Center for Clean Air Policy's "Urban Leaders Adaptation Initiative" (Lowe et al. 2009) and ICLEI – Local Governments for Sustainability's Climate Resilient Communities™ Program (ICLEI Local Governments for Sustainability 2009; see also Snover et al. 2007).

Some business alliances are also emerging as important actors in the adaptation arena. For example, a coalition of the eight largest US water utilities launched the Water Utility Climate Alliance (WUCA) in 2007, which supports impacts and adaptation research and lobbies for federal adaptation policy (http://www. wucaonline.org/html/). In March 2009, the National Association of Insurance Commissioners adopted a mandatory "climate change risk disclosure" requirement, demanding that insurance companies across the United States report on the financial risks they face from climate change – a move expected to strongly impact at-risk construction and development (http://www.ceres.org).

Many states and communities draw on a cadre of local scientists and consultants for technical assistance and for facilitating adaptation planning processes. For example, the Center for Climate Strategies has guided and facilitated adaptation processes in Maryland, Alaska, and other states (http://www.climatestrategies.us). Expertise in consulting firms is quite variable, however. Most desperately missing is in-depth knowledge of the (social science) vulnerability and adaptation literature, reflecting, again, the general lack of this type of expertise as communities and states begin to manage the impacts of climate change.

Conclusion: Entering into a "Period of Consequences"

This chapter has taken a cursory survey of adaptation-related activity currently underway in the United States. From the emergence in the news media to legislative action in Congress and nascent operational changes in federal agencies, to state and local government activities with considerable engagement of NGOs, scientists, and consultants, it is apparent that adaptation has finally, and explosively, emerged on the political agenda. At the same time, the current policy rush is not underlain by widespread public engagement and mobilization nor does it rest on a solid research foundation. It appears that the largest "bottleneck" to increasing US preparedness still lies in social scientific vulnerability and adaptation-specific expertise. This bottleneck stems from various reasons, including the limited US-specific research literature on vulnerability and adaptation; the lack of research funding and data gathering; the dearth of social science expertise in government agencies; a comparatively small number of university scientists working in this area and/or willing to engage in policy-relevant work; and the limited capacity in this area of specialization in consulting firms. The above-mentioned America's Climate Choices studies addressed the country's adaptation options and related research and capacity needs specifically. In the meantime, the burgeoning US adaptation efforts are at once a welcome development and at-risk of being done hastily and without the benefit of the appropriate expertise at the table.

A balance must be struck now where adaptation planning is initiated: small, commonsense, and meaningful policy and programmatic commitments should be made, while untenable promises and over-commitments of resources for ill-advised actions should be avoided. Changing scientific understanding and environmental conditions will require considerable policy flexibility, debate over difficult challenges, and resolving painful trade-offs. Meanwhile, a serious commitment at the highest levels is required to substantially expand vulnerability and adaptation research, and build technical and decision support capabilities. Without such a commitment, there is considerable danger that America will engage in countless expensive and damaging maladaptations, or that sectors and communities will prepare insufficiently for climate change, creating liabilities far more costly than the investment called for now.

Acknowledgments This chapter was prepared in part with support from the California Energy Commission (CEC). An adapted, more detailed version was published – with Springer Verlag's consent – by CEC and NOAA's Coastal Services Center. Vicki Arroyo, Terri Cruce, Josh Foster, Gregg Garfin, Patty Glick, Peter LaFontaine, an anonymous reviewer, and the editors provided helpful comments on an earlier draft of this chapter. The findings and opinions expressed do not reflect the views of any of CEC or those of the reviewers, and any remaining omissions, mistakes, and interpretations are mine alone.

References

AAAS (2009) Climate change research flat in 2008 budget. http://www.aaas.org/spp/rd/ccsp08p. htm. Cited 15 Jul 2009

Abramovitz JN, Starke L (2001) Unnatural disasters. Worldwatch Institute, Washington, DC

Adger WN, Agrawala S, Mirza MMQ et al (2007) Assessment of adaptation practices, options, constraints and capacity. In: Parry ML, Canziani OF, Palutikof JP et al (eds) Climate change 2007: impacts, adaptation and vulnerability – contribution of working group II to the fourth assessment report of the Intergovernmental Panel on Climate Change. Cambridge University Press, Cambridge

Alexander A, Owers J, Carveth R et al (eds) (2004) Media economics: theory and practice. Lawrence Erlbaum, Philadelphia

Anderson KE, Cahoon DR, Gill SK et al (2009) Coastal elevations and sensitivity to sea level rise. A report by the US Climate Change Science Program and the Subcommittee on Global Change Research, Synthesis and Assessment Product 4.1, Climate Change Science Program, Washington, DC

Association of State Floodplain Managers (2007) National flood programs and policies in review 2007. http://www.floods.org/. Cited 15 Jul 2009

Backlund P, Janetos A, Schimel D et al (2008) The effects of climate change on agriculture, land resources, water resources, and biodiversity in the United States. A report by the US Climate Change Science Program and the Subcommittee on Global Change Research, Synthesis and Assessment Product 4.3, Climate Change Science Program, Washington, DC

Bader DC, Covey C, Gutowski Jr WJ et al (2008) Climate models: an assessment of strengths and limitations. A report by the US Climate Change Science Program and the Subcommittee on Global Change Research, Synthesis and Assessment Product 3.1, Department of Energy, Office of Biological and Environmental Research, Washington, DC

Beller-Simms N, Ingram H, Feldman D et al (2008) Decision-support experiments and evaluations using seasonal-to-interannual forecasts and observational data: a focus on water resources. A report by the US Climate Change Science Program and the Subcommittee on Global Change Research, Synthesis and Assessment Product 5.3, NOAA National Climatic Data Center, Asheville

Boykoff MT (2007a) Flogging a dead norm? Newspaper coverage of anthropogenic climate change in the United States and United Kingdom from 2003 to 2006. Area 39(4):470–481

Boykoff MT (2007b) From convergence to contention: United States mass media representations of anthropogenic climate change science. Trans Inst Br Geogr 32(4):477–489

Boykoff MT, Boykoff JM (2004) Balance as bias: global warming and the US prestige press. Glob Environ Change 14(2):125–136

Burton I (1994) Deconstructing adaptation . . . and reconstructing. Delta 5(1):14–15

California DWR (2006) Progress on incorporating climate change into management of California's water resources. Technical Memorandum Report, California DWR, Sacramento

California DWR (2008) Managing an uncertain future: climate change adaptation strategies for California's water. California DWR, Sacramento

California Department of Fish and Game (2005) California wildlife: conservation challenges. http://www.dfg.ca.gov/wildlife/wap/report.html. Cited 15 Jul 2009

California Department of Water Resources [California DWR] (2005) California water plan update 2005: a framework for action. California DWR, Sacramento

Cannon T (1994) Vulnerability analysis and the explanation of 'natural' disasters. In: Varley A (ed) Disasters, development and environment. Wiley, New York

CCSP (2008) Uses and limitations of observations, data, forecasts, and other projections in decision support for selected sectors and regions. A report by the US Climate Change Science Program and the Subcommittee on Global Change Research Synthesis and Assessment Product 5.1, NASA, Washington, DC

Churchill WS (1936) The locust years speech before the house of commons on November 12, 1936. The Churchill Society. http://www.churchill-society-london.org.uk/Locusts.html. Cited 15 Jul 2009

City of New York (2007) PlaNYC: A greener, greater New York. City of New York. 156 pages. http://nytelecom.vo.llnwd.net/o15/agencies/planyc2030/pdf/full_report_2007.pdf

Clark PU, Weaver AJ, Brook E et al (2008) Abrupt climate change. A report by the US Climate Change Science Program and the Subcommittee on Global Change Research, Synthesis and Assessment Product 3.4, Climate Change Science Program, Washington, DC

Compaine BM, Gomery D (2000) Who owns the media? Competition and concentration in the mass media industry, 3rd edn. Lawrence Erlbaum Associates, Philadelphia

Croteau D, Hoynes W (2006) The business of media: corporate media and the public interest, 2nd edn. Pine Forge Press, Thousand Oaks

CSO (2007) The role of coastal zone management programs in adaptation to climate change. Final report of the CSO Climate Change Work Group, CSO, Washington, DC

CSO (2008) The role of coastal zone management programs in adaptation to climate change. Second annual report of the CSO Climate Change Work Group, CSO, Washington, DC

Cutter SL (1996) Vulnerability to environmental hazards. Prog Hum Geogr 20:529–539

Dinse K, Read J, Scavia D (2009) Preparing for climate change in the Great Lakes region. Michigan Sea Grant, Ann Arbor

Dumanoski D (1988) Coastal regions urged to prepare for rising sea level: planners decry apathy toward greenhouse effect's threat. The Boston Globe (13 Dec 1988), A1

Easterling WE, Hurd BH, Smith JB (2004) Coping with climate change: the role of adaptation in the United States. Pew Centre on Global Climate Change, Arlington

Fagre DB, Charles CW, Allen CD et al (2008) Thresholds of climate change in ecosystems. A report by the US Climate Change Science Program and the Subcommittee on Global Change Research, Synthesis and Assessment Product 4.2, Climate Change Science Program, Washington, DC

Galbraith H (2008) Confronting climate change – the challenge of adaptation [PowerPoint]. Presented to the California Climate Adaptation Working Groups, Climate Change Initiative, Manomet Center for Conservation Science, Sacramento, 23 Oct 2008

Gamble JL, Ebi KL, Sussman FG et al (2008) Analyses of the effects of global change on human health and welfare and human systems. A report by the US Climate Change Science Program and the Subcommittee on Global Change Research, Synthesis and Assessment Product 4.6, Climate Change Science Program, Washington, DC

GAO (2007a) Climate change: agencies should develop guidance for addressing the effects on federal land and water resources. GAO, Washington, DC

GAO (2007b) Climate change: financial risks to federal and private insurers in coming decades are potentially significant. GAO, Washington, DC

GAO (2009) Climate change: observations on federal efforts to adapt to a changing climate. Testimony before the Subcommittee on Energy and Environment, Committee on Energy, House of Representatives, Washington, DC

Garfin G, Jacobs K, Buizer J (2008) Beyond brainstorming: exploring climate change adaptation strategies. EOS Trans Am Geophys Union 89(25):227

ICLEI Local Governments for Sustainability (2009) Climate adaptation: introducing ICLEI's climate resilient communities program. http://www.icleiusa.org/programs/climate/Climate_Adaptation/adaptation. Cited 15 Jul 2009

Julius SH, West JM, Baron JS et al (2008) Preliminary review of adaptation options for climate-sensitive ecosystems and resources. A report by the US Climate Change Science Program and the Subcommittee on Global Change Research, Synthesis and Assessment Product 4.4, Environmental Protection Agency, Washington, DC

Karl TR, Meehl GA, Miller CD et al (2008) Weather and climate extremes in a changing climate – regions of focus: North America, Hawaii, Caribbean, and US Pacific Islands. A report by the US Climate Change Science Program and the Subcommittee on Global Change Research, Synthesis and Assessment Product 3.3, NOAA, National Climatic Data Center, Asheville

King JS (2008) Call for ideas on living in a warming world: competition asks designers to 'climate-proof' Bay Area. The San Francisco Chronicle (19 Dec 2008), B1

Knuuti K (2002) Planning for sea level rise: US Army Corps of engineers policy. In: Ewing L, Wallendorf L (eds) Solutions to coastal disasters '02. American Society of Civil Engineers, Reston

Kumagai Y, Edwards J, Carroll MS (2006) Why are natural disasters not "natural" for victims? Environ Impact Assess 26:106–119

Lowe A, Foster J, Winkelman S (2009) Ask the climate question: adapting to climate change impacts in urban regions. A report of the CCAP's Urban Leaders Adaptation Initiative, CCAP, Washington, DC

McMasters P (1989) Topic: the ozone shield – nations must join hands against a global threat. USA Today (4 Jan 1989), 11A

Miller HI (2007) Global warming resilience required. The Washington Times (16 Mar 2007), A18

Morgan G, Dowlatabadi H, Henrion M et al (2009) Best practice approaches for characterizing, communicating, and incorporating scientific uncertainty in decisionmaking. A report by the US Climate Change Science Program and the Subcommittee on Global Change Research, Synthesis and Assessment Product 5.2, NOAA, Washington, DC

Moser SC (2005) Impact assessments and policy responses to sea-level rise in three US states: an exploration of human-dimension uncertainties. Glob Environ Change 15(4):353–369

Moser SC (2006) Climate change and sea-level rise in Maine and Hawai'i: the changing tides of an issue domain. In: Mitchell RB, Clark WC, Cash DW et al (eds) Global environmental assessments: Information, institutions, and influence. MIT, Cambridge, MA

Moser SC (2009) Good morning, America! The explosive US awakening to the need for adaptation. California Energy Commission, Sacramento CA and NOAA Coastal Services Center, Charleston

Moser SC (2010) Now more than ever: The need for more societally relevant research on vulnerability and adaptation to climate change. Appl Georg 30:464–474

Moser SC, Luers AL (2008) Managing climate risks in California: the need to engage resource managers for successful adaptation to change. Clim Change 87(Suppl 1):S309–S322

Mote PW, Parson EA, Hamlet AF et al (2003) Preparing for climatic change: the water, salmon, and forests of the Pacific Northwest. Clim Change 61(1–2):45–88

National Assessment Synthesis Team (2000) Climate change impacts on the United States: the potential consequences of climate variability and change – foundation. Cambridge University Press, New York

New Mexico Drought Task Force (2006) New Mexico drought plan. New Mexico Office of the State Engineer, Albuquerque

NOAA Coastal Services Center (2009) Local strategies for addressing climate change. NOAA CSC, Charleston

NRC (2003) Planning climate and global change research: a review of the draft US Climate Change Science Program strategic plan. National Academy Press, Washington, DC

NRC (2004) Implementing climate and global change research: a review of the final US Climate Change Science Program strategic plan. National Academy Press, Washington, DC

NRC (2007) Evaluating progress of the US Climate Change Science Program: methods and preliminary results. National Academies Press, Washington, DC

NRC (2009a) Restructuring federal climate research to meet the challenges of climate change. National Academies Press, Washington, DC

NRC (2009b) Informing decisions in a changing climate. National Academies Press, Washington, DC

NRC (2010) America's climate choices: adapting to the impacts of climate change. National Academies Press, Washington, DC

Parry ML, Canziani OF, Palutikof JP et al (eds) (2007) Climate change 2007: impacts, adaptation, and vulnerability – contribution of working group II to the fourth assessment report of the Intergovernmental Panel on Climate Change. Cambridge University Press, Cambridge

Parzen J (ed) (2008) Chicago area climate change quick guide for municipalities and other organizations: adapting to the physical impacts of climate change. City of Chicago

Pendleton EA, Williams SJ, Thieler ER (2004) Coastal vulnerability assessment of Assateague Island National Seashore (ASIS) to sea-level rise. US Geological Survey, Reston

Pew Center on Global Climate Change (2008) Adaptation planning – what US states and localities are doing (2008 update). Pew Center for Global Climate Change, Arlington

Pew Research Center for the People & the Press (2008) Internet overtakes newspapers as news outlet. http://pewresearch.org/pubs/1066/internet-overtakes-newspapers-as-news-source. Cited 15 Jul 2009

Pielke RA Jr, Prins G, Rayner S et al (2007) Lifting the taboo on adaptation. Nature 445:597–598

Repetto R (2008) The climate crisis and the adaptation myth. Yale School of Forestry & Environmental Studies, New Haven

Rosenzweig C, Major DC, Demong K et al (2007) Managing climate change risks in New York City's water system: assessment and adaptation planning. Miti Adapt Strateg Glob Change 12(8):1391–1409

Rubinoff P, Vinhateiro ND, Piecuch C (2008) Summary of coastal program initiatives that address sea level rise as a result of global climate change. Rhode Island Sea Grant/Coastal Resources Center, University of Rhode Island, Narragansett

Savonis MJ, Burkett VR, Potter JR (2008) Impacts of climate change and variability on transportation systems and infrastructure: Gulf Coast study, phase I. A report by the US Climate Change Science Program and the Subcommittee on Global Change Research, Synthesis and Assessment Product 4.7, Climate Change Science Program, Washington, DC

Smith JB, Tirpak D (eds) (1989) The potential effects of global climate change on the United States: report to Congress. Environmental Protection Agency, Office of Policy, Planning and Evaluation, Office of Research and Development, Washington, DC

Snover AK, Miles EL, Hamlet AF (2003) Learning from and adapting to climate variability in the Pacific Northwest. Paper presented at 'Insights and tools for adaptation: learning from climate variability' Workshop, Washington, DC, 18–20 Nov 2003

Snover AK, Binder LW, Lopez J et al (2007) Preparing for climate change: a guidebook for local, regional, and state governments. ICLEI–Local Governments for Sustainability, Oakland

Solomon S, Qin D, Manning M et al (eds) (2007) Climate change 2007: the physical science basis – contribution of working group I to the fourth assessment report of the Intergovernmental Panel on Climate Change. Cambridge University Press, Cambridge

Stern PC, Wilbanks TJ (2008) Fundamental research priorities to improve the understanding of human dimensions of global change: a discussion paper prepared for the National Research Council's Committee on Strategic Advice to the US Climate Change Science Program. National Academy of Sciences, Washington, DC

Turner BL II, Kasperson RE, Matson PA et al (2003) A framework for vulnerability analysis in sustainability science. Proc Natl Acad Sci USA 100(14):8074–8079

USACE (2009) Water resource policies and authorities incorporating sea-level change considerations in civil works programs. Circular No. 1165-2-211 of July 1, 2009, Department of the Army, US Army Corps of Engineers, Washington, DC

White GF (ed) (1974) Natural hazards: local, national, global. Oxford University Press, New York

Wilbanks TJ, Bhatt V, Bilello DE et al (2007) Effects of climate change on energy production and use in the United States. A report by the US Climate Change Science Program and the Subcommittee on Global Change Research, Synthesis and Assessment Product 4.5, Climate Change Science Program, Washington, DC

Chapter 4
Perspectives on Adaptation to Climate Change in Europe

Stéphane Isoard

Abstract Many regions in Europe are vulnerable to climate change impacts and these have already been observed in many human and natural systems. There is therefore a need for all European countries to adapt to climate change. However, climate change does not pose a threat at all levels of change, nor in a similar way across all economic sectors and regions. Impacts of climate change vary by region, with the Mediterranean basin, north western and central-eastern Europe, and the Arctic, together with many coastal zones and other areas prone to river floods, mountains, and cities being particularly vulnerable. Consequently, adaptation options have to be tailor-made to local geographic conditions in terms of vulnerable landscape types and sectors involved. This is why European and national adaptation strategies must be complemented by regional approaches that address implementation issues at this scale. The European Union (EU) Adaptation White Paper together with national and regional adaptation strategies provide key steps toward European frameworks for adaptation measures and policies to strengthen resilience to climate change impacts. Monitoring and evaluation methods must be further developed, so that adaptation practices can be assessed across time and space. A preliminary set of success factors and barriers to adaptation is also important to identify and assess the determinants of good practices and assist stakeholders in developing robust adaptation strategies that can make Europe more resilient to climate change. Potential conflicts, synergies, and trade-offs between adaptation and mitigation have also to be identified early in the decision-making process, and adaptation and mitigation efforts should be coordinated.

Keywords Climate change • Adaptation • Impacts • Vulnerability • Resilience • Human and natural systems • Adaptive capacity • Costs • European Union • Europe

S. Isoard (✉)
European Environment Agency, Kongens Nytorv 6, Copenhagen DK-1050, Denmark
e-mail: stephane.isoard@eea.europa.eu

J.D. Ford and L. Berrang-Ford (eds.), *Climate Change Adaptation in Developed Nations: From Theory to Practice*, Advances in Global Change Research 42, DOI 10.1007/978-94-007-0567-8_4, © Springer Science+Business Media B.V. 2011

Europe Needs to Adapt

The European Union (EU) has agreed to aim to limit the long-term increase of global mean temperature to 2°C above preindustrial levels, as reflected in the recent Cancún Agreement. Even if this goal is achieved through stringent worldwide mitigation actions to stabilize global greenhouse gas concentrations, some impacts will remain, at least in the short and medium terms, making adaptation imperative to reduce vulnerability and enhance resilience.

Many regions in Europe are vulnerable to climate change impacts and these have already been observed in many human and natural systems, as shown in Fig. 4.1. Most of the impacts are adverse and are generally projected to worsen, certainly beyond a few decades (Parry et al. 2007, 2009; EEA 2010a). Portions of northern Europe might benefit to some extent in the short and medium terms from climate

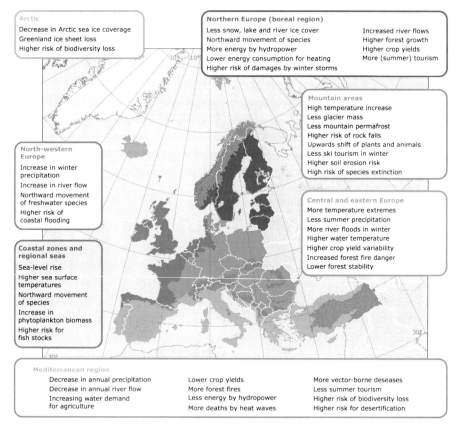

Fig. 4.1 Key past and projected impacts of climate change and effects on sectors for the main bio-geographic regions of Europe (EEA 2010a)

change (e.g., increasing crop yields and forest growth, increased tourism demand). However, more frequent and intense extreme events in the long term are projected to lead to adverse effects.

Climate change impacts affect, directly and/or indirectly, virtually all socioeconomic sectors in Europe. Those particularly affected include energy supply, health, water management, agriculture, forestry, tourism, transport, biodiversity and ecosystem goods and services, and fisheries. There is therefore a need for all European countries to adapt to climate change. Adaptation policies will have to pay due attention to cross-sectoral dimensions (e.g., competition for the same scarce resources) to take into account the variety of drivers, success factors, and barriers to good practices.

Adaptation to climate change is defined by the Intergovernmental Panel on Climate Change (IPCC) as "adjustment in natural or human systems in response to actual or expected climatic stimuli or their effects, which moderates harm or exploits beneficial opportunities." There are various types of adaptation decision-making processes (Parry et al. 2007, 2009). Planned adaptation, which can be anticipatory/proactive or reactive, aims at taking measures to counteract expected impacts of climate change – before these are observed. They are the result of a deliberate decision, based on an awareness that conditions have changed or are about to change and that action is required to return to, maintain, or achieve a desired state. Autonomous adaptation is a reactive response to a variety of factors and is triggered by changes in natural and human systems, including the climatic stimuli and market forces. The focus here is mainly on planned adaptation, even though it is challenging in practice to disentangle and systematically distinguish the various types of adaptation, since planned adaptation packages may also facilitate autonomous adaptation. Adaptation aims at managing the unavoidable impacts, while mitigation aims at avoiding the unmanageable impacts.

Adaptation aims at increasing the resilience of natural and human systems to current and future impacts of climate change. Depending on the context, adaptation refers to measures taken or strategies developed at the European, national, regional, and local levels. The implementation of adaptation measures takes place primarily across boundaries (e.g., river catchments) and various sub-national levels, and therefore involves many levels of decision-making. Examples of adaptation measures include: heat-related health action plans, vaccination, health-system planning, flood-risk planning (early warning systems), drought and water scarcity risk management, water demand management, coastal and flood defenses, coastal zone management, economic diversification, natural hazard monitoring, reinforcing infrastructure (such as roads, bridges, and electric wires), land-use management, and greening of cities.

On the other hand, adaptation strategies are usually developed at national or regional levels for setting strategic directions and dealing with issues across the board (e.g., building adaptive capacity, transboundary issues). The transboundary nature of climate change and associated adaptation responses, together with the subsidiary principle, are important to consider when developing and implementing adaptation strategies and measures.

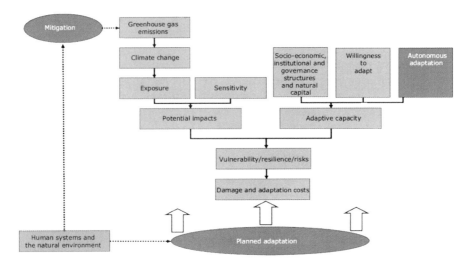

Fig. 4.2 Conceptual framework for climate change impacts, vulnerability, disaster risks, and adaptation options (Based on Isoard et al. 2008; IPCC 2007; Füssel and Klein 2006)

Climate change does not pose a threat at all levels of change, nor in a similar way across all economic sectors and regions. Consequently, adaptation options have to be tailor-made to local geographic conditions, given the vulnerability of landscape types and sectors involved, with the goal of implementing measures at the appropriate level of decision-making. Different systems (e.g., regions, landscape types, sectors) at different geographic levels will require different approaches and levels of intervention. However, tailor-made adaptation measures must be consistent with broader strategies at larger scales.

Climate change may also provide opportunities for innovation in technology and governance in various economic sectors (e.g., tourism, energy supply, water management, health, construction, and shipping). Increasing resilience and taking advantage of these opportunities require proactive adaptation, which calls for integrated analysis and tools (e.g., spatial planning, vulnerability mapping, and territorial analytical frameworks) covering interconnected issues ranging from demographic shifts to biodiversity (Uhel and Isoard 2008). There is now significantly improved understanding of the relationship between impacts, which forms the basis for climate change concerns, and the concept of vulnerability (see Fig. 4.2 and Grothmann and Patt 2005). Adaptation aims precisely at addressing the determinants of vulnerability (or resilience).

Many adaptation options are already identified and have been implemented in some cases. Nowadays, adaptation is seldom undertaken for the sake of climate change alone; rather, adaptation is generally integrated into other crosscutting and precautionary policy actions, such as disaster preparedness, coastal zone management, rural development, health services, spatial planning and regional development, and ecosystems and water management. Adaptation is very much about managing the risks associated with future climate change impacts and in many cases indeed a link with disaster management will be appropriate. An increasing consideration of adaptation issues in decision-making is expected to lead to the development of new assessment tools and more integrated adaptation measures.

Successful adaptation requires strategic approaches that recognize that appropriate measures today may be inappropriate or insufficient in the future. The uncertainty inherent to climate change scenarios or vulnerability assessments calls for flexible adaptation decisions that are easy to adjust, if required. Continued evaluations are therefore needed to make the necessary adjustments, and incorporate new scientific information that reduces uncertainty as it becomes available. Effective adaptation needs to contemplate all possible climate conditions during the horizon of a policy decision. It should therefore consider no-regret measures (suitable under every plausible scenario) and a broad variety of adaptation options (i.e., gray or technological measures; green/ecosystem-based measures; and soft measures addressing behavior, management and policies). The overall aim of adaptation measures and strategies is indeed to ensure that vulnerable human and natural systems "adapt well to climate change" (as opposed to "are well adapted") in a dynamic and iterative process that properly articulates determinants of success. However, there are limits to adaptation (Parry et al. 2007, 2009; EEA 2008, 2009), which are usually identified for a specific system.

There is also a clear lack of information across Europe on impacts and vulnerability assessment at regional and local levels, and on adaptation activities and measures planned or currently being implemented by countries (Parry et al. 2007, 2009; EEA 2010a; Swart et al. 2009). Enhancing information sharing in a concerted manner therefore appears critical to inform the many levels of decision-making involved in practical adaptation responses. This activity should be based on a careful analysis of the added value for those whom the information is supposed to reach.

Linkages between adaptation and mitigation must also be considered, particularly to align and coordinate future actions (Swart and Raes 2007). Potential synergies, conflicts, and trade-offs between adaptation and mitigation have to be identified early in the decision-making process, and these must be addressed. Some adaptation options can be developed in synergy with mitigation, for example in the land and water management sectors. The development of mitigation measures also needs to consider vulnerabilities and adaptation options. Identifying possible conflicts between mitigation and adaptation is key for avoiding maladaptation or unsustainable solutions.

There is also a need to better understand maladaptation, which occurs when adaptation measures do not increase resilience/adaptive capacity or reduce vulnerability, for example inappropriate, not proportionate or cost-ineffective solutions; are environmentally unsustainable; or conflict with other long-term policy objectives, for example, artificial snow making or air conditioning that conflict with mitigation targets (EEA 2009; IPCC 2007). To avoid maladaptation both climatic and socio economic factors of vulnerabilities have to be considered when developing policy responses.

Europe has to adapt to climate change, and also has a moral obligation to assist developing countries, as their communities, economic sectors, and ecosystems are most vulnerable. The assistance to developing countries should be informed, among others, by: the 5-year Nairobi program of work on impacts, vulnerability, and adaptation to climate change (UNFCCC 2006); the National Adaptation Plans of Actions (NAPAs); the Bali Action Plan (UNFCCC 2007c); and the Cancún Adaptation Framework (December 2010). A number of developing countries have prepared NAPAs using the United Nations Development Programme (UNDP) Adaptation Policy Framework (UNDP 2004). Furthermore, the Bali Action plan, resulting from the December 2007 Conference and Meeting of the Parties (COP/MOP) meetings, recognizes that adaptation will need to be explicitly included in a global post-2012 climate change agreement. This will facilitate support for adaptation measures in developing countries, and enhance the sharing of information and experiences between those and the developed nations.

The European, National and Regional Policy Frameworks for Adaptation

Recognizing the necessity for Europe to adapt to climate change impacts, the European Commission adopted a White Paper in April 2009 (EC 2009a) to build upon the June 2007 Green Paper. The EU Adaptation White Paper, together with national and regional adaptation strategies, provides key steps toward European frameworks for adaptation measures and policies to increase resilience. The knowledge base and governance structures are important issues to consider for developing policy options that ensure effective and robust adaptation measures.

The European Union Should Provide the Overall Policy Frameworks

The European dimensions of climate change impacts are many-fold. Impacts of climate change vary by region, with the Mediterranean basin, north western and central-eastern Europe, and the Arctic, together with many coastal zones and other areas prone to river floods, mountains, and cities being particularly vulnerable. This is why most adaptation measures will be implemented at national, regional, or local levels. The European Union should support these efforts through an integrated and coordinated approach, particularly in connection with cross-border and regional solidarity issues, as well as EU policy areas (e.g., agriculture, water, biodiversity, fisheries, and energy). The vulnerability of ecosystems, economic sectors, landscapes, and communities in the long term will also require: mainstreaming adaptation into EU sectoral policies and external relations; aligning EU funding mechanisms (i.e., structural/cohesion funds); fostering research; and involving stakeholders. The EU also has a role in coordinating and enhancing information sharing on adaptation best practices, and current and planned adaptation activities in Europe, and in providing methodological guidance to encourage the implementation of appropriate, proportionate, and cost-effective adaptation measures at all levels of decision-making. Climate change adaptation will eventually need to be embedded in all EU policies.

The EU Adaptation White Paper, published by the European Commission on April 1, 2009, is a key step toward a framework for adaptation measures and policies to strengthen EU resilience to climate change impacts (EC 2009a). It stresses the need to develop the knowledge base further and to integrate adaptation into EU policies. The White Paper also prominently recognizes that the impacts of climate change will have varying regional implications, which means that most adaptation measures will need to address local needs. The White Paper is framed to complement and ensure synergies with actions by Member States and focus on four pillars to reduce the EU's vulnerability and improve its resilience:

- Develop and improve the knowledge base at regional level on climate change impacts, vulnerabilities mapping, costs and benefits of adaptation measures to inform policies at all levels of decision-making
- Integrate or mainstream adaptation into EU policies
- Use a combination of policy instruments – market-based instruments, guidelines, and public–private partnerships – to ensure effective delivery of adaptation
- Work in partnership with the Member States and strengthen international cooperation on adaptation by mainstreaming adaptation into the EU's external policies

The EU aims at an integrated approach with top-down policy strategies for mainstreaming adaptation into sectoral policies together with bottom-up activities building adaptive capacity and implementing actions. The pillars of action support adaptation initiatives at all levels of decision-making across sectors. Concrete initiatives have started to integrate adaptation into EU sectoral policies (Pillar 2), in particular with water and flood risk management, agriculture and rural development, health, and nature protection and biodiversity.

The EU framework for action presented in the White Paper sets out a two-phase strategy that complements actions taken by Member States through an integrated and coordinated approach. The first phase, which runs until 2012, will be dedicated to preparing a comprehensive EU adaptation strategy from 2013 onward. The first phase will focus primarily on improving the knowledge base on climate change, possible adaptation measures, and means of embedding adaptation in EU policies.

The EU is well placed to facilitate the implementation of the first phase of the framework through a series of actions (EEA 2009) including: supporting monitoring and data collection networks to expand the knowledge base; developing analytical tools and assessments that report on the adaptive capacity and vulnerability of natural and human systems (some existing management tools consider climate change (e.g., analytical frameworks, downscaled scenarios, cost-benefit analyses and good practices) but are usually largely unknown among local and regional stakeholders, a fact that underlines the need for dissemination and outreach); developing information platforms on climate change impacts, risks, and adaptation options at the local and regional levels; facilitating exchange of good practices between Member States to further support capacity building of regional and local authorities, and encourage those that have not yet prepared national and/or regional adaptation strategies to do so; and fostering stakeholder participation in research projects to bridge the gap between scientists, policymakers, and all other relevant parties.

To address the current lack of knowledge-sharing and management, the EU Adaptation White Paper proposed to develop a European Clearinghouse on climate change impacts, vulnerability, and adaptation to maintain a wide range of information at European, national, regional, and sectoral levels on climatology and impacts, vulnerability assessments, good adaptation practices, and policy frameworks. The EU Adaptation Clearinghouse for Europe (ACE) would also link to other similar or related initiatives such as the Biodiversity Information System for Europe (BISE)/European Community Biodiversity Clearing House Mechanism, the Water Information System for Europe (WISE), and the Global Monitoring for Environment and Security (GMES). To further ensure synergies and avoid duplication of work, the clearinghouse would also link to initiatives and knowledge platforms in European regions, which would provide regional nodes to the knowledge base. The European Environment Agency will be hosting the clearinghouse as of 2012 and will maintain it thereafter.

The Water Framework Directive establishes a legal framework to protect and restore clean water across Europe by 2015 and ensure the long-term sustainable use of water. The 1st River Basin Management Plans for 2009–2015, which should since December 2009 be available in all River Basin Districts across the EU, take into account the impacts of climate change. In addition, climate change must also be properly integrated in the implementation of the Floods Directive. There are, however, serious delays in some parts of the EU – in several countries as of June 2010 consultations are still continuing – or river basin management plans have not yet been established.

Water directors of EU Member States issued a Common Implementation Strategy (CIS) guidance document (EC 2009b) that addresses the integration of climate change impacts and adaptation in the implementation of the Water Framework Directive, the Floods Directive and the Strategy on Water Scarcity and Droughts, particularly in view of Member States' second (2015) and third (2021) river basin management plans. The EEA contributed to the EC's guidance document and is currently working on examples of good practices for the integration of climate change adaptation into water management (e.g., management plans and information systems). The United Nations Economic Commission for Europe (UNECE) (through its Water Convention) also released guidance on water and adaptation to climate change, particularly in connection with developing adaptation strategies and cooperation in transboundary basins (UNECE 2009). Reviews of the Water Framework and Floods Directives, foreseen for 2012, will provide an opportunity to assess how climate change should be further integrated into EU water policies.

To address water scarcity, the EC will assess the need to further regulate the standards of water-using equipment and water performance in agriculture, households, and buildings. When reviewing the implementation of the Water Scarcity and Droughts Strategy in 2012, options for boosting the water-storage capacity of ecosystems to increase drought resilience and reduce flood risks will be evaluated.

The forthcoming Common Agricultural Policy reform and the new rural development policy plan (Pillar 2) will specifically take adaptation into account with, for example, requirements for more efficient water consumption.

By 2011 adaptation to climate change should also become part of the EU health strategy through the development of guidelines. The Adaptation White Paper included a dedicated Staff working document on "Human, animal, and plant health impacts of climate change" and DG SANCO has established a task force to implement it. In addition the European Centre for Disease Prevention and Control developed a handbook for national vulnerability, impact, and adaptation assessments that addresses specifically climate change and communicable diseases in the EU Member States (ECDC 2010). WHO Europe works to identify policy options to help prevent, prepare for, and respond to the health effects of climate change, and its products include various guidance documents (WHO-Europe 2010c). The Parma Ministerial conference in March 2010 also adopted a European Regional Framework for Action (WHO-Europe 2010a).

The impacts of climate change must also be factored into the management of the Natura 2000 network and Habitats Directives to ensure the diversity and connectivity of natural areas and allow species migration and survival when climate conditions change. The EC has developed a Discussion Paper (EC 2009c) within the EU Ad Hoc Expert Working Group on Biodiversity and Climate Change which showcases the links and interdependency between biodiversity and ecosystems, ecosystem services and climate change. Ecosystem-based adaptation measures have the potential to lead to win-win situations as they both provide adequate responses to climate change challenges and sustain ecosystems functions in the long term. In 2010 the EC also published a Communication that sets out possible future options for a long-term (2050) EU vision on biodiversity policy and mid-term (2020) targets beyond 2010 (EC 2010a), also addressing climate change, and is planning to further address green infrastructure in 2011. The United Nations Convention on Biological Diversity (CBD) and UNFCCC have set up an experts group addressing the issues from a global perspective and formulating recommendations. The links and dependencies between solving the climate change and biodiversity loss challenges are at the heart of developing complementary policies that simultaneously promote biodiversity conservation and support climate change mitigation and adaptation objectives.

In 2010, the EC also launched a Green Paper to open the debate on options for an EU approach to forest protection and information systems in the framework of the EU Forest Action Plan, as announced by the Adaptation White Paper (EC 2010c). The Green Paper sets out the main challenges facing Europe's forests, reviews existing forest information systems and the tools available to protect forests, and raises a series of questions relevant to the development of future policy options.

The European Commission undertook a study in 2009 on the design of guidelines for the elaboration of regional climate change adaptations strategies that highlights, amongst other issues, the importance of existing EU regional funding instruments for mainstreaming adaptation (EC 2009d). In future, EU Structural Funds might constitute an essential instrument to support regions and cities in allocating adequate funds for mainstreaming climate change adaptation in building, water or energy policies. The establishment of adaptation strategies requires indeed strategic long-term decisions and funding opportunities to efficiently adjust management practices. The ESPACE project also addressed governance issues by stressing the importance of adaptive management structures, and combining change and risk management approaches for integrating adaptation into spatial planning (ESPACE 2007). In this context mainstreaming adaptation also refers to integrating climate change into instruments such as the Environmental Impact Assessment (EIA) and Strategic Environmental Assessment (SEA) and the related directives (EC 2010b).

National and Regional Adaptation Strategies

The EEA keeps a regularly updated overview of progress toward the development and implementation of national adaptation strategies (NAS) online (EEA 2010b). Eleven European countries – Denmark, Finland, France, Germany, Hungary, the Netherlands, Norway, Portugal, Spain, Sweden, and the United Kingdom – have adopted NAS so far, while several others are expected to adopt such a strategy in the next few years – Austria by 2011, Belgium by 2012, Estonia, Ireland, Latvia, and Switzerland by 2011. In addition countries have also submitted information on their adaptation plans in their fifth National Communication to the UNFCCC due on January 1, 2010. National audit offices are also increasingly involved in undertaking reviews of adaptation policies for national parliaments, and guidelines for auditing responses to climate change are being developed (NAO 2009).

Adaptation initiatives are already being implemented through policies, invest-ments, and changes in behaviors, such as no-regret measures that are relevant under all plausible future scenarios. These have recently been reviewed and evaluated for some EU member states (Biesbroek et al. 2010; Swart et al. 2009). Key findings from these studies show that in general countries are aware of the need to adapt to climate change. Nevertheless, compared to climate change mitigation, many are only at an early stage of developing policy frameworks. Many adaptation initiatives are not undertaken as stand-alone actions, but are embedded within broader sectoral measures, for example water-resource management and coastal defense strategies.

In many countries adaptation started with assessing needs, setting-up research programs to increase the knowledge base, identifying policy and financial in-struments and fields of cooperation. The implementation of national adaptation strategies is only starting, due partly to the complexity in translating existing knowledge into policy packages. Developing implementation programs is therefore a policy challenge in many countries in connection with plans for mainstreaming adaptation within existing policies and economic instruments and working out reporting/monitoring mechanisms. In addition only a small number of adaptation measures are found to explicitly consider scenarios of future climate change.

The integration of adaptation into sectoral policies is considered as crucial to avoid contradictions between policies and benefit from synergies, though the number of sectors included differs widely between countries. Some countries, especially in north-western and northern Europe, have also acknowledged possible positive impacts of climate change and defined policies to take advantage of them. There is also a strong need for communication tools that enhance information sharing, which differs greatly between countries, to address the lack of awareness about adaptation issues and the multidimensional aspect of the topic – multiple sectors, scales, regions, communities, and stakeholders.

Country comparisons show that many existing policies for coping with weather-related events already contribute to climate change adaptation in most countries, and that priority sectors differ widely due to historical, climatic, and geographical circumstances. With regard to disaster-risk reduction at the national level, one major activity has been the establishment of national strategies and national platforms for disaster-risk reduction. So far, 11 European countries have established such a platform and many more have established official Hyogo Framework Focal Points (UNISDR 2009b; PreventionWeb 2010b).

Participative processes for developing NAS have mainly included ministries. Participation of other stakeholders, for example representatives of other decision-making levels, environmental and business NGOs, scientists, has so far been limited although it has been recognized as important.

Various factors that motivate, trigger, and facilitate the development of adaptation policies have also been identified, as have a variety of weaknesses and threats that hinder the process (Biesbroek et al. 2010; Swart et al. 2009; Per et al. 2009).

Various organizations or countries have developed tools and guidance for screening adaptation options to cope with events such as floods, droughts, heat-waves, and sea-level rise, and setting priorities, such as the Adaptation Wizard (UKCIP), the Adaptation Decision Explorer (weADAPT), or the Digital Adaptation Compendium (EU ADAM project 2009). A multi-criteria analysis is commonly used for assessing adaptation options. The Netherlands has developed a routeplanner to assess and rank adaptation options depending on criteria such as feasibility and cost-benefit considerations (van Ierland et al. 2006), while in France, criteria for ranking sectoral options with a long-term planning horizon have been developed (Hallegatte et al. 2008). The process of ranking adaptation options should involve appropriate stakeholders at the regional, national, or sectoral level to reflect the specificities of the vulnerable region, sector, or community, both in spatial and temporal terms.

Adaptation will to a large extent occur at a decentralized level, so the efficiency of individual adaptation measures depends on local conditions and the ability to take these into consideration. Regional and sub-national adaptation strategies and processes support this and offer solution-oriented and stakeholder-specific perspectives. Regions and municipalities are increasingly concerned about their vulnerability, and some have started developing and implementing their own adaptation strategies. At least 29 regional and local adaptation initiatives and strategies in the EU have recently been identified (Swart et al. 2009). Key sectors included are landscape and spatial planning, water, health, and biodiversity. Andalucia, Spain; North-Rhine Westphalia, Germany (NRW 2009); and Rhônes-Alpes, France have been the first regions to embark on developing and adopting regional adaptation strategies. Since 75% of the European population lives in urban areas, some cities, including Barcelona, Copenhagen, London, and Rotterdam, have also started to develop adaptation plans and are getting increasingly organized (e.g., C40 Cities – Climate Leadership Group) to tackle this environmental and governance challenge (Rotterdam City 2007; London City 2010). Various trans-national initiatives and strategies also exist, such as the Action Plan on Climate Change of the Alpine Convention.

The objectives of these plans differ widely, ranging from increasing public awareness and reducing vulnerability to increasing coping capacity to extreme events, but the implementation of practical measures has so far been given little attention. However, the 2007 floods in the United Kingdom revealed that some of the most effective management measures may be taken by local authorities and individual households (Pitt 2008).

Success Factors for and Barriers to Adaptation

The determinants of successful (or failed) adaptation should be identified and assessed to better inform policymakers about good practices. The drivers discussed below are drawn from regional case studies in a recent EEA report addressing climate change adaptation and water-related issues in the Alps (EEA 2009), and a literature review.

With regard to success factors, political support is a key catalyst for initiating, driving, and coordinating adaptation to climate change, providing a strategic framework for effective action. Policies have generally been responses to extreme events or natural hazards that motivate demand for action by public authorities. Once initiated, adaptation measures rely on a broad variety of factors for their success, relating primarily to institutional and governance structures, as well as organizational settings:

- Measures are generally more accepted and successful when they promote other goals, including economic gains.
- A sound legal framework is a crucial complement to political support.
- Technological measures play a major part in adaptation measures.
- An increasing number of initiatives consider complementary "soft" actions on the demand management side (e.g., behavioral adaptation, full participation of stakeholders).

- Introducing market-based economic incentives and financial support is also helpful in encouraging proactive and innovative adaptation measures.
- Raising stakeholder awareness about the need for anticipatory adaptation actions is vital, especially in sectors with long lead times (e.g., forestry and power generation).
- Other social factors, in particular local practices and social networks, are also key.

The case studies and the literature review also reveal various barriers to adaptation:

- Limited scientific knowledge and uncertainty about future local climate change impacts.
- The lack of long-term planning strategies, coordination, and use of management tools that consider climate change at the level of region or river basin, and across sectors.
- The failure to consider climate change explicitly in management plans.

Obviously, local and regional situations differ considerably within Europe. Significant variance exists in demography, climate and environmental impacts, economic structures, cultures and values, land uses (urban, agriculture, pasture), and patterns of public and private partnerships. These local conditions, which are key to be considered, make it challenging to transfer lessons learned from successful adaptation measures across time and space.

Nonetheless, the determinants highlighted above and lessons learnt from practical experiences provide guidance for designing successful and robust adaptation strategies. As noted earlier, key elements of successful strategies include:

- Tailoring measures to specific regional climate conditions, sectors, and political and socioeconomic contexts.
- Ensuring dialogue between stakeholders through cooperative structures and knowledge transfers.
- Monitoring progress to support regular reviews of policy objectives and the inclusion of new scientific information when becoming available.

These elements would help deliver decision-making processes that take climate change into account more systematically and follow a proactive, precautionary, and cross-sectoral approach.

The Cost of Adaptation

Watkiss and Hunt (2010) reviewed the information on the costs and benefits of adaptation in Europe, covering European sectoral, regional, and national studies. The review found that the coverage of adaptation cost estimates is limited, though the evidence base is now growing as a result of such on-going Europe-wide studies as the EU FP7 ClimateCost project. The existing information has a very uneven distribution, the largest number of studies, and those with most advanced and wide range of methods used, being for the coastal zone. There are also a number of cross-sectoral studies emerging, which look at the indirect effects of coastal flooding on, for example, health and tourism. For other sectors the coverage is more limited.

These studies use a range of methods and metrics for different time periods and with different assumptions, and therefore it is challenging to compare estimates, undertake a systematic review and build up a coherent picture of the overall costs of adaptation. Nevertheless, the information provides some early context and highlights important issues.

Estimates, in global aggregated studies, of adaptation costs for Europe include the UNFCCC (2007a, b) review with an indicative assessment of USD 3–19 billion/year by 2030 (EUR 2.5–16 billion/year) for the infrastructure and coastal zones, while estimates given in the Stern Review (2006) based on costs of 0.05–0.5% GDP imply adaptation costs for infrastructure for Europe of EUR 4–60 billion/year. In addition to investment and financial flow assessments, economic IAMs also report a wide range of estimates: the PAGE model, using the Stern analysis findings, assumes adaptation costs for the EU15 of USD 25–60 billion/year (EUR 21–50 billion/year) with a mean estimate of USD 45 billion/year (EUR 37 billion/year); other IAM studies report much lower adaptation costs, such as the adaptation and mitigation strategies (ADAM) project (Aaheim et al. 2010) which reports adaptation costs in western Europe in 2020 estimated at 0.04% of GDP (USD 5 billion or EUR 4.1 billion) rising to USD 35 billion in 2050 (EUR 29 billion or 0.13% of GDP) under a 2°C scenario; OECD (de Bruin et al. 2009) estimates total weighed adaptation cost for a 2.5°C temperature rise at 0.64% of total output for Europe and 0.14% for eastern Europe.

The sectoral and national studies provide some comparative information on adaptation costs. The PESETA coastal study (Richards and Nicholls 2009) and Hinkel et al. (2009, 2010) reports costs of EUR 0.25–1.7 billion/year in the period 2010–2040 and EUR 0.3–3.5 billion/year in the period 2070–2100 across a range of temperature increase and sea-level rise scenarios. These studies also report that the economic benefits of adaptation options in reducing costs of inaction far outweigh the costs. For health, there are estimates of the costs of adaptation for diarrhoeal disease in Europe, based on costs of health interventions (Ebi 2008; Markandya and Chiabai 2009) which report annual adaptation costs in the period up to 2030 for Europe at USD 12–260 million/year for a range of scenarios and assumptions (EUR 10–215 million/year). These estimates indicate potentially large adaptation costs in Europe – billions of Euro per year in the short-term, potentially tens of billions per year in the longer-term.

A number of national studies have undertaken more comprehensive adaptation cost assessments, though these also have only partial coverage and, in general, much less information is available for the EU-12 than for EU-15. The most detailed national information is currently available for the Netherlands, Sweden, and the United Kingdom but a large number of other European national studies will be published over the next few years, many of which will assess the costs of adaptation.

The individual national studies also imply large adaptation costs, particularly for flood protection. For example, the United Kingdom Foresight study estimated the total adaptation investment needed to address flooding – coastal, river, and intra-urban – over the next 80 years at between GBP 22 billion and GBP 75 billion for a portfolio of responses, depending on the scenario, implying average annual costs of up to EUR one billion per year. Similarly, a recently conducted assessment on flood protection and flood risk management in the Netherlands estimates that the implementation of a comprehensive set of adaptation measures will cost EUR 1.2–1.6 billion/year up to 2050 and EUR 0.9–1.5 billion/year during the period 2050–2100. The recent Swedish evaluation estimated potentially large investment costs for adaptation to climate change across a wider range of sectors, including transport, energy, water treatment, infrastructure, flood protection, and agriculture, at up to a total of EUR ten billion for the period 2010–2100.

When scaled up to the European level, these national studies imply potentially higher adaptation costs (even for individual risks such as flooding) than found in many of the more aggregated sectoral studies, and certainly higher than many of the IAM assessments. At the EU level, they suggest costs of tens of billions/year when scaled to all countries and all sectors.

The lack of information about full adaptation costs hinders to some extent the implementation of adaptation measures across Europe. Substantial work in this area is therefore required to inform further integrated policy-making.

Acknowledgments The author gratefully acknowledges the helpful comments on an earlier draft of Ronan Uhel (EEA), André Jol (EEA), and the anonymous reviewers. The opinions expressed here do not necessarily reflect those of the European Environment Agency and are the sole responsibility of the author.

References

Aaheim A, Dokken T, Hochrainer S et al (2010) National responsibilities for adaptation strategies: lessons from four modelling frameworks. In: Hulme M, Neufeldt H (eds) Making climate change work for us. European perspectives of adaptation and mitigation strategies 1: 4. Cambridge University Press, Cambridge

Biesbroek GR, Swart RJ, Carter T et al (2010) Europe adapts to climate change: comparing national adaptation strategies. Global Environmental Change 20(August):440–450

de Bruin, K, Dellink, R, Agrawala S (2009). Economic aspects of adaptation to climate change: integrated assessment modelling of adaptation costs and benefits. OECD Environment Working Papers No. 6, OECD publishing. doi:10.1787/225282538105

Ebi KL (2008) Adaptation costs for climate change-related cases of diarrhoeal disease, malnutrition, and malaria in 2030. Global Health 4:9

EC (2009a) Adapting to climate change: towards a European framework for action. COM (2009) 147/4, European Commission, Brussels

EC (2009b) Common implementation strategy for the water Framework directive (2000/60/EC): work programme 2010-2012 – Supporting the implementation of the first river basin management plans. European Commission, Brussels

EC (2009c) Towards a strategy on climate change, ecosystem services and biodiversity, a discussion paper prepared by the EU Ad Hoc Expert Working Group on Biodiversity and Climate Change. http://ec.europa.eu/environment/nature/pdf/discussion_paper_climate_change. pdf

EC (2009d) Design of guidelines for the elaboration of Regional Climate Change Adaptations Strategies. European Commission – DG Environment – Tender DG ENV. G.1/ETU/2008/0093r. Ribeiro M, Losenno C, Dworak T, Massey E, Swart, R, Benzie M, Laaser C. Ecologic Institute, Vienna. http://ec.europa.eu/environment/water/adaptation/pdf/ RAS%20Final%20Report.pdf

EC (2010a) Options for an EU vision and target for biodiversity beyond 2010. COM(2010)4 final. http://ec.europa.eu/environment/nature/biodiversity/policy/index_en.htm; http://ec.europa. eu/environment/nature/biodiversity/conference/index_en.htm. Accessed 13 Sep 2010

EC (2010b) Environmental assessment. http://ec.europa.eu/environment/eia/home.htm. Accessed 27 Aug 2010

EC (2010c) Green paper on forest protection and information in the EU: Preparing forests for climate change. COM(2010) 66 final. http://ec.europa.eu/environment/forests/pdf/green_paper. pdf

ECDC (2010) Climate change and communicable diseases in the EU member states – Handbook for national vulnerability, impact and adaptation assessments. European Centre for Disease Prevention and Control, Stockholm. www.ecdc.europa.eu/en/publications/Publications/1003_ TED_handbook_climatechange.pdf

EEA (2008) Impacts of Europe's changing climate – 2008 indicator-based assessment. EEA report no 4/2008. European Environment Agency, Copenhagen

EEA (2009) Regional climate change and adaptation – the Alps facing the challenge of changing water resources. EEA report no 8/2009. European Environment Agency, Copenhagen

EEA (2010a) The European environment – state and outlook 2010: adapting to climate change. European Environment Agency, Copenhagen. http://www.eea.europa.eu/soer/europe/adapting-to-climate-change

EEA (2010b) National adaptation strategies. www.eea.europa.eu/themes/climate/national-adaptation-strategies. Accessed 13 Sept 2010

ESPACE (2007) ESPACE – Planning in a changing climate – putting adaptation at the heart of spatial planning. www.espace-project.org/part1/publications/ESPACE%20Stategy%20Final.pdf

Füssel HM, Klein R (2006) Climate change vulnerability assessments: an evolution of conceptual thinking. Climatic Change 75:301–329

Grothmann T, Patt A (2005) Adaptive capacity and human cognition: the process of individual adaptation to climate change. Glob Environ Change 15(3):199–213

Hallegatte S, Patmore N, Mestre O et al (2008) Assessing climate change impacts, sea level rise and storm surge risk in port cities: a case study on Copenhagen. OECD Environment Working Papers No. 3, OECD. doi:10.1787/236018165623

Hinkel J, Nicholls RJ, Vafeidis AT et al (2009) The vulnerability of European coastal areas to sea level rise and storm surge. Contribution to the EEA SOER 2010 report, Potsdam Institute for Climate Impact Research (PIK)

Hinkel J, Nicholls R, Vafeidis A et al (2010) Assessing risk of and adaptation to sea-level rise in the European Union: an application of DIVA. Mitigation and Adaptation Strategies for Global Change, 1–17. Published online. doi:10.1007/s11027-010-9237-y

IPCC (2007) Climate change 2007: impacts, adaptation and vulnerability. Contribution of working group II to the fourth assessment report of the Intergovernmental Panel on Climate Change. Cambridge University Press, Cambridge

Isoard S, Grothmann T, Zebisch M (2008) Climate change impacts, vulnerability and adaptation: theory and concepts. Paper presented at workshop 'Climate Change Impacts and Adaptation in the European Alps: Focus Water', UBA, Vienna

London City (2010) www.london.gov.uk/climatechange/sites/climatechange/staticdocs/Climiate_ change_adaptation.pdf

Markandya A, Chiabai A (2009) Valuing climate change impacts on human health: empirical evidence from the literature. Int J Environ Res Public Health 6(2):759–786

NAO (2009) Adapting to climate change – a review or the environmental audit committee. National Audit Office, United Kingdom. www.nao.org.uk/publications/0809/adapting_to_climate_change.aspx

NRW (2009) North-Rhine Westphalia. www.umwelt.nrw.de/umwelt/pdf/klimawandel/Klimawandel_Anpassungsstrategie_Gesamt.pdf

Parry ML, Canziani OF, Palutikof JP et al (eds) (2007) Climate change 2007: impacts adaptation and vulnerability – contribution of working group II to the fourth assessment report of the Intergovernmental Panel on Climate Change. Cambridge University Press, Cambridge

Parry M, Arnell N, Berry P et al (2009) Assessing the costs of adaptation to climate change: review of the UNFCCC and other recent estimates. International Institute for Environment and Development and Grantham Institute for Climate Change, London

Per M, Aix F, Beck S et al (2009) Climate policy integration, coherence and governance. PEER report no 2. Partnership for European Environmental Research, Helsinki. http://peer-initiative.org/media/m235_PEER_Report2.pdf

Pitt M (2008) Learning lessons from the 2007 floods. Full report. http://archive.cabinetoffice.gov.uk/pittreview/thepittreview/final_report.html. Accessed 20 Sept 2009

PreventionWeb (2010b) National platforms. www.preventionweb.net/english/hyogo/national/?pid:3&pil:1. Accessed 27 Aug 2010

Richards J, Nicholls RJ (2009) Impacts of climate change in coastal systems in Europe. PESETA-Coastal systems study. JRC scientific and technical reports

Rotterdam City (2007) http://kennisvoorklimaat.klimaatonderzoeknederland.nl/gfx_content/documents/regio%20rotterdam/algemeen/Brochure%20Rotterdam%20Climate%20Proof%20The%20Rotterdam%20Challenge%20on%20Water%20and%20Climate%20Adaptation.pdf

Stern N (2006) The economics of climate change. Cabinet Office – HM Treasury. Cambridge University Press

Swart R, Raes F (2007) Making integration of adaptation and mitigation work: mainstreaming into sustainable development policies? Clim Policy 7(4):288–303

Swart R, Biesbroek R, Binnerup S et al (2009) Europe adapts to climate change: comparing national adaptation strategies. PEER report no 1. Partnership for European Environmental Research, Helsinki. http://peer-initiative.org, http://peer-initiative.org/media/m256_PEER_Report1.pdf

Uhel R, Isoard S (2008) Regional adaptation to climate change: a European spatial planning challenge. InfoRegio Panorama, March 2008 No 25 'Regional policy, sustainable development and climate change', European Commission DG REGIO

UNDP (2004) Adaptation policy frameworks for climate change: developing strategies, policies and measures. UNDP, New York

UNECE (2009) Guidance on water and adaptation to climate change, convention on the protection and use of transboundary watercourses and international lakes. www.unece.org/env/documents/2009/Wat/mp_wat/ECE_MP.WAT_30_E.pdf

UNFCCC (2006) Five-year programme of work on impacts, vulnerability and adaptation to climate change. Report of the Subsidiary Body for Scientific and Technological Advice on its twenty-fifth session, Nairobi, 6–14 Nov 2006

UNFCCC (2007a) The analysis of existing and potential investment and financial flows relevant to the development of an effective and appropriate international response to climate change. UNFCCC, Bonn

UNFCCC (2007b) Investment and financial flows to address climate change. Background paper. UNFCCC, Bonn

UNFCCC (2007c) Decision 1/CP.13, Bali Action Plan. UNFCCC, Bonn

UNISDR (2009b) Implementing the hyogo framework for action in Europe: advances and challenges – report for the period 2007–2009. www.unisdr.org/preventionweb/files/9452_V114.05HFABLEU7.pdf

van Ierland EC, de Bruin K, Dellink RB et al (2006) A qualitative assessment of climate change adaptation options and some estimates of adaptation costs. Routeplanner naar een klimaatbestendig Nederland Adaptatiestrategiën. Study performed within the framework of the Netherlands Policy Programme ARK as Routeplanner projects 3, 4 & 5

Watkiss P, Hunt A (2010) Review of Adaptation Costs and Benefits Estimates in Europe for SOER 2010. Contribution to the EEA SOER 2010. Report to the European Environment Agency

WHO (2010a) Protecting health in an environment challenged by climate change: European Regional Framework for Action. Fifth Ministerial Conference on Environment and Health, Parma, Italy, 10–12 March 2010. www.euro.who.int/en/what-we-publish/information-for-the-media/sections/latest-press-releases/european-governments-adopt-comprehensive-plan-to-reduce-environmental-risks-to-health-by-2020. Accessed 27 Aug 2010

WHO-Europe (2010c) www.euro.who.int/en/what-we-do/health-topics/environmental-health/Climate-change. Accessed 27 Aug 2010

Chapter 5
The Australian Experience

Timothy Frederick Smith, Dana C. Thomsen, and Noni Keys

Abstract The past focus of climate change action in Australia was dominated by mitigation initiatives and these remain a critical and urgent issue. However, the global imperative for planning and action to adapt to likely climate change impacts articulated by the scientific community has now been adopted as a key imperative for governments, industries, and communities alike. While it is often difficult to separate climate change adaptation initiatives from existing risk management or sustainability initiatives, over recent years there has been clear identification of new climate change adaptation policies and programs. These initiatives range from local-scale planning to reduce vulnerability, to national research programs such as the National Climate Change Adaptation Research Facility and the Commonwealth Scientific and Industrial Research Organisation (CSIRO) Climate Adaptation Flagship. The Australian Government has also created a Department of Climate Change to coordinate and support activities at the national level. One of the key challenges that remains is mainstreaming the understanding of vulnerability to climate change impacts and associated adaptation initiatives, across and within sectors. For example, the emphasis of research and action has been based on assessments of exposure, with only a limited number of past research projects focused on the understanding of sensitivity and adaptive capacity. Furthermore, climate change adaptation is usually framed within an economic rationalist paradigm, rather than a quality of life paradigm, and consequently there are challenges and potential paradoxes associated with achieving overriding goals such as short-term economic growth. Nevertheless, Australia continues to build upon existing mitigation and risk management initiatives and has now embraced a range of policies, strategies, and promising actions to enhance climate change adaptation.

T.F. Smith (✉) • D.C. Thomsen • N. Keys
Sustainability Research Centre, University of the Sunshine Coast, Maroochydore, DC 4558 Australia
e-mail: Tim.Smith@usc.edu.au; dthomsen@usc.edu.au; noni.keys@gmail.com

J.D. Ford and L. Berrang-Ford (eds.), *Climate Change Adaptation in Developed Nations: From Theory to Practice*, Advances in Global Change Research 42, DOI 10.1007/978-94-007-0567-8_5, © Springer Science+Business Media B.V. 2011

Keywords Australia • Adaptation policy • Adaptation research • Australian Government • Local adaptation • CSIRO • National Climate Change Adaptation Research Facility • Climate Adaptation National Research Flagship • Local Adaptation Pathways Program

Introduction

> I love a sunburnt country ... of droughts and flooding rains.
>
> (Dorothea Mackellar 1911)

Climatic conditions have been prevalent in the Australian psyche and popular culture since European colonization, and earlier as part of the Indigenous Australian dreamtime. Drought, fire, floods, and other climate-related natural hazards have been frequent and intense for many parts of Australia, and often following each other in rapid succession. The discussion of potential impacts of climate change to many Australians is therefore no big surprise, as many have seen examples of the impacts of climate variability either first hand or through extensive media coverage. While there is some (declining) contention over the causes of contemporary climate change, there is a general acceptance of the devastation that can be wrought by climatic events.

While a distinction between adaptation and mitigation has been made in the literature and is reflected in various policy documents internationally, mitigation is essentially a form of adaptation. However, for the purposes of this chapter, we focus on the popularized interpretation of adaptation, which excludes activities aimed at addressing the root causes of climate change. Thus, adaptation in an Australian context (similar to many other parts of the world) is defined as actions taken to reduce the severity of impacts, or capitalize on the opportunities, resulting from climate change. This Australian definition is largely consistent with several international definitions (e.g., United Nations Framework Convention on Climate Change (UNFCCC), Intergovernmental Panel on Climate Change (IPCC), United Nations Development Programme (UNDP), and UK Climate Impact Programme) (Levina and Tirpak 2006). We begin by describing the context for climate change in Australia; second, we outline the relevant institutional arrangements; third, we present the key initiatives and activities that have shaped the evolution of climate change adaptation in Australia; fourth, we outline the current Australian climate change adaptation policies and research initiatives; and finally, we discuss future research needs and policy directions.

The Context for Climate Change in Australia

Many Australians are acutely aware of their role in contributing to climate change. For example, while Australia accounts for less than 6% of total global coal extraction, Australia is the largest coal exporter in the world, at over 31% of the global market share (IEA 2009). Coal also represents Australia's largest commodity export; export earnings for black coal were valued at approximately $54.6 billion (AUD) in the 2008–2009 financial year (ABARE 2009). The remaining 22.5% of Australian coal not exported is used primarily for electricity to supply, for example, residential and industrial consumers. Hence, from about 1988 to 2005 the dominance of Australian discourse to climate change had focused primarily on mitigation, as demonstrated by the public pressure exerted on the Australian Government to eventually commit to the Kyoto Protocol for reducing greenhouse gas emissions.

Australia's perceived reliance on coal exports to sustain its economic well-being highlights an ill-fated trade-off between economics and ecology – favoring economic growth over functioning socio-ecological systems. These competing goals, values, and worldviews are often dominated by anthropocentric creations, such as markets, that may or may not manifest themselves with devastating consequences over ecological phenomena. For example, from early 2008 to early 2009, the climate change crisis was overshadowed, in both media exposure and political discourse, by the global economic "crisis" – just as it was gaining considerable widespread public support for action. As Preston and Stafford-Smith (2009) suggest, one of the critical challenges to adaptation is the complexity of social interactions involved in adaptation decision making. Since the return of the climate change debate to media favor, the skeptics have again emerged but are currently a considerable minority, particularly in political discourse. Similarly, the Australian Government has acknowledged the vulnerability of many sectors of the Australian economy to climate change, in particular the agricultural sector (e.g., impacts from drought and intense storm events), fisheries (e.g., impacts from changes in ocean temperature), and tourism (e.g., impacts on iconic tourist destinations such as the Great Barrier Reef). The fundamental point of debate now relates to the economic cost of action and/or inaction, highlighting that the challenge will be in moving from rhetoric to action for both mitigation and adaptation.

Overview of Australian Institutional Arrangements

In order to discuss the Australian response to climate change it is important to first explain the interactions of the various tiers of the government. Then, this section will overview national, state, and local arrangements related to climate change adaptation.

In 1901, when the nation-state of Australia was proclaimed, the powers of the national government were outlined in the Commonwealth of Australia Constitution Act (1900) (the Constitution). These powers largely relate to taxes, trade, government, and judicial power, and any powers not specified in the Constitution were deemed to be the responsibility of the States of Australia. Similar to many other environmental issues, climate change was not mentioned in the Constitution. Hence, while the Australian Government may develop climate change adaptation policy, it is not binding to the States. However, the Australian Government has one distinct advantage over the States in its ability to influence adoption: the ability to raise taxes. The common forum for agreement occurs through the Council of Australian Government (COAG), which often involves Australian Government funds transfers to the States as incentive payments. In 2007 the Australian Government Department of Climate Change was established to focus and build on previous functions and roles transferred from the former Department of the Environment and Water Resources (including the Australian Greenhouse Office), the Department of Foreign Affairs and Trade, and the Department of the Prime Minister and Cabinet.

Next, as mentioned above, individual States have been vested with powers to form and enact some legislation and policies, including those related to climate change adaptation. Each State has taken a somewhat distinct approach to the way it has responded to this adaptation. While many States have taken proactive steps in the formation of various Departments of Climate Change over recent years, they operate largely in isolation of one another. Some have even adopted different climate change projections, such as for sea level rise as a basis for land use planning (although in some cases this is justified because they aim to predict local changes, which may vary between States due to a range of geo-physical parameters such as bathymetry, dominant wind directions and speed, and ocean currents).

The third tier of Australian government is local government, which is a direct function of that of the State. That is, local governments are created through State government legislation, and the States vary both the boundaries and responsibilities of local governments from time to time (e.g., the recent regionalization of local government in Queensland through amalgamation). For this reason, the roles and responsibilities of local government vary considerably between States. Similarly, spatial and population sizes of local governments also vary considerably within and between States. In all States, local governments act as a delivery arm from the State government in one form or another. However, while local governments can raise revenue through rates, for example, the mechanisms for raising revenues are controlled by the State governments. However, considerable community pressure has initiated the adoption of a number of de facto responsibilities for environmental management (Smith et al. 2001), and this is likely to be the case for climate change adaptation in coming years. The de facto nature of environmental roles and responsibilities means that they are fragmented and uncoordinated across scales and sectors. This is partly driven by: (1) the urgency of the imperative to adapt; (2) an absence of deep conceptual understanding of climate change; and (3) entrenched (mandated) sectoral interests of government agencies between and within the various tiers of government.

The Evolution of Climate Change Adaptation in Australia

Australia's experience with adaptation to climate change has followed from its involvement in international research and negotiations through the World Meteorological Organization (WMO), United Nations Environment Programme (UNEP), and

IPCC, leading to the UN Framework Convention on Climate Change (UNFCCC). To simply cite domestic climate change-related landmark initiatives and achievements is a credit to the ongoing effort of climate change scientists to contribute to policy development in Australia, but does not reflect the magnitude of political controversy and industry influence surrounding domestic responses to climate change, including the development of adaptation policy. Although the reduction of greenhouse gases (mitigation) and adaptation to climate change impacts are often discussed as two separate response strategies, their evolution as policy responses in Australia are inextricably linked (Table 5.1).

Following Australia's participation in the World Conference on the Changing Atmosphere in Toronto in 1988, the Australian government adopted an ambiguously worded "Interim Planning Target" to reduce greenhouse gases by 20% by 2005, based on 1988 levels, unless such action caused "harm" to the Australian economy (Kay 1997; Pearse 2007). Subsequent policy response to climate change was thus framed as a debate over costs to the economy of taking action to reduce greenhouse emissions versus costs to the economy, human health, coastal settlements, and ecosystems of not taking action, i.e., of impacts from climate change (Bulkeley 2001). With Australia's domestic electricity generation and a substantial portion of export revenue dependent on relatively cheap coal reserves, it was in the interests of what Bulkeley (2001) terms "the resource-based discourse coalition" to downplay the magnitude and extent of climate impacts and the link between costly extreme weather events and climate change. This position was supported by past Australian governments, making it politically dissonant to allocate resources to adaptation programs.

However, as a result of increasing information about the likely impacts of unmitigated climate change at least at larger scales, and Australia's continuing involvement in the IPCC and international climate negotiations, domestic climate change policy processes were initiated. In 1990, two separate processes were implemented for assessing the costs, benefits, and options for reducing greenhouse gases: (1) an extensive stakeholder consultation process established to determine strategies for Ecologically Sustainable Development (ESD) and (2) a more narrowly referenced Industry Commission inquiry into the costs and benefits of stabilizing greenhouse gas emissions. Of significance for subsequent adaptation policy development, the industry inquiry emphasized uncertainties in climate change science as an argument against taking national action (Taplin 1994). According to one of the commissioners, "economists cannot measure impacts if scientists cannot tell them what the impacts are likely to be" (Hundloe 1992, p. 476).

Following these two inquiries, the Australian government, in collaboration with State and local government, produced the National Greenhouse Response Strategy (Taplin 1994). Unlike the broad-ranging policy recommendations emanating from the cross-sectoral ESD process, the Strategy took a minimalist "no-regrets" approach and focused on a review of State and local government initiatives in addressing climate change (Taplin 1995, p. 18). Although the National Greenhouse Response Strategy included references to research and adaptation, and was adopted

Table 5.1 A summary of key climate change-related initiatives in Australia

	National	State	Local
1949	CSIRO Meteorological Physics Research Section established	–	–
1988	CSIRO Meteorological Physics Research Section, after several incarnations, becomes the CSIRO Division of Atmospheric Research	–	–
	Australia participates in World Conference on the Changing Atmosphere		
1989	Australian Climate Change Science Program established	–	–
1990	Australian Government adopts interim target to reduce emissions by 20%	–	–
	Ecologically Sustainable Development working groups to report on options for meeting interim target to reduce greenhouse gas emissions		
	Industry Commission inquiry into the costs and benefits of stabilizing greenhouse gas emissions		
1992	Australia signs UNFCCC at Rio Earth Summit	COAG adopt National Greenhouse Response Strategy	Agenda 21 local government sustainability planning commences
	National Greenhouse Response Strategy produced		
1997	–	–	ICLEI Cities for Climate Protection pilot program commences in Australia
1998	National Greenhouse Strategy published, which includes a national framework for adaptation to climate change	–	–
1999–2009	–	–	Australian Government funds ICLEI Cities for Climate Protection Oceania Head Office
2004	National Climate Change Adaptation Program created	–	–

Year		
2005	CSIRO Division of Marine Research merges with the Division of Atmospheric Research to form CSIRO Marine and Atmospheric Research (CMAR) Allen Consulting report to Australian Government: Climate Change Risk and Vulnerability	New South Wales launches Greenhouse Action Plan
2006	—	New South Wales Sustainability Action Statement
2007	COAG endorses the National Climate Change Adaptation Framework (13 April) Australian Government Department of Climate Change established (3 December) *Climate Change Adaptation Actions for Local Government* report released Review of Australian Climate Change Research (Solomon & Steffen) Australia ratifies the Kyoto Protocol in the UNFCCC Bali negotiations	Queensland Climate Smart Adaptation Action Plan
2007–2008	CSIRO launches the $44 million (AUD) Climate Adaptation National Research Flagship Australian Government invests a further $126 million (AUD) over 5 years for adaptation, including the National Climate Change Adaptation Research Facility Garnaut Climate Change Review	—
2008	(Wilkins) Strategic Review of Australian Climate Change Programs Commonwealth Water Act commences (establishes new governance arrangements including sustainable limits on water extraction (3 March))	—
2010	Proposed introduction date for the Australian Government Emissions Trading Scheme	—
2011	Revised introduction date for the Carbon Pollution rReduction Scheme	—

by the Council of Australian Governments (COAG), by 1995 there had been little implementation of any actions (Bulkeley 2001, p. 161).

In 1998 a revised National Greenhouse Strategy was published (Australian Government 1998). It described a national framework for adaptation to climate change "to assist policy makers and industries plan for and adapt to the potential impacts of climate change in a cost-effective and timely manner" (p. 90). The Australian government was to assume a coordinating and research support role in collaboration with the States and Territories, and relevant research organizations. Priorities for investigation included identifying mechanisms for incorporating adaptation strategies into existing planning mechanisms (p. 90).

Meanwhile, some advances were made in considering the impacts of climate change to ecosystems and sectors reliant on natural resources, such as agriculture, forestry, coastal regions, and human health. However, government attention for ensuing years remained focused on the economic costs and benefits of mitigation (Bulkeley 2001; Head 2009). In its formal communication to the UNFCCC in 2005, the Australian Government reported that:

> [most] of the funding under the recently developed Climate Change Strategy is for policies and measures aimed at limiting and reducing greenhouse gas emissions. [However,] the Australian Government is investing in measures to identify, assess and adapt to the unavoidable consequences of climate change. (AGO 2005a, p. 36)

In 2004, the National Climate Change Adaptation Program was created, and a report was commissioned to identify priorities for adaptation (Allen Consulting 2005, p. iii), indicating a shift to more serious consideration of the impacts of climate change by the Australian government.

The evolution of adaptation policy in Australia has continued to be marked by a series of inquiries and commissioned reports by experts outside of the formal government process. This reflects the complexity of both the emerging scientific knowledge, and different attitudes to the costs and benefits of responding to climate change assumed by different government departments (Taplin 1994; Bulkeley 2001). For example, an inquiry into the Australian Climate Change Science Program found that climate change is already influencing vulnerable ecosystems and urbanized coastlines in Australia, with increasing effects expected in the coming years (Solomon and Steffen 2007, p. 2). Significantly, the report noted that funding for climate change science that underpins adaptation had remained static over the previous 5 years (p. 3).

A detailed tracing of national budget allocations to adaptation programs over the period of Australia's participation in climate change policy over the past two decades is beyond the scope of this discussion. However, in 2007 one estimate of government spending on adaptation was $14.6 million (AUD) or three cents per Australian per year (Pearse 2007, pp. 120–121). In the same year Cyclone Larry cost $239 million (AUD). At that point the then government had also provided $2.3 billion (AUD) in relief to farmers affected by the continuing drought (Pearse 2007, p. 121). Spending on adaptation was increased in 2007 with the announcement of funding of $170 million (AUD) over 5 years for a National Adaptation Framework to be developed with State governments through the Council of Australian Ministers (COAG) (Solomon and Steffan 2007, p. 3).

Under the COAG National Adaptation Framework, the Australian government funds research and planning. This includes assistance for local governments to identify and implement adaptation in relation to: infrastructure and property; recreation facilities; health services; planning and development approvals; natural resource management; and water and sewerage (Council of Australian Governments 2007). The Framework aims to reduce sectoral and regional vulnerability to climate change and to build adaptive capacity. Since 2007, every State and Territory has initiated research and planning for adaptation, with some States building on previous sustainability programs developed as a result of Agenda 21 commitments (Head 2009). Western Australia launched a strategy in 2004 (Western Australian Greenhouse Taskforce 2004), and New South Wales and Victoria in 2005, with other States and Territories following (Head 2009). In addition, more than 60 local governments, most located in coastal and urban areas, have received funding to carry out climate change risk assessments and develop adaptation plans (DCCEE 2008a). The stated aim of the assistance is to build adaptive capacity through the provision of scale-relevant information, the engagement of stakeholders, and assistance to integrate climate change impacts and adaptation considerations into key policies and programs. However, a commitment to develop an implementation plan for the Framework detailing "partnerships, milestones and resources ... the roles and responsibilities of the different levels of government in advancing the actions [and] monitoring and evaluation," (Council of Australian Governments 2007, p. 4) has yet to be fully enacted.

The Australian government's role in providing research and information to support adaptation decisions by State and local governments was reinforced in a review of all climate change programs carried out after the change of government in late 2007. The Wilkins review (2008) produced a set of principles to assist the new Australian government in streamlining all programs to be complementary to the government's perceived centerpiece of climate change policy: an emissions trading system. It advised that while the Australian government should focus on its national strategy for emissions reduction, States and local government should deal with issues of adaptation. In this regard, the report remarked that "currently governments are spending too much time, energy and resources on reduction and not enough on adaptation" (p. 6). As of 2008, it was estimated that 6% of the Australian government climate change budget was spent on adaptation (Wilkins 2008). However, it is recognized that ongoing involvement by the Australian government in issues of high complexity and magnitude – such as improving urban water supplies through desalination, recycling, and stormwater harvesting, which was allocated AU$1 billion over 6 years in 2008 (Wong 2008), and improving flows in the Murray–Darling river system – will continue to require national leadership and funding.

Despite recent inquiries and reports related to adaptation policy stressing the interrelationships with mitigation (Allen Consulting 2005; Solomon and Steffan 2007; Wilkins 2008; Garnaut 2008), concern remains about a lack of direction and complementarity between the aims and implementation of the two policy areas. For example, during recent Senate committee hearings, one Senator suggested that farmers accessing government training assistance for managing climate change impacts were "only getting bits of the picture," due to the government's reluctance to direct individuals toward sustainable production options such as renewable energy generation and sale (Milne 2008).

Current Australian Climate Change Adaptation Policies and Research Initiatives

Adaptation Policies

At the national level, the *Climate Change Budget Overview 2009–2010* details that the present Australian Government has allocated a total of over $15 billion (AUD) for programs and initiatives to address issues related to climate change (DCCEE 2009). This is a substantial increase in funding from the initial $2.3 billion (AUD) allocated in the 2008–2009 budget, and supports a three-way strategy focused on mitigation, adaptation, and fostering a global response. The 2009–2010 Climate Change Budget is weighted toward mitigation strategies with the development and implementation of the Carbon Pollution Reduction Scheme (a recent iteration of the former Emissions Trading Scheme) and funding for clean businesses, clean energy, and sustainable homes. For example, $4.5 billion (AUD) will be spent on the Clean Energy Initiative, supporting renewable energy and carbon sequestration; $4 billion (AUD) will fund energy efficiency initiatives; and $11.8 million (AUD) will contribute to the ongoing development and implementation of the Carbon Pollution Reduction Scheme (DCCEE 2009).

Funding for adaptation focuses on an integrated approach between the different levels of government and the provision of scientific research to direct or enable future adaptations. Activities at the national level are coordinated by the National Climate Change Adaptation Framework that was endorsed by the Council of Australian Governments (COAG) in April 2007 as part of its *Plan of Collaborative Action on Climate Change* (DCCEE 2008b). The Framework is designed to facilitate collaboration between National, State, and Territory governments in order to develop adaptive capacity and reduce vulnerability to climate change in key sectors and regions. It is also designed to ensure that climate change is considered in policy development and decision making across all scales and jurisdictions in the medium term (5–7 years) (DCCEE 2008b).

The key element of the Australian Government adaptation strategy is the ongoing $126 million (AUD) Climate Change Adaptation Program designed to provide guidelines, planning tools, and information regarding climate change (AGO 2005b). For example, $20 million (AUD) will be spent establishing a National Climate Change Adaptation Research Facility and $30 million (AUD) will assist in the development of research plans prioritizing the information needs for adaptation across sectors (DCCEE 2009). Several Australian Government Departments are also undertaking initiatives and programs to support adaptation. For example the Department of Innovation, Industry, Science and Research is responsible for a $387.7 million (AUD) investment over 5 years to conduct integrated marine and climate research in order to ensure appropriate adaptation strategies, through the Marine and Climate Super Science Initiative; The Department of Climate Change has $31.2 million (AUD) for the Australian Climate Change Science Program; the Department of Agriculture, Fisheries and Forestry will conduct research and training and provide financial assistance to support adaptation in the agricultural sector through the Australia's Farming Future Initiative; and the Department of the Environment, Water, Heritage and the Arts supports several initiatives including Reef Rescue and Water for the Future (DCCEE 2009).

The third part of the national strategy focuses on global solutions, including funding toward adaptation measures. From 2008 to 2011 AusAID, the Australian Government's Aid Program, will spend $150 million (AUD) on a new International Climate Change Adaptation Program to assist economically poorer nations to adapt to the impacts of climate change through the provision of scientific information and technical expertise, funding support for adaptive responses, and contributions to multilateral adaptation funds (AusAID 2009). This program centers on Pacific Island Nations and East Timor. Initiatives in this program build on other sustainability-related projects such as those addressing water resource management and longer term climate change-related impacts in the region (e.g., construction of rainwater tanks in Tuvalu and support for solar desalination stills in Vanuatu, see AusAID 2009).

Enacting national-level strategies and undertaking on-ground works, local governments are perceived as being at the "forefront of managing the impacts of climate change" (DCCEE 2008a). Indeed, the Australian Government is funding a Local Adaptation Pathways Program (LAPP) to assist local governments to identify risks and develop appropriate responses to climate change threats (DCCEE 2008a). Possible actions for local government are identified in *Climate Change Adaptation Actions for Local Government* (SMEC Australia 2009). This report clearly articulates that local governments should focus on mitigation and adaptation strategies simultaneously and that local government has a regulatory "duty of care" with regard to the health and safety of its community. There are regional differences between the responsibilities of different government levels for climate change adaptation. Nevertheless, the report identifies that all local governments can play a leadership role in implementing adaptation strategies through education and awareness raising, changes to approval processes, and setting appropriate examples through the conduct of its own operations. For example, the City of Melville in Western Australia has developed sustainable residential design guidelines. These serve as an educational tool and were also incorporated into council policy governing the approval of residential developments (SMEC Australia 2009). Similarly, the Sunshine Coast Regional Council and the Moreton Bay Regional Council are both supporting the Living Smart Homes program, a free educational initiative.

However, many local governments are only beginning to shift their attention toward adaptation initiatives (Smith et al. 2009), and undertaking these initiatives are complicated by a range of other constraints such as the lack of clearly defined roles and responsibilities of the various tiers of government (see Smith et al. 2009 for other examples).

Most existing climate change-related policies focus on mitigation. Nevertheless, there are some policies that encourage adaptation including, for example: (1) the Commonwealth Water Act (3 March 2008) that sets out new water governance arrangements and new limits on water extraction, given the unsustainable use of current resources such as those of the Murray–Darling Basin and (2) guidance on sea level rise in Victoria and New South Wales that emerged in 2009.

Adaptation Research Initiatives

Since 2007 there has been a massive increase in funding to support climate change adaptation research initiatives. Most of these resources are being invested by the Australian Government; initiatives at other tiers of government represent only a fraction of the total investment in research effort. This chapter discusses major research initiatives by the Department of Climate Change (including the National Climate Change Adaptation Research Facility) and the Commonwealth Scientific and Industrial Research Organisation (CSIRO) Climate Adaptation National Research Flagship, which represent a combined investment in excess of $170 million (AUD) over the next 5 years.

Adaptation research initiatives were funded by the Australian Government prior to 2007 (e.g., the Urban Integrated Assessment Programme of the Australian Greenhouse Office). In 2007, several significant research initiatives were announced and dwarfed past investment in this domain. The most significant announcement was that of the formation of the National Climate Change Adaptation Research Facility (NCCARF). The NCCARF forms a catalyst for university research endeavors across eight research themes (although we recognize that numerous other university centers and consortiums related to climate change adaptation also exist independently from NCCARF): (1) terrestrial biodiversity; (2) human health; (3) marine biodiversity and resources; (4) water resources and freshwater biodiversity; (5) settlements and infrastructure; (6) social, economic, and institutional dimensions; (7) emergency management; and (8) primary industries. Each theme has a research network (often hosted by non-NCCARF university partners), and is developing or has developed a National Adaptation Research Plan (NARP). The eight NARPs set the funding priorities for climate change adaptation research through NCCARF. The Human Health NARP is the only NARP to be finalized, and it has attracted significant co-investment through a special funding program within the National Health and Medical Research Council competitive grant scheme. It is likely that similar co-investment strategies will be adopted for the remaining seven NARPs that are expected to be released progressively over the next 3–12 months. Indigenous issues were included in many of the NARPs; however, at the time of writing, there were suggestions that given the unique and cumulative pressures facing Indigenous Australians, a dedicated network and NARP should be formed. However, perhaps other disadvantaged communities – such as those living below the poverty line and without adequate resources to enhance their adaptive capacity – have not been adequately included within the research priority areas.

Also in 2007, the CSIRO, Australia's national science agency, launched the $44 million (AUD) Climate Adaptation National Research Flagship. This initiative has four research themes, including: (1) pathways to adaptation: positioning Australia to deal effectively with climate change; (2) sustainable cities and coasts; (3) managing species and natural ecosystems in a changing climate; and (4) adaptive primary industries, enterprises, and communities (CSIRO 2007). While the CSIRO has significant capacity and expertise in climate change research, they have also expanded their research partnerships for climate adaptation research through several Collaboration Fund projects. For example, in early 2009 the South East Queensland Climate Adaptation Research Initiative (SEQ CARI) began. SEQ CARI consists of a partnership between the CSIRO, Queensland Government, Australian Department of Climate Change, Griffith University, University of the Sunshine Coast, and the University of Queensland. SEQ CARI represents a $14 million (AUD) research initiative over 3 years aimed at building the resilience of the region through understanding vulnerability, developing adaptation strategies, and building adaptive capacity (CSIRO 2009). This is Australia's largest regional adaptation research initiative. While the CSIRO have actively supported climate change research for more than half a century (e.g., the CSIRO Meteorological Physics Research Section was established in 1949), significant investment in research directly related to adaptation occurred in 2007 with the establishment of the CAF.

Adaptation research in Australia has also begun to be recognized in national awards. For example, up until 2009 the Australian Museum Eureka Prize (Australia's pre-eminent science awards) in the category of Innovative Solutions to Climate Change was awarded to research focused on mitigation. However, in 2009, the prize was awarded to climate change adaptation research in the Sydney region (Australian Museum 2009).

A significant barrier to the facilitation of future climate change adaptation research relates to the nature, not amount, of research and university funding in Australia (which is also a similar issue in many other parts of the world). Australia is currently restructuring its research incentives toward one that favors quality over quantity, through the Excellence for Research in Australia framework, and associated changes in funding arrangements to the university research block grants. While this is certainly a desirable goal, the way in which quality is judged creates a strong bias toward rewarding disciplinary-specific researchers, over those working in transdisciplinary or emerging fields, such as climate change. Thus, while there are emerging significant funding programs for climate change adaptation research over the coming 3–4 years, Australia is creating disincentives for the scientific revolution (Kuhn 1962) that is needed to provide sustained support for research that would facilitate the radical breakthroughs needed in these emerging fields.

Future Research Needs and Policy Directions

Australia is well advanced in anticipatory approaches to adaptation planning at all tiers of government, within industry, and increasingly within communities (both communities of place and communities of interest). Australia also continues to invest significant resources into climate change adaptation research. Both planning and research are critical to a strategic and proactive approach to climate change adaptation. While there remain key knowledge gaps – particularly in the under-standing of context-specific vulnerability and adaptive capacity – the primary issue for Australia is not necessarily one of improvements in research or policy directives, more so, one of taking direct action. In other words, how does all of this investment in research get translated into reduced vulnerability, and are funds being invested in the right research to achieve this end?

Conclusion

Australians are acutely aware of the impacts of climate variability and thus the potential impacts of climate change. All tiers of government, along with industries and communities, have acknowledged the need for climate change adaptation. While significant investment has gone into climate change adaptation policies, strategies, and research, these initiatives remain subservient to the short-term economic rationalism that grips western societies. A defining trait of the Australian experience is preparing for the battle against climate change through research, capacity building, policies and strategies; however, the battle began long ago and we are still a long way from engaging at the frontline.

Acknowledgments We thank the two anonymous reviewers for their valuable comments and suggestions for improving the Chapter.

References

ABARE (2009) Australian mineral statistics 2009 (June Quarter). Australian Government, Canberra
AGO (2005a) Australia's fourth national communication on climate change: a report under the United Nations Framework Convention on Climate Change. Australian Government, Canberra
AGO (2005b) National Climate Change Adaptation Program. Australian Government. www. climatechange.gov.au/impacts/publications/pubs/nccap.pdf. Cited 27 Jun 2009
Allen Consulting (2005) Climate change risk and vulnerability. Australian Greenhouse Office, Australian Government, Canberra
AusAID (2009) Climate change adaptation. Australian Government. www.ausaid.gov.au/keyaid/ adaptation.cfm. Cited 12 Jul 2009

Australian Government (1998) National Greenhouse Strategy. Australian Government Publishing Service, Canberra

Australian Museum (2009) Australian museum eureka prizes: innovative solutions to climate change. http://eureka.australianmuseum.net.au/eureka-prize/innovative-solutions-to-climate-change2. Cited 14 Sep 2009

Bulkeley H (2001) No regrets? Economy and environment in Australia's domestic climate change policy process. Glob Environ Change 11(2):155–169

Council of Australian Governments (2007) National climate change adaptation framework. Commonwealth of Australia, Canberra

CSIRO (2007) Climate adaptation flagship. http://www.csiro.au/org/ClimateAdaptationFlagshipOverview.html. Cited 21 Aug 2009

CSIRO (2009) South East Queensland climate adaptation research initiative. http://www.csiro.au/partnerships/seqcari.html. Cited 14 Sep 2009

DCCEE (2008a) Local government. Australian Government. http://www.climatechange.gov.au/impacts/localgovernment/index.html. Cited 11 Jul 2009e

DCCEE (2008b) Key initiatives. Australian Government. http://www.climatechange.gov.au/impacts/about.html. Cited 11 Jul 2009

DCCEE (2009) Climate change budget overview 2009–10. Australian Government, Canberra

Garnaut R (2008) The Garnaut climate change review. Cambridge University Press, Melbourne

Head B (2009) Why adaptation matters: beyond the emissions trading debate. Paper presented at the public policy network conference, Australian National University, Canberra, 29–30 Jan 2009

Hundloe T (1992) The role of the industry commission in relation to the environment and sustainable development. Aust J Public Adm 51(4):476–489

IEA (2009) Key world statistics. Organisation for Economic Co-operation and Development, Paris

Kay P (1997) Australia and Greenhouse Policy: a chronology – background paper 4. Parliament of Australia, Parliamentary Library, Canberra

Kuhn TS (1962) The structure of scientific revolutions. University of Chicago Press, Chicago

Levina E, Tirpak D (2006) Key adaptation concepts and terms. Organisation for Economic Co-operation and Development, Paris

Mackellar D (1911) My country. In: The closed door and other verses. Specialty Press, Melbourne (First published in 1908 as 'Core of My Heart' in the London *Spectator*)

Milne C (2008) Farmers and climate change: adaptation and reducing emissions – Senate estimates transcripts, agriculture, fisheries and forestry portfolio – Climate Change, Canberra, available online: http://christine-milne.greensmps.org.au. Cited 10 Aug 2009

Pearse G (2007) High and dry: John Howard – climate change and the selling of Australia's future. Viking, Camberwell

Preston BL, Stafford-Smith M (2009) Framing vulnerability and adaptive capacity assessment: discussion paper. CSIRO climate adaptation Flagship working paper No. 2. http://www.csiro.au/org/ClimateAdaptationFlagship.html. Cited 10 May 2010

SMEC Australia (2009) Climate change adaptation actions for Local Government. A report prepared for the Australian Department of Climate Change. Australian Government, Canberra

Smith TF, Sant M, Thom B (2001) Australian estuaries: a framework for management. Cooperative Research Centre for Coastal Zone, Estuary and Waterway Management, Brisbane

Smith TF, Brooke C, Measham TG et al (2009) Case studies of adaptive capacity: systems approach to regional climate change adaptation strategies. Prepared for the Sydney Coastal Councils Group and the DCCEE

Solomon S, Steffan W (2007) Australian climate change research: perspectives on successes, challenges and future directions. DCCEE, Australian Government, Canberra

Taplin R (1994) Greenhouse: an overview of Australian policy and practice. Aust J Environ Manag 1(3):142–155

Taplin R (1995) International co-operation on climate change and Australia's role. Aust Geogr 26(1):16–22

Western Australian Greenhouse Taskforce (2004) Western Australian greenhouse strategy. Government of Western Australia, Perth

Wilkins R (2008) Strategic review of Australian Government climate change programs. Australian Government, Canberra

Wong P (2008) AU$6 million to help manage Murray-Darling Basin water resources. Media release, Parliament House, Canberra, 9 Oct 2008. http://www.climatechange.gov.au/minister/wong/2008/media-releases/October/mr20081009b.aspx. Cited 16 Sep 2009

Chapter 6
Leading the UK Adaptation Agenda: A Landscape of Stakeholders and Networked Organizations for Adaptation to Climate Change

Emily Boyd, Roger Street, Megan Gawith, Kate Lonsdale, Laurie Newton, Kay Johnstone, and Gerry Metcalf

Abstract This chapter is a review of the landscape of adaptation science and policy stakeholders and networks in the United Kingdom. The aim is to broadly map the key stakeholders and the activities in the science and policy sectors, and across regions and the Devolved Administrations. This is done to gain a better understanding of the range and scope of adaptation actions in the United Kingdom to date. The chapter makes advances in understanding the evolution of the UK adaptation landscape and demonstrates how adaptation policy and science are building adaptive capacity across institutions and networks. We conclude that adaptation policy and actions represent a change to living with climate change futures. The chapter also highlights that there remain significant challenges ahead, in particular in the realm of science.

Keywords Adaptation • Mapping • UK stakeholders • Institutions • Science • Policy • United Kingdom • Stakeholders • UKCIP • Adaptation legislation

Introduction

The United Kingdom has taken a leadership role in developing adaptation institutions and science over the past 10 years at both domestic and international levels. Since the 1980s, adaptation governance in the United Kingdom has evolved from a small set of research led institutions to a much broader governance landscape, which consists of multiple level actors and decision-making processes, including

E. Boyd (✉)
School of Earth and Environment, University of Leeds, Leeds, LS2 9JT UK
e-mail: e.boyd@leeds.ac.uk

R. Street • M. Gawith • K. Lonsdale • L. Newton • K. Johnstone • G. Metcalf
UK Climate Impacts Programme, Environmental Change Institute, Oxford, UK
e-mail: roger.street@ukcip.org.uk; megan.gawith@ukcip.org.uk; kate.lonsdale@ukcip.org.uk; laurie.newton@ukcip.org.uk; kay.johnstone@ukcip.org.uk; Gerry.metcalf@ukcip.org.uk

J.D. Ford and L. Berrang-Ford (eds.), *Climate Change Adaptation in Developed Nations: From Theory to Practice*, Advances in Global Change Research 42, DOI 10.1007/978-94-007-0567-8_6, © Springer Science+Business Media B.V. 2011

national level departments, Devolved Administrations, regional partnerships, local authorities, and private and voluntary organizations. The original focus on impacts has moved on and adaptation is now part of the political agenda, including through the Climate Change Act 2008 (Her Majesty's Government 2008) and the Climate Change (Scotland) Act 2009 (Scottish Government 2009a). In this vein, the focus on adaptation has shifted to the recognition that both adaptation and mitigation are needed to tackle climate change. Over time the linkages between adaptation and development have also become more pronounced, and adaptation actions have emerged in public, private, and community sectors.

Under devolution in the United Kingdom, the Westminster Parliament has devolved some of its powers, including responsibility for adapting to climate change, to the national authorities in Scotland, Wales, and Northern Ireland. In this chapter, we trace the evolution of adaptation science, practice, and policy over the past 10–12 years. We highlight the significant activities that are currently underway in this field not only with reference to the United Kingdom as a whole, but also with some reference to specific activities in England and the Devolved Administrations. We show that work on the climate change issue has matured throughout most of the United Kingdom from one of solely identifying and raising awareness of climate change impacts to one where the capacity of individual institutions, partnerships, and networks to adapt to climate change risks is rapidly increasing. Transitions to living with climate change are evident in the United Kingdom and illustrative examples are presented.

We show further that while capacity to adapt is increasing and delivery of adaptation actions is taking place, this is fairly patchy and challenges to widespread delivery of adaptation remain.

The chapter is set out in the following manner. We start by describing a brief history of adaptation in the United Kingdom and the status of adaptation research. Examples of adaptation in practice across various scales and sectors are then described and analyzed. The concluding discussion reflects on what adaptation in practice means and reflects on the challenges that lie ahead.

A Brief History of Adaptation in the United Kingdom

This section provides a brief synthesis of adaptation history in the United Kingdom. This is presented in three distinct phases: impacts led adaptation (1980–1996), encouragement of adaptation policy and practice (1997–2007), and requirements and national legislation (2008–present) (see Table 6.1).

Impacts Led Adaptation: 1980–1996

With the establishment of the Intergovernmental Panel on Climate Change (IPCC) in 1989, the UK Government considered it imperative to understand more fully what a

Table 6.1 Summary of stages in the development of the adaptation landscape in the United Kingdom

Time line	Focus	Action	Observations
1980–1996	Impacts led adaptation	Research	Focus on establishing the evidence base for mitigation
		CCIRG review 1991, 1996 not being used to inform adaptation	Adaptation seen as distraction from mitigation
1997–2007	Encouragement of adaptation policy and practice	Engagement of stakeholders	Adaptation seen as an alternative to mitigation
		Impact description – > adaptation responses	Link between adaptation and development recognized
		Working with the willing and learning about adaptation (early adaptors)	
2008– present	Requirement through national legislation	Climate Change Act 2008, Climate Change (Scotland) Act 2009	Economic, social, and environmental risk and impact are important
		National Adaptation Programme established, Climate Change Risk Assessment and Adaptation Reporting Powers	Need for both adaptation and sustainable development
		Lessons from encouragement	
		UK "Adapting well to climate risks"	

changing climate would mean for the United Kingdom. To this end, the Department of the Environment, Transport and Regions (DETR) commissioned a group of independent experts, the Climate Change Impacts Review Group (CCIRG), to assess the potential impacts of climate change on the United Kingdom. The resulting CCIRG's reports (CCIRG 1991, 1996) were groundbreaking at the time, but had limitations. First, the sector-based approach meant knock-on effects across sectors were not considered in an integrated manner. Second, most of the research reviewed had not engaged stakeholders and did not necessarily provide the information they would need to respond to climate change. Third, the research drawn on by the CCIRG used different climate change scenarios, different data, different scales, and different assumptions making it difficult to build up a composite picture of UK impacts (West and Gawith 2005).

The DETR sought to stimulate work that was user driven and based on common scenarios and datasets to address the limitations of previous assessments. The

UK Climate Impacts Programme (UKCIP) was thus established in April 1997 to facilitate an integrated, stakeholder led assessment of climate change impacts in the United Kingdom (UKCIP 2010). It was based on two main premises: (1) climate impacts research driven by stakeholders would provide information that would meet their needs and help them plan how to adapt to climate change and (2) by providing an integrative framework within which studies are undertaken, individual sectors would obtain a more realistic assessment of climate change impacts (McKenzie Hedger et al. 2000). Other institutions established in this phase include the Environmental Change Unit (ECU), now Environmental Change Institute (ECI) at Oxford University; and the Hadley Centre for Climate Prediction and Research housed within the UK Met Office. One of the main focuses of the ECU then was on modeling and assessing what the future impacts of climate change might be (e.g., Carter et al. 1994). The Hadley Centre's mandate was to develop the United Kingdom's climate modeling and prediction capability (Hulme and Turnpenny 2004), with most of its funding coming from the UK Government.

Encouragement of Adaptation Policy and Practice: 1998–2007

UKCIP was a primary focus for encouraging adaptation practice within the United Kingdom and did so by adopting a modular structure with individual studies funded by stakeholder partnerships and set in an integrative framework. Two main categories of study were conducted to provide the required knowledge and experience to inform adaptation decision-making. First, regional studies considered multiple sectors within a given region. These studies delivered information that was relevant to local decision-making, facilitated the inclusion of climate risks into spatial planning, and helped develop climate adaptation strategies (West and Gawith 2005). Second, sector-focused studies were typically undertaken at the national level, tended to be quantitative in nature, and informed decision-making on climate impacts and adaptation at the local, regional, and national scale for given sectors (West and Gawith 2005). In its early years, UKCIP was concerned primarily with raising awareness and engaging organizations in assessing how climate change might affect them. With time, recognition that organizations also need internal capacity to act on, or respond to, information on climate change impacts resulted in a shift in emphasis to more explicitly helping stakeholders to build their capacity to deliver adaptation actions. This included the delivery of a set of climate projections (UKCIP98, UKCIP02) and tools such as the Risk, Uncertainty and Decision-Making Framework, and UKCIP Adaptation Wizard (see http://www.ukcip.org.uk).

As a focus for delivery of the regional studies, UKCIP worked with stakeholders in setting up Climate Change Impacts Partnerships which now flourish in all English Regions (and in the Devolved Administrations). The distinctive feature of these partnerships is that they were and continue to be "stakeholder led."

Initially this meant that they were self-funded and consisted of regional governance organizations, local authorities, business-facing organizations (i.e., organizations that advise, regulate, or represent business), and nongovernmental organizations (NGOs). The first task for each partnership was to commission a Regional Scoping Study, which was typically published as both a Technical Report and a Summary Report (for further information, see http://www.ukcip.org.uk). The partnerships have created a wealth of information on impacts and risks, and built extensive networks in public and private sectors with their respective regions. Hitherto, The Department for Environment, Food and Rural Affairs (Defra) Adapting to Climate Change has recently recognized the potential of these partnerships and provided useful additional funds which have enabled and enhanced coordinating/secretariat functions and some valued stakeholder led regional projects that occurred regionally, inter-regionally, nationally, and internationally.

These partnerships benefit from continued and increased support as they have created and sustain powerful but independent networks of agencies in public and private sectors that are committed to dealing with resilience to a changing climate at a regional scale. The outputs from these partnerships are increasingly valuable and they are now cooperating, under the name "Climate UK," to share their increasing understanding and work jointly on some projects. The regional work of these self-appointed partnerships is particularly important as the arrangements for formal regional governance in England are still very tentative and subject to radical change or even removal under the new coalition government (for further information on the nature and work of these partnerships see http://www.ukcip.org.uk and click the tab "Who we work with").

Other major milestones in this period include the establishment of the Tyndall Centre for Climate Change Research in the autumn of 2000. It was the first large-scale multidisciplinary research center funded by three UK research councils – Natural Environment Research Council (NERC), Engineering and Physical Sciences Research Council (EPSRC), and Economic and Social Research Council (ESRC) (Tyndall Centre 2001). Its headquarters are at the University of East Anglia, with a consortium of nine academic research groups across the United Kingdom. Its aim was to conduct integrated research to identify and evaluate sustainable responses to climate change (Hulme and Turnpenny 2004). One example of Tyndall Centre support to policy-making was the Defra funded adaptation inventory conducted by Tompkins et al. (2009). Moreover, there were about 6–10 contributing authors to the IPCC report from the Tyndall Centre (Parry et al. 2007). In the mid-2000s, there was a need for further expansion on the economic and policy dimensions of climate change adaptation. The ESRC funded the Climate Change Centre for Economics and Policy (the "Stern Centre"), which was established to conduct for climate change research. This Centre was jointly hosted by the Grantham Institute at London School of Economic and University of Leeds.

National Level Requirements: 2008–Present

England

While the Government had already set out significant steps to strengthen the domestic program on climate change, it became clear that the urgency of the need to tackle climate change required further concerted action. In October 2006 the Government announced its intention to publish legislation on climate change, and a draft Climate Change Bill was published for public consultation and pre-legislative scrutiny in March 2007.

The revised Bill as introduced into Parliament on November 14, 2007 aimed to take into account findings from the parliamentary scrutiny and public consultation processes. One of the key results of these findings was the strengthening of the Bill's consideration of adaptation. Originally a miscellaneous provision, impacts of and adaptation to climate change became a full-fledged part of the Bill and the resulting Act.

The Climate Change Act 2008 (or the Act) was enacted by Royal Assent in November 2008 (Her Majesty's Government 2008). In addition to a legally binding long-term framework to cut carbon emissions, the Act established a statutory framework for work on climate change adaptation. Included within this statutory framework is the requirement to undertake a UK-wide Climate Change Risk Assessment (CCRA) and report on it within 3 years of the Act coming into force (i.e., in January 2012) and then every 5 years thereafter. It also set out a statutory National Adaptation Programme to address the most pressing climate change risks; this is required as soon as practically possible after publication of the CCRA with the requirement to report on progress at two yearly intervals.

The Act gives the Secretary of State the power to direct reporting authorities – over 100,000 eligible organizations with functions of a public nature and statutory undertakers – to produce reports on: the current and future predicted impacts of climate change on their organization and proposals for adapting to climate change.

When reporting, an authority must have regard to Statutory Guidance from the Secretary of State and must then consider its report when carrying out its functions (for further information, refer to http://www.defra.gov.uk/environment/climate/legislation/guidance.htm). A list of priority reporting authorities has been prepared with the first set of reports from authorities to be published at the end of 2011 (see http://www.defra.gov.uk/environment/climate/documents/rp-list.pdf).

Devolved Administrations

The governance arrangements vary among the devolved administration with Scotland, Wales, and Northern Ireland each having different degrees of autonomy.

Each administration began its adaptation work with a scoping study from which each has gone on to approach adaptation in a different way (for further information, see http://www.ukcip.org.uk).

For example, the Climate Change (Scotland) Act, which received Royal Assent in August 2009 (Scottish Government 2009a), created a legislative framework to drive actions to both reduce emissions to limit future changes and to adapt to the changes. Although heavily focused on mitigation, it does include provisions related to adaptation. Included are requirements for reports and a program of action before Parliament.

In support of this action, a Climate Change Adaptation Framework was published in 2009 setting out the strategic direction for Scottish Government actions (Scottish Government 2009b). It presents a national, coordinated approach intended to ensure that Scotland understands the risks and opportunities these challenges present and is adapting in a sustainable way. To do so, the Framework aims to lead planned adaptation across all sectors through three pillars:

- Improving the understanding of the consequences of a changing climate and both the challenges and opportunities it presents.
- Equipping stakeholders with the skills and tools needed to adapt to a changing climate.
- Integrating adaptation into wider regulation and public policy so that it is a help, not a hindrance, to addressing climate change issues.

The Welsh Assembly Government (WAG 2010) has recently published its Climate Change Strategy that includes the outline of an Adaptation Framework. The main WAG body responsible for climate change – Climate Change Commission for Wales – has tended to focus on mitigation issues but recently created an Adaptation Task and Finish Group.

Northern Ireland commissioned a follow-up scoping study (Arkell et al. 2007). Amongst this study's recommendations is the creation of a Climate Change Impacts Partnership, which has now been established and is beginning to work effectively.

It appears that the prevailing culture of the wider political relationships between Westminster and the Devolved Administrations (DAs) is challenging the potential for cooperation on adaptation work. Adaptation is a new topic for many, so, alternative approaches can provide real opportunities for experiment, comparison, and shared understanding. Such opportunities exist in the adaptation work of the Devolved Administrations, both between the three DAs themselves and in both directions between the UK Government and the DAs. UKCIP is being encouraged by Defra and the DAs to ensure that tools and guidance are developed which are applicable in all settings and that examples of current practice are widely drawn. There is further scope here for cooperation at government level to exploit the different adaptation experiences derived from the four different administrations.

Adaptation in Practice Across Scales and Sectors

Across the United Kingdom

Evidence collected by UKCIP (West and Gawith 2005), the Environmental Audit Committee, and the Royal Commission on Environmental Pollution demonstrates that adaptation is being practiced within the United Kingdom. This includes action primarily focusing on building adaptive capacity and also delivering adaptive actions. There is also recognition that these efforts are insufficient to achieving the goal of a United Kingdom that is adapting well. The following are highlights of progress.

UK Government Programs

Government departments published their "Departmental Adaptation Plans" in March 2010 joining up with Carbon Reduction Delivery Plans as part of a comprehensive Government response to the challenges of climate change (for further information, see Defra 2010a). Each of the 16 departments published a plan which highlights the main climate risks affecting their policies, programs, and estates and the actions being taken to adapt. Quality assurance, key crosscutting issues, and policy/programs were discussed using a peer review process that involved senior officials from across Whitehall. To accompany the plans, Defra also published a supplement which looks at the impact climate change will have on natural resources and the services derived from them; and seeks to stimulate a debate about how adaptation to climate change in the natural environment should be approached.

A cross-government Adapting to Climate Change Programme (ACC Programme) was established in 2008 to bring together and drive forward work in Government and the wider public sector on adaptation in England and the United Kingdom for reserved matters (reserved matters are the areas of government policy where Parliament has kept the power [jurisdiction] to make laws [legislate] in Scotland, Northern Ireland, and Wales). A Programme Board directs the ACC Programme with senior representatives from most central government departments. Defra provides the ACC Programme delivery team, but responsibility for embedding adaptation into individual government policies is the responsibility of the relevant government department.

The ACC Programme is currently undertaking the groundwork for the statutory National Adaptation Programme to be put in place by 2012, and the second phase of its work from 2012 will be to implement this National Adaptation Programme. The groundwork involves:

- Developing a more robust and comprehensive evidence base about the impacts and consequences of climate change on the United Kingdom
- Raising awareness of the need to take action now and help others to take action

- Working across Government to embed adaptation into Government policies, program, and systems
- Measuring success and taking steps to ensure effective delivery

UK Government Agencies

The leadership and strong roles being played by the Environment Agency and Natural England reflect other efforts within the United Kingdom.

The Environment Agency, which operates in England and Wales, was established in 1996 to bring together previously disparate responsibilities for protecting the environment and to contribute to sustainable development. The Agency is a Non-Departmental Public Body (NDPB) responsible to the Secretary of State for Defra, and is an Assembly Government Sponsored Body (AGSB) responsible to the Welsh Assembly Government (WAG). The Agency works closely and in partnership with the nature conservation agencies Natural England and the Countryside Council for Wales, as well as with local authorities. The Agency's work includes: regulation of major industry, flood and coastal risk management, waste management, agriculture, navigation, fisheries, contaminated land, conservation and ecology, water quality and resources, and climate change.

The Agency sees itself taking a leadership role on adaptation, publishing its first formal adaptation strategy in 2005 and an updated version in 2008. The Agency's approach has evolved over time with the intention of implementing an integrated adaptation program. This program is based on (1) an assessment that identified the risks that climate change poses to each of the Agency's main functions and (2) plans that were developed to address these risks. The program is intended to inform investments in priority areas and in exploiting synergies in adaptive responses (e.g., opportunities to combine flood defense with provision of connected habitat for wildlife).

Natural England is a statutory body created in 2006, charged with the responsibility to ensure that England's unique natural environment is protected, managed and improved for the benefit of present and future generations. Since its inception, Natural England has advocated placing sustainable adaptation at the heart of its own operational policy, decision-making, and advice. Natural England has noted a shift in thinking from only biodiversity conversation and dealing with the impacts of climate change on biodiversity to considering the role that biodiversity conservation can play in supporting societal adaptation (and mitigation). There has also been a shift to exploring adaptation at a landscape scale that recognizes the nested nature of action, and that seeks to work across spatial and administrative boundaries. This latter area of work is being done through a series of studies looking across a subset of England's "National Character Areas" (for further information see http://www.natural-england.org.uk/regions/east_midlands/ourwork/backgroundandmethodology.aspx). Natural England has developed an approach in which the overall landscape and ecosystem services provide a framework encompassing a more detailed assessment of measurable assets such as biodiversity, historic environment, natural resources, and access and recreation opportunities.

Local Government

England's local authorities have been engaged with tackling climate change for a significant period, but until recently their responses have been dominated by mitigation policies. As the importance of adaptation has been increasingly recognized, local authorities are becoming seen as one of the primary means of delivery. Two initiatives that reflect this interest and level of engagement originated from local rather than central government.

An initial signal was the Nottingham Declaration (2000). This Declaration was a voluntary pledge that English councils could sign to demonstrate commitment to tackling the challenges of climate change. Although the original Declaration focused primarily on mitigation, a revised version launched in 2005 gave equal weight to the challenges of adapting to the impacts of a changing climate. More than 95% of English councils have signed up to the Nottingham Declaration.

In 2007, the Local Government Association undertook a commission of enquiry on climate change that included adaptation (Local Government Association 2007). The enquiry concluded that local government is uniquely placed to tackle climate change and recommended action on adaptation focusing on providing leadership, building capacity, and engagement with their local communities.

Central government included an indicator on "Planning to adapt to climate change" (National Indicator 188 – NI188) in the local government performance framework (April 2008). NI188 requires authorities to report progress over the subsequent 3 years against five levels of achievement based on the assessment and management of changing climate risks to their own operations and to the wider local community.

The indicator has proved a useful stimulus to encourage local authorities to consider adaptations to the risks of a changing climate, although the levels of response remain variable. In particular, many smaller district councils lack the financial and staffing resources to respond effectively. Supporting uptake and delivery of this indicator, UKCIP developed a workshop program focusing on developing an adaptation strategy and action plan. Further support is needed on the threats and opportunities for local authority service areas and can be based on similar workshops piloted with relevant professional bodies like the Transport Planners and the Emergency Planners Society (2009).

Private Sector

Awareness and action on adaptation within the private sector has been growing rapidly over the past couple of years but remains at a low level relative to the public sector. Much of the activity is still in the realm of building adaptive capacity rather than delivering specific adaptation actions, with organizations that support, regulate, and represent components of the private sector (e.g., professional bodies, regulators, and trade associations) playing an increasingly important role. These types of organizations are well placed to drive forward the agenda through sector-based studies, networks, guidance, and awareness-raising activities.

Adaptation efforts within the private sector are varied across different types of organizations and sectors. UKCIP is working with a number of businesses and business-facing organizations to facilitate the adaptation efforts of the private sector. For example, the National Association of Cider Makers carried out a modest industry led study (NACM 2008) with the intention of providing advice and guidance to members on the risks of climate change and background information on potential opportunities. In terms of regulatory bodies, Ofwat (The Water Services Regulation Authority) has and continues to consider adaptation (and mitigation) within the water and sewerage sectors for England and Wales, including as part of the water resources planning process. Ofwat (2010) have issued a policy statement describing how they will regulate in a way that takes account of climate change, and encourage and enable water companies to respond to climate change. The interest and seriousness within the water industry community related to adaptation is further reflected by the work underway within Water UK (2010) and UK Water Industry Research (2010).

Exemplar among the activities of professional bodies is work by the Institute of Mechanical Engineers (IMechE), the Association of British Insurers (ABI), and the Confederation of British Industry (CBI). IMechE commissioned research exploring the engineering implications of climate change over the next 1,000 years and recommends what engineers need to do to adapt to the future world (IMechE 2008). The ABI has a large portfolio of work on adaptation including ClimateWise (2010), a collaborative initiative for their members. Also of significance and demonstrating the rising profile of the adaptation agenda, the CBI, the premier lobbying organization for UK business, has been steadily increasing the level of attention devoted to adaptation and have recently published a report on adaptation with recommendations for business and government (CBI 2010). Since 2006, their corporate leaders' group on climate change has begun to extend its work on mitigation to include adaptation as reflected in recent publications (UK Met Office 2007; CBI 2009).

While adaptation continues to evade the plans of most individual companies and particularly small and medium enterprises, some large UK companies are also beginning to take on the agenda. This increased interest is reflected in related efforts within UKCIP and Defra's Adaptation to Climate Change Programme (Defra 2010b).

The growing need for advice on adaptation has been reflected in the growth of private consultancy capabilities with the United Kingdom to meet this available market. New consultancies have been established and many existing consultancies have been strengthening their capabilities and portfolios to include adaptation in the context of climate change. The Environmental Audit Committee heard evidence that the adaptation consultancy community within the United Kingdom is "growing up and that internationally many recognize that UK consultancies are further along on providing services related to adaptation than consultancies in their home countries" (Environmental Audit Committee 2010, EV 141 Q305 Hilary Benn).

The Royal Commission on Environmental Pollution (RCEP) in its report recognized the importance of maintaining quality control for those offering consultancy services in relation to adaptation, in particular, with reference to the use of climate projections. It recommended that governments and relevant bodies build climate change adaptation into existing personal and corporate accreditation schemes to help ensure that climate projection information is being interpreted and used appropriately (RCEP 2010, paragraph 5.56).

Voluntary Sector

Questions such as "Who is likely to be most vulnerable?" "Who is responsible for bearing the cost of adaptation?" and "Who makes adaptation decisions?" are now starting to be asked in the United Kingdom. These questions are being asked in response to direct experience of extreme weather events, and also in anticipation of climate change and greater climate risks in the future. With this recognition, there has been an increase in interest in the climate adaptation agenda from both environmental and non-environmental voluntary organizations, often in connection with work on the mitigation agenda.

Early work by CAG Consultants emphasized how climate impacts would have a disproportionately heavy impact on more deprived communities and households. People in the United Kingdom most likely to be vulnerable to climate change are those that:

> are already deprived by their health, the quality of their homes and mobility; as well as people who lack awareness of the risks of climate change, the capacity to adapt and who are less well supported by families, friends and agencies.

> (CAG Consultants 2009)

Voluntary organizations have long been involved with the climate change agenda through regional and local climate change partnerships and initiatives, collaboration in research partnerships such as the EPSRC funded Adaptation and Resilience for a Changing Climate (ARCC) programme (ARCC 2010), and a number of new voluntary organization specific initiatives that have been developed recently. For example, the Baring Foundation launched a set of four projects in July 2008 focusing on climate change and refugee organizations, community anchor organizations (i.e., local, multipurpose community organizations), children and youth organizations, and organizations that support vulnerable communities.

In 2009, the Joseph Rowntree Foundation funded six projects to investigate the social implications of climate change through their Climate Change and Poverty Programme (Joseph Rowntree Foundation 2010) (http://www.jrf.org.uk/work/workarea/climate-change-and-poverty). This continuing work seeks to ensure that people or places facing poverty and disadvantage are not disproportionately affected by climate change, or by policy or practice responses to it. The Programme aims to do this by providing evidence on the social impact of climate change in the United Kingdom; raising awareness of the consequences of climate change for vulnerable people and places; and supporting the development of fair responses to climate change among policymakers, practitioners, and communities undertaking mitigation and adaptation activity at a national and local level.

In addition, the Third Sector Task Force on Climate Change was launched in May 2009 to raise the profile of climate change in the voluntary sector. As shown by this Task Force, as well as the initiatives described above, there is a growing group of voluntary organizations in the United Kingdom exploring the implications of climate change from a social justice perspective. This movement is creating new and innovative collaborations between the climate change academics and practitioners and social justice and poverty academics and practitioners.

Challenges Ahead: The Environmental Audit Committee and the Royal Commission for Environmental Pollution

Two recent assessments of the role of government and other institutions in adaptation have highlighted advancements and the need for an increased focus on adaptation move from "talking to doing." The Environmental Audit Committee (appointed by the House of Commons) examined the extent to which the Government is embedding climate change adaptation and management of risks from future climate change impacts into government programs, policies, and decision-making and into those of the wider public and private sectors. The Committee also examined how well the Government is tackling the key questions on adaptation. Evidence gathered and assessed by the Audit Committee allowed it to recommend actions to be taken by government to further support adaptation (Environmental Audit Committee 2010). The Committee concluded that there has been good progress on adaptation since 2008, and that the Government should build and support multiple levels of government, business, and civil society to respond to the risks of climate change. Uncertainty is not a reason for delay in action, and the Government should move rapidly to strengthen its new policy framework. While this is an important conclusion, the extent to which the new coalition government will act upon these recommendations remains to be discovered.

The Royal Commission for Environmental Pollution (RCEP) has reported to parliament on the challenges facing institutions (i.e., organizations, and practical arrangements for implementing policies, or legal, regulatory and policy frameworks) in adapting to climate change. Their report, entitled *Adapting Institutions to Climate Change* suggests that although there are institutions that are seriously considering how they will adapt, many are poorly prepared and many have simply not started to consider it. The Commission concluded that framing the problem of adapting, organizational learning, and implementation are specific components underlying building of adaptive capacity. They went on to put forward a series of recommendations designed to help institutions develop their capacity. These were grouped into the policy framework, specific institutional arrangements, and resources to build capacity, equity, and public engagement. Included amongst the recommendations is an adaptation test to be applied to new policies and programs. While the conclusions of the report are important, there remain gaps in understanding how the evaluation of a range of adaptation actions will be conducted (RCEP 2010).

Discussion: UK Leadership on Adaptation Domestically and Worldwide

We suggest that the United Kingdom is in transition from primarily building adaptive capacity to delivering adaptation through practice, learning from policy, and science networks and stakeholders. As suggested by the RCEP report (RCEP 2010) and Environmental Audit Committee report (EAC 2010), this transition is essential and efforts will need to be enhanced and supported.

The UK policy framework is motivated by the understanding that climate change is real and what is required now are practical ways to adapt. There is encouraging evidence that many actions are occurring at different scales from local, regional, national to international.

The United Kingdom appears to be a leader in terms of scientific initiatives, which has important implications for policy. The science supports, and in turn is supported by, the administrative structures and key leaders in adaptation (the Prime Minister's office, government ministers in Defra and Department of Energy and Climate Change [DECC], cross-Whitehall programmes and efforts, and the UK Research Councils [RC UK]). UK climate science is dominated, as elsewhere, by climate impacts research, but the focus on adaptation, vulnerability and resilience is rapidly strengthening.

The UKCIP continues to play an important role in linking theory, practice and policy in climate adaptation work, thereby combining bottom-up and top-down approaches. It too, though, needs to be able to adapt to the changing social, economic, and political circumstances in which it operates to be able to continue to fulfill its vital function of helping UK organizations adapt to climate change.

There is evidence in the United Kingdom of bottom-up and interdisciplinary applied research and funding: RC UK; Living With Environmental Change (LWEC) Programme; continuing support within individual research councils (e.g., EPSRC and NERC); investment in applied, interdisciplinary science by academic institutions (e.g., Universities of Reading, East Anglia, Oxford); and buy-in from businesses and local authorities.

International leadership is based on the foundations of key institutions and how they collaborate and interact with the national and international policy context through fora such as the IPCC and the United Nations Framework Convention on Climate Change (UNFCCC). Some brief examples of UK initiatives with international importance include:

- The Met Office is a primary source of information, for example, PRECIS supplied projections to developing countries, including China and India and 50 other countries.
- Australia and New Zealand used UK Met Office climate models.
- The Tyndall Centre for Climate Change Research UK focuses on international dimensions of climate change.
- The UK plays a role in the IPCC.

Internationally, the UKCIP has been an important source of information and many of its tools and guidance documents are being picked up, translated, and used as the basis for local and regional guidance. For example, universities and consultancies in Australia, the United States, Finland, and Germany have picked up on the UKCIP Adaptation Wizard (UKCIP 2008) and are working to develop local variations of the tool. UKCIP's Risk and Uncertainty and Decision-Making Framework (Willows and Connell 2003) is highly regarded and has been referenced widely both in the United Kingdom (e.g., Stern 2006) and abroad (e.g., UNFCCC 2008; EEA 2004). *Identifying Adaptation Options* (UKCIP 2007) has been emulated in the Philippines.

The UK international development position on adaptation has grown significantly (e.g., see Boyd and Osbahr 2010) on resilient responses in networked development organizations in the United Kingdom. The UK coalition government in their program for partnership government has agreed that they will work toward an ambitious global climate deal that will limit emissions and explore the creation of new international sources of funding for the purpose of climate change adaptation and mitigation (Her Majesty's Government 2010).

Key challenges and gaps remain. These include, for example, an agreed upon framework for adaptation, measuring adaptive capacity, measuring progress in adaptation, and identifying priorities for UK adaptation science. A greater focus is also required on social dimensions of climate change, flexible and adaptive policy cycles, regional capacity and research, and mechanisms for linking practice in local authorities and businesses with a vulnerability perspective.

Swart et al. (2009) observe that a close look at organizational architectures for coherent adaptation strategies across countries illustrate divergences between political priorities, resource availability, scale of programs, institutions, and organizations that are in competition with external pressures from public and private organizations. This observation is also true of the United Kingdom, but by focusing resources in the areas identified above, the United Kingdom could continue to make real progress on adaptation and retain its position as a global leader in this field.

Conclusion

In conclusion, this chapter set out to highlight the development of the adaptation landscape in the United Kingdom and to provide insights into the various adaptation actions in practice. The chapter illustrates that in many ways, the United Kingdom is leading on adaptation science, stakeholder engagement, and in terms of legislation. However, there also remain gaps and challenges – such as evaluating and funding adaptation – that still require further consideration.

Acknowledgments Thanks to anonymous reviewers. Special thanks to Professor Mike Hulme, Phil Irving, Caroline Cowan, Nick MacGregor, and Richenda Connell.

References

Adaptation and Resilience to a Changing Climate (2010) ACN – A coordinating research network. http://www.ukcip-arcc.org.uk. Cited 15 Sept 2010

Arkell B, Darch G, McEntee P (eds) (2007) Preparing for a changing climate in Northern Ireland – Project UKCC13. Scotland and Northern Ireland Forum for Environmental Research. http://www.doeni.gov.uk/preparing_for_a_climate_change_in_northern_ireland_executive_summary.pdf. Cited 15 Sept 2010

Boyd E, Osbahr H (2010) Resilient responses to climate change: exploring organisational learning in networked development organisations. Environ Educ Res (forthcoming)

CAG Consultants (2009) Differential social impacts of climate change in the UK – Project UKCC22. Scotland and Northern Ireland Forum for Environmental Research. http://www.sniffer.org.uk/Resources/UKCC22/Layout_Default/12.aspx?backurl=. Cited 15 Sept 2010

Carter TR, Parry ML, Harasawa H et al (1994) IPCC technical guidelines for assessing climate change impacts and adaptations. Department of Geography, University College London, UK and Center for Global Environmental Research, National Institute for Environmental Studies, Tsukuba, Japan

CBI (2010) Whatever the weather: managing the risks from a changing climate. http://climatechange.cbi.org.uk/reports/00419/?utm_source=feedburner&utm_medium=feed&utm_campaign=Feed%3A+ClimateChangeFeed+%28CBI+Climate+Change++RSS+Feed%29. Cited 15 Sept 2010

CBI Climate Change Board (2009) Future Proof: preparing your business for a changing climate. Confederation of British Industry. http://climatechange.cbi.org.uk/reports/00230/. Cited 15 Sept 2010

CCIRG (1991) The potential effects of climate change in the United Kingdom. Department of the Environment, Her Majesty's Government, London

CCIRG (1996) Review of the potential effects of climate change in the United Kingdom. Department of the Environment, Her Majesty's Government, London

ClimateWise (2010) http://www.climatewise.org.uk. Cited 15 Sept 2010

Defra (2010a) Adaptation across Government. Department for Environment, Food and Rural Affairs, Her Majesty's Government. http://www.defra.gov.uk/environment/climate/programme/across-government.htm. Cited 15 Sept 2010

Defra (2010b) An update from the Adaptation to Climate Change Programme. http://www.defra.gov.uk/environment/climate/documents/1003-newsletter.pdf. Cited 15 Sept 2010

EEA (2004) Impacts of Europe's changing climate: an indicator-based assessment – EEA Report no.2/2004. EEA, Copenhagen

Emergency Planning Society (2009) Emergency planning: Adapting to Climate Change. http://www.climatesoutheast.org.uk/images/uploads/Emergency_planning_-_adapting_to_climate_change.pdf. Cited 15 Sept 2010

Environmental Audit Committee (2010) Adapting to climate change – Sixth Report of Session 2009–10, HC 113. House of Commons, UK Parliament. http://www.publications.parliament.uk/pa/cm200910/cmselect/cmenvaud/113/113.pdf. Cited 15 Sept 2010

Her Majesty's Government (2008) Climate Change Act 2008. Office of Public Sector Information. http://www.opsi.gov.uk/acts/acts2008/ukpga_20080027_en_1. Cited 15 Sept 2010

Her Majesty's Government (2010) The Coalition: our programme for government. http://programmeforgovernment.hmg.gov.uk/files/2010/05/coalition-programme.pdf. Cited 15 Sept 2010

Hulme M, Turnpenny J (2004) Understanding and managing climate change: the UK experience. Geogr J 170(2):105–115

IMechE (2008) Climate change: Adapting to the inevitable?. http://www.imeche.org/NR/rdonlyres/D72D38FF-FECF-480F-BBDB-6720130C1AAF/0/Adaptation_Report.PDF. Cited 15 Sept 2010

Joseph Rowntree Foundation (2010) http://www.jrf.org.uk/work/workarea/climate-change-and-poverty. Cited 15 Sept 2010

Local Government Association (2007) A climate of change, final report of the LGA Climate Change Commission. http://www.lga.gov.uk/lga/aio/20631. Cited 15 Sept 2010

McKenzie Hedger M, Gawith M, Brown I et al (eds) (2000) Climate change: assessing the impacts – identifying responses, the first three years of the U.K. Climate Impacts Programme. UKCIP and Department for Environment Transport Regions, Her Majesty's Government, Oxford

NACM (2008) Industry prepares for climate change. http://cideruk.com/cider_news/view/climate_change_story1/. Cited 15 Sept 2010

Nottingham Declaration (2000) http://www.energysavingtrust.org.uk/nottingham/Nottingham-Declaration/The-Declaration/About-the-Declaration. Cited 15 Sept 2010

Ofwat (2010) Preparing for the future – Ofwat's climate change policy statement http://www.ofwat.gov.uk/sustainability/climatechange/pap_pos_climatechange.pdf. Cited 15 Sept 2010

Parry ML, Canziani OF, Palutikof JP et al (eds) (2007) Climate change 2007: impacts adaptation and vulnerability – contribution of working group II to the fourth assessment report of the Intergovernmental Panel on Climate Change. Cambridge University Press, Cambridge

RCEP (2010) Twenty-eighth report: Adapting institutions to climate change. Her Majesty's Government. http://www.rcep.org.uk/reports/28-adaptation/28-adaptation.htm. Cited 15 Sept 2010

Scottish Government (2009a) Climate Change Scotland Act 2009. http://www.scotland.gov.uk/Topics/Environment/climatechange/scotlands-action/climatechangeact. Cited 15 Sept 2010

Scottish Government (2009b) Scotland's climate change adaptation framework. http://www.scotland.gov.uk/Topics/Environment/climatechange/scotlands-action/adaptation/AdaptaitonFramework/TheFramework. Cited 15 Sept 2010

Stern N (2006) The Stern review on the economics of climate change. http://www.hm-treasury.gov.uk/sternreview_index.htm. Cited 15 Sept 2010

Swart R, Biesbroek R, Binnerup S et al (2009) Europe adapts to climate change: comparing national adaptation strategies – PEER report no. 1. Partnership for European Environmental Research, Helsinki

Tompkins EL, Boyd E, Nicholson-Cole SA et al (2009) An inventory of adaptation to climate change in the UK: challenges and findings – Tyndall working paper 135. Tyndall Centre for Climate Change Research, Norwich

Tyndall Centre (2001) The Tyndall Centre research strategy. Tyndall Centre for Climate Change Research, Norwich

UK Met Office (2007) Climate change adaptation for UK businesses: report for the confederation of British industry task force on climate change. Her Majesty's Government. http://www.metoffice.gov.uk/consulting/CBI_TFCC.pdf. Cited 15 Sept 2010

UK Water Industry Research (2010) http://www.ukwir.org/site/web/content/home. Cited 15 Sept 2010

UKCIP (2007) Identifying adaptation options. http://www.ukcip.org.uk/images/stories/Tools_pdfs/ID_Adapt_options.pdf. Cited 15 Sept 2010

UKCIP (2008) The UKCIP adaptation wizard v 2.0. http://pisd-pak.org/ClimateChangePDF/CC_Adaptation_Wizard.pdf. Cited 15 Sept 2010

UKCIP (2010) UKCIP timeline: dates of key events. http://www.ukcip.org.uk/index.php?option=com_content&task=view&id=422&Itemid=9. Cited 15 Sept 2010

UNFCCC (2008) Compendium on methods and tools to evaluate impacts of, vulnerability and adaptation to, climate change. http://unfccc.int/adaptation/nairobi_work_programme/knowledge_resources_and_publications/items/2674.php. Cited 15 Sept 2010

WAG (2010) Wales climate change strategy. http://wales.gov.uk/topics/environmentcountryside/climatechange/tacklingchange/strategy/walesstrategy/;jsessionid=gJ68LY3ZNGkPksV5BJC9MjyDRcszLCJSTJR0ybtPrvcR37DdYQzJ!686978193?lang=en. Cited 15 Sept 2010

Water UK (2010) Climate change: adaptation to climate change http://www.water.org.uk/home/policy/climate-change/adaptation-briefing. Cited 15 Sept 2010

West C, Gawith M (2005) Measuring progress: preparing for climate change impacts through the UK Climate Impacts Programme. UKCIP, Oxford

Willows R, Connell R (eds) (2003) Climate adaptation: risk uncertainty and decision-making – technical report. UKCIP, Oxford

Chapter 7
Adaptation to Climate Change in Canada: A Multi-level Mosaic

Thea Dickinson and Ian Burton

Abstract The necessity for adaptation is now widely recognized in Canada. However, the developing pattern of response is an expanding mosaic. Individual pieces – i.e., initiatives at the provincial, territorial, and municipal levels – are visible, but the overall strategic design is lacking clarity and cohesion. This is likely due, in part, to Canada's federalism, and to the conceptualization of adaptation in the United Nations Framework Convention on Climate Change (UNFCCC). The negotiations leading to the UNFCCC conceived of adaptation as largely a place-based and local matter; of concern only to those most vulnerable communities and countries. In consequence, a bottom-up approach was viewed as the preferred option. Over the life of the UNFCCC, adaptation has grown in significance and has come to be seen as requiring top-down strategic approaches. A major challenge now facing Canada – and indeed all Parties to the Convention – will be the effective and simultaneous management and coordination of both top-down and bottom-up approaches. Currently, in Canada, the blend has been allowed to evolve almost unguided, with modest encouragement from the federal government. Leadership has emerged at both provincial and municipal levels across the country. But it is not clear what the consequences of such an approach will be.

Keywords Canada • Adaptation • Climate change • Top-down • UNFCCC • National assessment • Province • Territory • Municipality • National strategy

T. Dickinson (✉)
Burton Dickinson Consulting, 600 Kingston Road, Suite 204, Toronto, ON M4E 1R1, Canada
e-mail: thea.dickinson@rogers.com

I. Burton
Emeritus, Meteorological Service of Canada, Environment Canada, Downsview, ON, Canada

Emeritus, University of Toronto, Toronto, ON, Canada
e-mail: Ian.Burton@ec.gc.ca

J.D. Ford and L. Berrang-Ford (eds.), *Climate Change Adaptation in Developed Nations: From Theory to Practice*, Advances in Global Change Research 42, DOI 10.1007/978-94-007-0567-8_7, © Springer Science+Business Media B.V. 2011

Adaptive Learning

The necessity for adaptation is now widely recognized in Canada. The developing pattern of response is an expanding mosaic. Individual pieces are visible, but the overall strategic design is lacking clarity and cohesion (Lemmen et al. 2007). This is not an unusual experience in Canada. Canada is geographically large, with a relatively small and scattered population. More significantly, Canada is a federal state with ten provinces and three territories. This constitutional arrangement involving a complex division of powers and allocation of responsibilities among three levels of government is constantly under question, shifting at the margins, especially in areas of so called shared jurisdiction (Gardner 1994; Morton 1996). When a new issue such as adaptation to climate change emerges, there is almost always some uncertainty about how the needed policies and actions will be identified, developed, and shared. Important parts of the climate change adaptation (and mitigation) debate still remain unanswered and even unaddressed: who will pay what share of the costs for adaptation of different kinds, in different places, and in relation to what risks? In addition, how will allocation of costs and responsibilities be decided for climate-resilient infrastructure (design of buildings, highways, drainage systems), agriculture (new cultivars and drought losses), control of invasive species (pine beetle spread and infestations), public health (heat episodes, and heat/health alerts), and the impacts of extreme climate-related weather events and many others?

Adaptation to the impacts of climate change is a new challenge for all countries. In some instances, a sense of urgency has led to the development of strong top-down-guided approaches. This is especially the case in the least developed countries, where support has been provided under the United Nations Framework Convention on Climate Change (UNFCCC) through the Global Environment Facility (GEF) for the preparation of National Adaptation Plans of Action (NAPAs) (Government of Canada 2010; COWI and IIED 2009). Support for national adaptation planning is now an expanding part of many bilateral aid programs, along with multilateral aid agencies, the World Bank, and several regional development banks. The World Bank, for example, has instituted a Pilot Programme for Climate Resilience (PPCR) that aims to help selected developing countries adopt a programmatic and "transformational" approach to adaptation (http://siteresources. worldbank.org/INTCC/Resources/progressreportPPCR.pdf). This approach is now spreading to many developed countries.

On the other hand, an overarching and top-down approach or framework has yet to emerge in Canada. Pieces are added to the multi-level mosaic in a spontaneous and loosely connected way. To date, there is no overall strategy or grand design. What is therefore happening can best be described as adaptive learning or learning to adapt. The adaptation that is occurring in Canada is place- and region-specific, involving a diverse range of risks, sectors, and ecosystems. To develop an overall strategic design from a Canadian perspective requires an Olympian vision. This is not to say that it is an impossible feat, but it does require far-sighted leadership, time, and resources. The last chapter of the national assessment (Lemmen et al. 2007), for example Chap. 10, *Moving Forward on Adaptation*, presents four relatively modest suggestions for "building the momentum" and four similarly modest "near-term steps" (see Box 7.1).

Box 7.1 Suggestions from the National Assessment for "Building the Momentum" and "Near-Term Steps"

Building the Momentum
1. Maintain and strengthen the knowledge base
2. Synthesize and share knowledge
3. Remove barriers to action
4. Review and contribute to international initiatives

Near-Term Steps
1. Broaden engagement and collaboration
2. Lead by example
3. Enhance institutional capacity
4. Promote and mandate adaptation measures

These suggestions and steps seem to be valid and useful as far as they go, but by most criteria fall well short of a national strategy of the sort required to meet the growing and future impacts of climate change.

Characteristics of the Canadian Approach

For much of the life of the United Nations Framework Convention on Climate Change, Canada, like other high income developed countries, has focused its attention on mitigation with less interest being given to domestic adaptation (several reasons are outlined in Dickinson 2007; Burton et al. 2007). At the federal level, Canada has two major adaptation groups: (1) Adaptation Impacts and Research Service (AIRS) housed in Environment Canada; and (2) the Climate Change Impacts and Adaptation Division (CCIAD) within Natural Resources Canada. In 1997, Environment Canada completed the first major Canadian climate change assessment, *The Canada Country Study* (CCS). In the last few years, beginning around 2005, there was a rapid international awakening about the need for adaptation in developed countries. It is widely recognized that we are entering the "period of consequences" (see Chap. 5 of this volume, by Susan Moser). Consequently or coincidentally, in 2007, Natural Resources Canada (NRCan) published the second national assessment, *From Impacts to Adaptation: Canada in a Changing Climate*. The report represents the most comprehensive assessment of the impacts of climate change in Canada to date. The report concludes that:

- Adaptive capacity is generally high, but is unevenly distributed between and within regions and populations.
- Some adaptation is occurring, both in response to, and in anticipation of, climate change impacts.

- Integrating climate change into existing planning processes, often using risk management methods, is an effective approach to adaptation.
- Barriers to adaptation need to be addressed, including limitations in awareness and availability of information and decision-support tools.
- Although further research will help to address specific knowledge gaps and adaptation planning needs, we have the knowledge necessary to start undertaking adaptation activities in most situations now.

These conclusions seem to be accurate but cautious, and lacking in a strong sense of direction or cohesion. The message seems to be "let the mosaic evolve."

Many developed countries have created or are in the process of creating national climate change adaptation plans, strategies, and/or programs. Implementation of these is well underway in the member countries of the European Union, including the United Kingdom. This is also true of Australia, Japan, and now, after some hesitation, the United States. In Canada, at the federal level, emphasis has not been on developing a national plan or strategy, but rather has been focused on climate change model and scenario development, as well as on providing research to the growing community of adaptation scientists and networks; the Canadian Climate Change Scenarios Network (CCCSN), for example, is Canada's state-of-the-art network that has contributed to the reports of the Intergovernmental Panel on Climate Change (IPCC) (IPCC 2007). Public servants in Environment Canada and Natural Resources Canada have been promising for at least the past 3 years that a national framework for adaptation will shortly be issued.

Abroad, Canada continues to respond to requests for financial support for vulnerability, impacts, and adaptation projects in developing countries. To illustrate, Canada was among the first contributors to the GEF-administered Least Developed Countries Fund (LDCF), and the Pilot Programme on Climate Resilience (PPCR) managed by the World Bank. Canada has provided project level support for the Canada–China Climate Change Cooperation Project (C5) to help increase the capacity of Chinese research institutions and government agencies to identify and assess the sensitivities and vulnerabilities associated with climate change. Owing to this collaboration, the Chinese government began to integrate adaptation strategies into development and planning initiatives. Canada is also supporting the building of adaptive capacity in Western Africa (Government of Canada 2010).

Nevertheless, top-down directed action based on a shared vision or framework has been slow to evolve as different levels of government look to each other for leadership. Canada, by any measure, has the capacity to adapt: the country possesses high national income per capita, highly skilled and educated human resources, effective and efficient public organizations and institutions, a strong private sector, and access to technology. But federal hesitation exists. The mantra "adaptation is local" still lives in the minds of those at the highest policy level, and, unlike mitigation, the notion persists that the benefits of adaptation fall largely to those who invest in it, at the place where the investment occurs. This idea has also slowed adaptation at the international level, and only recently has a more strategic view of adaptation gained ground – one requiring planning, policy, and action at national levels. Such an approach is needed for Canada, one where actions at the federal level support both provincial and municipal adaptation.

The Multi-level Evolving Mosaic

Although the federal government has yet to develop a national plan or strategy, a number of Canada-wide initiatives have been supported on an interim or temporary basis. Following the release of the 2007 national assessment, NRCan announced a \$35 million project to establish six Regional Adaptation Collaboratives (RACs) across Canada (with the requirement of matching provincial funding). Prior to the RAC program, in 2001, Natural Resources Canada developed Canadian Climate Impacts and Adaptation Research Network (C-CIARN), with funding that was sustained until June 30, 2007. From the perspective of the major users of these funds, their short-term and temporary nature creates difficulties for the development of longer-term strategic adaptation planning or capacity building. The recent RAC program fits this pattern: this program is aimed at coordinating sustained action to reduce vulnerability to climate change by advancing adaptation planning and decision-making – yet is funded for just over 2 years.

The lack of sustained federal leadership has forced (perhaps intentionally) a multitude of provincial, municipal, and nongovernmental players to develop their own plans, strategies, and programs. Over the past decade, several climate change consortiums have cropped up at the provincial level: Ouranos in Québec; Pacific Climate Impacts Consortium (PCIC) in British Columbia; and Ontario Climate Change Impacts and Adaptation Research group (OCCIAR) in Ontario. However, while these and several other climate change consortiums exist across Canada, there is no mandate to exchange information or lessons learned, and neither are there mechanisms in place for formal knowledge transfer. The Council of the Federation (COF) hopes to change this with the development of the Climate Change Adaptation Community of Practice (CoP), proposed during a 2008 Climate Change and Adaptation Forum in Vancouver. The online community, set to launch in August 2010, will attempt to bridge this disconnect by promoting knowledge transfer between provinces and territories across Canada. Similar to other such initiatives, committed funding is only available for 2 years and beyond that the support remains unclear.

In spite of this ad hoc federal support, strategic thinking has been demonstrated in several provinces in Canada. In Ontario, the Minister of the Environment established the Ontario Expert Panel on Climate Change Adaptation, which submitted its report in November 2009. The report contains a total of 59 recommendations, most of them specific action items addressed to a wide diversity of government departments. Most recommendations are directed to situations where immediate action in response to or in anticipation of climate change can be undertaken. In addition, and more importantly for this discussion, the report makes five major recommendations:

1. Launch a province-wide climate change adaptation action plan.
2. Establish a Climate Change Adaptation Directorate (CCAD).

3. Ensure that the CCAD has ongoing access to expertise.
4. Enhance climate change science and modeling capacity.
5. Identify dedicated funding for climate change adaptation initiatives.

Provinces across Canada are also undertaking similar initiatives as illustrated in Table 7.1.

More recently, at the municipal level, the Federation of Canadian Municipalities (FCM) has worked closely with ICLEI to promote mitigation and, of late, adaptation at the local level (FCM 2009). Their Partners for Climate Protection (PCP) program has 180 municipal governments actively involved in their 5-milestone framework for reducing greenhouse gas emissions. The need for leadership is felt at the municipal level, with officials asking FCM for more information, resources, and tools to aid them in adapting to climate change. FCM has called upon the federal government to establish a municipal adaptation fund to assist municipal governments in responding to climate change. Additionally, in the fall of 2010, ICLEI Canada is planning on launching a new adaptation initiative to help cities develop climate change adaptation plans. This comes after several communities across Canada have initiated their own adaptation activities, including:

Calgary, Alberta
Capital Regional District, British Columbia
Clyde River, Nunavut
Dawson City, Yukon
Delta, British Columbia
Edmonton, Alberta
Halifax, Nova Scotia
Iqaluit, Nunavut
London, Ontario
Metro Vancouver, British Columbia
Montréal, Québec
Oakville, Ontario
Ottawa, Ontario
Peel Region, Ontario
Pickering, Ontario
Port Alberni, British Columbia
Prince George, British Columbia
Portage la Prairie, Manitoba
Richmond, British Columbia
St. John's, Newfoundland
Sudbury, Ontario
Toronto, Ontario
Vancouver, British Columbia
Yellowknife, Northwest Territories
York Region, Ontario
(FCM 2009)

Table 7.1 Overview of Provincial climate change plans and adaptation characteristics

Province/territory	Summary/status of plan	Selected characteristics of plan or planned actions (unless otherwise indicated below)
Northwest Territories		
Climate change adaptation plan	In 2008 the Northwest Territories completed a climate change impacts and adaptation report in which the government stated it is working toward developing a climate change adaptation plan for the government and a territorial plan for the territory.	• Repair and replace foundations damaged by ground movement or water accumulation under buildings. • Rehabilitate runways in Inuvik and Yellowknife due to damage caused by permafrost degradation. • Award contracts to supply vendors 1 month earlier as a result of shorter winters. • Use climate data from communities in the territory with the highest amounts of snow, rain, and wind as outlined in the National Building Code of Canada. • Integrate climate change into school curriculum for kindergarten to grade 12 students. (Northwest Territories Government 2008)
Yukon		
Climate change action plan	In February 2009, Yukon released its Climate Change Action Plan focusing on forests, water, permafrost and infrastructure, climate change scenarios and community information and needs assessments. Currently, work is underway in the areas of land-use planning, human health, community adaptation planning, emergency response planning, agriculture, and building standards.	• Complete a Yukon infrastructure risk and vulnerability assessment and determine adaptation strategies in response. • Develop an inventory of permafrost information for use in decision-making. • Complete a Yukon water resources risk and vulnerability assessment. • Create a tool to facilitate the collection and distribution of water quantity and quality data. • Conduct a Yukon forest health-risk assessment and tree species and vulnerability assessment. • Enhance knowledge and understanding of climate change. • Establish a Yukon Research Centre of Excellence. (Yukon Government 2009)

(continued)

Table 7.1 (continued)

Province/territory	Summary/status of plan	Selected characteristics of plan or planned actions (unless otherwise indicated below)
Nunavut		
Climate change adaptation plan	In 2003, the Department of Sustainable Development released the Nunavut Climate Change Strategy. A Climate Change Adaptation Plan is currently being developed with several pilot adaptation projects already underway. The project "Addressing Climate Change Adaptation: A Collaborative Approach in Support of the Nunavut Climate Change Adaptation Plan" is scheduled to be completed by March 2011.	• Support community-based climate change adaptation planning. • Conduct feasibility studies and pilot projects for wind, solar, and hydroelectric energy being developed across the territory. • Develop energy efficiency projects throughout Nunavut, including residual heating projects, new efficient building practices, and efficient housing retrofits. • Work to develop climate change science and Inuit traditional knowledge research and monitoring initiatives in partnership with federal government, universities, institutes, and Inuit organizations. (Nunavut 2003)
British Columbia		
Preparing for Climate Change: British Columbia's Adaptation Strategy;British Columbia Regional Adaptation Collaborative (BC RAC)	British Columbia was the first province to announce its Regional Adaptation Collaborative, which will include forest, watershed, and floodplain management. In 2009, the province released Preparing for Climate Change: British Columbia's Adaptation Strategy, which outlines several intended goals for adapting to climate change.	• Build a strong foundation of knowledge and tools to help public and private decision-makers across British Columbia prepare for a changing climate. • Make adaptation a part of the Government of British Columbia's business, ensuring that climate change impacts are considered in planning and decision-making across government. • Assess risks and implement priority adaptation actions in key climate-sensitive sectors. • Develop initiatives including Mountain Pine Beetle Action Plan; Drought Response Plan; and B.C.'s Fire Smart initiative. (British Columbia Government 2009)

Manitoba

Climate change action plan Partner in the Prairie Regional Adaptation Collaborative; Manitoba Rural Adaptation Council (MRAC)

In 2001, Manitoba established the Manitoba Climate Change Task Force. Manitoba's current Climate Change Action Plan contains several adaptation initiatives. The province is integrating climate change adaptation into environmental assessments and land use instruments and is relocating winter ice roads to land. Manitoba is planning on developing a separate Climate Change Adaptation Strategy. Manitoba's Climate Change Action Plan contains over 60 initiatives, of which 10% relate to adaptation.

- Continue developing integrated watershed management plans to address water budgeting and water conservation.
- Improve flood protection throughout the province, including upgrading the Red River Floodway from protection against a 1-in-90 year spring flood to a 1-in-700 year spring flood.
- Continue expanding Manitoba's hydrometric network.
- Continue introducing incentives such as the Riparian Tax Credit and a Nutrient Management Regulation to protect lakes and rivers.
- Work with municipalities to establish local emergency management plans to prepare for extreme weather events, and increase the speed and effectiveness of local and regional emergency response measures.
- Invest in the realignment of winter roads and in improved river and stream crossings, along with work on all-weather roads.

(Manitoba Government 2008)

Saskatchewan

Partner in the Prairie Regional Adaptation collaborative (PRAC)

In 2009, Saskatchewan introduced greenhouse gas emissions legislation that includes adaptation. The province is currently working toward development of a provincial adaptation strategy. Saskatchewan is a partner in the Prairie Regional Adaptation Collaborative, which includes plans for effective drought management, water conservation, and adaptation planning.

Expected outcomes include:

- Overview of the range of probable future climates, which will provide the basis for appropriate adaptation actions.
- Report describing Saskatchewan biophysical assessment scenarios based on potential future climate scenarios.
- Report describing Saskatchewan climate change impacts and potential adaptation strategies.
- Recommendations to enhance Prairies water management.
- Recommendations to enhance forest adaptation to climate change.
- Web-based Wizard Adaptation Tool that includes future climate scenarios and the key climate variables that will enable users to make informed decisions adjusted to Saskatchewan's future climate.
- Coordinated programming under the Prairie Regional Adaptation Collaborative.

(Saskatchewan Government 2009)

(continued)

Table 7.1 (continued)

Province/territory	Summary/status of plan	Selected characteristics of plan or planned actions (unless otherwise indicated below)
Alberta Adaptation strategy Partner in the Prairie Regional Adaptation Collaborative (PRAC)	In 2008, a report on Climate Change Vulnerability Assessment for Alberta was released. In 2009, Alberta released Phase 1 of its Adaptation Strategy, which focuses on risk assessment, capacity building, taking action, evaluating outcomes, and strategic integration. Alberta is a partner in the Prairie Regional Adaptation Collaborative.	• Develop a provincial Climate Change Adaptation Strategy to provide overall direction, identify measures, and indicators of climate change, provide information about the impacts, and identify risks and vulnerabilities. Planned actions include: • Coordinate policy and research on adaptation. • Communicate and inform Albertans on the potential impacts of climate change. • Develop appropriate responses to adapt to climate change. (Alberta Government 2008)
Ontario Climate change action plan Expert panel on climate change adaptation; Ontario Regional Adaptation Collaborative (ORAC)	Ontario's Climate Change Action Plan was released in 2005. In December 2007, Ontario established an external Expert Panel on Climate Change Adaptation. The province has also established the Ontario Regional Adaptation Collaborative to be launched in 2010.	Actions already taken: • Create new rules for green energy projects to improve the climate resiliency of the province's energy grid. • Consider the impacts of climate change on protecting sources of Ontario's drinking water. • Increase public awareness of health hazards associated with extreme weather. • Consider climate change vulnerabilities and risk when developing the Northern Growth Plan. • Set up research centers to look into developing pest- and drought-resistant crops. • Consult on a plan to develop a water conservation and efficiency strategy for Ontario. (Ontario Government 2007)

Newfoundland and Labrador Climate Change Action Plan; Regional Adaptation Collaborative (RAC) for Atlantic Canada	Newfoundland and Labrador released its Climate Change Action Plan in 2005. Recently, the province provided $1.3 million toward the development of the necessary tools, policies, and strategies to help communities assess and adapt to climate change impacts. The province is also partner in the Regional Adaptation Collaborative (RAC) for Atlantic Canada.	• Organize a workshop on climate change impacts and adaptation for local municipalities. • Require that infrastructure projects receiving public funds meet a standard set of criteria with respect to climate change. • Initiate dialogue with Fisheries and Oceans Canada, industry, and stakeholders on climate change mitigation and adaptation. • Promote the consideration of climate change impacts in areas of the province that have initiated efforts toward Integrated Coastal Zone Management Planning. • Include climate change considerations in its Sustainable Development Strategy. • Report annually on the provincial Climate Change Action Plan. (Newfoundland and Labrador Government 2005)
Prince Edward Island 2008 climate change strategy	A risk-based approach to adaptation is included in the province's 2008 Climate Change Strategy, which establishes a provincial, interdepartmental working group to identify and manage current and projected climate-related risks.	• Incorporate climate change outcomes into the environmental impact assessment process. • Conduct coastal erosion sensitivity mapping • Including risk assessment. • Improve the ability to withstand future climatic conditions by designing better bridges, roads, dams, water supplies, sewers, and buildings. • Improve coastal communities' preparedness to cope with severe weather events. • Avoid decisions that make adaptation harder or that increase vulnerability, by preventing development in areas at risk from increased flooding, storm surges, and catastrophic erosion events. (Prince Edward Island Government 2008)

(continued)

Table 7.1 (continued)

Province/territory	Summary/status of plan	Selected characteristics of plan or planned actions (unless otherwise indicated below)
Nova Scotia 2009 climate change action plan	In 2009, Nova Scotia released its Climate Change Action Plan that contains over a dozen actions related to adaptation, including the development of an adaptation fund, departmental adaptation planning, land-use planning guidelines, a wetlands strategy and a water resource management strategy.	• Create an Adaptation Fund within Nova Scotia Environment to encourage adaptation research and development. • Develop statements of provincial interest on adaptation to provide guidance on land-use planning. • Establish criteria for the consideration of climate change during Nova Scotia Environment's environmental assessment process, and develop a guide to climate change for project proponents. • Launch a web-based clearinghouse of information and tools to support adaptation to climate change in Nova Scotia. • Update biannually a report that provides latest climate research, and review critical information gaps and policy direction for the province. (Nova Scotia Government 2009)
New Brunswick Climate change action plan 2007–2012	In 2007, New Brunswick released its Climate Change Action Plan 2007–2012. Since then it has released two progress reports. The Plan outlines numerous adaptation actions.	• Establish a formalized roundtable process with municipal associations, to promote and encourage regular dialogue and the exchange of ideas. • Adopt smart growth community-planning principles that consider climate change impacts/adaptation and emissions reductions. • Develop and implement a comprehensive provincial water management strategy. • Assist the tourism industry to make informed decisions and mainstream adaptation. • Incorporate vulnerability considerations into departmental decision-making processes involving economic, social, and environmental considerations in support of the public and private sectors' development and adaptation needs. (New Brunswick Government 2007)

Québec
Climate change action plan 2006–2012

Québec has included adaptation in its Climate Change Plan 2006–2012. The province is focusing on health monitoring and warnings, improved water and air quality management, and research and monitoring related to coastal erosion, water, transportation infrastructure, forests, agriculture, permafrost, and biodiversity.

- Set up mechanisms to prevent and mitigate the impact of climate change on health and public safety, with an investment of $34 million.
- Perform various evaluations and research related to permafrost thawing, coastline erosion, and adapting to the impacts of these climate changes, with a $6.6 million investment.
- Determine the vulnerability of Québec forests and the forest sector to climate change and incorporate the anticipated effect of these changes into forest management. This component will be supported with investment of $6 million.

(Québec Government 2008)

A Strategic Approach to Adaptation

As first conceived in the negotiations leading to the UNFCCC, adaptation was thought of as largely a place-based and local matter; of concern only to those most vulnerable communities and countries. In consequence, a bottom-up approach was viewed as the preferred option – with financial and technical assistance being made available from higher levels of government – supported where necessary by the international donor community. Over the life of the UNFCCC, adaptation has grown in significance and has come to be seen as requiring top-down strategic approaches. A major challenge now facing all Parties to the Convention will be the effective and simultaneous management and coordination of both top-down and bottom-up approaches. Currently, in Canada, the blend has been allowed to evolve almost unguided, with modest encouragement from the federal government. Leadership has emerged at both provincial and municipal levels across the country. But it is not clear what the consequences of such an approach will be.

At present there is a growing and relatively unconnected multi-level mosaic of adaptation activities. It is very much a learning-by-doing approach that could lead to a lack of sufficient preparedness and failure to create the needed climate resilience across regions and sectors for a variety of climate risks. The evolving mosaic has the potential to leave particular regions of Canada overexposed. Also, an uncoordinated effort may fail to take maximum advantage of new opportunities that climate change will bring to a cold northern country. Surely, Canada cannot continue to avoid the development of federally sponsored nation-wide climate change adaptation strategy for much longer. Such an activity would not obviate the current multi-level mosaic approach. On the contrary, it would give it focus and enable Canadians to see the full picture and facilitate greater awareness and stronger action in both the near and the long term.

References

Alberta Government (2008) Climate change strategy: responsibility, leadership and action. http://environment.gov.ab.ca/info/library/7894.pdf. Cited 5 July 2010

British Columbia Government (2009) Preparing for climate change: British Columbia's adaptation strategy. http://www.livesmartbc.ca/attachments/Adaptation_Strategy.pdf. Cited 5 July 2010

Burton I, Bizikova L, Dickinson T et al (2007) Integrating adaptation into policy: upscaling evidence from local to global. Clim Policy 7(4):371–376

COWI, IIED (2009) Joint external evaluation: evaluation of the operation of the least developed countries fund for adaptation to climate change. Evaluation Department, Ministry of Foreign Affairs, Denmark

Dickinson T (2007) The compendium of adaptation models for climate change, 1st edn. Adaptation and Impacts Research Division, Environment Canada, Downsview

Environment Canada (1997) The Canada country study: climate impacts and adaptation. Adaptation and Impacts Research Group, Downsview

FCM (2009) Municipal resources for adapting to climate change. http://www.sustainablecommunities.ca/files/Capacity_Building_-_PCP/PCP_Resources/Mun-Re-_Adapting-Climate-Change-e.pdf. Cited 5 July 2010

Gardner A (1994) Federal intergovernmental cooperation on environmental management: a comparison of developments in Australia and Canada. Canadian Environmental Protection Act Office, Hull

Government of Canada (2010) Fifth national communication on climate change: actions to meet commitments under the United Nations framework convention on climate change. Submitted to the UNFCCC Secretariat 12 Feb 2010. http://unfccc.int/resource/docs/natc/can_nc5.pdf. Cited 5 July 2010.

IPCC (2007) Climate change 2007: synthesis report – contribution of working group I, II and III to the fourth assessment report of the Intergovernmental Pannel on Climate Change [IPCC]. IPCC, Geneva

Lemmen DS, Warren FJ, Lacroix J et al (eds) (2007) From impacts to adaptation: Canada in a changing climate 2007. NRCan, Government of Canada, Ottawa

Manitoba Government (2008) Adapting to climate change: preparing for the future. http://www.gov.mb.ca/asset_library/en/beyond_kyoto/adapting_to_climate_change.pdf. Cited 5 July 2010

Morton F (1996) The constitutional division of powers with respect to the environment in Canada. In: Holland KM, Morton FL, Galligan B (eds) Federalism and the environment: environmental policymaking in Australia. Greenwood, Westport

New Brunswick Government (2007) Climate change action plan 2007–2012. Department of Environment. http://www.gnb.ca/0009/0369/0015/0001-e.pdf. Cited 5 July 2010

Newfoundland and Labrador Government (2005) Climate change action plan. Department of Environment and Conservation. http://www.env.gov.nl.ca/env/climate_change/govt_action/climatechangeplanfinal.pdf. Cited 5 July 2010

Northwest Territories Government (2008) Climate change impacts and adaptation report. Environment and Natural Resources. http://www.enr.gov.nt.ca/_live/documents/content/NWT_Climate_Change_Impacts_and_Adaptation_Report.pdf. Cited 5 July 2010

Nova Scotia Government (2009) Toward a greener future Nova Scotia's Climate Change Action Plan. Department of Environment. http://www.gov.ns.ca/energy/resources/spps/energy-strategy/Climate-Change-Action-Plan-2009.pdf. Cited 5 July 2010

Nunavut Government, 2003. Nunavut climate change strategy. Department of Sustainable Development. http://www.gov.nu.ca/env/Climate%20Change%20Full%20English%20low.pdf. Cited 5 July 2010

Ontario Government (2007) Ontario's action plan on climate. Ministry of the Environment. http://www.ene.gov.on.ca/publications/6445e.pdf. Cited 5 July 2010

Prince Edward Island Government (2008) Prince Edward Island and climate change a strategy for reducing the impacts of global warming. Department of Environment, Energy and Forestry. http://www.gov.pe.ca/photos/original/env_globalstr.pdf. Cited 5 July 2010

Québec Government (2008) Québec and climate change: a challenge for the future. http://www.mddep.gouv.qc.ca/changements/plan_action/2006-2012_en.pdf. Cited 5 July 2010

Saskatchewan Government (2009) Climate change adaptation research. http://www.parc.ca/saskadapt. Cited 5 July 2010

Yukon Government (2009) Climate change action plan. Environment Yukon. http://www.environmentyukon.gov.yk.ca/pdf/YG_Climate_Change_Action_Plan.pdf. Cited 5 July 2010

Part II
Adaptation in the Public Health Sector

Chapter 8
Overview: Adaptive Management for the Health Risks of Climate Change

Kristie L. Ebi

Abstract Climate change is expected to increase health risks in all countries. Although public health agencies and organizations have impressive records of controlling the burden of climate-sensitive health outcomes, current and planned programs and activities may need to be modified to address the additional risks of climate change. Programs and activities need to take an iterative risk management approach if they are to maintain current levels of health burdens as diseases change their geographic range and incidence in response to changing temperature and precipitation patterns and as the risks of adverse health outcomes from extreme weather events increase. Public health can learn from the experiences in ecosystem management with adaptive management, a structured and iterative process of decision-making in the face of imperfect information, with an aim of reducing uncertainty through monitoring and evaluation. Although many of the steps in adaptive management are familiar to public health, key differences include: a stronger emphasis on stakeholder engagement; taking a systems-based approach; developing interventions based on models of future impacts; and a strong and explicit focus on iterative management that can facilitate the capacity for further adaptation. Incorporating these elements into public health programs and activities will increase their effectiveness to address the health risks of climate change.

Keywords Climate change • Public health • Adaptation • Adaptive management • Systems approach • Stakeholder engagement • Risk management • Ecohealth • Learning by doing • Health risks

K.L. Ebi (✉)
ClimAdapt LLC, 424 Tyndall Street, Los Altos, CA 94022, USA

Department of Global Ecology, Stanford University, Stanford, CA, USA
e-mail: krisebi@essllc.org; krisebi@stanford.edu

J.D. Ford and L. Berrang-Ford (eds.), *Climate Change Adaptation in Developed Nations: From Theory to Practice*, Advances in Global Change Research 42, DOI 10.1007/978-94-007-0567-8_8, © Springer Science+Business Media B.V. 2011

Introduction

Climate change presents current risks to population health, with the risks expected to increase with additional climate change (Confalonieri et al. 2007). As discussed in the companion chapters in this volume, public health institutions and organizations are beginning to identify and implement programs and activities to avoid, prepare for, and respond to these risks. The focus has been on interventions that are the responsibility of national and sub-national public health agencies, such as developing and deploying early warning systems for heat waves and other extreme weather events, and improving the quality and extent of surveillance and control programs for infectious diseases (Ebi 2009; Frumkin et al. 2008; Jackson and Shield 2008), although Semenza (this volume) rightly notes that effective adaptation requires a comprehensive approach that includes community-based participation. A key question is how to most effectively incorporate options and processes into current and planned programs that address risks that are changing over time and space, while integrating top-down and bottom-up approaches. Furthermore, these risks arise not just from a changing climate, but also from changes in other factors that determine the distribution and incidence of climate-sensitive health outcomes.

Recent Developments: From Theory to Practice

A key conclusion from the *Synthesis Report* of the Intergovernmental Panel on Climate Change *Fourth Assessment Report* was:

> Responding to climate change involves an iterative risk management process that includes both adaptation and mitigation and takes into account actual and avoided climate change damages, co-benefits, sustainability, equity, and attitudes to risk.

(IPCC 2007)

Current public health programs to address the risks associated with extreme weather events, poor air quality, malnutrition, and infectious diseases were generally not designed and implemented within an iterative risk management framework. Adaptive management is an approach used in natural resource management that addresses many of the issues identified as important for adapting to climate change; lessons learned from adaptive management can inform approaches to modify existing and create new programs to address the health risks of climate change.

Adaptive management is a structured, iterative process of decision-making in the face of imperfect information, with an aim of reducing uncertainty through monitoring and evaluation. Adaptive management is currently used in fishery management and waterfowl harvest management (Williams et al. 2009), and is designed to enhance scientific knowledge and reduce uncertainties that arise from natural variability, the stochastic behavior of ecosystems, and social and economic changes, in order to improve future decisions (NRC 2004). It does not postpone actions until enough information is known. Adaptive management aims to create policies that can help organizations, managers, and other stakeholders respond to and take advantage of unanticipated events (Holling 1978; Walters 1986). Instead of seeking precise predictions of future conditions, adaptive management recognizes the uncertainties associated with projecting future outcomes and calls for consideration of a range of possible future outcomes (Walters 1986). Management policies and measures are designed to be flexible and are subject to adjustment in an iterative, social learning process.

Adaptive management is intended to increase the ability to fashion timely responses in the face of new information, in a situation where there are multiple stakeholder objectives and preferences (NRC 2004). It encourages stakeholder engagement in decision-making, and aims to reduce decision-making gridlock by making it clear that decisions are provisional, that there is often no "right" or "wrong" decision, and that modifications are expected. It explores alternative ways to meet management objectives, predicts the outcomes of alternatives based on the current state of knowledge, implements one or more of these alternatives, monitors to learn about the impacts of management actions, and then uses the results to update knowledge and adjust management actions (Williams et al. 2009).

A distinction is often made between "passive" and "active" adaptive management approaches, although both involve iterative decision-making, feedback between surveillance and decisions, explicit characterization of system uncertainty, and embracing risk and uncertainty as a way of building understanding.

- Passive adaptive management uses predictive modeling to select a single course of action. Outcomes are monitored to inform adjustment of subsequent decisions, and to update models. This approach contributes to learning and to more effective management, but it is limited in its ability to enhance scientific and management capabilities for conditions outside the course of action selected.
- Active adaptive management approach evaluates a range of competing, alternative models of system dynamics and responses to identify management options. Learning is achieved by observing system responses to these actions. The goal of active adaptive management is to test new hypotheses to determine the best management strategy. A lack of concordance between observation and expectation should lead to a revised model(s) of how the system functions and to revised management options and actions.

The process of adaptive management is relatively new and is still being defined. Key elements of adaptive management that have been identified in theories and practice include (NRC 2004; Williams et al. 2009):

- Identify the key problem(s) to be addressed in collaboration with all relevant stakeholders, to reflect social and political preferences. Involvement of stakeholders from the beginning increases management effectiveness and the likelihood of achieving the project goals. Not all stakeholders will give each problem the same priority, so there should be explicit consideration of objectives and tradeoffs that capture the values of stakeholders.

- Assess the problem and the range of interventions to help solve it. Stakeholders will be helpful in identifying the causes of the problem and the range of options to address the problem.
- Project the possible current and future consequences of the different decisions. This requires an explicit baseline understanding of and assumptions about the system being managed to develop a foundation for learning (Holling 1978; Walters 1986). A model(s) should characterize the state of the system and its possible responses to management actions, as well as identify gaps in and the limits of knowledge. Model complexity should be tailored to the decision at hand. Mathematical models are often developed to help understand systems behavior. But in poorly understood systems, or when the scale or risks of the actions being considered do not justify the expense of rigorous models, simple schematic diagrams can serve as useful conceptual models. Each activity is evaluated to determine the extent to which it is likely to achieve the stated objectives and to generate new information or foreclose future choices. This step also should evaluate stakeholders' tolerance for the potential consequences of decisions.
- Identify key uncertainties and constraints that could affect future management decisions, and the options to address these issues.
- Monitor and evaluate outcomes to compare the results of management decisions. Stakeholders can help implement and monitor these activities, and participate in their evaluation. Monitoring should focus on significant indicators of progress toward objectives. However, monitoring does not ensure progress and should not be equated with adaptive management. Monitoring programs and results should improve understanding of systems and models, to evaluate the outcomes of management decisions, and to provide a basis for better decision-making. Monitoring systems should be an integral part of program design at the outset and not simply added post hoc after implementation (Holling 1978); a mechanism(s) for incorporating learning into future decisions. The political will to act upon that information must also exist. Adaptive management organizations must have the flexibility to adjust operations in light of new information, environmental changes, and shifting social and economic conditions and preferences.

Figure 8.1 shows general steps in an adaptive management program: assess the problem; design the intervention(s) (including projecting the consequences of different actions); implement; monitor; evaluate; and adjust based on lessons learned (Williams et al. 2009). Stakeholders should be engaged in all steps.

Adapting to the risks of climate change follows a process that is fundamentally similar to adaptive management. Similar steps were identified for developing adaptation strategies, policies, and measures in the Adaptation Policy Framework (Lim et al. 2005) and in the process of community-based adaptation (Ebi and Semenza 2008). Both adaptation processes and adaptive management take systems-based approaches, are iterative, explicitly incorporate learning by doing (informed by monitoring and evaluation), and place a strong emphasis on stakeholder engagement.

Fig. 8.1 General steps in an adaptive management program (Williams et al. 2009)

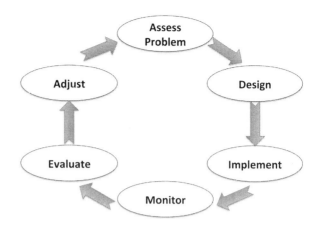

Key Challenges

Effectively applying adaptive management approaches within the public health sector to promote climate change adaptation will be a key challenge in the coming decades. The elements of adaptive management and the steps in an adaptive management process are already familiar in public health, although key differences include:

- A stronger emphasis on stakeholder engagement
- Taking a systems-based approach
- Developing interventions based on models of future impacts, as well as modeling future benefits and harms of public health interventions under different environmental and socioeconomic scenarios
- A strong and explicit focus on iterative management that will facilitate the capacity for further adaptation as the climate continues to change

Incorporating these elements into public health programs and activities will be challenging, yet will increase the effectiveness of these programs and activities in addressing the health risks of climate change.

Stakeholder Engagement

Effective adaptation processes include a broad range of stakeholders, including representatives of those who may be affected by climate change or the intervention and those who will implement identified adaptation options (Lim et al. 2005). Substantial literature exists on stakeholder engagement, including approaches and principles of effective engagement (e.g., Lim et al. 2005; Semenza this volume). Stakeholders contribute significantly to understanding current vulnerability and to identifying necessary public health and health care interventions. At the same time, their involvement in the assessment process will educate them about the risks of climate change and motivate them to continue the adaptation process after the assessment. Ensuring effective stakeholder engagement requires: identifying stakeholders for inclusion in the process; specifying their roles and responsibilities; and ensuring their continued involvement (Lim et al. 2005).

Ideally, stakeholders should be included who represent the programs that deal with the health outcome; organizations and institutions knowledgeable about climate change and development plans; local, regional, and national policymakers; and the most vulnerable groups. If, for example, water-borne diseases are a priority issue, then stakeholders could include representatives from the department(s) in the ministry of health that deal with water-borne diseases, the ministries of the environment (assuming they are the primary ministry dealing with climate change) and finance (assuming they oversee development infrastructure planning), water managers, university scientists involved in water-related issues, and community leaders and others who understand patterns of water use and misuse in their community. When identifying possible stakeholders, consideration should be given to stakeholders who will be involved with the effective design, implementation, and monitoring of public health and health care interventions.

Systems-Based Approaches

Public health moved away from systems-based approaches over the past several decades, with increased focus on individual-level risks. A new research framework is needed for effectively preparing for and responding to the health risks of climate and other global environmental changes (McMichael 2001). Ecohealth is an emerging field of study focusing on how changes in the earth's ecosystems are affecting human health (Corvalan et al. 2005; Lebel 2003). Ecohealth includes researchers and other specialists, community members, and decision-makers in the study of changes in the biological, physical, social, and economic environments. An ecohealth approach is based on three pillars: transdisciplinarity, participation, and equity.

- Transdisciplinarity refers to taking a systems-based approach that includes all relevant scientific disciplines, as well as community stakeholders and decision-makers. The challenges of global environmental changes transcend traditional research boundaries. Therefore, not including all relevant scientific disciplines limits the ability to understand the potential magnitude and extent of health impacts, as well as of interventions to efficiently and effectively reduce risks where the main drivers are within other sectors. For example, changes in temperature and precipitation patterns are affecting freshwater resources that can alter the prevalence of water-borne diseases, vector-borne diseases (through changes in standing water for breeding, etc.), and malnutrition (through water availability affecting agricultural productivity). Further, choices made in other sectors can ameliorate or exacerbate climate change-related health risks. Understanding and addressing these complex risks requires coordination across disciplines and organizations.

- Participation aims to achieve consensus and cooperation, within and among the community, scientific, and decision-making groups. The importance of wide stakeholder participation is discussed in Semenza (this volume).
- Equity involves analyzing the roles of men and women, and those of various social groups. Because men and women often have different responsibilities and different degrees of influence on decisions, it is important to take gender into account when considering the design and implementation of interventions. Various castes, ethnic groups, and social classes can have proscribed roles that affect their health and interactions with health and other agencies. Further, the impacts of climate change are not equally distributed across population groups and regions (Confalonieri et al. 2007). Socioeconomic factors play a critical role in altering vulnerability and sensitivity, by interacting with biological factors that mediate risk and/or lead to differences in the ability to adapt or respond to exposures. Taking equity into account facilitates identifying more effective interventions. For example, infants and children are particularly sensitive to heat stress, ozone air pollution, water-borne and food-borne illnesses, Lyme disease, and dengue fever (Balbus and Malina 2009).

Ecohealth provides a research framework that facilitates long-term sustainability. The health risks of climate change are part of a larger agenda that requires sustainable community-driven development (Corvalan et al. 2005; Lebel 2003). Building long-term sustainability requires strengthening formal and informal local community institutions, and building diversified financial mechanisms that are less reliant on a single donor. Ecohealth approaches provide an opportunity to empower communities, build crosscutting cooperation, and gain knowledge through projects.

Modeling Future Impacts and the Possible Consequences of Interventions

Current climate change-related risks are expected to alter as temperatures increase, and as precipitation patterns and the frequency and intensity of extreme events change. The actual impacts experienced in a particular location over a particular time period will depend not only on the climate change experienced, but also on specific vulnerabilities in the region and the actions taken within and outside the health sector to address the risks and vulnerabilities. Population vulnerabilities will change with changing demographics, economics, technological development, etc. Insights into how these changes could alter future risks are important for developing and evaluating adaptation options. Understanding how these various factors could interact over time under different climate and socioeconomic scenarios is best understood through modeling (Ebi 2008). For example, the impact of a heat wave depends not just on the size of the temperature anomaly but also on population acclimatization, urban infrastructure, the presence of an early warning system, and other factors, as discussed by Honda and Semenza (this volume).

The basic components of a model include: (1) estimates of baseline health burdens; (2) exposure–response relationships between changes in weather variables and disease outcomes, taking into consideration key non-climatic drivers of the health outcome; (3) projections of changes in temperature and precipitation under different socioeconomic and climate scenarios; and (4) approaches to estimate how adaptation could reduce projected health burdens (Ebi 2008). Projections of the potential burden of adverse health outcomes for vulnerable populations and locations can be used by local and regional risk managers to modify or implement new effective and timely adaptation policies and measures to reduce projected negative impacts, and can be used by policymakers to identify mitigation targets and the possible health consequences of approaches to meet those targets. In addition, models can provide regulators and lawmakers with more comprehensive evaluations of the full costs and benefits of various policy pathways. Models also can provide insights into the possible consequences of various management choices; this can be used in conjunction with adaptive management approaches to establish appropriate monitoring and evaluation programs.

Iterative Management

Public health has a long and impressive history of controlling major public health threats. Within the challenges of climate change are opportunities to identify where and when climate-sensitive health outcomes could change their distribution and/or incidence, thus facilitating the proactive modification of programs and activities to reduce outbreaks when they occur. The complexities of disease transmission systems and the inherent uncertainties with projections of future health impacts under different scenarios can be viewed as barriers to action. However, adaptive and iterative management is designed to take these complexities and uncertainties into account, in part by formally managing programs using a "learning by doing" approach. This approach requires increased emphasis on monitoring and evaluation to provide early information on modifications that could increase program efficiency and effectiveness under different environmental conditions.

For example, the increasing number of heat wave early warning systems provides an opportunity for understanding not just whether the overall systems are effective, but which elements of the response system contribute the most to system effectiveness (Kovats and Ebi 2006). This information is needed to design new systems that incorporate necessary elements without spending scarce public health funds on unnecessary activities. It is important to understand the degree to which effectiveness depends on the local context (i.e., governance and other factors) to understand how best to modify early warning systems for new locations. Various combinations of components can be tested in various locations, to quickly identify best practices. Further, experience is needed in how to adjust the weather conditions that trigger an early warning as the climate continues to change. This requires developing an indication of population acclimatization to higher temperatures, to understand when current triggers no longer indicate conditions that are particularly hazardous. Using iterative management approaches, national and international organizations can design and fund different communities to try different approaches to determine which is more efficient and effective, and can then disseminate the lessons learned.

Opportunities and Future Directions

Avoiding, preparing for, and effectively responding to climate change require moving beyond business-as-usual public health. Effectively addressing the health risks of climate variability and change requires coordination across local to national agencies and organizations. Because the health risks of and public health responses to climate change cover a broad range of issues, and because the risk and responses will change over temporal and spatial scales, national coordination of local programs and activities will be most effective. Programs and activities designed to address climate change and health issues should be created within agencies and organizations whose mission mandate includes human health. Key public health activities – including surveillance and monitoring; field, laboratory, and epidemiologic research; model development; development of decision support tools; and education and capacity building of the public and public health and health care professionals – should consider how climate change could alter the effectiveness of current and planned activities. In addition, formal coordination mechanisms should be established across all relevant ministries and organizations, including those dealing with environment, water resources, agriculture, transport, and urban planning. Entraining systems-based and iterative management approaches into local and national programs and activities will facilitate creating the flexibility and creativity needed to proactively prevent avoidable morbidity and mortality in a changing climate.

Case Studies

The case studies in this volume highlight various aspects of adaptive management. Several chapters recognize the importance of *stakeholder involvement*, particularly the involvement of those targeted by adaptation actions. Two chapters by Hutton and Berry and colleagues investigate the behavioral motivations and perceptions of adapting to climate change in Canada. The authors recognize that a participatory approach is necessary to identify people's needs and priorities, and thus increase their responsiveness to public health messaging and programming.

Many case studies also show how *transdisciplinarity* is useful in a health context. Lindgren and colleagues discuss the Swedish Government's Commission on Climate and Vulnerability (2005–2007), which assessed local vulnerability to climate change and adaptation needs. This chapter summarizes the section of the Commission's report dealing with water-related health impacts. This topic was explored in a "much broader context than is usually done," thanks to collaboration between the workgroups on health and water. Additional examples of collaboration are found in the chapter by Ogden and colleagues, which explores infectious diseases in Canada. The authors recognize that actors from diverse sectors and disciplines must collaborate with public health officials to manage disease, including: pharmacists who may be involved in innovative disease detection programs; experts in risk modeling and geographic information systems (GIS); decision-makers involved in watershed management, water treatment facilities, and sewer systems (to manage water-borne diseases); professional associations; and many others. Likewise, the case study by Semenza, which focuses on heat waves in urban environments, recognizes transdisciplinary cooperation and community-based participation in what the author terms "lateral public health."

These case studies also show the potential for applying adaptive management in addressing health issues. For instance, Ogden and colleagues acknowledge the uncertainty surrounding the effect of climate change on infectious diseases. Limited information implies the need for *iterative management*, under which adaptation strategies may be adjusted as more information becomes available. Another example is found in the chapter by Honda and colleagues that addresses the impacts of heat in Japan. This chapter lays the groundwork for *modeling future impacts* of heat in a changing climate, by discussing parameters such as the temperature–mortality relationship and population vulnerability (e.g., associated with age). In this way, the case studies in this volume demonstrate how adaptive management may be applied in addressing the health impacts of climate change.

References

Balbus JM, Malina C (2009) Identifying vulnerable subpopulations for climate change health effects in the United States. J Occup Environ Med 51(1):33–37

Confalonieri U, Menne B, Akhtar R et al (2007) Human health. In: Parry ML, Canziani OF, Palutikof JP et al (eds) Climate change 2007: impacts adaptation and vulnerability – contribution of working group II to the fourth assessment report of the Intergovernmental Panel on Climate Change. Cambridge University Press, Cambridge

Corvalan C, Hales S, McMichael A (2005) Ecosystems and human well-being: health synthesis – a report of the millenium ecosystem assessment. World Health Organization, Geneva

Ebi KL (2008) Healthy people 2100: modeling population health impacts of climate change. Clim Change 88:5–19. doi:10.1007/s10584-006-9233-0

Ebi KL (2009) Public health responses to the risks of climate variability and change in the United States. J Occup Environ Med 51(1):4–12

Ebi KL, Semenza J (2008) Community-based adaptation to the health impacts of climate change. Am J Prev Med 35(5):501–507

Frumkin H, Hess J, Luber G et al (2008) Climate change: the public health response. Am J Public Health. doi:10.2105/AJPH.2007.119362

Holling CS (ed) (1978) Adaptive environmental assessment and management. Wiley, New York

IPCC (2007) Climate change 2007: synthesis report – contribution of working group I, II and III to the fourth assessment report of the Intergovernmental Pannel on Climate Change [IPCC]. IPCC, Geneva

Jackson R, Shield KN (2008) Preparing the US health community for climate change. Am Rev Public Health 29:57–73

Kovats RS, Ebi KL (2006) Heatwaves and public health in Europe. Eur J Public Health. doi:10.1093/eurpub/ck1049

Lebel J (2003) Health: an ecosystem approach. International Development Research Centre, Ottawa

Lim B, Spanger-Siegfried E, Burton I et al (2005) Adaptation policy frameworks. Cambridge University Press, Cambridge

McMichael T (2001) Human frontiers, environments and disease. Cambridge University Press, New York

NRC (2004) Adaptive management for water resources project planning. Panel on adaptive management for resource stewardship, committee to assess the U.S. Army Corps of Engineers methods of analysis and peer review for water resources project planning, National Research Council, Washington, DC

Walters C (1986) Adaptive management of renewable resources. Blackburn, Caldwell

Williams CK, Szaro RC, Shapiro CD (2009) Adaptive management: the U.S. Department of the Interior Technical Guide. Adaptive Management Working Group, U.S. Department of the Interior, Washington, DC

Chapter 9
Behavioral Health and Risk Perception: Factors in Strengthening Community Resiliency and Emergency Preparedness

David Hutton

Abstract This chapter examines the relationship between people's general sense of psychosocial well-being and their capacity to respond to climate change and extreme weather events. Many people, while aware and concerned about climate change, are preoccupied with more immediate worries such as financial and health concerns as well as crime and other neighborhood conditions. Should environmental and emergency preparedness programming be effective in promoting adaptive behavior, it is important that these programs take into account how people's daily needs, priorities, attitudes, and values influence their motivation to prepare for climate change impacts. As such, there must be flexibility in the delivery of information and programming. This cannot be effectively achieved without greater emphasis on a participatory and consultative approach aimed at identifying those individual and community needs and priorities that shape people's responsiveness to public messaging and programs. This in turn increases the likelihood that people will not only perceive greater relevance in community activities aimed at enhancing climate change awareness and personal emergency preparedness, but will allow people to see themselves as part of the solution to climate change and other risks.

Keywords Adaptation • Climate change • Coping • Emergency preparedness • Psychosocial • Resiliency • Vulnerability • Well-being • Stress • Stressors • Behavioral health • Risk perception • Community resiliency

D. Hutton (✉)
United Nations Relief and Works Agency (UNRWA), West Bank Field Office,
Jerusalem 97200, Israel
e-mail: d.hutton@unrwa.org; huttondavid@hotmail.com

J.D. Ford and L. Berrang-Ford (eds.), *Climate Change Adaptation in Developed Nations: From Theory to Practice*, Advances in Global Change Research 42, DOI 10.1007/978-94-007-0567-8_9, © Springer Science+Business Media B.V. 2011

Introduction

The research presented in this chapter is intended to identify individual and community mechanisms which can enhance the coping and adaptive capacity of Canadian communities to the impacts of climate change and severe weather events especially. The chapter draws upon a 2007 research report by the Canadian Coalition for Immunization Awareness and Protection (CCIAP), *Impact of Climate Change and Extreme Events on Psychosocial Well-Being of Individuals and Community, and Consequent Vulnerability: Mitigation and Adaptation by Strengthening Community and Health-Risk Management Capacity.* This research was collaboratively undertaken by the Natural Resources Institute, the University of Manitoba and the Public Health Agency of Canada. The specific objectives of the project were to: (1) assess the degree of psychosocial stress associated with climate change across three communities in Western Canada; (2) identify key indicators which might influence people's perceptions of climate change; (3) determine the means and methods of integrating of psychosocial aspects of climate change into risk management, preparedness, and response policies and practices; and (4) determine options to enhance citizen perception of, and participation in, climate change and emergency preparedness initiatives.

This research is distinct in that it considers the influence of psychosocial factors on people's responsiveness to the possible impacts of climate change and extreme events, rather than considering psychosocial factors as consequences of climate change impacts. Traditionally, the psychological aspects of natural hazards and disasters have been contextualized in terms of individual distress and trauma arising from disaster events. At the same time, emergency management activities related to extreme weather events has focused mainly on promoting household preparedness through public awareness and education, with nominal consideration as to how psychosocial and behavioral factors such as daily worries and stress may limit the intended outcome of this approach. In light of these perspectives, this research project endeavored to link capacity enhancement to mitigate, prepare, and respond to climate change and extreme weather to underlying psychosocial indicators which influence the coping and adaptive capacity of people.

The research was undertaken between October 2005 and March 2007. This focused on three municipalities in southern Manitoba: the ward of North Kildonan in Winnipeg and the rural municipalities of Stuartburn and Cornwallis. North Kildonan is an urban ward located in the city of Winnipeg, with a population of approximately 37,000 persons (City of Winnipeg 2008). This is a mainly middle class community, with an average household income of $66,000. The most common types of employment are manufacturing (13%), health care and social services (12%), and retail trade (11%). The rural municipality of Stuartburn is located in southeastern Manitoba, approximately 120 km south of the city of Winnipeg. This is primarily a farming community, with 37% of the population deriving their livelihoods from agriculture. The average household income is approximately $37,600. The rural municipality of Cornwallis is located in southwestern Manitoba, surrounding the city of Brandon. Originally an agricultural community, the proportion of the population deriving their livelihoods from farming has steadily dropped. Today, agriculture accounts for only 12% of people's livelihoods; the government services industry is the largest employee (35%). The average income in Cornwallis is $66,100 (Manitoba Bureau of Statistics 2008a, b).

Methodologically, the project used a staged research strategy, employing a combination of quantitative and qualitative methods. This was built into a Community Participatory Framework (CPF) designed to assess factors influencing participants' perceptions, beliefs, and attitudes toward climate change and emergency preparedness. This mixed methodology was particularly useful in that it allowed the research team to attain information on community-level organizations and activities, as well as affording more in-depth analysis of individual coping and adaptation patterns.

At the outset of the project, the research team conducted an expert consultation workshop with provincial and regional health and emergency preparedness organizations. The purpose of the workshop was to solicit expert knowledge and opinion on the impacts of climate change in Manitoba. The main issues for consideration included: (1) the perceived scope and impacts of climate change in Manitoba, (2) the role of health and emergency management in addressing climate change, (3) strategies to facilitate coping and adaptation, and (4) key issues to be considered in planning for climate change.

Within each of the communities, focus groups were conducted to identify both individual and community-level perspectives and insights related to climate change. Particular attention was given to: identifying participants' beliefs and attitudes toward climate change and severe weather events, the degree to which these are perceived to be currently affecting people, and potential options to enhance people's capacity to cope with and adjust to related impacts. The focus groups were variably composed of community leaders, members of local interest and social groups, and key community informants.

Finally, random household surveys were conducted in each of the target communities. This totaled 269 respondents in North Kildonan ($N = 70$), Stuartburn ($N = 94$), and Cornwallis ($N = 105$). Information was collected on the respondents' background (for example, age, gender, education, income, length of residence) in order to examine the influence of individual characteristics on coping and adaptation processes. Respondents' level of awareness, knowledge, and concern related to climate change was also assessed, in addition to their attitudes around community life and neighborhood conditions. Finally, the respondent's own sense of well-being was assessed in terms of daily worries and stress, as well as the type of stressors they felt to be most prevalent in their lives.

Research Findings

The key objective of the research was to examine the degree to which psychosocial factors, including respondents' daily worries and resultant stress, influence their willingness and capacity to adjust to the risks posed by climate change. The results

Table 9.1 Level of concern and interest about climate change by community (%)		None/a little	Somewhat	Very
	Concern			
	North Kildonan	26.1	33.3	40.5
	Stuartburn	38.7	29.0	32.3
	Cornwallis	26.0	35.6	38.4
	Interest			
	North Kildonan	43.5	24.6	31.9
	Stuartburn	45.1	31.2	23.7
	Cornwallis	35.9	27.2	36.9

from both individual surveys and the community consultations indicated that the majority of respondents, while aware of and concerned about climate change and severe weather events, did not consider these as major concerns in their lives.

Across the three municipalities, 80.2% of the surveyed respondents acknowledged that climate change was underway, although only 13.5% indicated they were *very* concerned. Almost one third (30.5%) reported that they were either *not at all* or only *a little* concerned, 32.7% were *somewhat* concerned, and 23.3% *quite a bit* concerned. In terms of climate change knowledge, only 29.4% of the respondents indicated they knew a *great deal* about it, 47.0% knew *something* about climate change, and 23.6% *very little*.

In comparison, surveyed respondents were markedly more concerned with daily issues. Some 21.5% of the sample had found the previous month *very stressful* while 56.3% reported it to be either *occasionally* or *sometimes stressful*. The *biggest worry* reported by respondents was personal or family health issues (31.8%) followed by financial (27.7%) and family (19.1%) issues. Only 3.6% of the respondents cited the climate or weather as a worry in their daily lives.

When asked about the concerns they had about their neighborhoods, nearly one half of the respondents (47.4%) cited crime. This was distantly followed by concerns about available health, social and education programs (18.8%) and road conditions and travel (10.2%). Merely 12 or 6.8% of the respondents indicated they were concerned about the environment.

As shown in Table 9.1, an unexpected pattern was observed across the sampled communities in regard to the respondents' concern and interest with climate. Respondents living in the Winnipeg ward of North Kildonan demonstrated the greatest concern in climate change, with 40.5% indicating they were *very* concerned. However, less than one third (31.9%) were *very* interested in learning more about it. In Cornwallis, 38.4% of the respondents were *quite* or *very concerned* about climate change, with 36.9% being *very interested* in learning more about it. In Stuartburn, 32.3% were *quite* or *very concerned*, while 23.7% were *very interested* in learning more.

An interesting pattern was also found in relation to the respondent's demographic characteristics (Table 9.2). Almost as many high school (38.1%) as university graduates (42.1%) were either *quite* or *very concerned* with climate change. Similarly, almost the same proportion of higher income ($65,000 plus) as low

Table 9.2 Level of concern with climate change by demographics (%)

	None/a little	Somewhat	Quite a bit	Very
Gender				
Male	39.8	26.6	22.4	11.2
Female	25.0	36.3	32.8	14.9
Age				
18–30 years	32.3	38.7	16.1	12.9
31–45 years	36.8	28.1	21.1	14.0
46–60 years	26.6	36.2	20.0	17.1
60 years plus	27.9	30.9	33.8	7.4
Education				
Below high school	29.7	36.3	19.1	14.9
High school	34.9	27.0	25.4	12.7
Some college	29.8	38.8	14.9	16.5
College/university	28.4	29.5	30.7	11.4
Income				
Less than $35,000	36.4	26.1	26.1	11.4
$35–65,000	39.5	36.1	21.6	16.5
$65,000 plus	28.9	34.6	25.0	11.5

income respondents ($35,000 or less) was *very* concerned with climate change, respectively (11.5% and 11.4%). In regard to age and gender, however, a positive relationship was found in all communities. Overall, concern with climate change increased from a low of 29% among the youngest respondents to 34.1% among 31–45 year olds, 37.1% among 46–60 year olds, and 41.2% among the oldest respondents. Across the three communities, almost 20% more women than men were *quite or very concerned* (47.7% in comparison 29.9%).

Although not tested, this may reflect older persons' greater attention to weather because it often compounds daily living activities such as shopping or visiting friends, as discussed later in this chapter. During interviews, older respondents often expressed growing difficulties with travel during winter months because of reduced mobility and transportation. The higher level of concern found among women is consistent with other research that has shown that women are less likely to accept risk from all sources and more concerned about issues of health and safety generally (Flynn et al. 1994; Monson 2000).

Clearly, one of the more significant barriers to personal preparedness to environmental threats is the low salience of these events in most people's lives. Many people simply do not see the purpose in preparing for an event that may not occur and, even when acknowledging the importance of personal preparedness. In this study, only 15.7% of the surveyed respondents believed that a disaster would *definitively* one day occur in the area they lived (52.6% thought a disaster *might* occur but this was not likely). Only one half (49.6%) indicated they had supplies at home that could be considered part of a 72-h household emergency kit (for example, canned goods, supply of water, first aid kit, candles).

The challenge, then, is how best to motivate and engage individuals and communities in activities which will enhance their capacity to adjust to unpredictable but potentially disastrous events. As an underlying premise, two assumptions might be considered. First, there needs to be acknowledgement that simple awareness or even understanding of a possible risk is not a sufficient condition for behavioral change. Indeed, Ronan and Johnston (2005) have observed that motivation, as opposed to information and education, may be the *sine qua non* of community preparedness. "Despite the fact that people may be aware of both risk as well as strategies that can mitigate that risk, it does not follow directly that they will take the necessary action … Motivation is the psychological factor that fuels interest, concern, and action" (p. 7).

Second, it is important to recognize that the motivating factor to promote adaptive behavior toward climate change, as well as other remote threats to people's health and safety, may lie outside the immediate scope of climate change and/or emergency preparedness itself. That is, given people do not perceive climate change and severe weather as a significant threat, in part because they are preoccupied with more immediate daily stressors, should coping and adaptation to climate change be framed within a broader strategy which aims to enhance the capacity of people to cope and adapt more effectively to living demands generally, not merely environmental and severe weather impacts?

Implications for Policy and Programming

These findings speak to a need to adopt more collaborative and creative methods of engaging and supporting communities to prepare for the potential impacts of climate change and severe weather events. Currently in Canada, efforts to inform and prepare the public continue to take the form of broad public information activities which overlook the need to contextualize messages to the needs and priorities of different societal groups. A 2004 study on risk communication practices in Canada found that over three-quarters of surveyed emergency management organizations did not specifically target vulnerable groups. Nine of ten relied on generic written materials – namely websites, brochures, and fact sheets – to encourage individual and household preparedness (Haque et al. 2004).

One of the shortcomings of such an approach is a failure to account for underlying beliefs, values, and attitudes that determine how people perceive and respond to public messaging and information. As shown in this study, a significant proportion of population may have poor understanding and varying levels of concern regarding environmental threats like climate change and severe weather events. Moreover, having understanding does not necessarily imply action. In a telling study of heat waves and seniors in Canada, Jacque (2005) found that 70% of respondents were aware of heat wave advisories and 73% knew at least one means of protection. However, 47% would not use air conditioning in their homes. The main reasons cited by the respondents included the purchase price (cited by 44%), noise (42%), excessive cooling (45%), and the belief that it is harmful (38%) or would aggravate health problems (43%).

For many years, denial and minimizing of threats was seen by many emergency managers as the main reason why individuals did not prepare for potential hazards and risks, even when living in areas prone to events such as earthquakes, flooding, and severe storms like tornados. Although research has shown that both these cognitive processes influence household preparedness (Mileti and O'Brien 1992), there are additional factors that need to be examined when considering how people perceive and react to hazards and risks. In this study, for example, senior-aged respondents living in North Kildonan were less concerned about the changing weather than its perceived effects on normal daily activities like shopping and visiting friends. Moreover, this was seen to be interdependent on changing neighborhoods and access to essential urban services like public transportation on which people rely for conducting daily living activities. As one respondent put it, "the weather may be changing, yes, but the real problem is that Winnipeg has changed. If I want to buy groceries now, I have to catch a bus downtown, and the buses don't run as often. Returning at night is also a problem because crime has increased and one doesn't feel safe anymore."

This highlights the importance of understanding why people regard certain events or phenomena as being important to them, rather than assuming that knowledge of the risk alone will prompt action. This takes on particular relevance in respect to climate change: both because it is not well understood by the public and also because the impacts of changing and severe weather events (as illustrated above) may not be contextualized by people within the context of climate change itself. Moreover, climate change is often seen as happening mostly in the future and having few immediate effects (Patchen 2006). This causes immediate difficulties in facilitating awareness and adaptive behaviors because people tend not to evaluate issues from long-term perspectives, rather they "overweight" or place undue emphasis on shorter term concerns that affect them more directly (Loewenstein and Thaler 1989). Weber (2006) has similarly observed that worry generally acts as a driver for risk management decisions, meaning that "when people fail to be alarmed about a risk or hazard, they do not take precautions" (p. 103).

These findings point to the importance of working closely with communities to ensure that messaging is contextualized to people's everyday concerns, thus increasing the probability it will be perceived to be relevant and in turn acted upon. This is not a "one-size-fit-all" approach. When respondents were asked how awareness of climate change information might be raised, taking into account people's relatively low interest, there was a general consensus that activities need to be aligned to people's priorities and values. As one respondent in Winnipeg observed, "You need to make it real for people. Talk about what matters most to them. Make them realize what will happen to their children in the future." In the rural municipality of Cornwallis, on the other hand, the value of providing practical information to the agricultural community was emphasized. "Awareness, I think that can be through forums or workshops to identify what some of the causes are. You can't prevent everything, you can't control the weather but you can educate people about zero till and minimum till."

This obviously cannot be achieved through the simple dissemination of information. Respondents emphasized the point that people are "overwhelmed" by the amount of information they now receive. "There's too much junk mail these days," stated one focus group participant, "so unless my name is on it, it goes in the garbage." Respondents also stressed the importance of raising the visibility of climate change through a mainstreaming approach. It was agreed that schools can be the most effective way of raising awareness and changing attitudes toward climate change, with participants noting that children can learn of environmental risks and personal preparedness just as they do of other subjects. At the community level, it was agreed that information might be introduced to and through different associations and organizations such as church groups and social clubs, many of which have regular meetings through which issues like climate change might be presented and discussed. It was also suggested that climate change and emergency preparedness be introduced through events like family fairs which might serve as natural draw for community members. Having fun activities for children which may also teach them of climate change and household preparedness as well as having free food and door prizes for adults to encourage their attendance were identified as two ways to increase participation.

Building on work of community groups and networks can also form the basis for a significantly more interactive and holistic system, providing outreach and raising awareness to the public, particularly hard-to-reach or socially invisible groups (for example, frail and isolated seniors, non-English speaking newcomers, and the poor and homeless). These groups are often not only the most knowledgeable of the distinct needs of their members – for example, the Canadian National Institute for the Blind would be most informed about sight impairment populations and their coping skills in crises – but usually have established networks of mutual aid, as well as channels of communication in which information can be readily disseminated.

This speaks to the importance of applying resources to a more community-based approach, should emergency management effectively utilize organizations to promote household preparedness. Emergency management organizations in Canada have long emphasized that emergency preparedness – to all hazards including weather-related events like heat waves and cold snaps – is a shared responsibility which begins with the individual and household. However, there has been little effort to engage and consult with communities and people in meaningful ways. This is reflected in the continuing use of generic written materials to encourage preparedness and an underutilization of community-based organizations to support emergency preparedness activities. In a 2006 study of voluntary sector organizations conducted on behalf of the Canadian Red Cross (Enarson and Walsh 2007) found that only 31% of the surveyed agencies participated in local emergency preparedness committees, although 87% indicated they were in position to disseminate information and 55% could provide support during a public health emergency.

This need not be resource intensive should greater attention be given to developing more collaborative approaches to build community resiliency and emergency preparedness. The fact that both climate change awareness and emergency preparedness activities have been introduced to children and youth by organizations as varied as the Canadian Red Cross (Expect the Unexpected Program) and both Scouts Canada (Climate Change Education and Action Program) and Girl Guides of Canada (Girl Guides Yukon Climate Change Challenge) speaks to the potential of mainstreaming these issues through existing organizations. Another notable example of this community-based approach is the Canadian Climate Change Action Fund-Public Education and Outreach Program, which encourages projects that raise public awareness and promote community actions that reduce greenhouse gas emissions. The project supports innovative projects that have included an award winning "live action" simulation workshop, education, and tools for school teachers, and a CO_2 calculator that guides the user through lifestyle choices to raise awareness of greenhouse gas emissions from everyday activities.

It is important, however, that mainstreaming is not reduced to portable awareness-raising tools. As shown in this research, it is essential that individuals and communities are engaged in a manner that takes into account their knowledge, beliefs, and ideas. While the majority of Canadians may well express concern with climate change and potentially disastrous events, they remain preoccupied with daily demands and stressors associated with their health, finances, and families. Consequently, it cannot and should not be assumed that individuals will place the same priority on climate change and personal preparedness as do decision makers, and they will not necessarily act on information and advice when it is received. Here, the need to identify people's motivation to change takes on particular relevance.

To this end, it is essential that the diversity of communities is taken into account in developing resources and projects aimed at promoting more adaptive behavior to climate change and extreme events. Clearly, the vulnerability of people to hazards and disasters is not equally distributed but varies across regions and takes different significance in different hazard zones. Equally, communities differ remarkably not only in their capacities to prepare for and respond to hazards but also in the way their members perceive, interpret, and respond to risks. As such, there must be flexibility in the delivery of information and programming, which cannot be effectively achieved without emphasis on a participatory and consultative approach aimed at identifying those individual and community priorities, values, and beliefs which shape people's responsiveness to public messaging and programs. In turn, this increases the likelihood that people will not only see the relevance of climate change in their daily lives, but also see themselves as part of the solution to climate change and other risks.

References

City of Winnipeg (2008) Winnipeg neighborhood profiles: North Kildonan ward. www.winnipeg. ca/Census1996/data/11-00.pdf. Cited 10 May 2007

Enarson E, Walsh S (2007) An integrated approach to emergency management and vulnerable populations: survey report and recommendations. Canadian Red Cross, Ottawa (Unpublished manuscript)

Flynn J, Slovic P, Mertz J (1994) Gender, race and perception of environmental health risks. Risk Anal 14:1101–1108

Haque E, Lindsay J, Lavery J et al (2004). Exploration into the relationship of vulnerability and perception to risk communication and behaviour: ideas for the development of tools for emergency management programs. Public Safety and Emergency Preparedness Canada, Government of Canada, Ottawa (Unpublished manuscript)

Jacque L (2005) Knowledge, attitude and behaviour of the elderly during oppressive heat and air pollution. Paper presented at adapting to climate change in Canada 2005: understanding risks and building capacity, Montreal, 5 May 2005

Loewenstein G, Thaler L (1989) Context effects in the judgements of causation. J Pers Soc Psychol 57(2):189–200

Manitoba Bureau of Statistics (2008a) 2006 census profile: Stuartburn RM. www.gov.mb.ca/asset_library/en/statistics/demographics/communities/stuartburn_rm.pdf. Cited 10 May 2007

Manitoba Bureau of Statistics (2008b) 2006 census profile: Cornwallis RM. www.gov.mb.ca/asset_library/en/statistics/demographics/communities/cornwallis_rm.pdf. Cited 10 May 2007

Mileti D, O'Brien P (1992) Warnings during disasters: normalizing communication risk. Soc Probl 29(1):40–57

Monson G (2000) Gender differences in environmental concerns and perception. J Geogr 99(2): 44–56

Patchen M (2006) Public attitudes and behaviour about climate change: what shapes them and how to influence them. Purdue Climate Change Research Center, Purdue University, West Lafayette

Ronan K, Johnston D (2005) Promoting community resilience in disasters: the role for schools, youth, and families. Springer, New York

Weber E (2006) Experience-based and description-based perceptions of long-term risks: why global warning does not scare us (yet). Clim Change 77(1–2):103–120

Chapter 10
Lateral Public Health: A Comprehensive Approach to Adaptation in Urban Environments

Jan C. Semenza

Abstract The unpredictable nature of climate change poses considerable challenges to public health because it acts as a multiplier on existing exposure pathways and thus exacerbates existing vulnerabilities. Urban settings are particularly susceptible to the impacts of extreme weather events due to high population densities with shared exposure pathways. Moreover, metropolitan areas tend to be at increased risk from heat waves because urban climates are often warmer than un-built surroundings. Three aspects of urban adaptation to climate change are addressed here: (1) social interventions that advance bonding, bridging, and linking social capital in order to enhance community capacity and resilience; (2) interventions that attenuate the negative consequences of climatic events by physically improving the built environment; and (3) social services interventions that integrate multiple sectors through emergency plans for risk reduction of vulnerable populations.

These adaptation strategies in urban environments illustrate the concept of lateral public health based on transdisciplinary cooperation and community-based participation. In order to mount an effective response, public health practitioners need to transcend the traditional disciplinary boundaries and embrace lateral public health. This framework farms out public health action to other sectors of society, as well as community members of at-risk populations, in order to promote sustainable adaptation.

Keywords Climate change • Public health • Heat waves • Heat-related mortality • Adaptation • Social capital • Urban • City • Vulnerability • Extreme weather events

J.C. Semenza (✉)
Head of Future Threats and Determinants Section, Scientific Advice Unit,
European Centre for Disease Prevention and Control (ECDC), Tomtebodavägen 11A,
Stockholm, S-171 83, Sweden
e-mail: jan.semenza@ecdc.europa.eu

J.D. Ford and L. Berrang-Ford (eds.), *Climate Change Adaptation in Developed Nations:* 143
From Theory to Practice, Advances in Global Change Research 42,
DOI 10.1007/978-94-007-0567-8_10, © Springer Science+Business Media B.V. 2011

Introduction

The hallmark of traditional public health lies in the prevention of disease, extension of life expectancy, and promotion of health through the concerted effort of state and local public health departments. Government public health agencies or ministries of health aim to implement public health interventions that prevent rather than treat disease. This approach has proven remarkably successful in reducing infectious diseases and controlling many of the chronic diseases (Omran 1971). However, the unpredictable nature of climate change poses new challenges to the discipline of public health. These long-term threats require a paradigm shift in the thinking of public health practitioners. In order to respond effectively to the predicament of climate change, an expanded view of public health needs to be adopted that lies outside the traditional confines of the discipline. Practicing public health "outside the box" requires lateral thinking (Fig. 10.1). The traditional discipline of public health operates within the boundaries of government and vertically administers programs to susceptible populations (Fig. 10.1a). In contrast, lateral public health aims to expand those boundaries; rather than operating within the traditional constituents, lateral public health aims to alter the dynamics between them. It is a transdisciplinary, grassroots approach to public health, grounded in community-based participation in the decision-making process (Fig. 10.1b). Since the frequency and intensity of climatic events are on the rise and mitigation efforts are slow acting at best, urban adaptation strategies require such a new approach with rapid implementation.

Today, half of the world's population lives in urban areas (UNDESA Population Division 2007). While in 1800, only 3% of the population lived in cities, this proportion has steadily increased over time (Galea and Vlahov 2005). In the middle of the last century, New York was the first city with a population surpassing ten million (Satterthaite 2000). The proportion of urban dwellers is the highest in North America and Europe where in 2000, 79% and 73% lived in cities, respectively. The number and size of cities has grown, particularly in developing countries, where the absolute number of urban dwellers far exceeds the number in developed countries, and in many cases city populations have exceeded 20 million. Metropolitan areas are large population centers with agglomerations that are economically and environmentally connected to the urban core through transportation corridors. Thus, multiple cities are now interconnected such as: the BosWash "megalopolis" comprising Boston, Providence, Hartford, New York City, Newark, Philadelphia, Wilmington, Baltimore, Washington, and vicinity; Los Angeles–San Diego–Tijuana; or Chicago–Milwaukee. In Europe, the concept of the "Blue Banana" was coined to denote the "megalopolis" that extends from London, through Paris, to Milan and includes approximately 70 million inhabitants.

The expansion and merging of highly urbanized zones creates challenges to public health practitioners charged with protecting the health and welfare of the public. The complexity of this phenomenon became painfully apparent when over 700 people died in Chicago during the record setting heat wave in 1995 or when

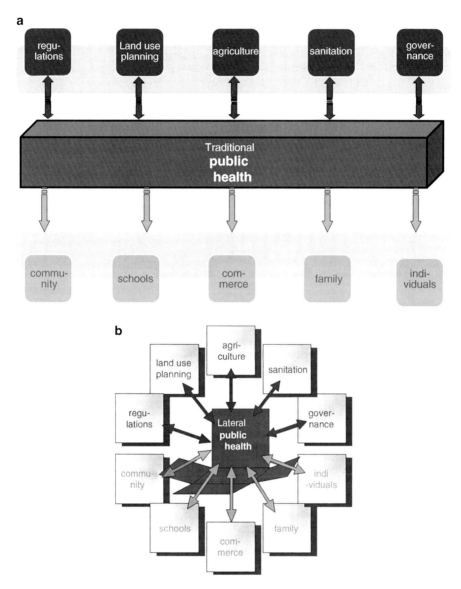

Fig. 10.1 (**a**) Traditional public health and (**b**) lateral public health. Public health aims at preventing disease, prolonging life and promoting health through the organized efforts and informed choices of society, organizations, public and private, communities, and individuals. The boxes in this diagram represent selected examples of different sectors of society involved in public health activities. Traditional public health (Fig. 10.1a) executes vertical functions while lateral public health (Fig. 10.1b) is based on interactive relationships. Thus, lateral public health aims to transcend the boundaries of traditional public health by acting outside the box and connecting directly with communities and other sectors of society

up to 70,000 people died in Europe during the 2003 heat wave (Semenza et al. 1996; Vandentorren et al. 2004; Conti et al. 2005; Johnson et al. 2005; Robine et al. 2008). Since the majority of individuals today live in urban environments, special consideration needs to be given to this particular form of living. Urbanization per se is a determinant of health that defines the conditions of living, working, and leisure time. Living in urban environments can thus impact health through three distinct pathways: the social environment, the physical environment, and access to social services (Galea and Vlahov 2005). The concept of lateral public health is applied here to urban interventions targeting these three pathways.

Social Environment

Adaptation to climate change can be enhanced by engaging urban communities in neighborhood renewal. Such community-based efforts have a twofold benefit: on one hand, they advance the adaptation capacity in urban settings by infrastructure improvements (see section below on physical environment) and on the other hand, they augment civic capacity which is beneficial to the overall well-being of the community. The bases for these adaptation efforts build social networks, which form a web of connections between people with different skills and backgrounds. These communal associations provide support and resources for problem solving and allow social engagement that advances social capital. The concept of social capital is the theoretical underpinning of the adaptation interventions in the social environment presented here.

Social capital, as detailed in Fig. 10.2, is defined as the potential embedded in social relationships that enable residents to coordinate community action to achieve shared goals (Bourdieu 1986; Coleman 1988; Putnam 1995). Social capital has two complementary facets: cognitive and structural social capital. Cognitive social capital includes norms, values, attitudes, and beliefs that emerge during community meetings and is defined as peoples' perception of level of interpersonal trust, sharing, and reciprocity. Community meetings and block parties, for example, increase cognitive social capital. In the process, social networks emerge, which is the basis for structural social capital. Structural social capital is inherent in social organizations of communities and can be described through these social nodes. Social capital relies on such networks for collaboration between residents to collectively address adaptation issues in the urban environment (Ziersch et al. 2005). These social relationships promote community participation and mutual cooperation in adaptation projects and are therefore not a characteristic of one particular individual, but rather a collective characteristic.

Social capital is a two-dimensional construct as described above and can also be portrayed as bonding (localized) and bridging/linking social capital (Hawe and Shiell 2000; Szreter 2002). Bonding social capital refers to the value assigned to social networks between homogeneous groups of people, and is inherent in existing social or religious groups; it is essential but not sufficient for neighborhood

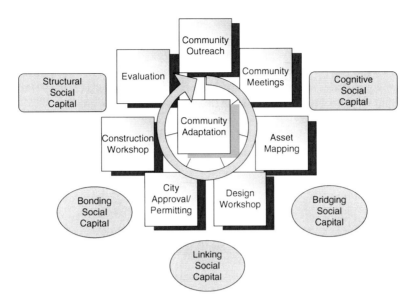

Fig. 10.2 The process of community-based adaptation to climate change. Cognitive social capital emerges as norms, values, attitudes, and beliefs during community meetings and is defined as peoples' perception of level of interpersonal trust, sharing, and reciprocity. Bridging social capital comes to light during asset mapping, when professional skills and talents are mapped and connected in the neighborhood. Linking social capital is developed after construction plans from the design workshop are submitted to city engineers for permitting. During the construction and installation of the projects, community members create friendship ties and increase bonding social capital. Extended social networks and civic engagement for the common good is part of structural social capital (Adopted with permission from Semenza and March 2009)

adaptation to climate change, because it may produce redundant information not applicable to adapting neighborhoods. Bridging social capital, on the other hand, connects different organizations and individuals and can reveal new information for problem solving and the creation of new opportunities. Linking social capital, an extension of bridging social capital, can connect parties unequal in power and access, such as residents with city officials (Szreter 2002).

Climate Change Adaptation of the Social Environment

Community-based adaptation to climate change can build on collective capacity inherent in social networks in order to increase resilience to extreme weather events (Keim 2008). Social networks and limited social support may predispose urban residents to poorer coping and adverse health outcomes (Crooks et al. 2008; Giles et al. 2005; Kawachi et al. 1999; Kawachi and Berkman 2001; McLeod and Kessler 1990). Conversely, social networks are important for positive health behaviors, including those related to climatic events such as heat waves (Semenza et al. 1996). Participating in group activities such as clubs, support groups, churches, etc. or having a friend close by is protective against heat-related mortality; conversely, those that live alone or do not have a pet to care for are at increased risk of succumbing to the heat during extreme weather events (Semenza et al. 1996).

In light of this close relationship between the social environment and climatic events, an intervention is presented here that aims to promote social interactions. These social interactions seem to advance collective efficacy that engage residents in direct social action, and in turn augment social capital in a sustainable manner. In Portland, Oregon a nonprofit organization entitled engages communities in a number of activities in order to strengthen social ties (http://www.cityrepair. org). Through a number of community events such as block parties, fairs and festivals, the organization fosters social interactions among urban residents. The City Repair Project is renowned for supporting communities in implementing urban renewal projects. In a collective effort, community members in Portland have created a number of green roofs, urban gardens, bioswales (water catchments designed to drain surface runoff to remove silt and pollution) and other features to counteract the urban heat island effect (see Table 10.1 and section on the physical environment) (Bradley 1995). By physically adapting and preparing the neighborhood for climate change, this process builds social networks among urban residents.

Staff members of The City Repair Project start with community outreach and organize community meetings as illustrated in Fig. 10.2 (Semenza and Krishnasamy 2007). These communal events among urban residents facilitate interpersonal trust, sharing and reciprocity and help to advance cognitive social capital (Semenza and March 2009). This proven strategy can be applied to the climate change field by informing residents about the challenges and threats from climate change and, in the process, building a common vision for the neighborhood (Ebi and Semenza 2008). Community organizers of The City Repair Project assist in the identification of resources in the neighborhood available to the interventions. These assets can be mapped out and made available to the project. As part of this process, bridging social capital is advanced by linking individuals with different talents and skill sets. During design workshops, urban features are developed with the support of landscape designers and other local talents, and subsequently circulated among residents for approval and signature. In Portland, this process has been institutionalized through city ordinances and has gained support from urban planners, politicians, and citizens (Semenza 2003). Construction plans are subject to review by engineers and architects prior to permitting by city engineers. This process is initiated by community members but encouraged and supported by a number of government officials. This grassroots effort is an example of lateral public health, which improves linking social capital in the community since it forces community members to interact with city administration (Fig. 10.1b). By bringing together the community with expert builders during construction workshops, the design plans are then realized in a collective effort, which builds bonding social capital (Fig. 10.2). Adaptation projects built by community members include green roofs to reduce rooftop temperatures,

Table 10.1 Community building strategies to advance community adaptation capacity

Interventions	Activities	Adaptation benefits
Urban forestry	Plant trees in parking strips or abandoned lots; adopt Friends of Trees programs (neighborhood tree plantings, restoration of green spaces, etc.)	Shading (preventing solar radiation from being absorbed by building materials)
		Ambient cooling from evapotranspiration
Urban vegetation	Build trellises for hanging gardens; planter boxes on street corners; plant in abandoned lots	Decreasing air-conditioning demand and peak energy consumption for cooling by shading buildings from solar radiation
		Ambient cooling from evapotranspiration
		Increasing property values through aesthetic enhancement
		Enhancing air quality by removing particulate pollutants from the air and by decreasing the emissions associated with air-conditioning energy demand
Urban gardens	Create community gardens in urban neighborhoods with elevated planter boxes if no soil available	Shading
		Ambient cooling from evapotranspiration
		Producing local food
Green roofs	Install green roofs on small structures (bike sheds, information kiosks, bus stops, etc.) or bigger public buildings as demonstration projects	Reducing rooftop temperatures and heat transfer to the surrounding air
		Decreasing summertime indoor temperatures, which reduces air-conditioning demand and peak energy consumption for cooling
		Lessening pressure on sewer systems through the absorption of rainwater
		Filtering pollution – including heavy metals and excess nutrients – through bio- and phytoremediation
		Protecting underlying roof material, reducing noise, providing a habitat for birds and other small animals, and improving the quality of life for building inhabitants
		Reducing the urban heat island effect by decreasing rooftop temperatures through evapotranspiration, which cools the surrounding air

(continued)

Table 10.1 (continued)

Interventions	Activities	Adaptation benefits
Cool roofs	Install roofs with high solar reflectance on houses in low socioeconomic status (SES) neighborhoods	Saving money on energy bills
		Reducing peak energy demand
		Reducing power plant emissions of air pollution and greenhouse gases
		Increasing indoor occupant comfort by lowering top-floor temperatures
		Contributing to urban heat island mitigation
Cool pavement	Install light-colored or permeable pavements in parking lots or school yards	Allowing water to percolate and evaporate through porous pavements
		Cooling the pavement surface and surrounding air
Window screens	Repair and install insect screens on houses	Protecting against emerging vector-borne diseases
Vector abatement	Eliminate standing water in depressions or objects	Preventing propagation of vector-borne diseases
Flood control	Restore barriers/dunes at river/beaches (e.g., with vegetation) to increase their resilience	Preventing flood waters from inundating inhabited areas
Habitat restoration	Plant native species and eliminate invasive plants	Decreasing dispersion of zoonoses through increased biodiversity
Rainwater storage	Increase rainwater storage (domestic water butts, unpaved gardens, etc.)	Preventing overflows in peak periods to avoid river contamination
Brown field development	Plant trees, gardens, vegetation, etc.	Shading
		Ambient cooling from evapotranspiration
		Local food production

Sources: http://www.epa.gov/heatisland/resources/faq.html#6
http://www.epa.gov/heatisland/strategies/community.html
http://www.friendsoftrees.org/about/index.php
http://www.growing-gardens.org/
http://www.permaculture.org/nm/index.php/site/index/

urban gardens to cool the air through transpiration, urban vegetation to increase shading of buildings, cool pavements to reduce heat absorbance of streets, etc. (Semenza 2005).

These interventions were subjected to program and process evaluation in 2003 (Semenza and Krishnasamy 2007; Semenza and March 2009). Portland residents were systematically sampled within a two-block radius of three intervention sites and interviewed before ($N = 325$) and after ($N = 349$) the interventions, of which, 265 individuals completed both surveys of a panel study. There was a significant change of sense of community between the first and the second survey ($p < 0.01$). For example, these data indicate that the participants appreciated their neighborhood as a good place to live and a good place for children to grow up significantly more after the improvements in the urban landscape had been implemented. Social interaction also displayed a consistent increase at all three sites, although the change was not statistically significant ($p = 0.06$). Social capital displayed a statistically significant increase after the intervention ($p = 0.04$). These data indicate that these community actions to improve the built urban environment led to an increase in social capital. In the context of climate change in urban settings, these findings are important in that they demonstrate that it is possible to improve the social environment, which augments the resilience of its residents. In summary, advancing community-based urban improvements can enhance the resilience of urban populations to climatic events.

Physical Environment

High population densities in urban centers where people work, live, and spend leisure time is the setting of a number of exposures with potential health consequences; urban dwellers are collectively exposed to the built environment, green spaces, drinking water, air quality, and weather conditions and are potentially at risk to be affected from climatic changes (Younger et al. 2008). For example, during hot summer days, metropolitan areas tend to heat up disproportionately compared to the surrounding areas, creating an urban heat island (Frumkin 2002). The thermal radiation of buildings and streets is determined by the heat capacity of concrete and asphalt, which can be responsible for up to 13°C higher temperatures in the urban core (Watkins et al. 2007). Furthermore, a limited number of trees and vegetation in urban areas limits the capacity to cool the air through transpiration (Table 10.1) (Jo 2002). At night, surface areas radiate the heat absorbed during the day, increasing the minimum temperatures. Elevated temperatures can increase the magnitude and duration of heat waves. Heat-related mortality and morbidity increase exponentially and peak 2 days after the maximum temperature surpasses a location-specific threshold (Semenza et al. 1996; Semenza et al. 1999). The heat island effect can exacerbate the impact of heat waves by escalating and prolonging heat exposure. This phenomenon can be particularly harmful when the relentless night-time exposure results in cumulative heat exposure. Residents of southern cities tend to be acclimated to warmer weather conditions and thus are less vulnerable; in contrast, urban populations living in mid and high latitudes with significant annual temperature variation are particularly at risk for illness and death.

We found Chicago residents living on the top floors of buildings to be at a fivefold increased risk for heat-related mortality during the 1995 heat wave. Living in a house with a flat roof or living in an apartment building were also risk factors. In order to save on energy bills, many of the roofs in Chicago are painted black to attract the sunlight during the cold winter months; however, during a heat wave this practice can pose a considerable risk for individuals living on the top floor. Furthermore, many of the windows had storm windows installed and were nailed shut, preventing aeration of hot residences. In this context, adaptation is defined as initiatives and measures to reduce the vulnerability of susceptible individuals to the effects of climate change, specifically the elderly, poor, and socially isolated. The interconnected nature of lateral public health is illustrated with urban interventions to adapt to climate change (Fig. 10.1b).

Climate Change Adaptation of the Physical Environment

Adaptation strategies in the urban environment can be combined with the community building interventions discussed above. These interventions are based on the notion that humans exhibit a strong sense of place (Altman and Low 1992; Brown and Perkins 1992; Stedman 2002), which increases with length of residence (Hay 1998). Attachment to place is rooted in culture, identity and history and is derived from psychological processes (Fullilove 1996). Sense of place is fundamental for adaptation to climate change because it is associated with sustainable practices and environmental conservation (Vorkinn and Riese 2001; Uzzell et al. 2002). Since local climatic conditions are fundamental to this sense of place, extreme weather events linked to climate change can disturb the psychological conception of place (Knez 2005). These events can do so by impacting the economic, societal, environmental, or cultural base and can have devastating consequences for the community, as documented by Hurricane Katrina. By disrupting the bonds people have with their physical environment, community health can be severely impacted with direct consequences to their sense of cultural identity (Kirsch 2001). Disrupting this sense of place can also impact community engagement and adaptation capacity (Adger et al. 2005; Hess et al. 2008). Conversely, if sense of place is linked to social networks, neighborhood attachment can motivate residents to take action in the interest of the common good (Lewicka 2005).

The interventions described above can help to decrease the heat island effect. Portland residents designed and built a number of urban projects with sustainable features that address this goal. For example, these community members have installed over 20 eco-roofs since 2001. An eco-roof is a lightweight, vegetated roof system that helps to reduce air temperatures and smog, captures and evaporates between 10 and 100% of precipitation and provides insulation. Through eco-roofs, urban temperatures are reduced through shading and transpiration; this in turn decreases cooling costs (Table 10.1) (Gill et al. 2007) and offers a number of other health benefits (Louv 2005). City parks can be up to 13°C cooler than the surrounding built urban environment and provide reprieve to overheated individuals during a heat wave (Watkins et al. 2007; Sponken-Smith and Oke 1998). Beyond these measures addressing the heat island effect, climate change adaptation in the physical environment of cities should also include vector abatement, flood control, urban habitat restoration, etc. A number of other steps can be taken as well to prepare households for an emergency or for gradual climatic change: increasing shading around the house, installing window screens, clearing out the gutters, etc. These efforts can further be enhanced through community efforts to advance adaptation in the neighborhood. For example urban agriculture in community gardens, vacant lots, or schoolyards is a local, small-scale and sustainable way of food production with a number of health benefits such as physical activity (Dixon et al. 2009). Through community-based efforts, these interventions also have multiple co-benefits for the social environment since building stronger relationships builds social capital.

However, many of these adaptation strategies are time consuming and might not yield immediate benefits to urban dwellers. Thus implementing short-term measures is an equally important aspect of public health practice. In the case of heat waves, the most effective way to reduce the risk of heat-related mortality is to have access to an air-conditioned environment. Interestingly, we found a fivefold decreased risk if there was an air conditioner in the household but an air-conditioned lobby of an apartment complex was equally protective (Semenza et al. 1996). Individuals with access to cooling shelters or other air-conditioned places are also at substantially reduced risk. Thus, in Europe or other parts of the world where the air conditioner penetration is relatively low, strategies need to be put in place to remove susceptible individuals from their homes with high ambient temperatures and provide access to cool churches, shopping centers, cinemas, etc.

Exacerbations of urban heat islands are not the only effects of climate change in metropolitan settings. Heavy precipitation can cause cryptosporidium outbreaks due to contamination of water treatment plants, but community water boil notices can contain the risk of infection (Semenza and Nichols 2007). Wildfires can threaten urban settings, and community-based risk assessments are crucial for hazard identification. Work parties to remove flammable brush in close proximity to buildings, implementing emergency plans, establishing evacuation routes, etc. can potentially attenuate the public health impact of wildfires. Communities should also prepare for floods by increasing awareness of flood zones, evacuation routes, and response plans; particular attention needs to be placed on vulnerable populations in hospitals, nursing homes, schools, and prisons. Landslides are another threat to communities that requires special emergency warning with evacuation plans. Elevated ambient temperatures have been linked to food-borne diseases (Kovats et al. 2004); health education/promotion interventions for safe food handling and storage during hot weather episodes can reduce this risk to communities. Changing environmental conditions has altered the distribution of vector-borne diseases, but bed nets, protective clothing, vaccination, etc. can minimize these risks (Semenza and Menne 2009). These examples illustrate the complexity of intervening in urban settings and the difficulty of orchestrating a comprehensive approach to climate change adaptation.

Social Services

Breaking the traditional confines of public health, lateral public health aims to reach out to different sectors in society not traditionally associated with health (Fig. 10.1b). As documented above, connecting community members with urban planners, architects, engineers, meteorologists, and government can have multiple co-benefits (Capon et al. 2009). Similarly, working with social workers on adaptation issues can be particularly beneficial for vulnerable groups such as the elderly and the poor. In many cities, wealth inequalities between groups exist in close proximity, and these groups differ also significantly in health indicators (Wilkinson 1992). These inequalities manifest as systematic discrepancies in morbidity and mortality, in which individuals with lower levels of education or socio-economic status die at earlier ages and suffer more illnesses (Semenza and Maty 2007). These vulnerable groups are also more susceptible to adverse weather events, and while erratic but recurrent heat waves cannot be avoided, the public health consequences of such extreme events on susceptible populations are entirely preventable. These groups suffer from inadequate access to health services compared to more affluent groups in society (Andrulis 2000). In light of these inequalities, an adaptation strategy targeting vulnerable populations is described below.

Climate Change Adaptation of Social Services

The institutional memories of the 1995 heat wave in Chicago led to a proactive strategy of attenuating the public health consequences of excessive environmental heat. Results from the field investigation identified political, mass media, environmental, societal, and behavioral risk factors for heat-related mortality (Semenza et al. 1996). Based on the epidemiological evidence, a macrosocial intervention was developed with a number of specific steps, including: meteorological monitoring of the weather conditions, defining the roles of agencies and organizations, preparing for a heat emergency, initiating emergency procedures during the heat wave, media engagement and outreach, and evaluation (Box 10.1) (Luber and McGeehin 2008). This comprehensive heat emergency response plan incorporated the findings from the field investigation in order to assure evidence-based action (Bernard and McGeehin 2004). Twenty-four hour hotlines were developed to provide information on the nearest cooling shelter, transportation and recommended treatment for heat stroke in the case of an emergency. "Heat Outlook" databases were compiled to facilitate contact of vulnerable individuals during a crisis, such as the elderly and the very isolated, either by phone or in person.

In 1999, a public health response was mounted at the dawn of a heat wave in Milwaukee, Wisconsin. Early dissemination of information through hotlines and media as well as regular status checks of family members, particularly the socially

Box 10.1 Components of a Proposed Model Heat Wave Emergency Response Plan

Measurements and monitoring:

1. Identify the recording station where meteorological factors will be monitored.
2. Establish heat index (or other measure) criteria for heat warning levels. The local sector of the National Weather Service may have designated indices that can be adopted.
3. Define heat-related outcomes for physicians and medical examiners.
4. Develop methods for monitoring heat-related outcomes.

Participating organizations:

5. Define the role and response of each participating organization. Designate and coordinate the response at each organization.
6. Establish an emergency communication system. Include all organization contacts (up-to-date telephone numbers and fax numbers).
7. Maintain a list of resources (e.g., potable water, air conditioners, fans) that should be available during a heat wave (specify quantity, who to contact, etc.).
8. Ensure that utility (power and water) companies have contingency plans and communication mechanisms in case of problems.

Preparations prior to the onset of excessive heat:

9. Identify at-risk populations.
10. Organize informative material for the public, media, health care workers, and volunteers.
11. Verify collaboration of participating agencies.
12. Procure resources necessary for a heat emergency (e.g., equipment, supplies, personnel, cooling sites).
13. Maintain and update a database of at-risk populations.
14. Identify temporary cooling centers (air-conditioned areas available for public use). Some should be wheelchair accessible.
15. Locate swimming pools and public sprinklers for public use.

During a heat wave:

16. Communicate with participating organizations and agencies to activate response plans.
17. Open temporary cooling centers.
18. Contact at-risk populations using the established database.

(continued)

Box 10.1 (continued)

19. Initiate communication of emergency information through the media and other organizations with public interaction.
20. Establish a telephone hotline for information on heat-related illnesses and relief tips.
21. Arrange transportation to and from cooling centers.
22. Arrange distribution of resources (fans, potable water).
23. Ensure that utility companies do not suspend services during the heat wave.

Following a heat wave:

24. Terminate the emergency response. Notify media and all other participating agencies.
25. Return unused resources and replenish resource stocks.
26. Evaluate emergency procedures and improve them if necessary.
27. Assess damage to municipal infrastructure.
28. Evaluate data.

Source: Heat Wave Emergency Response: A review by the National Center for Environmental Health of the Centers for Disease Control and Prevention, Atlanta, Georgia (Bernard and McGeehin 2004).

isolated, proved to be crucial in saving lives. Improved public health response in 1999 resulted in fewer heat-related deaths than expected (Weisskopf et al. 2002). There were 17% and 51% reductions in heat-related deaths and emergency medical service runs, respectively, compared to 1995, and those adverse health outcomes that did arise were not the result of differences in heat levels alone.

The absence of extreme temperature alert systems and prevention measures became evident during the historic European heat wave of 2003 that resulted in tens of thousands of excess deaths (Vandentorren et al. 2004; Conti et al. 2005; Johnson et al. 2005; Robine et al. 2008). At the time, only two cities in Europe, Rome and Lisbon, had sophisticated heat wave alert plans (Koppe et al. 2004). However, now virtually all European cities have prevention plans in place in the eventuality of hot weather that direct messages through the most effective channels to vulnerable groups. Chicago's public health calamity was the starting point in the development and execution of heat preparedness plans, which proved to be effective at reducing heat-related deaths. Thus, multi-sectoral preparedness and response activities are essential to reduce vulnerability and increase resilience to climatic hazards (Keim 2008).

Conclusion

Lateral public health aims to transcend the boundaries of traditional public health by connecting directly with communities and reaching out to other sectors in society. This approach is particularly important in urban settings where high population densities and multiple exposures threaten vulnerable populations. Moreover, funding for adaptation to climate change has been limited, and this lateral public health strategy offers an effective approach to meet some of the impending challenges by empowering officials, agencies, and local communities (Bouwer and Aerts 2006; Dovers 2009). Engaging different actors in lateral public health will help the process of mainstreaming adaptation to climate change into a range of other programs and sectors with co-benefits for the health of the public (Halsnæs and Trærup 2009; Mertz et al. 2009; St Louis and Hess 2008). Public health practitioners can mount an effective adaptation response to climate change if they reach out to all stakeholders, including those not traditionally associated with their discipline, such as urban planners, and integrate governmental with community activities.

References

Adger WN, Hughes TP, Folke C et al (2005) Sociological resilience to coastal disasters. Science 309:1036–1039

Altman I, Low S (1992) Place attachment. Plenum, New York

Andrulis DP (2000) Community, service, and policy strategies to improve health care access in the changing urban environment. Am J Public Health 90(6):858–862

Bernard SM, McGeehin MA (2004) Municipal heat wave response plans. Am J Public Health 94(9):1520–1522

Bourdieu P (1986) The forms of capital. In: Richardson J (ed) Handbook of theory and research for the sociology of education. Macmillan, New York

Bouwer L, Aerts J (2006) Financing climate change adaptation. Disasters 30(1):49–63

Bradley GA (ed) (1995) Urban forest landscapes: integrating multidisciplinary perspectives. University of Washington Press, Seattle

Brown B, Perkins D (1992) Disruptions in place attachment. Hum Behav Environ Adv Theory Res 12:279–304

Capon AG, Synnott ES, Holliday S (2009) Urbanism, climate change and health: systems approaches to governance. NSW Public Health Bull 20(1–2):24–28

Coleman J (1988) Social capital in the creation of human capital. Am J Sociol 94(Suppl):S95–S120

Conti S, Meli P, Minelli G et al (2005) Epidemiologic study of mortality during the summer 2003 heat wave in Italy. Environ Res 98(3):390–399

Crooks VC, Lubben J, Petitti DB et al (2008) Social network, cognitive function, and dementia incidence among elderly women. Am J Public Health 98(7):1221–1227

Dixon JM, Donati KJ, Pike LL et al (2009) Functional foods and urban agriculture: two responses to climate change-related food insecurity. NSW Public Health Bull 20(1–2):14–18

Dovers S (2009) Normalizing Adaptation. Glob Environ Change 19(1):4–6

Ebi KL, Semenza JC (2008) Community-based adaptation to the health impacts of climate change. Am J Prev Med 35(5):501–507

Frumkin H (2002) Urban sprawl and public health. Public Health Rep 117(3):201–217

Fullilove MT (1996) Psychiatric implications of displacement: contributions from the psychology of place. Am J Psychiatry 153:1516–1523

Galea S, Vlahov D (2005) Chapter 1 – Urban health: populations, methods, and practice. In: Handbook of urban health: populations, methods and practice. Springer, New York

Giles LC, Glonek GF, Luszcz MA et al (2005) Effect of social networks on 10 year survival in very old Australians: the Australian longitudinal study of aging. J Epidemiol Community Health 59(7):574–579

Gill SE, Handley JF, Ennos AR et al (2007) Adapting cities for climate change: the role of the green infrastructure. Built Environ 33(1):115–133

Halsnæs K, Trærup S (2009) Development and climate change: a mainstreaming approach for assessing economic, social, and environmental impacts of adaptation measures. Environ Manage 43(5):765–778

Hawe P, Shiell A (2000) Social capital and health promotion: a review. Soc Sci Med 51(6):871–885

Hay RB (1998) A rooted sense of place in cross-cultural perspective. Can Geogr 42(3):245–266

Hess JJ, Malilay JN, Parkinson AJ (2008) Climate change: the importance of place. Am J Prev Med 35(5):468–478

Jo HK (2002) Impacts of urban greenspace on offsetting carbon emissions for middle Korea. J Environ Manage 64(2):115–126

Johnson H, Kovats RS, MacGregor G et al (2005) The impact of the 2003 heat wave on mortality and hospital admissions in England. Health Stat Q 25:6–11

Kawachi I, Berkman LF (2001) Social ties and mental health. J Urban Health 78(3):458–467

Kawachi I, Kennedy BP, Glass R (1999) Social capital and self-rated health: a contextual analysis. Am J Public Health 89(8):1187–1193

Keim ME (2008) Building human resilience: the role of public health preparedness and response as an adaptation to climate change. Am J Prev Med 35(5):508–516

Kirsch S (2001) Lost words: environmental disaster, "cultural loss" and the law. Curr Anthropol 42:167–178

Knez I (2005) Attachment and identity are related to place and its perceived climate. J Environ Psychol 25(2):207–218

Koppe C, Kovats S, Jendritzky G et al (2004) Health and global environmental change, vol 2, Waves: risks and responses. World Health Organization, Copenhagen

Kovats RS, Edwards SJ, Hajat S et al (2004) The effect of temperature on food poisoning: a time-series analysis of salmonellosis in ten European countries. Epidemiol Infect 132(3):443–453

Lewicka M (2005) Ways to make people active: the role of place attachment, cultural capital, and neighborhood ties. J Environ Psychol 25(4):381–395

Louv R (2005) Last child in the words: saving our children from nature-deficit disorder. Algonquin Books, Chapel Hill

Luber G, McGeehin M (2008) Climate change and extreme heat events. Am J Prev Med 35(5): 429–435

McLeod JD, Kessler RC (1990) Socioeconomic status differences in vulnerability to undesirable life events. J Health Soc Behav 31(2):162–172

Mertz O, Halsnæs K, Olesen JE et al (2009) Adaptation to climate change in developing countries. Environ Manage 43(5):743–752

Omran AR (1971) The epidemiologic transition; a theory of the epidemiology of population change. Milbank Mem Fund Q 49:509–538

Putnam R (1995) Bowling alone: America's declining social capital. J Democr 6(1):65–78

Robine JM, Cheung SL, Le Roy S et al (2008) Death toll exceeded 70,000 in Europe during the summer of 2003. C R Biol 331(2):171–178

Satterthaite D (2000) Will most people live in cities? Br Med J 312(7269):1143–1145

Semenza JC (2003) The intersection of urban planning, art, and public health: the sunnyside piazza. Am J Public Health 93(9):1439–1441

Semenza JC (2005) Chapter 23 – Building healthy cities: a focus on interventions. In: Vlahov D, Galea S (eds) Handbook of urban health: populations, methods and practice. Springer, New York

Semenza JC, Krishnasamy PV (2007) Design of a health-promoting neighborhood intervention. Health Promot Pract 8(3):243–256

Semenza JC, March TL (2009) An urban community-based intervention to advance social interactions. Environ Behav 41(1):22–42

Semenza JC, Maty S (2007) Chapter 21 – Acting upon the macrosocial environment to improve health: a framework for intervention. In: Galea S (ed) Macrosocial determinants of population health. Springer, New York

Semenza JC, Menne B (2009) Climate change and infectious diseases in Europe. Lancet Infect Dis 9(6):365–375

Semenza JC, Nichols G (2007) Cryptosporidiosis surveillance and water-borne outbreaks in Europe. Euro Surveill 12(5):E13–E14

Semenza JC, Rubin HC, Falter KH et al (1996) Risk factors for heat-related mortality during the July 1995 heat wave in Chicago. N Engl J Med 335(2):84–90

Semenza JC, McCullough J, Flanders DW et al (1999) Excess hospital admissions during the 1995 heat wave in Chicago. Am J Prev Med 16(4):269–277

Sponken-Smith RA, Oke TR (1998) The thermal regime of urban parks in two cities with different summer climates. Int J Remote Sens 19:2085–2104

St Louis ME, Hess JJ (2008) Climate change: impacts on and implications for global health. Am J Prev Med 35(5):527–538

Stedman R (2002) Toward a social psychology of place: predicting behavior from place-based cognitions, attitudes, and identity. Environ Behav 34:561–581

Szreter S (2002) The state of social capital: bringing back in power, politics and history. Theory Soc 31(5):573–621

UNDESA Population Division (2007) Urban agglomerations 2007. http://www.un.org/esa/population/publications/wup2007/2007_urban_agglomerations_chart.pdf. Cited 17 April 2009

Uzzell D, Pol E, Badenas D (2002) Place identification, social cohesion, and environmental sustainability. Environ Behav 34:26–53

Vandentorren S, Suzan F, Medina S et al (2004) Mortality in 13 French cities during the August 2003 heat wave. Am J Public Health 94(9):1518–1520

Vorkinn M, Riese H (2001) Environmental concern in a local context: the significance of place attachment. Environ Behav 33:249–263

Watkins R, Palmer J, Kolokotroni M (2007) Increased temperatures and intensification of the urban heat island: implications for human comfort and urban design. Built Environ 33:85–96

Weisskopf MG, Anderson HA, Foldy S et al (2002) Heat wave morbidity and mortality, Milwaukee, Wisconsin, 1999 vs. 1995: An improved response? Am J Public Health 92(5):830–833

Wilkinson RG (1992) Income distribution and life expectancy. Br Med J 304:165–168

Younger M, Morrow-Almeida HR, Vindigni SM et al (2008) The built environment, climate change, and health: opportunities for co-benefits. Am J Prev Med 35(5):517–526

Ziersch AM, Baum FE, Macdougall C et al (2005) Neighborhood life and social capital: the implications for health. Soc Sci Med 60(1):71–86

Chapter 11
Public Health in Canada and Adaptation to Infectious Disease Risks of Climate Change: Are We Planning or Just Keeping Our Fingers Crossed?

Nicholas Hume Ogden, Paul Sockett, and Manon Fleury

Abstract Climate change is expected to increase the health risks for Canadians from infectious diseases from our environment, including vector-borne, water-borne, and food-borne diseases. Adaptation efforts will be important to reduce the impact of these risks. Public health systems are in place in Canada to control many disease risks but there are still knowledge gaps on, and modifications needed to, existing approaches to protecting the population from endemic diseases and new or emerging pathogens. This chapter addresses five key questions on whether public health is on track to helping communities adapt to changing risks. The questions address adaptation to disease risk of climate change by exploring the following: assessments of disease risks, methods for adaptation, responsibility, resources, and public action and societal will. Overall, with these increasing risks to the health of Canadians, all sectors of society will need to participate in the adaptive response, while federal, provincial, and community public health bodies will need to work together to identify and communicate risk and promote and coordinate adaptation responses.

N.H. Ogden (✉)
Zoonoses Division, Centre for Food-borne, Environmental & Zoonotic Infectious Diseases, Public Health Agency of Canada, Jeanne Mance Building, 200 Eglantine, Tunney's Pasture, AL 1906B, Ottawa, ON K1A 0K9, Canada
e-mail: Nicholas.ogden@phac-aspc.gc.ca

P. Sockett
Communicable Disease Control Division, Primary Health Care and Public Health Directorate, First Nations and Inuit Health Branch, Health Canada, Ottawa, ON, Canada
e-mail: paul_sockett@hc-sc.gc.ca

M. Fleury
Environmental Issues Division, Centre for Food-borne, Environmental & Zoonotic Infectious Diseases, Public Health Agency of Canada, Guelph, ON, Canada
e-mail: manon_d_fleury@phac-aspc.gc.ca

J.D. Ford and L. Berrang-Ford (eds.), *Climate Change Adaptation in Developed Nations: From Theory to Practice*, Advances in Global Change Research 42, DOI 10.1007/978-94-007-0567-8_11, © Springer Science+Business Media B.V. 2011

Keywords Climate change • Adaptation • Infectious diseases • Vector-borne diseases • Water-borne diseases • Food-borne diseases • Canada • Public health • Health • Disease risk

Introduction

Climate change is expected to increase the health risks of Canadians from extreme weather events and infectious diseases found in nature. Studies in Canada and elsewhere highlight the potential impact of climate change on food-borne, water-borne, and vector-borne diseases, for which environment and climate are important determinants of risk. These determinants vary from increased temperature resulting in increased survival of pathogens through changes in the ecology and geographic range of disease reservoir and vector species. Adaptation to increased risks requires modification of existing approaches to protecting the population from familiar diseases and identification of new or emerging pathogens. Adaptation approaches will likely entail both dedicated interventions such as water treatment and vector control, and changes in human behavior to reduce risk of exposure.

This chapter addresses five key questions on whether public health is on track to help communities adapt to changing risks: (1) Have we made appropriate assessments of disease risks for developing adaptation methods? (2) What methods are available for adaptation to changing risks from food-borne, water-borne, and vector-borne diseases? (3) What levels of jurisdiction are responsible for adaptation? (4) What key resources are needed to establish effective adaptation? (5) Is adaptation dependent on public action and to what extent is it a product of societal will?

In addressing these questions, we discuss both general and specific methods of adaptation to food-borne, water-borne, and vector-borne disease risks. We conclude with a pragmatic look at the prospects for public health to respond to increased infectious disease risks due to climate change.

Have We Made Appropriate Assessments of Disease Risks for Developing Adaptation Methods?

There is concern that climate change will drive the emergence or reemergence of infectious diseases via various mechanisms (Table 11.1). These processes could be impacted by declining investment in control programs related to diseases previously controlled (Gubler and Wilson 2005). Food-borne, water-borne, and vector-borne diseases likely pose the most immediate increased risk to human health from altered climate conditions as they are linked to environmental and climate factors.

Prioritization of infectious disease risk must account for uncertainty about the quality of data available, the likelihood and extent of climate change effects, projected impacts in terms of pathogenicity, projected numbers of cases, and the potential costs and benefits of adaptation approaches (Watson et al. 2005).

Table 11.1 Mechanisms of emergence of infectious diseases and suggested links to climate change

Mechanism	Example	Direct effect of climate change	Indirect effect of climate change	Possible evidence for climate change effects to date
1. Emerging ability to recognize or diagnose an infectious disease	All emerging infectious diseases	No	No	–
2. Emergence due to range expansion of endemic areas	Lyme disease and raccoon rabies in Canada; Bluetongue virus in the Mediterranean	Changed geographic footprint of suitable temperature/humidity for vector survival and bacterial or viral multiplication	Increased habitat suitability or reservoir host abundance	Bluetongue virus in the Mediterranean (Purse et al. 2005); Lyme disease in Canada (Ogden et al. 2008a); malaria in Kenya (Pascual et al. 2006)
3. Emergence due to accidental introduction	West Nile virus in North America; SARS in Canada; airport malaria worldwide	Enhanced survival of introduced vectors and pathogens	Increased abundance of vectors and pathogens in endemic areas; increased or changed human movements and migrations	No
4. Emergence due to evolution of new pathogenic microorganisms	Highly pathogenic avian influenza	No	Effects on ecological dynamics affecting fitness of different genetic variants	No
5. Reemergence of endemic disease and, for zoonoses, "spill-over" from animal to human hosts	Food-borne and water-borne diseases; hantavirus in Western United States; *E. coli* O157 in Canada	Increased temperatures affecting pathogen multiplication; heavy rainfall enhancing human–pathogen contact	Enhanced abundance of pathogens in the environment via effects on ecological dynamics	No

Studies in Canada have explored relationships between disease or vector occurrence and climate variables through qualitative assessments and simple statistical or mathematical models. These are then applied to climate scenarios to attempt to quantify possible effects of climate change on disease occurrence (Ontario Forest Research 2003; Fleury et al. 2006; Thomas et al. 2006; Ogden et al. 2005, 2006). There is considerable knowledge of drivers of endemic diseases and this knowledge may be sufficient to allow adaptation to changing degrees of risk. The relation between ecosystems, infectious diseases, and global climate change are less intuitive in developed countries where water treatment, vector control, and higher quality housing partly mitigate infectious disease threats (Greer et al. 2008). However, to be sure, we need further analyses on the complex interactions of climate (and extreme weather events) on microbial or vector replication and survival, the natural environment, and on the human–environmental interaction, which includes human population growth and movement, habitat disruption, and modification.

The need for broader assessment of climate change on infectious diseases is illustrated by recent assessments of how climate change may affect transmission of vector-borne diseases. Vector-borne diseases are transmitted by arthropods (ticks and flies), from human to human (often by mosquitoes, e.g., malaria, dengue), or from animal reservoirs (mostly wildlife) to humans (vector-borne zoonoses). The latter are transmitted by a variety of arthropod species: West Nile virus is transmitted from birds to humans by mosquitoes, Lyme disease is transmitted from rodents to humans by ticks, and Bartonellosis is transmitted from rodents by fleas.

There is a global debate about the extent to which climate change may affect vector-borne disease ecology and disease risk. The debate has centered on key human-to-human transmitted vector-borne diseases such as malaria and dengue, and this debate has significance for how seriously threats of vector-borne diseases are considered in developed countries such as Canada. The two opposing points of view are reviewed in Box 11.1.

Box 11.1 Vector-Borne Diseases and Climate Change: The Debate

1. Climate change will have a large impact on global vector-borne disease risks:

   Strengths: Inherent, known, links between vector biology and climate (such as temperature influences on the "extrinsic incubation period": Rogers and Randolph 2006) mean that cautious risk assessments may be advisable (Martens et al. 1995; Patz et al. 1998). *Weaknesses*: Assessments have been based on simplified mathematical models that are likely very inaccurate (by overestimation) at identifying where vector-borne disease risk occurs (Rogers and Randolph 2006). Simplified models ignore climate-independent determinants of risk including habitat, biodiversity, and factors unrelated to the environment such as resistance to antimalaria drugs and societal wealth and health (Reiter 2001; Reiter et al. 2004; Charron et al. 2008).

2. Climate change will have a small impact on global vector-borne disease risks:

   Strengths: Assessments are based on statistical models that associate current known vector-borne disease occurrence with climate, and sometimes habitat variables, to identify current and future environmental footprints for vector-borne diseases (Rogers and Randolph 2000, 2006).
   Weaknesses: Observed statistical observations do not necessarily imply causality, and the statistical models have attempted to identify an "ecological niche" when the distribution of human-to-human transmitted vector-borne diseases is likely partly determined by societal factors not currently investigated in statistical models (Berrang-Ford et al. 2009).

Of significance to the debate is that: (1) we have limited information on the real effects of climate, relative to other factors, on the biology of vectors and vector-borne disease, except in a few examples such as Lyme disease (Ogden et al. 2004, 2005, 2006), and on which to base quantitative assessments of climate change impacts on vector occurrence and vector-borne disease incidence; (2) efforts to synthesize existing data on vector and vector-borne disease occurrence and ecology have been limited to date; and (3) with the exception of Bluetongue virus (Purse et al. 2005), perhaps malaria in a focal region (Pascual et al. 2006), and Lyme disease vectors in southeastern Canada (Ogden et al. 2008a), our recent and current efforts to attribute climate and climate change effects, or effects of alternative factors, rest on imperfect data. Thus, we have limited ability to determine climate effects on vector-borne disease occurrence and to predict and identify any climate change effects. It would be expected, however, that estimation of climatic influences and prediction of climate change effects on vector-borne zoonoses would be more robust than for diseases maintained in a human–vector–human transmission cycle, as the former are relatively independent of the occurrence of human cases.

Similarly, it remains difficult to assess the true burden of illness from enteric pathogens. We have started addressing the impact of climate change on food-borne and water-borne diseases in Canada by linking precipitation and temperature to enteric diseases and water-borne disease outbreaks (Charron et al. 2004; Fleury et al. 2006; Thomas et al. 2006). However, transmission routes include person to person, water, food, and the environment. Understanding these dynamics is difficult because investigation of the source of individual cases is not routine (Mead et al. 1999) and enteric symptoms may relate to non-enteric causes (Powell et al. 2007). Systems to ensure food safety, from the farm, through processing, to retail and consumption are established in Canada. Since the Walkerton outbreak, where excess rainfall resulted in contamination of the water supply, there have been advances in water safety regulation and water quality monitoring in Canada (Auld et al. 2004).

There are, however, key knowledge gaps on: (1) the ecology and epidemiology of vector-borne, food-borne, and water-borne diseases; (2) understanding of how human social and societal factors interact with the ecology of these diseases; and (3) synthesis of existing information on disease ecology, epidemiology, and societal factors that convert potential risks into real risks for public health, and investigations on how climate change may indirectly affect these risks via effects on socioeconomic conditions. These gaps hamper precise quantified risk assessments and we argue that, at present, we must use available risk assessment methods to identify potential risks, but interpret the assessments using the precautionary principle.

What Methods Are Available for Adaptation to Changing Risks from Food-Borne and Vector-Borne Diseases?

To meet the public health challenge of climate-change-induced changes to the distribution and frequency of endemic diseases, emergence of new diseases, and importation of diseases and disease vectors into Canada, a coordinated response of public health and environmental health infrastructures will be essential. Four key elements to response are presented in Box 11.2.

Box 11.2 Elements of Response
The key elements to this response involve:

1. Disease surveillance capable of detecting the geographic and temporal occurrence of emerging and reemerging diseases
2. Surveillance data analysis and communication
3. Networking amongst public health jurisdictions
4. Effective interventions to prevent disease or reduce incidence

---

Timely detection of disease events implies a surveillance infrastructure capable of identifying both routine and unusual events via regular data analysis. Surveillance programs in Canada are comprehensive for common food-borne and water-borne diseases and some vector-borne zoonoses. They include: direct reporting of laboratory confirmed infections at the local and provincial levels; national, provincial, and territorial reporting of notifiable diseases and selected laboratory identifications of enteric pathogens; as well as provision of laboratory reference services both provincially and nationally. In some cases, nontraditional indicators of health and innovative approaches may be used in disease detection programs. For example,

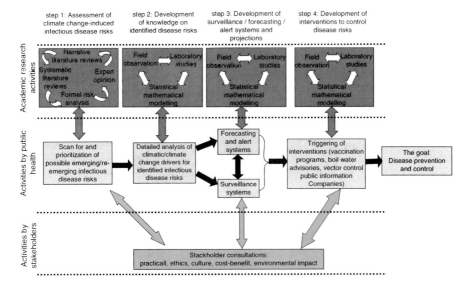

Fig. 11.1 A flow chart of adaptation activities required to identify and respond to emerging infectious disease threats driven by climate change. *Arrows* indicate the flow of adaptation activities from initial identification of climate-change-driven disease risks through knowledge acquisition and development of surveillance, forecasting or alert systems and the interventions they trigger to prevent or control disease, as well as collaborations with science research institutions and stakeholders

pharmacy data on over-the-counter sales, emergency room visits, and Telehealth calls can provide early warning for local disease events (Edge et al. 2004; Edge et al. 2006). "Environmental" surveillance to detect infection in sentinel animals or vectors and forecasting (Teklehaimanot et al. 2004; Bouden et al. 2008) to detect disease risks before they occur in humans will be crucial to minimizing risks from emerging zoonoses. Canada has relatively comprehensive programs for monitoring trends in food-borne and water-borne diseases and some vector-borne zoonoses. However, as in other developed countries, many diseases are underreported and national statistics represent trends in disease patterns rather than incidence or prevalence of disease (Flint et al. 2005; Majowicz et al. 2005). While current surveillance approaches in Canada successfully report trends for common diseases, these programs may be less effective for monitoring rarer events, new diseases, and diseases not captured in national statistics.

Methods to detect new or unusual diseases should be based on detailed comprehension of pathogen ecology and yield defensible threshold values for risk which trigger interventions. Ideally, feedback in the system will incorporate analysis of surveillance data and comparison with forecasts/projections, and evaluation of the efficacy of interventions. This "ideal" system is summarized in Fig. 11.1. Current initiatives in Canada, such as the C-EnterNet pilot study in Waterloo, Ontario, integrate current information on human enteric disease with levels of pathogen exposure from food, animal, and water sources to detect changes in trends and to contribute to the risk management process (Public Health Agency of Canada 2007).

Sharing data and information within networks and between jurisdictions facilitates risk identification and contributes to defining the scope and spread of risk. The public health response to West Nile virus in Canada provides an example of how networking of surveillance information is feasible in large countries with multiple jurisdictions and mandates. Although the technology exists for rapid information sharing, privacy and confidentiality requirements challenge surveillance objectives. The range of diseases that climate change may affect suggests that these issues must be addressed thorough planning of data information systems and by identifying the needs and information required for timely data sharing. Timely analysis of surveillance data to identify where and when disease risks are occurring and presentation of these analyses in a way meaningful for public health professionals are essential to effective disease control. Thus, capacity in surveillance, epidemiology, risk modeling, geographic information systems (GIS), and communications will be important to adaptation.

Interventions to prevent or limit many predictable disease risks in Canada include: water treatment, food safety through hazard control, and vector-borne disease control programs. Regulation of imported foods, animals, and animal products also limits the potential for the introduction and establishment of animal diseases. Regulations and guidelines are relatively easily adapted to changing disease risks but imply an established infrastructure to monitor their effectiveness and to respond when issues arise. Canada's regulatory requirements are designed to protect human and animal health interests within an international travel and trading environment. This includes the import–export requirements of trading partners and international obligations of reporting and control of diseases under the World Health Organization (WHO) International Health Regulations and the World Trade Organization (WTO) Sanitary and Phytosanitary (SPS) Measures (WHO 2005).

Some risks may require development of new interventions. For example, capacity to respond effectively to tick-borne zoonoses is limited and under development (Piesman and Eisen 2008). For emerging diseases, information for medical professionals and the public will be important in reducing human exposure and ensuring rapid diagnosis and treatment (Public Health Agency of Canada 2005; Ogden et al. 2008b). Furthermore, Canada's experience with West Nile virus emphasized the need for increased availability of expertise in entomology and vector-borne disease control and demonstrated the need to develop capacity in public health. For water-borne diseases there is a need to enhance watershed management, and update water treatment facilities and sewer systems built to cope with past rainfall projections that may not have enough capacity for future extreme weather events linked to climate change.

What Levels of Jurisdiction Are Responsible for Adaptation?

In Canada, national standards and policy and regulatory frameworks focus on supporting safe food supplies, preventing the importation of contaminated foods, and maintaining the health of economically important animals. Enforcement and implementation of much of these regulations are delivered by the provincial, territorial, and municipal levels. Changes in public health risks resulting from climate change will therefore impact multiple levels in Canada's political infrastructure. However, climate impacts will vary across the country and some effects may be relatively local. For these reasons, the municipality is likely to play a major role in adaptation alongside provincial or territorial and national authorities (Ebi and Semenza 2008). This implies that local health authorities recognize the issue and make adaptation to changing disease risks a priority. At the provincial and municipal levels, these responsibilities can be addressed by multiple sectors such as agriculture, health, sustainable development, and natural resources. At the national level, the organizations likely to play a major role in adaptation include federal departments and agencies with responsibilities for agriculture and food, environment, and public health. The Canadian response to West Nile virus offers an interesting model in which initial federal government leadership to address the incursion of this new infectious disease in North America has devolved to provincial and municipal decision-making surveillance, and prevention and control activities. This approach permits local needs and concerns to be addressed, including the acceptability of some measures (see Box 11.3).

Box 11.3 Case Study of West Nile Virus
The arrival of West Nile virus in North America in 1999 and the subsequent spread of the virus west across the continent provided a model for future adaptation to an emerging infectious disease. The federal government response was coordinated across three departments, which linked surveillance, blood safety, and communications activities both horizontally across departments and vertically with provinces and territories and also linked internationally with the United States and the WHO. A national multi-jurisdictional collaboration was developed to monitor for dead corvids and mosquitoes infected with West Nile virus. The monitoring program itself was managed by a wildlife health NGO, the Canadian Cooperative Wildlife Health Centre, which in turn coordinated diagnostic and surveillance activities within Canada's veterinary colleges. In the absence of a human vaccine, communications advice on limiting exposure to infected mosquitoes continues to be a major component of the control and prevention activity. Information on virus activity has played a major role in local decision-making about mosquito control and targeting of health protection advice, and over time much of the monitoring and control activity has devolved to the provincial and municipal levels.

In responding to the impacts of climate change, the importance of nongovernmental organizations (NGOs) should not be underestimated. Whilst these organizations have no mandate to develop policy and implement regulations, they play important roles. For example, university research programs provide information essential to understanding the science related to climate change. National professional bodies such as the Canadian Public Health Association help develop public health and research priorities, can improve both professional and public awareness of issues, and provide an independent scientific opinion to decision-makers. The Canadian Red Cross plays a major role in responding to severe events and disasters. Industry-based associations promote response to climate-linked issues through improving awareness, development of industry-based guidelines and codes of practice. Finally, public advocacy groups play a role in identifying issues and scrutinizing industry and government response to problems.

What Key Resources Are Needed to Establish Effective Adaptation?

Climate change has major implications for the awareness, preparedness, analysis, and action responsibilities of public health. Awareness requires a clear understanding of the risks, while preparedness, analysis, and action mean development of a surveillance and response strategy that addresses the anticipated evolving risk from infectious diseases with a sustainable long-term program (Frumkin and McMichael 2008). These require public buy-in to actions that may impact personal lifestyle at one extreme, to future municipal and provincial resource allocation and planning at the other. Communication with the public and across organizational boundaries of both risks and consumer incentives to address them will be essential in obtaining support for development of cross-sector adaptive strategies (Semenza et al. 2008).

Public health preparedness has two components: ability to assess immediate and evolving risk to communities and populations and the ability to respond to emergency events. In the context of climate change, this includes changes that will occur over decades as well as sudden disaster or near disaster events. Canada is relatively well prepared for disaster scenarios with organized multilevel public health and disaster response infrastructure and planning. However, Canada is just beginning to address the longer term implications of climate change in health, including infectious diseases. Action to date has focused on reporting the implications of climate change and investment in developing approaches to adaptation (Charron et al. 2008). However, municipalities need to be prepared for projected future climate change effects (on an unprecedentedly long timescale) that integrate other pressures on infrastructure, such as population expansion and demand for potable water that is protected from contamination in the most environmentally sustainable way.

Key to development of adaptive approaches will be the ability to predict and detect current, emerging, or new health events. These requirements are predicated on the ability to access appropriate data and information and to apply appropriate analysis to provide clear understanding of the current incidence or prevalence of a disease and to model future trends.

Public health action to address infectious disease risk has four important requirements. First is the effective and balanced communication of climate change-related health issues, which identifies the type of risk, its likelihood to occur, the time period over which it may occur, and the alternative actions of the individual or society which may be needed to reduce risk.

Second is to promote the value and strength of public health and population health approaches to identifying and addressing risks. By addressing such risks, the whole population benefits from broadly applied risk reduction strategies. Some interventions, such as the application of population-wide vaccine programs to prevent or control infectious diseases have been in place now for several decades. The effectiveness of such strategies is amply demonstrated by temporal reductions in such diseases as measles, rubella, diphtheria, and pertussis.

Third is the utilization of accurate and timely information to inform public health policy. As with specific public health interventions such as vaccination programs, Canada has a long history of policy activity designed to protect individuals and populations from harm. This can be an adaptive process that recognizes new expectations and the ability to respond to new threats. For new and emerging diseases, challenges related to food supply globalization and rapid travel to exotic locations require policy and regulation capable of responding to new risks while at the same time accommodating economic realities.

Last, the promotion of multidisciplinary and collegial approaches to the analysis of data and information from multiple sources, and the assessment and analysis of risk will facilitate development of policy and action. This component depends on the availability of essential and appropriate data, opportunity and willingness to share data and information, to form strategic research alliances, and to develop innovative use of forecasting, surveillance, and intervention.

These responses are predicated on the availability of appropriate resources and leadership, and where the responsibilities for these lie remains to be decided. Clearly, essential resources extend beyond purely financial components to include technologies, infrastructure, expertise, research, multi-sectoral collaboration, and corporate and individual responsibility.

Is Adaptation Dependent on Public Action and to What Extent Is It a Product of Societal Will?

The most recent report of the Intergovernmental Panel on Climate Change (IPCC) confirms that health impacts linked to climate change are being detected world-wide and that some regions and populations (e.g., the young, elderly, aboriginal populations, and those living in polar and coastal regions) will experience greater impacts (IPCC 2007, p. 104). In general, public health professionals recognize the importance of developing adaptive capacity and approaches to adaptation now, but investment in public health preparedness for climate change-related risks will depend on public and societal support. Thus, the ultimate test of awareness of the impacts of climate change and their challenge to health may be a societal response that demands effective public health action. We suggest four requirements for support for public health action on climate change in Box 11.4.

Box 11.4 Four Requirements for Support for Public Health Action on Climate Change

1. Promotion of the values and strengths of population-based approaches to surveillance and interventions to preserve and improve health through concerted societal action. The public health and communicable diseases arena has a strong history of adaptation to infectious disease risk, and a multi-sectoral armament of responses involving individual and population-based action.
2. A multi-sectoral, multidisciplinary, collegial approach to analysis, assessment, and response to complex public health threats.
3. Public health policy and communications that are informed and directed by the best available scientific evidence.
4. Effective communications on climate change-related health issues that are scaled according to the audience, whether health professionals, the public, vulnerable groups, etc., and should provide culturally appropriate, transparent, and balanced information about health risk. The potential for communication approaches to promote changes in behavior at the individual and population levels should not be underestimated (Maibach et al. 2008).

The key elements of a societal approach to adaptation include: consideration of the need for accurate assessment of potential threats; recognition of the timelines over which health threats will occur; the social and economic costs and benefits of doing something, or nothing; and consideration of the societal impacts of adaptation (Roberts 2006; Stern 2007). Stern (2007) makes three key points about adaptation that are crucial to understanding an adaptation framework: (1) adaptation is essentially unavoidable, (2) adaptation cannot solve the impacts of climate change but will contribute to limiting those impacts, and (3) the benefits of adaptation will be largely local and can have relatively rapid results.

This chapter has sought to clarify some of the issues posed by the current and future predicted impacts of climate change on infectious disease risks, particularly food-borne, water-borne, and vector-borne agents. Addressing these risks through adaptation approaches is a broad objective of public health which will engage all levels of government, research, industry, and the public in making choices. Government response will inevitably promote a horizontal approach incorporating issues of trade, travel, and animal health in assessing human health and economic risks. Not all choices will be financially costly but may require individual lifestyle changes, such as regular use of insect repellent or education of doctors to include emerging diseases such as Lyme disease in their differential diagnoses. Other strategies would be more expensive with benefits realized over decades, as in the case of safe drinking water infrastructure investment. In the longer term reduced vulnerability to climate change effects, including altered risks to infectious diseases, may have more to do with improved "social and material conditions of life," and leveling of "inequalities within and between populations" at risk (McMichael and Kovats 2000). What is clear is that all sectors of society will be involved in the response and that government has a leadership role in identifying and communicating risk and promoting and coordinating, and sometimes funding, response.

References

Auld H, MacIver D, Klaassen J (2004) Heavy rainfall and waterborne disease outbreaks: the Walkerton example. J Toxicol Environ Health A 67(20–21):1879–1887

Berrang-Ford L, MacLean JD, Gyorkos TW et al (2009) Climate change and malaria in Canada: a systems approach. Interdiscip Perspect Infect Dis. doi:10.1155/2009/385487

Bouden M, Moulin B, Gosselin P (2008) The geosimulation of West Nile virus propagation: a multi-agent and climate sensitive tool for risk management in public health. Int J Health Geogr 7:35

Charron DF, Thomas MK, Waltner-Toews D et al (2004) Vulnerability of waterborne diseases to climate change in Canada: a review. J Toxicol Environ Health A 67:1667–1677

Charron DF et al (2008) The impacts of climate change on foodborne, waterborne, rodent-borne and vector-borne zoonoses. In: Séguin J (ed) Human health in a changing climate: a Canadian assessment of vulnerabilities and adaptive capacity. Health Canada, Ottawa

Ebi KL, Semenza JC (2008) Community-based adaptation to the health impacts of climate change. Am J Prev Med 35(5):501–507

Edge VL, Pollari F, Lim G et al (2004) Syndromic surveillance of gastrointestinal illness using over-the-counter sales: a retrospective study of waterborne outbreaks in Saskatchewan and Ontario. Can J Public Health 95(6):446–450

Edge VL, Pollari F, Ng L-K et al (2006) Syndromic surveillance of norovirus using over the counter sales of medications related to gastrointestinal illness. Can J Infect Dis Med Microbiol 17(4):235–241

Fleury M, Charron DF, Holt JD et al (2006) A time series analysis of the relationship of ambient temperature and common bacterial enteric infections in two Canadian provinces. Int J Biometeorol 50(6):385–391

Flint JA, Van Duynhoven YT, Angulo FJ et al (2005) Estimating the burden of acute gastroenteritis, foodborne disease and pathogens commonly transmitted by food: an international review. Clin Infect Dis 41:698–704

Frumkin H, McMichael AJ (2008) Climate change and public health thinking, communicating, acting. Am J Prev Med 35:403–409

Greer A, Ng V, Fisman D (2008) Climate change and infectious diseases in North America: the road ahead. Can Med Assoc J 178(6):715–722

Gubler DJ, Wilson ML (2005) The global resurgence of vector-borne diseases: lessons learned from successful and failed adaptation. In: Ebi KL, Smith JB, Burton I (eds) Integration of public health and adaptation to climate change: lessons learned and new directions. Taylor and Francis, London

IPCC (2007) Climate change 2007: synthesis report – contribution of Working Group I, II and III to the Fourth Assessment Report of the Intergovernmental Panel on Climate Change. IPCC, Geneva

Maibach EW, Chadwick A, McBride D et al (2008) Climate change and local public health in the United States: preparedness, programs and perceptions of local public health department directors. PLoS ONE 3(7):e2838. doi:10.1371/journal.pone0002838

Majowicz SE, Edge VL, Fazil A et al (2005) Estimating the under-reporting rate for infectious gastrointestinal illness in Ontario. Can J Public Health 96(3):178–181

Martens WJ, Niessen LW, Rotmans J et al (1995) Potential impact of global climate change on malaria risk. Environ Health Perspect 103:458–464

McMichael AJ, Kovats SR (2000) Climate change and climate variability: adaptations to reduce adverse health impacts. Environ Monit Assess 61(1):49–64

Mead PS, Slutsker L, Dietz V et al (1999) Food-related illness and death in the United States. Emerg Infect Dis 5(5):607–625

Ogden NH, Lindsay LR, Beauchamp G et al (2004) Investigation of relationships between temperature and developmental rates of tick Ixodes scapularis (Acari: Ixodidae) in the laboratory and field. J Med Entomol 41:622–633

Ogden NH, Bigras-Poulin M, O'Callaghan CJ et al (2005) A dynamic population model to investigate effects of climate on geographic range and seasonality of the tick *Ixodes scapularis*. Int J Parasitol 35:375–389

Ogden NH, Maarouf A, Barker IK et al (2006) Climate change and the potential for range expansion of the Lyme disease vector *Ixodes scapularis* in Canada. Int J Parasitol 36:63–70

Ogden NH, St-Onge L, Barker IK et al (2008a) Risk maps for range expansion of the Lyme disease vector, Ixodes scapularis, in Canada now and with climate change. Int J Health Geogr 7:24

Ogden NH, Artsob H, Lindsay LR et al (2008b) Lyme disease: a zoonotic disease of increasing importance to Canadians. Can Fam Physician 54:1381–1384

Ontario Forest Research Institute (2003) A synopsis of known and potential diseases and parasites associated with climate change – forest research information paper no. 154. Ontario Government, Ministry of Natural Resources, Sault Ste. Marie

Pascual M, Ahumada JA, Chaves LF et al (2006) Malaria resurgence in the East African highlands: temperature trends revisited. Proc Natl Acad Sci USA 103:5829–5834

Patz JA, Martens WJ, Focks DA et al (1998) Dengue fever epidemic potential as projected by general circulation models of global climate change. Environ Health Perspect 1106:147–153

Piesman J, Eisen L (2008) Prevention of tick-borne diseases. Annu Rev Entomol 53:323–343

Powell N, Benedict H, Beech T et al (2007) Increased prevalence of gastrointestinal symptoms in patients with allergic disease. Postgrad Med J 83:182–186

Public Health Agency of Canada (2005) Managing patients with West Nile virus: guidelines for health care providers. Can Commun Dis Rep 31S4:1–10

Public Health Agency of Canada (2007) Canadian National Enteric Pathogen Surveillance System (C EnterNet) 2006. Government of Canada. http://www.phac-aspc.gc.ca/publicat/2007/c-enternet06/areport06-eng.php. Cited 26 May 2009

Purse BV, Mellor PS, Baylis M (2005) Climate change and the recent emergence of bluetongue in Europe. Nat Rev Microbiol 3:171–181

Reiter P (2001) Climate change and mosquito-borne disease. Environ Health Perspect 109 (Suppl 1):141–161

Reiter P, Thomas CJ, Atkinson PM et al (2004) Global warming and malaria: a call for accuracy. Lancet Infect Dis 4:323–324

Roberts JA (ed) (2006) The economics of infectious disease. Oxford University Press, New York

Rogers DJ, Randolph SE (2000) The global spread of malaria in a future, warmer world. Science 289:1763–1766

Rogers DJ, Randolph SE (2006) Climate change and vector-borne diseases. Adv Parasitol 62: 345–381

Semenza JC, Hall DE, Wilson DJ et al (2008) Public perception of climate change: voluntary mitigation and barriers to behaviour change. Am J Prev Med 35(5):479–487

Stern N (2007) The economics or climate change. Cambridge University Press, Cambridge

Teklehaimanot HD, Schwartz J, Teklehaimanot A et al (2004) Weather-based prediction of *Plasmodium falciparum* malaria in epidemic-prone regions of Ethiopia II: weather-based prediction systems perform comparably to early detection systems in identifying times for interventions. Malar J 3:44

Thomas KM, Charron DF, Waltner-Toews D et al (2006) A role of high impact weather events in waterborne disease outbreaks in Canada, 1975–2001. Int J Environ Health Res 16(3):167–180

Watson RT, Patz J, Gubler DJ et al (2005) Environmental health implications of global climate change. J Environ Monit 7(9):834–843

WHO (2005) International Health Regulations, 2nd edn. World Health Organization, Geneva. http://www.who.int/csr/ihr/en/. Cited 26 May 2009

Chapter 12
Climate Change, Water-Related Health Impacts, and Adaptation: Highlights from the Swedish Government's Commission on Climate and Vulnerability

Elisabet Lindgren, Ann Albihn, and Yvonne Andersson

Abstract The Swedish Government's Commission on Climate and Vulnerability (2005–2007) assessed local vulnerability to climate change and adaptation needs within the main sectors of society for the periods 2011–2040, 2041–2070, and 2071–2100. The workgroup on Health (human and animal) worked closely with that on water. This chapter presents the main findings of the section on water-related health consequences from climate change. Heavy rain, floods, landslides, and increases in water flows may directly or indirectly impact infrastructure, buildings, public services, water sources, etc., which could cause health consequences ranging from deaths, injuries, and outbreaks of infectious diseases, to exposure to toxic compounds, and allergic reactions.

The Commission's recommendations to different governmental authorities on issues connected to water-related health consequences are presented, together with additional adaptive measures that were suggested by the Health and Water workgroups. The value of incorporating the added risks of climate change into existing programs and planning was emphasized. In response to the recommendations, the government has now created new funding opportunities for climate change-related scientific research, and for knowledge building within its agencies. At the local level,

E. Lindgren (✉)
Division of Global Health / IHCAR, Department of Public Health, Karolinska Institute, SE-171 77 Stockholm, Sweden
e-mail: Elisabet.Lindgren@ki.se

A. Albihn
Section of Environment and Biosecurity, National Veterinary Institute, Uppsala, Sweden
e-mail: ann.albihn@sva.se

Y. Andersson
Department of Epidemiology, Swedish Institute for Infectious Disease Control, Stockholm, Sweden
e-mail: yvonne.andersson@smi.se

J.D. Ford and L. Berrang-Ford (eds.), *Climate Change Adaptation in Developed Nations: From Theory to Practice*, Advances in Global Change Research 42, DOI 10.1007/978-94-007-0567-8_12, © Springer Science+Business Media B.V. 2011

some county and municipality boards have themselves initiated vulnerability and adaptation assessments focusing on local conditions, as bases for further planning and responses.

Keywords Climate change • National assessment • Adaptation • Vulnerability • Health • Water runoff • Floods • Flooding • Infectious diseases • Epidemic outbreaks • Water-borne diseases • Food-borne diseases • Water treatment • Disaster management • Medical responses • Government agencies

The Swedish Government's Commission on Climate and Vulnerability

A severe storm hit southern Sweden in January 2005. It caused the most extensive storm-felling of trees in at least a 100 years, and damaged almost 30,000 km of power lines, leaving 660,000 households without electricity, for up to 45 days in some rural locations. The direct costs were estimated to be about SEK 21 billion (2.23 billion €). Then in June 2005, the Swedish Government appointed The Commission on Climate and Vulnerability (hereafter referred to as "the Commission") to assess Swedish society's vulnerability to extreme weather events and long-term climate change, and to evaluate the need, and costs, for various sectors of society to adapt to a changing climate. Areas of particular interest were infrastructure, agriculture, fishing and forestry, water supply and sewage systems, and human and animal health and biodiversity. The full report was released in October 2007 (Ministry of the Environment 2007).

Workgroup 3 consisted of two subgroups: one on health (human and animal) issues, and one on water resources and management (Box 12.1). Parts of the work on health issues were done in collaboration between the two subgroups.

Box 12.1 Vulnerability Assessments

Sector- or area-specific analyses were carried out within three main working groups, plus subgroups to these:

1. Technical infrastructure, spatial planning, and buildings
2. Agriculture, forestry, and the natural environment
3. Health and water resources

The working groups included participants with expert knowledge from central and regional agencies, municipalities, businesses and organizations, as well as research institutions.

The Swedish Government's Commission on Climate and Vulnerability, 2005–2007

The Health workgroup concluded (Lindgren et al. 2008) that the main climate change-related concerns for human health in Sweden are: (1) consequences due to heat waves (e.g., indoor temperatures may soar depending on current adaptation of buildings to summer heat); and (2) changes in the risk of infectious diseases. The latter includes changes in the geographical distribution, the seasonality, and the incidence of several diseases transmitted by rodents and arthropods, such as ticks transmitting Lyme borreliosis and tick-borne encephalitis, but also changes in the frequency and intensity of outbreaks of food- and water-borne diseases, and the risk of new diseases to become established in Sweden.

The health report (Lindgren et al. 2007) includes a section that focuses on water-related health impacts in a much broader context than is usually done. This chapter summarizes the main findings of this section, including recommended adaptive strategies and measures that will directly or indirectly affect the risk of water-related health consequences.

Changes in precipitation, water runoff, floods, and landslides will have various effects on the environment as well as on different sectors of society. Some of these effects may in turn, directly or indirectly, cause consequences for human health. This chapter presents an overview of the background knowledge, the methods used for the vulnerability assessments, and the approaches taken to evaluate which adaptive measures and strategies are needed to counteract, or at least lessen the burden of water-related health impacts of climate change in Sweden.

Methods

The regional climate models used were developed by the Rossby Centre at the Swedish Meteorological and Hydrological Institute (SMHI) (e.g., Jones et al. 2004; Kjellström et al. 2005). They were based on the Intergovernmental Panel on Climate Change (IPCC) scenarios A2 and B2 for greenhouse gas emissions (IPCC 2000), and scaled down from two global climate models, HadAM3H by the Hadley Centre and ECHAM4/OPYC3 by the Max Planck Institute for Meteorology, for the following time frames: 2011–2040, 2041–2070, and 2071–2100. The reference period was 1961–1990. In consultation with various internal and external experts, 40 specific climate indices were developed as the foundation for assessing the future vulnerability within the chosen sectors of society. A total of over 10,000 climate maps showing the development of the indices were then made.

The Health workgroup based their theoretical risk and vulnerability projections of how climate change may cause health consequences in different regions of Sweden in the coming decades on a combination of factors:

1. Current conditions of interest. These depended on geographical location (county-level), and included demography, epidemiology, prevalence of disease vectors, land cover and land use, ecosystems, infrastructure, health care sector, disaster planning and responses, etc.

2. Known relationships between different climate variables and health effects/ diseases. The scientific literature was reviewed and evaluated, in conjunction with empirical experiences.
3. Outcomes of different climate scenarios. The local scenarios were based on climate variables and specific indices of interest for the assessments, e.g., monthly water and air temperatures, monthly precipitation, length of the vegetation season, date of the first frost day, number of days with snow cover, maximal precipitation during a 7-day period, number of days with precipitation greater than 25 mm, water runoff, etc.

Predicted Changes in Regional Climate

Sweden is situated between 69.4°N and 55.2°N. The proximity to the North Atlantic and the predominantly south-westerly to westerly winds give it a less harsh winter climate, given its latitude. The climate ranges from cold temperate to warm temperate going from north to south, with polar climate at the highest altitudes in the north, and semi-arid conditions on the island of Öland in the south-east. In general, the western regions of the country receive more precipitation than the east.

Temperatures are predicted to increase more in Sweden and Scandinavia than the global mean, with an average rise in annual mean temperature by 3–5° by 2071–2100 in comparison with the period 1961–1990, according to the regional models by the Rossby Centre. Winter temperature may increase by 7° in northern Sweden. The length of the vegetation season may increase from 6 to 7 months to as much as 10–11 months in Stockholm County by the end of this century. The climate in the Stockholm area in the 2080s will, in terms of temperature, resemble the climate in northern France today.

Precipitation is projected to increase during the autumn, winter, and spring seasons in most of the country, with an increase in the number of days with heavy precipitation. Summers will become drier, except for the northern regions. However, local heavy rainfall and downpours, which mostly occur during the summer months, will increase in intensity throughout the country. Runoff will increase in most parts of the country. In the west, high flows with a return period averaging 100 years, known as 100 year flow, will increase sharply. Spring floods will, however, decrease as a result of less snow cover. The risk of landslides will increase in known risk areas.

Water-Related Health Consequences

Deaths from extreme weather-related events like flash floods and landslides are not very likely to occur in Sweden. However, the Health workgroup identified a range of possible health consequences that may be linked to changes in water flows and water-related events. They are presented in Table 12.1. Extreme precipitation events

Table 12.1 Water-related health consequences of climate change

	Chain of causation	Health consequences
Humans (direct)	Flooding Landslides	Injuries Death (from trauma, drowning) Psychological effects: acute and from living in disaster prone areas
Transport sector	Damage to roads and railways Interruption of public services	Injuries Death Aggravation of disease
Energy plants and electricity supply	Prolonged power outages caused by landslides or flooding of electricity substations and plants	Water-borne disease outbreaks Food-borne disease outbreaks Aggravation of diseases
Industrial plants, service stations, and urban environments	Leakage of toxic compounds into soil and water sources due to increases in runoff and floods Contaminated standing water after floods	Infectious disease outbreaks Exposures to toxic compounds Skin infections Systemic diseases Exposures to toxic compounds
Buildings	Water leakages, including sewage and stormwater into basements	Allergy Respiratory disorders Infectious diseases Exposures to toxic compounds
Water treatment plants and water pipes	Floods, landslides, and increased water runoff: Flooding of treatment plants. Back flush of sewage into water system. Water pipe fractures. Contamination of water sources Discoloration (humus) leading to decreases in water intake during summer months	Water-borne disease outbreaks Intoxications Dehydration Kidney stones
Sewage, stormwater, and urban runoff	Leakages into soil, arable land, pastures, water sources, water pipes, contamination of agricultural products	Infectious disease outbreaks Exposures to toxic compounds
Land	Leakages from landfills (toxic compounds and microorganisms) Increased water runoff in soil	Exposures to toxic compounds Infectious disease outbreaks Infectious disease outbreaks Exposures to toxic compounds

(continued)

Table 12.1 (continued)

	Chain of causation	Health consequences
Agriculture and animal sectors	Leakages of micro-organisms and chemicals (pesticides, fertilizers) from agriculture, and from the animal sector (pastures, stables, dung heaps, etc.)	Infectious disease outbreaks Exposure to toxic compounds
Ecosystems	Long-term changes in water flows	Changes in prevalence of zoonoses and other vector-borne diseases
	Standing water after flooding	Insect invasions/allergy to bites Insect-borne diseases Leptospirosis

Source: Modified from the health report of The Swedish Government's Commission on climate and vulnerability (Lindgren et al. 2007)

may cause damage to infrastructure, the transport sector, and buildings, which in turn may lead to a range of health consequences depending on local conditions. Tap water is the major source of drinking water in Sweden due to its high quality. Water quality and availability may be affected negatively in several ways. Operations of water treatment plants may be interrupted, water pipes fractured, water sources and tap water contaminated, etc.

Heavy precipitation and increased flows in watercourses, as well as raised and variable groundwater levels, will increase the risk of landslides in current risk areas. Floods and landslides may damage buildings and infrastructure such as roads and power plants. Power outages may indirectly cause a range of health hazards, caused by shutdown of water treatment plants, cold storages for food, heating/cooling systems, etc. Road damages may interrupt ambulance car and other services, and leave vulnerable groups stranded.

Local downpours causing surface water and sewer systems to flood are already a problem in parts of Sweden today. Problems with basements being flooded and sewage being discharged will become even more serious in the future. Repeated flooding and increased precipitation and humidity may increase the risk of respiratory disorders and mould allergies. Increased interest in lakeside living will mean more homes being built in areas threatened by floods.

Increased water runoff may pick up contaminants, which in turn may contaminate drinking water sources, as well as water courses used for recreational and irrigation purposes. Such contaminants could consist of: petroleum and toxic compounds from industrial areas and urban runoff; pesticides or fertilizers from the agriculture sector; pathogens from the animal sector; and soil pathogens.

Epidemic outbreaks of water-borne diseases have been reported after heavy rains and flooding from both Europe and North America (Box 12.2). In Sweden, outbreaks of water-related infections are underreported as not all of the diseases are notifiable by law, and people do not seek medical attention for minor diarrheal problems, but even if they do laboratory tests are seldom undertaken. However, some episodes have been brought to attention. In 1980, 82% of the population in a southern municipality in Sweden got diarrhoeal symptoms after severe flooding caused leakages of contaminated water into the drinking water system. In 2005 an epidemic outbreak occurred in the south western parts due to salad that had been contaminated with verocytotoxin-producing *Escherichia coli* (VTEC). Prior to packaging, the salad had been watered with water taken from a nearby stream. Heavy rain had caused runoff of pathogens from nearby livestock into the stream (SMI 2006).

Box 12.2 Examples of Epidemic Outbreaks of Water-Borne Diseases Due to Heavy Rain and Flooding

- Walkerton, Canada, 2000: 48% of the population got sick after heavy rain caused contaminated water runoff (with *Escherichia coli* /VTEC and *Campylobacter jejuni*) from stables to flow into nearby groundwater source (Auld et al. 2004).
- Milwaukee, United States,1993: 67% of the population fell ill from drinking water contaminated with *Cryptosporidium parvum* due to repeated heavy rains and spring floods that "overloaded" the water treatment system. The parasite is not sensitive to conventional chlorine treatment (MacKenzie et al. 1995).
- Central Europe, 2002: following severe floods, outbreaks of *Shigella* infection were reported (Tuffs and Bosch 2002).
- Russia: outbreaks of diarrhoeal diseases and Hepatitis A have been reported on several occasions in connection with flooding (Kalashnikov et al. 2003).
- Finland: outbreaks of water-borne infections caused by caliciviruses, rotaviruses, and *Campylobacter jejuni* have been reported after high spring floods (Kukkula et al. 1997; Miettinen et al. 2001).

Bloodsucking insects may flourish in the aftermath of heavy rain and floods (Kriz 1998; Hubalek et al. 2004), and contaminated standing water can become hazardous to health. Outbreaks of leptospirosis are not uncommon worldwide after flooding, and are usually caused by exposure to water contaminated with the urine of infected animals. Cattle, pigs, horses, dogs, rodents, and various wild animals are the reservoirs of the pathogen. Outbreaks have been reported on several occasions in Europe (e.g., Zitek and Benes 2005).

In a long-term perspective, changes in precipitations patterns and seasonal climate will cause alterations of local ecosystems and biodiversity. This may open up opportunities for disease-transmitting vectors and reservoir species to become established in new locations.

Adaptive Measures and Strategies

The list of possible water-related health consequences from climate change, presented in Table 12.1, is long. Many of these effects would of course only occur during extraordinary circumstances. However, the increased risk of heavy rainfall and floods in Sweden is significant, as is the risk of increases in water runoff in many places. This will force many sectors of society to adapt in various ways. The measures could be implemented immediately or later, and be general or more specific, like flood-proofing electricity substations to prevent power outages.

The Commission proposed some general adaptive strategies to reduce vulnerability to flooding, landslides, and erosion:

- Adaptations of current infrastructure to a changing climate should be included in the objectives of current policies. Funds should be earmarked for these measures.
- Physical planning should be adapted to future risks. The government should provide information and produce material in support of planning and preventive measures.
- Special climate adaptation panels should be established at the county administrative boards.
- Information and training should be made available for staff in municipalities.
- Early warning systems should be developed. Registers of local data should be improved.
- A special climate adaptation fund should be created for greater investments aimed at reducing vulnerability to extreme weather events and long-term climate change. Priority should be given to projects aimed at preventing floods, landslides, and erosion.

The Commission recommended specific tasks to be given to different government agencies to lessen the health burden of climate change in the decades to come. In Sweden, government agencies are state-controlled organizations that act independently to carry out the policies of the Swedish Government. Recommendations and suggestions directly and indirectly related to water-related health impacts were made to several of the agencies. First, The National Board of Housing, Building and Planning (Boverket) should evaluate regulations to ensure safe planning in flood- and landslide-prone areas, and healthy buildings. In regard to the latter, focus should be on mould-resistant materials, or ventilation in areas with increased precipitation and risk of flooding.

Second, two agencies focusing on infectious disease received several recommendations: (1) The National Board of Health and Welfare, which is responsible for communicable disease prevention, control, and epidemiology; and (2) the expert agency, The Swedish Institute for Infectious Disease Control (SMI), which monitors the epidemiological situation for infectious diseases in humans, and is responsible for promoting protection against such diseases. Recommendations were given to these agencies to ensure surveillance and monitoring of infectious diseases that may be affected by climate change and to initiate control measures when they are needed. In the case of outbreaks of zoonoses, collaboration is already established with the National Veterinary Institute (SVA), an expert authority under the Ministry of Agriculture.

Third, The Swedish Civil Contingencies Agency has the overall responsibility for the country's emergency preparedness and responses. The climate change aspect needs to be acknowledged in risk assessments and in response planning for all sectors of the society involved.

Fourth, recommendations were given to the National Food Administration, to evaluate food regulations in light of climate change consequences, and to inform the general public of increased risks posed by climate change.

Fifth, recommendations were made to various groups responsible for managing drinking water. In Sweden, several authorities divide the responsibility for managing water sources, surface water, groundwater, infrastructure like water mains, water pipes, and water treatment plants, private water sources, the quality of recreational outdoor baths; and implementing the European Union (EU) Water Framework Directive. These authorities include the National Food Administration, Boverket, the Swedish Environmental Protection Agency, the Geological Survey of Sweden, the National Board of Health and Welfare, and the five Swedish River Basin District Authorities. It was recommended that potential consequences of climate change for water quality should be included in current programs and planned strategies of the different authorities involved.

Research and knowledge capacity building are the areas where most action has been taken (as of spring 2009) since the Commission's recommendations were released in November 2007. The Swedish Government has created new funds for scientific climate change-related research starting from 2009. Several of the governmental authorities, like SVA, have received governmental funding to set up knowledge centres on climate change-related topics within their respective fields of expertise.

In addition to the recommendations stated above, the Health and Water work-groups highlighted the need for adaptive measures within several areas. The most important are presented below:

Information to the general public: Information should be given both in general and to high-risk groups/regions. Risks of particular concern include those associated with basement flooding, standing water after heavy rain, and the contamination of private water sources from increased runoff after heavy rains. In Sweden summer cottages with private water sources are common.

Information to farmers: Avoid watering vegetables with water from streams and watercourses close to pastures after heavy rain.

Education of staff: Education about changes in risks and of potential adaptive measures should be given to persons involved in areas that might be affected, such as the health care and public health sectors, disease surveillance and monitoring units, laboratories handling pathogens, fire and rescue departments and other emergency units, as well as local policymakers and other stakeholders of interest.

Surveillance and monitoring of diseases: This is already being done by SMI and SVA, but emphasis should be on changes in outbreak patterns, and causes of epidemics. Collaboration at the EU level is now being facilitated by the European Centre of Disease Control and Prevention (ECDC), which began operating in 2005.

Research: Focus should be on filling existing scientific knowledge gaps.

Risk group identification: Primary health care units/GPs should keep a list of individuals that may need extra help in case of climate change-related disasters, like the elderly, physically or mentally handicapped, especially if living alone.

Landfills: Registers of geographical locations of landfills (pathogens and toxic compounds) should be kept at a county level. Landfills located in flood or landslide-prone areas, should be marked.

Water sector: Water treatment plants and water pipes need to be adapted to increases in water volumes and higher temperatures, and to increased concentrations of microorganisms, humus levels, and toxic algal bloom in water sources. The costs for additional water treatment due to climate change-related increases in pathogens during the period 2011–2040 were estimated to be about 130 million €.

Many of the recommendations will be evaluated and implemented at the county and municipality levels. It remains unclear – about 1.5 years after the recommendations were released from the Commission – which groups will bear the costs and be responsible for the execution and monitoring of some adaptive measures. Some municipalities and counties have now, on their own initiative, started risk and adaptation assessments to evaluate the need for local adaptation to climate change for different sectors. One such example is Sundsvall municipality, with nearly 50,000 inhabitants, located at the northern Baltic coast line. Sundsvall recently started a 2-year climate change and adaptation assessment project that involves local stakeholders. Depending on the outcomes of such local assessments, further planning and initiation of adaptive measures at the local level will be taken. However, it is often unnecessary to develop new programs or infrastructure to handle the increased risk of water-related health consequences from climate change. Climate change risks should be incorporated into existing planning.

Conclusion

This chapter summarizes the findings and recommendations of the Health and Water workgroups of the Swedish Commission on Climate and Vulnerability on water-related health impacts of climate change. The Rossby Centre's regional climate scenarios point to significant increases in the risk of heavy rain, high flows, and floods. The risk of landslides is likely to increase in current risk areas. Health consequences from water-related impacts on infrastructure, buildings, the environment, and public services include injuries, epidemics, allergies, and aggravation of chronic diseases. Outbreaks of infectious diseases and exposures to toxic compounds could be linked to contamination of water sources, irrigation water, tap water, recreational waters, and still standing waters. Increases in precipitation may create more breeding sites for insects. However, health risks vary considerably between locations depending on local vulnerability and resilience, which has been shown in other studies and national assessments as well (e.g., Ebi et al. 2006).

The adaptive evaluations and recommendations made by the Commission focused mainly on tasks for the public sector. However, a main task of society is to provide information in order to promote adaptive measures also within industry and business, and to increase knowledge about climate change through research. The recommendations of the workgroups on Health and Water issues included both general and detailed adaptive strategies and measures. The importance of education of staff in affected sectors, and of information to risk groups and the general public were underlined. The Health workgroup concluded that local risk assessments and adaptation evaluations on water-related health consequences of climate change require crosscutting collaboration between scientific disciplines as well as between different sectors of society, and should from the start include different experts in the field, government agencies, local authorities, as well as local stakeholders. The value of incorporating the added risks of climate change into existing planning at both national and local levels was emphasized throughout the recommendations.

Acknowledgments This chapter is based on the work done by the Swedish Government's Commission on Climate and Vulnerability. We would like to thank in particular Mats Bergmark at MittSverige Vatten, and Per Ericsson at the Northern Water Board.

References

Auld H, MacIver D, Klaassen J (2004) Heavy rainfall and waterborne disease outbreaks: the Walkerton example. J Toxicol Environ Health A 67(20–22):1879–1887

Ebi KL, Mills DM, Smith JB et al (2006) Climate change and human health impacts in the United States: an update on the results of the U.S. national assessment. Environ Health Perspect 114(9):1318–1324

Hubalek Z, Zeman P, Halouzka J et al (2004) Antibodies against mosquito-borne viruses in human population of an area of Central Bohemia affected by the flood of 2002. Epidemiol Mikrobiol Imunol 53(3):112–120 [in Czech]

IPCC (2000) IPCC special report – emissions scenarios: summary for policy makers. A special report of Working Group III of the IPCC. Cambridge University Press, Cambridge

Jones CG, Willén U, Ullerstig A et al (2004) The Rossby Centre regional atmospheric climate model part I: model climatology and performance for the present climate over Europe. AMBIO J Hum Environ 33(4–5):199–210

Kalashnikov IA, Mkrtchan MO, Shevyreva TV et al (2003) Prevention of acute enteric infections and viral hepatitis A in the Krasnodar Territory appearing in connection with a natural disaster in 2002. Zh Mikrobiol Epidemiol Immunobiol 6:101–104 [in Russian]

Kjellström E, Bärring L, Gollvik S et al (2005) A 140-year simulation of the European climate with the new version of the Rossby Centre regional atmospheric climate model (RCA3). Swedish Meteorological and Hydrological Institute reports, No.108

Kriz B (1998) Infectious disease consequences of the massive 1997 summer floods in the Czech Republic. Working group paper. EHRO 020502/12

Kukkula M, Arstila P, Klossner ML et al (1997) Waterborne outbreak of viral gastroenteritis. Scand J Infect Dis 29(4):415–418

Lindgren E, Albihn A, Andersson Y (2007) Report of the working group on human and animal health. Ministry of the environment, The commission of climate and vulnerability. Swedish government official report SOU 2007:60, Appendix 34. Stockholm

Lindgren E, Albihn A, Andersson Y et al (2008) Consequences of climate changes for the health status in Sweden. Heat waves and disease transmission most alarming. Lakartidningen 105 (28–29):2018–2023 [in Swedish]

MacKenzie WR, Schell WL, Blair KA et al (1995) Massive outbreak of waterborne *Cryptosporidium* infection in Milwaukee, Wisconsin: recurrence of illness and risk of secondary transmission. Clin Infect Dis 21(1):57–62

Miettinen IT, Zacheus O, von Bonsdorff CH et al (2001) Waterborne epidemics in Finland in 1998–1999. Water Sci Technol 43(12):67–71

Ministry of the Environment (2007) Sweden facing climate change – threats and opportunities. Swedish Government Official Report SOU 2007:60. The commission on climate and vulnerability, Stockholm

SMI (2006) Annual epidemiological report on communicable diseases in Sweden, 2005. The Swedish Institute for Infectious Disease Control, Stockholm [in Swedish]

Tuffs A, Bosch X (2002) Health authorities on alert after extensive flooding in Europe. Br Med J 325:405

Zitek K, Benes C (2005) Longitudinal epidemiology of leptospirosis in the Czech Republic (1963–2003). Epidemiol Mikrobiol Imunol 54(1):21–26 [in Czech]

Chapter 13
Adaptation to the Heat-Related Health Impact of Climate Change in Japan

Yasushi Honda, Masaji Ono, and Kristie L. Ebi

Abstract High ambient temperatures are a cause of preventable morbidity and mortality in Japan. First, we describe the health impacts of heat waves, which include not only heatstroke, but also mortality and morbidity due to indirect effects of high temperatures. Older adults, young children, and persons with chronic medical conditions are particularly susceptible. Second, we discuss public health adaptation to projected increases in the frequency, intensity, and duration of heat waves due to climate change in Japan. These adaptation measures include those taken by the Ministry of Environment, and various local governments in response to local factors. Third, we discuss key challenges to adapting to high temperatures and minimizing heat-related health impacts. We finish by discussing the adaptation measures used by the government of Japan and some local cities to increase population resilience to heat waves. Critical issues include that measures should be suitable for the target populations (the vulnerable subgroups) and comprehensive.

Keywords Japan • Temperature • Heat wave • Heatstroke • Public health

Y. Honda (✉)
Graduate School of Comprehensive Human Sciences, University of Tsukuba,
Tsukuba, Ibaraki, Japan
e-mail: honda@taiiku.tsukuba.ac.jp

M. Ono
Association of International Research Initiatives for Environmental Studies, Tokyo, Japan
e-mail: ono@airies.or.jp

K.L. Ebi
ClimAdapt LLC, 424 Tyndall Street, Los Altos, CA 94022, USA

Department of Global Ecology, Stanford University, Stanford, CA, USA
e-mail: krisebi@essllc.org; krisebi@stanford.edu

J.D. Ford and L. Berrang-Ford (eds.), *Climate Change Adaptation in Developed Nations: From Theory to Practice*, Advances in Global Change Research 42, DOI 10.1007/978-94-007-0567-8_13, © Springer Science+Business Media B.V. 2011

Introduction

High ambient temperatures are a leading cause of weather-related mortality in developed countries. Although the numbers of heat-related deaths are fairly small relative to other causes of death, all these deaths are potentially preventable. Ministries of Health and other public health agencies are increasingly recognizing the population burden of heat waves, and consequently designing and implementing early warning systems to reduce premature mortality. The importance of these activities is increasing with projections of increases in the frequency, intensity, and duration of heat waves with climate change that suggest heat-related mortality could increase if no public health actions are taken.

This chapter first discusses the impacts of heat waves in Japan, including those on vulnerable population groups. This is followed by a discussion of public health adaptation to heat waves in Japan, including those taken by the government of Japan and some local cities. Finally, we discuss key challenges to adapting to high temperatures to minimize heat-related health impacts.

Impacts of Heat Waves

Heat waves are periods of consecutive extremely hot days. Authors of heat-wave-related papers define them on the basis of their intensity and duration. For example, Anderson and Bell (2009) developed an approach that considers six heat wave types: periods of 2 or more, or 4 or more days of continuous temperatures occurring in more than the 98.5th, 99th, or 99.5th percentile of the community's temperature distribution.

Exposure to high ambient temperatures can cause heat-related illnesses (heat cramps, heat exhaustion, heat syncope, or heatstroke) as well as contribute to mortality from a wide range of causes. Heat exhaustion is the most common response to prolonged exposure to high outdoor temperature. Heat exhaustion is characterized by intense thirst, heavy sweating, dizziness, fatigue, fainting, nausea or vomiting, and headache. If unrecognized and untreated, heat exhaustion can progress to heatstroke, a severe illness with a rapid onset that can result in delirium, convulsions, coma, and death (Lugo-Amador et al. 2004). Heatstroke has a high fatality rate, although even nonfatal heatstroke can lead to long-term effects. For example, severe impairment was observed in 35% of 58 patients admitted with heatstroke during the 1995 Chicago heat event, with no improvement after 1 year in those still alive (Dematte et al. 1998).

The rate at which adverse health effects appear depends on an individual's physical fitness, health, access to medical care, and other factors (Havenith et al. 1995). People usually try to adapt to extreme weather by altering clothing, gaining access to air-conditioned spaces, or staying in places with less harsh temperature.

Overall, older adults, young children, and persons with chronic medical conditions are particularly susceptible to heat-related illnesses and are at high risk for heat-related mortality (Basu and Samet 2002).

Temperature–Mortality Relationships in Japan

The relationship between temperature and mortality is complex. First, when evaluating this relationship for a particular heat wave, it is impossible to determine at the individual level which deaths were heat related (except for heatstroke deaths), because high ambient temperatures can affect several organ systems that can lead to death. Furthermore, heatstroke is only a small portion of all heat-related deaths: the total number of deaths in Tokyo during 1972–1994 was 38,331; only 50 deaths, or 0.13%, were due to "excessive heat." In this way, heat can indirectly lead to mortality. Many deaths during heat waves, such as those by cardiorespiratory diseases, suicide, and other external causes are heat related even if the official cause of death is not specifically heat (Page et al. 2007; Honda et al. 1995; Ishigami et al. 2008). For example, Honda et al. (1995) found a two-fold increase in all-cause mortality among children on days over 33°C, compared with days with lower temperatures. Because the increased risk was mostly due to accidents and was higher on weekends than on weekdays (Honda 2007), the excess deaths may be due to a "risk shift," i.e., increased mortality from higher risk activities such as swimming when the temperature is very high. These indirect effects of temperature on mortality may be important in a warmer climate as people change behavioral patterns to adapt to warmer temperatures.

Population-Level Heat-Related Impact and Adaptation

Mortality

At the population level in Japan, individual susceptibility to temperature results in the basic V-shape relationship between temperature and mortality (Honda et al. 1995), as shown in Fig. 13.1. This pattern has been reported from many areas in the world, including Europe, the Middle East, and the United States (e.g., Bull and Morton 1978; Kunst et al. 1993; Saez et al. 1995; Keatinge et al. 2000; El-Zein et al. 2004; Curriero et al. 2002). Although this pattern is a result of collective individual susceptibilities, this pattern can also be considered as a human response to a certain thermal environment at the population level. Using this pattern, we can assess the impact of the thermal environment. Based on studies in Japan and other countries, older adults are more vulnerable to high ambient temperatures (Bull and Morton 1978; Saez et al. 1995; Honda et al. 1995).

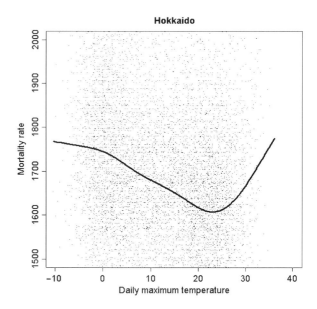

Fig. 13.1 Relation between daily maximum temperature and mortality rate in Hokkaido, Japan (1972–1994). Each *dot* represents data for a certain day. The *curve* is a nonparametric regression line, i.e., a smoothing spline curve ($df = 6$)

As shown in Fig. 13.1, the bottom of the V-shaped curve represents the temperature at which mortality is lowest. This can be called the "optimum temperature" (OT), because this temperature is optimal for the population. OT tends to be higher when the climate is warmer (Keatinge et al. 2000; Curriero et al. 2002; Honda et al. 1998). This relationship is important in projecting the impacts of climate change, because this pattern suggests that the OT will become higher as populations acclimatize (including changing infrastructure) to higher temperatures, even without explicit adaptation measures undertaken at the population level (for example, heat wave early warning and response systems).

Given this relationship between OT and climate, which weather factors best predict OT? At first glance, one might think that if area A is warmer than area B, and the long-term average temperature is higher in area A than area B, then this difference determines the OT level. However, this is not the case. As reported in Honda et al. (2007) and shown in Fig. 13.2, Okinawa is an outlier in the relationship between long-term average temperature and OT in Japan. Okinawa has a maritime subtropical climate; the long-term average temperature is very high, but there are fewer extremely hot days than in other prefectures. We analyzed Japanese data from 1972 to 1995 to determine which weather variables best predict mortality. Using daily mortality data and daily meteorological data, we evaluated the relationship between a wide range of weather factors and the OTs for 47 Japanese prefectures. The factors we explored were 0, 5, 10, … , 95, and 100 percentile values of daily minimum, average, and maximum temperature. Based on these analyses, the best factor that predicted OT was the 80–85 percentile value of daily maximum temperature (Fig. 13.3) (Honda et al. 2007). In this analysis, Okinawa was not an outlier. Further, the correlation coefficient was higher compared with that between long-term average temperature and OT.

Fig. 13.2 Relation between long-term average temperature (1972–1995) and optimum temperature (OT) for 47 prefectures in Japan (Modified from Honda et al. 2007)

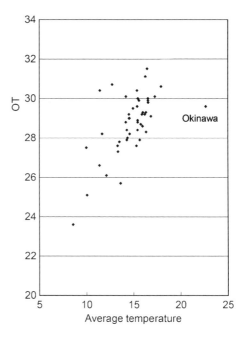

Fig. 13.3 Relation between 85 percentile value of daily maximum temperature ($T_{max}85$) and optimum temperature (OT) in Japan. *Solid line* is the regression line and *dotted line* indicates $T_{max}85 = OT$ (Honda et al. 2007)

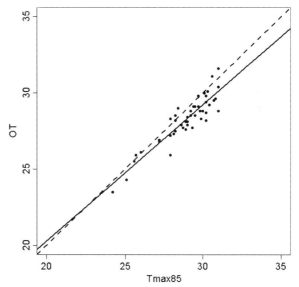

The above relationship can be considered a population-level acclimatization in that the OT will eventually become approximately 2°C higher as the 85 percentile value of daily maximum temperature increases by 2°C. This implies that, even when the average temperature becomes only 2°C higher, the OT will become 5°C higher when the 85 percentile of the daily maximum temperature becomes 5°C.

Table 13.1 Age and sex
distribution of heat disorder
patients among 18 cities in
Japan (2007)

Age (years)	Male	Female	Total
0 ~ 6	29	25	54
7 ~ 18	560	268	828
19 ~ 39	889	216	1,105
40 ~ 64	987	215	1,202
65+	1,065	848	1,913
Total	3,530	1,572	5,102

Note that the numbers were taken from the records of
ambulance visits in each city

This finding can be used to improve projections of the impacts of climate change on heat-related mortality, which is important for policymaking.

Morbidity

In Japan, there is no systematic surveillance system for morbidity, but some local governments have record systems for ambulance visits. We collected data for 18 cities, including Sapporo, Tokyo, and Osaka. Table 13.1 shows the age and sex distribution of the patients. Although the Japanese population is aging, the proportion of older subjects in this table was much higher than that in the total population. Also, young to middle-aged males had much higher incidence than females. Work, sports, and outdoor activity occupied a large proportion of young to middle-aged patients' locations or occasions of heat disorder events; in contrast, about half of the heat disorder in patients 65 years of age and over occurred at home (Fig. 13.4).

In terms of the relationship between heat and the incidence of heat disorders, wet bulb globe temperature (WBGT) – which consists of temperature, humidity, and radiation – was the best predictor of incidence (Fig. 13.5), suggesting WBGT should be used in studies of heat disorder. Of course, separate evaluations of temperature, humidity, and radiation may be necessary in studies of other heat-related health outcomes.

Preventing Heat-Related Impacts

Heat-related morbidity and mortality can be avoided if individuals at risk take appropriate steps to ensure their body temperature remains within a safe range. One effective approach is to provide air-conditioned shelters for residents during a heat wave. However, although the percentage of houses equipped with air-conditioners has been rising, evidence from other countries indicates that not all older adults at risk use air-conditioning even when they have access to it (e.g., Sheridan 2007). Figure 13.6 is an example (Kaido et al. 2007): as outdoor temperature decreased, bedroom temperatures decreased; but at outdoor temperatures less than

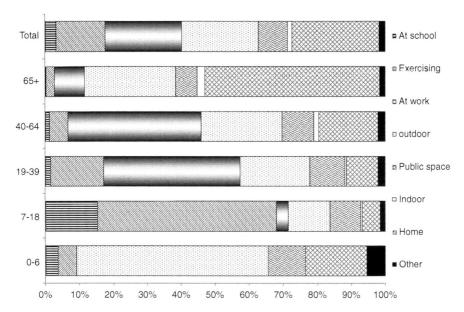

Fig. 13.4 Location or occasion of occurrence of heat disorder by age group (18 cities combined, Japan)

20°C, the bedroom temperature appeared stable at around 20°C and never dropped below 10°C. On the other hand, when the outdoor temperature was high, room temperatures also became high even beyond 30°C, although the house in question was equipped with an air-conditioner. The reason for this behavioral pattern was partly due to the belief that artificial cooling is not good for health; Japanese houses used to be equipped with only one air-conditioner per house, and going in and out of the air-conditioned room frequently was thought to be harmful for the autonomic nervous system. This implies that additional educational actions are needed to reduce heat-related mortality.

Adaptation Practice by Ministry of Environment

The urban heat island phenomenon and climate change have increased heat stress. The number of heatstroke patients has been rising in conjunction with these warming trends (Fig. 13.7). Recognizing these trends and that heatstroke is an avoidable risk, the Japanese government initiated an action plan for heatstroke prevention. Because this is an environmental health problem, the Ministry of Health, Welfare and Labour or the Ministry of Environment could be a responsible ministry. In Japan, the Ministry of Environment has traditionally managed environmental health issues such as air pollution, Minamata disease, and cedar pollen disease, and so has taken the lead.

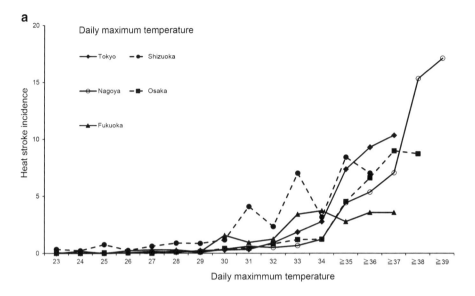

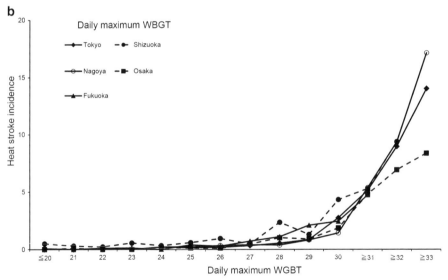

Fig. 13.5 Relation between heat disorder patients and index of heat: (**a**) daily maximum temperature and (**b**) wet bulb globe temperature

The Ministry of Environment action plan consists of the following (A) to (F):

(A) Preparation and distribution of heatstroke manual

The 2008 version of this heatstroke manual covers the following:

1. Pathophysiological outline of heatstroke
2. Treatment of heatstroke

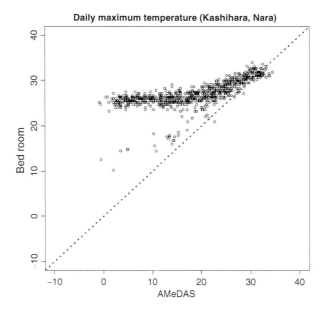

Fig. 13.6 The relation between outdoor temperature measured with Automated Meteorological Data Acquisition System (AMeDAS) and bedroom temperature in a house in Nara prefecture, Japan (Kaido et al. 2007)

3. Preventive measures to avoid heatstroke
4. Instructions for public health professionals
5. Sources of further information
6. Comprehensive actions by the government

The manual is revised annually. The manual is very useful for health workers, such as public health nurses at local health centers, but also includes illustrations so that information is accessible to the general public. For example, Fig. 13.8 shows how heatstroke can be prevented and be treated.

(B) Preparation and distribution of awareness-raising posters for heatstroke prevention.

Posters are distributed to local health centers and so-called mini-mini posters (business card size) are distributed to the public so that they can carry and refer to them whenever necessary.

(C) Government advertising

Each year, when the summer season begins, the government airs television advertisements about heatstroke prevention.

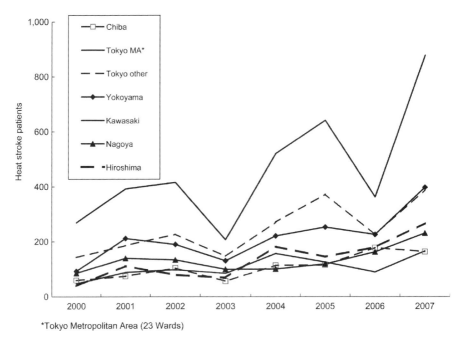

*Tokyo Metropolitan Area (23 Wards)

Fig. 13.7 Trend of heat disorder patients in Japan

(D) Creation of websites for heatstroke prevention

Websites for heatstroke include those by the Ministry of Environment, Japan
Weather Association, Japanese Society of Biometeorology, Japan Sports Association, Tokyo Fire Department, and Japan Red Cross. The site by Ministry
of Environment provides the trend of WBGT for the previous week, as
well as today's WBGT measurements every 3 h, and levels of heatstroke
risk (http://www.nies.go.jp/health/HeatStroke/index.html). Also, the above-
mentioned heatstroke manual in PDF format and some statistics of emergency
visits of heatstroke patients are available (http://www.env.go.jp/chemi/heat_
stroke/manual.html). Due to increasing numbers of heat waves and heatstroke
casualty reports on television or in newspapers, the number of access counts to
this site has increased and exceeded two million in 2007.

(E) Committee for Climate Change Impact Evaluation and Adaptation

The Ministry initiated a top-down impact assessment of the impacts of climate
change on key sectors, including water, ecosystems, and health. The health
impacts in the assessment include heat-related mortality, heatstroke, diseases
related to air pollution, and vector-borne infectious diseases. The core members
of the project discussed the climate change impacts and possible adaptations,
and reported the findings. Also, the Committee published a book, *Wise
Adaptation to Climate Change,* in 2008. This book has an English version and

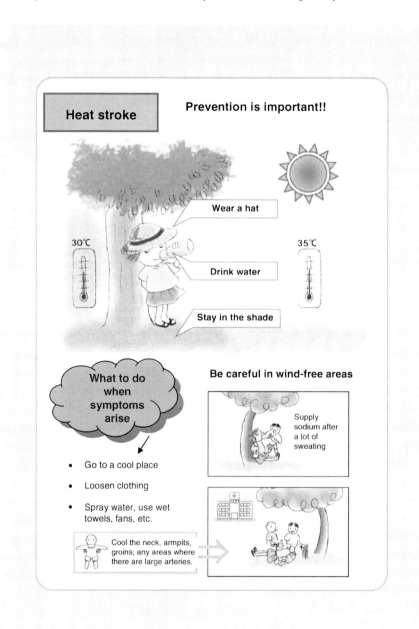

Fig. 13.8 An illustration in the heatstroke manual by Ministry of Environment (2008), translated from Japanese by the authors of this chapter

Table 13.2 Characteristics of heat warning systems of local governments in Japan

City	Target	Location of wet bulb globe temperature (WBGT) measurement	Media for dissemination	Other activities
Kusatsu	All residents	Elementary schools	E-mail	–
Kumagaya	All residents	Thirty elementary schools and city government building	E-mail, leaflet for elderly living alone	–
Tajimi	All residents	A kindergarten	E-mail, bulletin board, FM radio	Green curtain (tree belt) campaign
Oubu[a]	All residents	The University	–	–
Machida	Children	All elementary and junior high schools	Manual	Lecture for teachers, large fans, freezers

[a]Cooperated with Chukyo Women's University

is available from the Ministry of Environment upon request. In brief, this book covers food, water, ecosystem, disaster prevention and coastal zones, health, citizen's life, and developing countries.

(F) Ministry Liaison Conference

The Liaison Conference includes: the Fire Defence Agency; Ministry of Education, Culture, Sports, Science and Technology; Ministry of Health, Welfare and Labour; Meteorological Agency; and Ministry of Environment. An example of the importance of the Liaison Conference is the use of information collected by the Fire Defence Agency and Meteorological Agency on the website for heatstroke maintained by the Ministry of Environment.

Adaptation Practice by Local Governments

Some local governments initiated programs for heatstroke prevention, with motivations varying from one government to another. For example, Machida experienced heatstroke casualties at a school; hence as shown in Table 13.2, Machida's actions target elementary and junior high schools. Meanwhile, Kumagaya and Tajimi have the highest daily maximum temperatures (exceeding 40°C). Due to these high temperatures, these governments are acutely aware of the risk of heat stress, so these cities have established heat wave early warning systems.

As shown in Table 13.2, two main features of heat warning systems are (1) local WBGT measurements, and (2) a heat wave warning mailing system that sends warnings and related information to people's PCs or cell phones. The mail includes recommendations for activity level and actions for preventing heatstroke.

Other local initiatives aim to reduce heat-related health impacts by addressing the heat island effect. Tajimi initiated the "green belt" campaign that covers walls of buildings and houses with plants to reduce radiation heat. Similarly, Kyoto and Tokyo have been covering school grounds with lawns, which reduces surface temperatures by about 8°C compared to paved ground (Tokyo Metropolitan Research Institute 2010). Tokyo plans to cover the grounds of 2,000 schools by 2017. These adaptation measures are important because they also reduce carbon emissions. For example, lawn coverings at schools capture CO_2, and reduce the need for air-conditioning, by reducing the heat island effect.

Challenges to Adaptation Practices

Although the above practices may have increased awareness of the risks of high temperatures for the general public, further improvements are needed to overcome major problems. First, heat-related health impacts are not restricted to heatstroke. Different adaptation strategies may be needed to address different heat-related causes of morbidity and mortality. For example, children's swimming accidents cannot be completely avoided using only heatstroke prevention measures. Second, our understanding of cardiovascular or respiratory diseases is insufficient for effective adaptation. For example, we do not know why mortality from respiratory diseases increases during hot days, or to what extent people with cardiovascular diseases can be acclimatized to high temperatures. Third, the effectiveness of Japanese adaptation measures may be fundamentally limited. Vulnerable subpopulations, such as the elderly living alone, may not be reached by messages via PCs or cell phones. Kumagaya has distributed leaflets to elderly living alone, an important practice that other local governments could emulate. However, distributing leaflets does not mean that the elderly understand and take appropriate actions to avoid risk. Because the number of people with cognitive problems increases with the aging of populations, it is necessary for public health nurses to visit the houses of the elderly and instruct them so that they can understand appropriate actions to take.

As mentioned above, some older adults do not turn on air-conditioners even when the indoor temperature exceeds 30°C. As Kovats and Jendritzky (2006) pointed out, using air-conditioners may hinder natural acclimatization and potentially increase the risk of heatstroke if it is used improperly. On the other hand, air-conditioning can be a valuable adaptation measure to prevent heatstroke during heat waves when used properly. Its use should be encouraged, especially for the elderly in Japan.

However, air-conditioning is not a perfect adaptation solution. First, some houses are not equipped with air-conditioners, particularly in northern prefectures in Japan. Second, heat waves may increase electricity consumption, which could cause blackouts. Given these limitations, it may be necessary to build large, air-conditioned public shelters. These shelters could be located in community halls for neighborhood associations. These shelters could run on solar energy, or other off-grid electricity sources, to avoid blackout risks during heat waves. Also, energy-efficient city planning should be taken into consideration, especially for cities. In Tokyo, skyscrapers were built on the edge of waterfront, and this construction was criticized for blocking sea breeze and exacerbating the heat island phenomenon.

In summary, much has been learned about the risks of heat waves and actions that can be taken to reduce those risks. However, there is much more that needs to be learned to increase population resilience to increases in the frequency, intensity, and duration of heat waves in a changing climate.

References

Anderson BG, Bell L (2009) Weather-related mortality: how heat, cold, and heat waves affect mortality in the United States. Epidemiol 20(2):205–213

Basu R, Samet J (2002) Relation between elevated ambient temperature and mortality: a review of the epidemiologic evidence. Epidemiol Rev 24(2):190–202

Bull GM, Morton J (1978) Environment, temperature and death rates. Age Ageing 7(4):210–224

Curriero FC, Heiner KS, Samet JM et al (2002) Temperature and mortality in 11 cities of the Eastern United States. Am J Epidemiol 155(1):80–87

Dematte JE, O'Mara K, Buescher J et al (1998) Near-fatal heat stroke during the 1995 heat wave in Chicago. Ann Intern Med 129(3):173–181

El-Zein A, Tewtel-Salem M, Nehme G (2004) A time-series analysis of mortality and air temperature in Greater Beirut. Sci Total Environ 330(1–3):71–80

Havenith G, Luttikholt VGM, Vrijkotte TGM (1995) The relative influence of body characteristics on humid heat stress response. Eur J Appl Physiol O 70(3):270–279

Honda Y (2007) Impact of climate change on human health in Asia and Japan. Glob Environ Res 11(1):33–38

Honda Y, Ono M, Sasaki A et al (1995) Relationship between daily high temperature and mortality in Kyushu, Japan. Nippon Koshu Eisei Zasshi 42(4):260–268 (in Japanese)

Honda Y, Ono M, Sasaki A et al (1998) Shift of the short-term temperature mortality relationship by a climate factor – some evidence necessary to take account of in estimating the health effect of global warming. J Risk Res 1(3):209–220

Honda Y, Kabuto M, Ono M et al (2007) Determination of optimum daily maximum temperature using climate data. Environ Health Prev Med 12(5):209–216

Ishigami A, Hajat S, Kovats RS et al (2008) An ecological time-series study of heat-related mortality in three European cities. Environ Health 7(5):1–7

Kaido T, Honda Y, Suzuki Y et al (2007) Weather factors related with the QOL and exacerbation of the respiratory disease of patients on home oxygen treatment. Report of the grant-in-aid for scientific research 2006. Ministry of Education, Culture, Sports, Science and Technology, Tokyo

Keatinge WR, Donaldson GC, Cordioli E et al (2000) Heat related mortality in warm and cold regions of Europe: observational study. Br Med J 321(7262):670–673

Kovats RS, Jendritzky G (2006) Heat-waves and human health. In: Menne B, Ebi K (eds) Climate change and adaptation strategies for human health. Springer, Steinkopf

Kunst AE, Looman CW, Mackenbach JP (1993) Outdoor air temperature and mortality in the Netherlands: a time-series analysis. Am J Epidemiol 137(3):331–341

Lugo-Amador NM, Rothenhaus T, Moyer P (2004) Heat-related illness. Emerg Med Clin North Am 22(2):315–327

Page LA, Hajat S, Kovats RS (2007) Relationship between daily suicide counts and temperature in England and Wales. Br J Psychiat 191(2):106–112

Saez M, Sunyer J, Castellsague J et al (1995) Relationship between weather temperature and mortality: a time series analysis approach in Barcelona. Int J Epidemiol 24(3):576–582

Sheridan SC (2007) A survey of public perception and response to heat warnings across four North American cities: an evaluation of municipal effectiveness. Int J Biometeorol 52(1):3–15

Tokyo Metropolitan Research Institute (2010) Heat island effect alleviation by covering school grounds with lawn. http://www2.kankyo.metro.tokyo.jp/kankyoken/research/heatiland/koteisibahuka/kouteisibahuka.pdf. Cited 10 Sep 2010

Chapter 14
Risk Perception, Health Communication, and Adaptation to the Health Impacts of Climate Change in Canada

Peter Berry, Kaila-Lea Clarke, Mark Pajot, and David Hutton

Abstract Climate change poses increasing risks to the health of Canadians, particularly those most vulnerable to the impacts. Effective adaptations are needed to help people safeguard their health and well-being. Governments and communities have an important role to play in protecting citizens from climate-related health risks. Individuals also have a central role in adapting to climate change. Information about the perceptions of climate-related health risks and current vulnerability to the impacts is limited, but this information is needed by public health and emergency management authorities to inform education and outreach programs to promote greater adaptation actions among Canadians. This case study reports on the results of a national survey that was conducted in spring 2008 to better understand how Canadians perceive risks to health from climate change. Canadians generally accept that the climate is changing and many are concerned about its impacts – 72% view climate change as at least a moderate risk to health, with 32% rating it as a major risk. However, few Canadians have knowledge of specific risks to health posed by climate change, and many are not responding to health messages encouraging them to take protective measures to reduce impacts from current climate-related hazards. Results of this case study suggest that greater efforts are needed to inform Canadians about specific health risks related to climate change and tailor messages to populations most vulnerable to the impacts, in order to facilitate the adoption of adaptive actions.

P. Berry (✉) • K.-L. Clarke
Climate Change and Health Office, Health Canada, Ottawa, ON, Canada
e-mail: peter_berry@hc-sc.gc.ca; kaila-lea_clarke@hc-sc.gc.ca

M. Pajot
MES Candidate, York University, Toronto, ON, Canada
e-mail: pajotm@peelregion.ca

D. Hutton
United Nations Relief and Works Agency (UNRWA), West Bank Field Office,
Jerusalem 97200, Israel
e-mail: d.hutton@unrwa.org; huttondavid@hotmail.com

J.D. Ford and L. Berrang-Ford (eds.), *Climate Change Adaptation in Developed Nations: From Theory to Practice*, Advances in Global Change Research 42, DOI 10.1007/978-94-007-0567-8_14, © Springer Science+Business Media B.V. 2011

Keywords Climate change and health • Risk perception • Adaptation • Extreme heat • Vulnerable populations • Risk communication • Public health • Emergency management • Adaptation barriers • Extreme weather event

Introduction

The association between specific climate conditions and mortality, illnesses, and discomfort is well documented (McMichael et al. 2003; Riedel 2004; Confalonieri et al. 2007; Ebi et al. 2008). The rate of climate change and its impacts will likely create new stresses on individual and community health and well-being, and increase vulnerability to existing environmental and social pressures (Ebi et al. 2008; Seguin and Berry 2008; Menne et al. 2008). The Intergovernmental Panel on Climate Change (Parry et al. 2007) reports that many of the global-level impacts of climate change are now being felt and will worsen, given that warming is expected to increase in the coming decades. Seguin and Clarke (2008) document the climate-related health risks many Canadians and communities face across the country, as well as how these are expected to evolve as the climate changes. Potential impacts include more illnesses and deaths related to poor air quality, water- and food-borne contamination, extreme weather events, including extreme heat events[1] and changing patterns of diseases spread by animals, ticks, and insects.

Effective adaptations are needed to reduce the vulnerability of Canadians to these potential health impacts in the face of projected health, demographic, and climate trends in Canada (Gosselin 2004; Riedel 2004; Seguin 2008). As Dr. Marc Danzon, World Health Organization (WHO) Regional Director for Europe, stated, "In the face of what we know about the serious threats posed by climate change to health, the question today is not whether public health action is necessary, but what to do and how to do it" (European Food Safety Authority 2008). Governments and communities have an important role to play in protecting citizens from climate-related health risks. Public health officials monitor and intervene to reduce risks to health associated with contaminated food and water supplies, infectious diseases, and air pollution. Emergency management officials increase the resilience of communities to a range of natural and man-made hazards through vulnerability assessments, disaster mitigation, and emergency preparedness activities. Local volunteer organizations (e.g., Canadian Red Cross, Salvation Army) help to maintain the social networks and resources that sustain strong and vibrant communities. A key responsibility for adapting to the potential health impacts of climate change falls to individuals. Climate change adaptation literature focuses primarily on the factors (e.g., infrastructure, economic resources, institutions, technology) that affect whether governmental and nongovernmental decision-makers formulate effective

---

[1] This document uses the term "extreme heat event" rather than "heat wave" unless reporting on the findings of a public opinion survey.

adaptations to the impacts of a changing climate (Grambsch and Menne 2003). However, few adaptation studies have examined the determinants of adaptive behaviors at the individual level, particularly related to efforts to reduce possible impacts on health and well-being.

The development of adaptation options requires an understanding of how individuals perceive climate change and its impacts (Sharma et al. 2008). Public health and emergency management decision-makers require information about how Canadians perceive health risks associated with climate change in order to inform effective risk communication approaches for motivating individuals to take adaptive actions. Such information should be used to tailor messages in education and outreach products to specific populations.

This case study examines the current knowledge and perceptions of Canadians regarding health risks related to climate change. It draws on the results of a national survey conducted in spring 2008 for this research project. Information from the survey and other sources is used to examine whether Canadians are responding to existing risk communication activities by government authorities to get people to adopt protective behaviors. The analysis is guided by the following questions: (1) Do Canadians perceive risks to their health from climate change? (2) Are they knowledgeable about the health risks? (3) Are they taking actions to protect themselves from current climate-related hazards? (4) What are the implications for risk communication activities aimed at promoting adaptive behaviors?

Individual Adaptations to Protect Health from Climate-Related Risks

Climate change scenarios project an increased risk of extreme weather and other climate-related events in Canada such as floods, droughts, forest fires, and extreme heat events – all of which increase health risks to Canadians (Cheng et al. 2005; McBean 2006; Berry et al. 2008). In addition, risks associated with respiratory and cardiovascular diseases are expected to increase in many Canadian communities due to climate change through increased smog formation, wildfires, pollen production, and greater emissions of air contaminants because of changed personal behaviors (e.g., use of air-conditioning) (Cheng et al. 2005; Garneau et al. 2005; Lamy and Bouchet 2008). Climate change is also likely to increase risks associated with some infectious diseases across the country, and may result in the emergence of diseases that are currently thought to be rare or exotic to Canada (Charron et al. 2008; Lemmen et al. 2008). Specific population groups such as seniors, the poor, immigrants, young children, people with disabilities, Aboriginal Canadians, and those with preexisting health conditions are expected to be more severely affected by climate change health impacts due to their physiological characteristics or social barriers to adaptation (Tamburlini 2002; Public Health Agency of Canada 2004; Health Canada 2005; Mensah et al. 2005; Kovats and Jendritzky 2006; Furgal 2008).

For example, people 75 years of age or older made up nearly 80% of the excess deaths that occurred in France during the extreme heat event that struck Europe in 2003 (Fouillet et al. 2006).

Adaptive actions that reduce health risks can save lives, reduce injuries, and lessen the suffering associated with potential climate change impacts. Individuals are responsible for a wide range of actions to reduce health risks associated with climate-related hazards. Effective actions by individuals require preventative measures that include efforts to reduce greenhouse gas emissions as well as adaptations to address existing risks and protect health (Box 14.1).

Risk communication can be an effective means for getting people to adopt behaviors that promote and sustain good health (Committee on Communication for Behavior Change in the 21st Century, 2002). Individual-level factors that influence the adoption of health behaviors, including those adaptations needed to reduce risks to health from climate change, have been identified through research in the fields of psychology, communications research, and epidemiology (Maibach et al. 2007). Perceptions of risks and of individual and community vulnerabilities are strongly linked to behaviors to reduce the impacts of hazards such as floods, storms, and extreme heat events (Smit et al. 2000; Haque et al. 2004; Glik 2007). Such perceptions influence how and when people will use existing tools and resources to address climate change impacts (Health Canada 2003). For example, "... increased risk perception of heat results in increased response to a warning. Different social factors such as sex, race, age and income all play an important role in determining whether or not people will respond to a warning" (Kalkstein and Sheridan 2006, p. 43).

Seguin and Berry (2008) identify important barriers to adaptation, including an incomplete knowledge of the possible impacts on health and well-being, a lack of urgency among Canadians and decision-makers to take protective actions, and insufficient capacity by some vulnerable populations to adapt. Public health and emergency management officials in government and volunteer sector organizations can help address these barriers by raising awareness of the expected impacts of climate change and needed measures to protect health.

Box 14.1 Examples of Preventative Measures to Reduce Health Risks

Mitigating Hazards

- Reducing greenhouse gas emissions to slow down climate change and its impacts
- Avoiding high-risk areas (e.g., floodplains, avalanche corridors, wildland–urban interface, unprotected coasts) when choosing place of residence
- Taking measures (e.g., planting trees, driving less) to reduce the urban heat island effect

Reducing Exposures

- Personal health practices such as reducing exposure to extreme heat, safe food preparation and storage, correct hand washing, safe driving practices, emergency preparedness plans and kits
- Learning about important risks to health (e.g., new infectious diseases) and measures that should be taken to protect health
- Obeying public health alerts and warnings to reduce health risks from imminent hazards (e.g., smog alerts, extreme heat warnings)

Managing Illnesses and Promoting Health

- Effective management of individual and family (e.g., children, seniors) illnesses and conditions (e.g., diabetes, disabilities, mental illnesses)
- Proper and informed use of medications to treat illnesses and injuries
- Observance of healthy living practices (e.g., no smoking, regular exercise, optimal diet)

Building Resiliency

- Buying locally to develop and maintain local food sources
- Participating in community level outreach activities on climate change issues
- Improving social networks to reduce vulnerability to disasters and public health emergencies

Canadian's Perceptions of Climate Change and Related Risks to Health

Canadians generally make the link between environmental degradation and risks to health and consider environmental pollution as an important concern (Health Canada 1997; McAllister 2008). Recent surveys have demonstrated that Canadians are concerned about climate change (Eko Research Associates Inc. 2005), including potential health impacts (Pollara 2006; CMA 2007). However, there is a lack of information about the nature of individual perceptions of climate change health risks, the level of knowledge that Canadians possess about these issues, and whether an increase in knowledge and concern about risks will lead to adaptive actions.

For this study, a survey of Canadians was conducted to identify current perceptions regarding key health risks and vulnerabilities associated with climate change, as well as current adaptive actions individuals are taking to reduce these risks.

The type of information collected from participants through the survey was derived from a literature review of climate change health impacts and proposed adaptations, which was also undertaken as part of this research study. Forty questions were asked of respondents to obtain information on: views about whether climate change is occurring; perceptions and knowledge of health risks related to climate change; feelings of vulnerability from the impacts such as weather-related disasters; and current actions to reduce climate-related health risks, including current levels of emergency preparedness. Given the broad range of climate change health risks and proposed adaptations, and the limits on survey time and length, survey questions on adaptation activities focused on the issue of emergency preparedness. Emergency preparedness included issues related to the use of media to obtain weather information, evidence of individuals or households with a 72-h supplies kit and an evacuation plan, and compliance with preventive advice issued during extreme climatic events. The survey also included questions tailored to better understand perceptions of risk and adaptations being taken by Canadian seniors and people living with chronic conditions that can be exacerbated by climate change impacts on food, air, or water quality (e.g., heart disease, respiratory disease, cancer).

Telephone interviews were conducted from February 12 to March 3, 2008 with 1,600 adult Canadians from all provinces and territories. Results are accurate to within plus or minus 2.4 percentage points in 19 out of 20 samples (larger margins of error apply for subgroups of this population). The survey also included an oversample of 203 seniors and 200 persons aged 18–64 self-identifying as having at least one of a number of chronic health conditions.

The survey found that most Canadians are aware of climate change and many are concerned about its impacts, but have little knowledge of specific risks to health that it poses (Environics 2008). In the survey, 39% of Canadians reported having noticed a change in their community that they think is definitely or likely the result of climate change and 72% viewed climate change as at least a moderate risk to health, with 32% rating it as a major risk (Environics 2008). This perceived level of risk is considerably below the ratings assigned to issues such as obesity (70% considered this a major risk), heart disease (65%), and air pollution (62%). However, it is higher than the rating assigned to West Nile virus (16%) and the quality of tap water (14%) (Environics 2008) (Fig. 14.1).

A large majority of Canadians viewed their community as being vulnerable to climate change impacts. Fully three-quarters who reported that climate change poses a risk to health either now or in the future (93% of all Canadians surveyed) said their community is either definitely or likely vulnerable. In addition, two-thirds reported feeling personally vulnerable and almost half (46%) said someone in their household is especially vulnerable to the health risks of climate change (Environics 2008). The strongest concern about the possible health impacts of climate change was expressed by those Canadians who were more aware of, or more sensitized to, the issue of climate change (as identified by indicators such as having noticed definite effects of climate change locally, believing that climate change is definitely happening and that it is happening now). Those sensitized to the issue, such as the

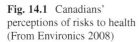

Fig. 14.1 Canadians' perceptions of risks to health (From Environics 2008)

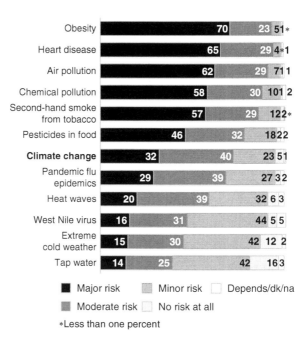

chronically ill and those with a high level of impairment, were more likely to believe that they are personally vulnerable to its negative health impacts (Environics 2008).

Seniors, on the other hand, despite being at potentially greater risk than younger Canadians, were no more likely than others to think of climate change as a major health risk. Seniors are also less knowledgeable about climate change health impacts as they were the least likely to be able to name at least one potential health risk without prompting and to think that climate change poses health risks today. Importantly, they were less likely than Canadians under age 65 to think that either they, or their community, are vulnerable. Seniors were also less likely (34% vs. 48% of those aged 18–64) to name their cohort as a population group that might be susceptible to the health risks of climate change (Environics 2008). This suggests that seniors are not making a direct connection between climate change and their personal vulnerability to its health effects. Given that many seniors – particularly those that are older and frail, isolated, or poor – are at significantly higher risk of illness and death, their limited knowledge of climate change health impacts and relatively lower perceptions of vulnerability should be a cause for concern for public health and emergency management officials.

The survey also revealed that most Canadians do not believe they will be personally affected by health impacts during a natural disaster. There seems to be a disconnect between views about climate change vulnerability and views about vulnerability to extreme weather events. Many Canadians did not view themselves as being at risk from extreme weather. The majority reported that if an extreme weather event occurred in their community, it would pose only a moderate (36%) or minor (38%) health risk to people living there. Few (11%) reported that the people in their community would suffer major impacts (Environics 2008). As well, only 7% of Canadians reported that they would have great difficulty in protecting themselves, or others, in the event of an extreme weather event such as a heat wave (Environics 2008). These findings are supported by other studies which suggest that many people do not view themselves as being vulnerable to health risks associated with extreme heat events (Kalkstein and Sheridan 2006; Sheridan 2006; Wolf et al. 2009). This perception that extreme weather events pose little threat is expressed well in the response of a Toronto homeowner upon receiving a flyer with his hydro bill warning him to keep 72 h worth of food and water on hand in case of an emergency. He discounted the need for action, "It sounds reasonable, but who's going to do that? It's a bit hard to imagine, really, having no supplies whatsoever for 3 days .... I think I would just go to the store" (Banks 2008).

Current Adaptations to Protect Health

Individuals who are prepared for an emergency know their risks, have developed an emergency plan for their household, and have set some basic supplies (flashlight, battery-operated radio, food, water, medication, cash, and blankets) aside in an emergency kit so that they are prepared to be self-sufficient for at least 72 h (Public Safety Canada 2008). Research suggests that many individuals are largely unprepared for emergencies (Tierney et al. 2001; Murphy 2004). In Canada, one-half to three-quarters of people are not prepared for emergencies (GPC Public Affairs 2005; Ekos Research Associates Inc. 2007). The survey conducted for this study found that only 42% claim to have a household emergency plan (Environics 2008). In addition, almost half of Canadians (45%) indicated that they do not have an emergency kit at home (Environics 2008). The low perception of vulnerability, as discussed in the earlier section, may be contributing to the failure of many high-risk individuals to take protective actions during extreme heat events. For example, in a study of the public response to extreme heat warnings in Phoenix, Arizona, it was found that while 93% of seniors (age 65 and over) reported that they had heard of heat warnings in the past, less than half reported changing their activities as a result of the warning (Kalkstein and Sheridan 2006). Similarly, studies undertaken in Canada indicate that although many seniors have the ability to eliminate risks from extreme heat through the use of air-conditioning, they often choose not to do so (Jacque 2005; Gosselin et al. 2008).

More generally, Canadians have little apparent knowledge of adaptation actions that can be taken to reduce health risks associated with climate change. This survey revealed that many people are taking some actions that they associate with reducing health risks of climate change, but most of these actions do not directly protect health. When asked, "Have you personally taken any steps in the past year to protect yourself or family members from the potential health impacts of climate change?" about half (51%) of Canadians reported to have recently taken at least

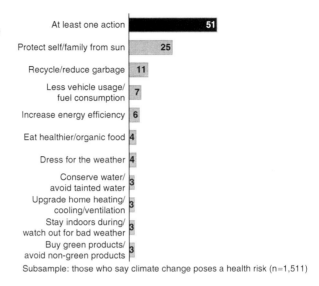

Fig. 14.2 Steps Canadians have taken in the past year to protect against climate change health impacts (% of respondents) (From Environics 2008)

Subsample: those who say climate change poses a health risk (n=1,511)

one step to protect themselves (Environics 2008). However, as Fig. 14.2 shows, many of the steps mentioned are actions to reduce environmental impacts in general (e.g., reducing garbage, vehicle use, and energy use) rather than to protect health (e.g., seeking cool areas during extreme heat events, safe food handling practices, planning for emergencies). This indicates that people understand they have a role to play in reducing the rate of climate change, but they are not making a direct connection with climate change health risks and the measures they should be taking to reduce them. For example, very few people mentioned staying indoors in bad weather (3%), or dressing for the weather (4%), as steps they have taken in order to protect themselves from the negative health effects of climate change (Environics 2008).

Discussion: Implications for Public Health Interventions

Concern About Climate Change Impacts Is Not Leading to Increased Preparedness

The finding that few Canadians are prepared for emergencies seems to be related to low perceptions of vulnerability and risk from emergency events. However, fully two-thirds of Canadians surveyed in 2008 feel personally vulnerable to climate change health impacts (Environics 2008). Therefore, while people believe that climate change is occurring and that they are vulnerable to the health impacts, it does not seem that this is being translated into increased preparedness for extreme weather such as extreme heat events. This may be because many people are not linking climate change with increases in extreme weather. People who indicated that they have noticed changes due to climate change generally did not cite extreme weather among those that they have noticed. Only 13% of Canadians identified more unpredictable or more extreme weather as changes they have noticed in this regard (Environics 2008).

The complacency displayed by many Canadians regarding the need to take protective measures during extreme weather events is a public health concern. For example, vulnerability to extreme heat events in some Canadian cities is high due to a growing risk of such occurrences, the urban heat island effect, an aging population, an increasing number of high rise apartments, and a growing reliance on air-conditioning for cooling. A prolonged and widespread power outage during such an event could result in a large number of heat-related illnesses and deaths among the affected populations and concomitant pressures on health and social services in the affected communities (Riedel 2004).

Greater Efforts Are Needed to Increase Knowledge About Climate Change Health Risks

Efforts to reduce the vulnerability of Canadians to health impacts from climate change through risk communication activities require an understanding by the public of specific risks to health that exist. While such knowledge, in and of itself, may not be sufficient to motivate adaptation, the current lack of understanding poses an important barrier to efforts by public health and emergency management officials to prepare people for climate change. Supported by the results of the survey, the general concern expressed by Canadians about risks to health posed by climate change means that the ground is ripe for behavioral change. People are aware that there is a problem, have some sense of vulnerability to the impacts, and believe human activity is the cause. Because many Canadians accept that climate change is either a health risk now or will be in the near future, there is the potential for strong uptake of messages which provide information on potential health impacts and highlight the need for the adoption of protective measures. However, education and outreach efforts to motivate adaptation will need to be complemented by actions that enhance the capacity of vulnerable populations to plan for, and respond to, the health risks (e.g., strengthening social networks, improving accessibility to services, removing barriers to communication).

Climate Change and Health Risk Communication Messages Need to Be Tailored to Specific Population Groups

To influence behavioral change, risk communication messages concerning the potential health impacts from climate change need to be tailored to specific population groups (Berry 2008). Kovats and Ebi (2006) recognize that heat health warning systems need to be linked to the active identification and outreach to high-risk individuals because passive dissemination of heat avoidance advice is unlikely to be effective. Differences in the perceptions and attitudes of specific populations (e.g., seniors, the poor, immigrants) may significantly impact how hazards are conceptualized and either addressed or ignored (Haque et al. 2004). Therefore, messages to seniors confirming the reality of climate change and outlining the potential for definite, short-term health effects may serve to enhance understanding and increase the participation of this population in preventive actions.

However, among the efforts that health sector and emergency management officials are taking in Canada to protect people from extreme heat events, natural disasters and other climate-related health risks, there is little evidence that the provision of information to, and tailored for, vulnerable populations is a key activity. Very little information geared to vulnerable populations is being disseminated by health authorities in Canada to reduce climate-related health risks (Haque et al. 2004; Paszkowski 2007). A review of 103 public health authority websites in Canada for information on a range of climate-related health risks, including extreme heat events, revealed that 95% of the communication products were tailored for the general public, while only 2% targeted workers or parents with children (Pajot 2008). Few materials were identified that were developed exclusively with the aim of providing advice to seniors or people with chronic diseases. In addition, only 4% of health authorities across Canada provided information specifically on climate change and its potential impacts (Pajot 2008). This may be contributing to the lack of knowledge displayed by most Canadians regarding climate-related health risks and the lack of awareness of actions they can take to reduce these risks. Because vulnerable populations bear a disproportionate burden of environmentally related ill-health on both local and national scales (Health Canada 2003), the lack of information being disseminated to address their needs is a cause for concern. Efforts by public health and emergency management officials to tailor messages to specific populations should also address barriers that prevent information about reducing risks to health from reaching these groups. For example, isolation experienced by many of the 700 seniors and poor people that died in the 1995 Chicago extreme heat event meant that their access to warnings and life-saving social interactions was limited (Hooke and Rogers 2005).

Conclusion

Climate change is expected to increase risks to the health and well-being of Canadians (Seguin and Berry 2008). The results of this case study indicate that Canadians are concerned about climate change and generally feel vulnerable to climate-related threats, but they have little knowledge about how climate change may impact their health. Public health and emergency management officials have an important role to play in communicating information about climate change health risks to Canadians. However, the findings also suggest that few health authorities in Canada are providing information to the public through their websites that addresses the issue of climate change and its expected risks.

Furthermore, many Canadians are not taking the needed actions to adapt to climate-related health risks (e.g., taking action to prepare for an increase in frequency and severity of extreme weather events). Of most concern are findings that indicate the current efforts to change the behaviors of some vulnerable populations, such as seniors, have limited efficacy. Consequently, greater actions are needed by health and emergency management authorities to raise awareness of the health risks related to climate change and promote effective adaptive behaviors. This can be accomplished by increasing the general understanding of current threats, susceptibility, severity and urgency related to climate-related health risks. Efforts are needed to address current gaps so that information is disseminated to Canadians in a format that is accessible, easily understood and retained. To this end, the following recommendations are made regarding climate change and health education, outreach, and research activities:

Education and Outreach

- Increase people's knowledge about the chances of a climate-related emergency or disaster (such as an extreme heat event) happening and its potential impacts on health, well-being, and quality of life.
- Develop information products targeted specifically for vulnerable populations and their caregivers which address differences in perceptions, knowledge, and abilities to take protective actions.
- Equip people with the skills (e.g., networking, protective behaviors, preparing for emergencies, seeking information about risks) they need to assess their own personal risk.

Research

- Increase knowledge related to perceptions of climate change and health issues held by other vulnerable populations in Canada such as the poor, immigrants, people with disabilities, parents with young children, and Aboriginal Canadians.
- Investigate how individuals receive and perceive emergency preparedness information and to what extent such materials change their perceptions of risk, attitudes, and behaviors.
- Assess the effectiveness of current education and outreach products (e.g., fact sheets) for a wider range of climate change and health risks (e.g., safe food handling, safe sun practices, preventing Lyme disease).
- Develop and share information and tools (e.g., best practices) for creating risk communication materials to assist public health and emergency management officials design, implement, and evaluate education campaigns.

References

Banks W (2008) Be prepared, be very, very prepared. Toronto Globe and Mail (26 Jul 2008). http://www.theglobeandmail.com%2Fservlet%2Fstory%2FLAC.20080726. EMERGENCY26%2FTPStory%2FTPEntertainment%2FOntario%2F&ord=44431395& brand=theglobeandmail&force_login=true. Cited 26 Jul 2008

Berry P (2008) Vulnerabilities, adaptation and adaptive capacity. In: Seguin J (ed) Human health in a changing climate: a Canadian assessment of vulnerabilities and adaptive capacity. Health Canada, Ottawa

Berry P, McBean G, Seguin J (2008) Vulnerabilities to natural hazards and extreme weather. In: Seguin J (ed) Human health in a changing climate: a Canadian assessment of vulnerabilities and adaptive capacity. Health Canada, Ottawa

Canada H (1997) Health and environment: partners for life. Health Canada, Ottawa

Canada H (2005) Your health and a changing climate: information for health professionals. Health Canada, Ottawa

Charron D, Fleury M, Lindsay LR et al (2008) The impacts of climate change on water-, food-, vector- and rodent-borne diseases. In: Seguin J (ed) Human health in a changing climate: a Canadian assessment of vulnerabilities and adaptive capacity. Health Canada, Ottawa

Cheng CS, Campbell M, Li Q et al (2005) Differential and combined impacts of winter and summer weather and air pollution due to global warming on human mortality in south-central Canada. Report prepared for Health Canada, Ottawa

CMA (2007) 7th annual national report card on health care. http://www.cma.ca/index.cfm?ci_id=46137&la_id=1&requiredfields=&q=Annual+REport+card+on+health&proxycustom=. Cited15 Oct 2008

Committee on Communication for Behavior Change in the 21st Century: Improving the Health of Diverse Populations (2002) Speaking of health: assessing health communication strategies for diverse populations. Institute of Medicine of the National Academies, National Academies Press, Washington, DC

Confalonieri U, Menne B, Akhtar R et al (2007) Human health. In: Parry ML, Canziani OF, Palutikof JP et al (eds) Climate change 2007: impacts adaptation and vulnerability – contribution of Working Group II to the Fourth assessment report of the Intergovernmental Panel on Climate Change. Cambridge University Press, Cambridge

Ebi K, Balbus J, Kinney PL et al (2008) Chapter 2: Effects of global change on human health. In: Gamble JL (ed) Analyses of the effects of global change on human health and welfare and human systems – Final report, synthesis and assessment product 4.6. U.S. Climate Change Science Program and the Subcommittee on Global Change Research. US EPA, Washington, DC

Ekos Research Associates Inc. (2005) Public perceptions of climate change: annual tracking – Spring 2005. Report prepared for Natural Resources Canada, Ottawa

Ekos Research Associates Inc (2007) Ekos Security Monitor 2006–2007 – public opinion research syndicated research summary. Health Canada, Ottawa

Environics Research Group Ltd. (2008) Assessing perceived health risks of climate change: Canadian public opinion – 2008. Report prepared for Health Canada, Ottawa

European Food Safety Authority (2008) World Food Day event to discuss health implications of climate change. http://www.efsa.europa.eu/EFSA/efsa_locale-1178620753812_1211902127685.htm. Cited 15 Oct 2008

Fouillet A, Rey G, Laurent F et al (2006) Excess mortality related to the August 2003 heat wave in France. Int Arch Occup Environ Health 80(1):16–24

Furgal C (2008) Health impacts of climate change in Canada's North. In: Seguin J (ed) Human health in a changing climate: a Canadian assessment of vulnerabilities and adaptive capacity. Health Canada, Ottawa

Garneau M, Guay F, Breton MC (2005) Modélisation des concentrations polliniques à partir de scénarios climatiques (Partie I). Consortium Ouranos, Montreal [in French]

Glik DC (2007) Risk communication for public health emergencies. Annu Rev Public Health 28(1):33–54

Gosselin P (2004) Health. In: DesJarlais C, Bourque A, Decoste R et al (eds) Adapting to climate change. Consortium Ouranos, Montreal

Gosselin P, Belanger D, Doyon B (2008) Health impacts of climate change in Quebec. In: Seguin J (ed) Human health in a changing climate: a Canadian assessment of vulnerabilities and adaptive capacity. Health Canada, Ottawa

GPC Public Affairs (2005) Are Canadians prepared for an emergency? An analysis of Canadians' preparedness for emergency situations and attitudes toward preparing for them. Report prepared for Public Safety and Emergency Preparedness Canada, Ottawa

Grambsch A, Menne B (2003) Adaptation and adaptive capacity in the public health context. In: McMichael AJ, Campbell-Lendrum DH, Corvalan CF et al (eds) Climate change and health: risks and responses. World Health Organization, Geneva

Haque CE, Lindsay J, Lavery J et al (2004) Exploration into the relationship of vulnerability and perception to risk communication and behaviour: ideas for the development of tools for emergency management programs. Office of Critical Infrastructure Protection and Emergency Preparedness, Ottawa

Health Canada (2003) Climate change and health: assessing Canada's capacity to address the health impacts of climate change. Report prepared for the Expert Advisory Workshop on Adaptive capacity, Ottawa

Hooke WH, Rogers PG (2005) Public health risks of disasters: communication, infrastructure, and preparedness – workshop summary. Institute of Medicine and National Research Council of the National Academies, Washington, DC

Jacque L (2005) Knowledge, attitude and behaviour of the elderly during oppressive heat and air pollution. Paper presented at Adapting to climate change in Canada 2005: understanding risks and building capacity. Montreal, 5 May 2005

Kalkstein AJ, Sheridan SC (2006) The social impacts of the heat – heat-health watch/warning system in Phoenix, Arizona: assessing the perceived risk and response of the public. Int J Biometeorol 52(1):43–55

Kovats S, Ebi K (2006) Heatwaves and public health in Europe. Eur J Public Health 16(6):592–599

Kovats RS, Jendritzky G (2006) Heat-waves and human health. In: Menne B, Ebi KL (eds) Climate change and adaptation strategies for human health. Steinkopff Verlag Darmstadt, Germany

Lamy S, Bouchet V (2008) Air quality, climate change and health. In: Seguin J (ed) Human health in a changing climate: a Canadian assessment of vulnerabilities and adaptive capacity. Health Canada, Ottawa

Lemmen DS, Warren FJ, Lacroix J et al (eds) (2008) From impacts to adaptation: Canada in a changing climate 2007. Natural Resources Canada, Ottawa

Maibach EW, Abroms LD, Marosits M (2007) Communication and marketing as tools to cultivate the public's health: a proposed "people and places" framework. BMC Public Health 7(88): 1–25. doi:10.1186/1471-2458-7-88

McAllister A (2008) The environmental monitor report – 2008 Canadian views on environment, health and sustainability. McAllister Opinion Research. Vancouver, Canada

McBean G (2006) Global change and threats to communities – disaster management. In: Fenech A, MacIver D, Auld H et al (eds) The Americas: building the adaptive capacity to global environmental change. Environment Canada, Toronto

McMichael AJ, Campbell-Lendrum DH, Corvalan CF et al (eds) (2003) Climate change and health: risks and responses. World Health Organization, Geneva

Menne B, Apfel F, Kovats S et al (2008) Protecting health in Europe from climate change. World Health Organization, Copenhagen

Mensah GA, Mokdad AH, Posner SF et al (2005) When chronic conditions become acute: prevention and control of chronic diseases and adverse health outcomes during natural disasters. Prev Chronic Dis 2(special issue):1–4. http://www.cdc.gov/pcd/issues/2005/nov/05_0201.htm. Cited 13 Feb 2008

Murphy BL (2004) Emergency management and the August 14th, 2003 blackout. Institute for Catastrophic Loss Reduction Research Paper Series – No. 40, Ottawa

Pajot, M (2008) Web-based inventory of Canadian risk communication materials related to the health impacts of climate change. Report prepared for Health Canada, Ottawa

Parry ML, Canziani OF, Palutikof JP et al (eds) (2007) Climate change 2007: impacts adaptation and vulnerability – contribution of Working Group II to the Fourth assessment report of the Intergovernmental Panel on Climate Change. Cambridge University Press, Cambridge

Paszkowski D (2007) Heat management in Canadian communities. Report prepared for Health Canada, Ottawa

Pollara (2006) Health care in Canada Survey 2006: a national survey of health care providers, managers, and the public. MediResource, Toronto

Public Health Agency of Canada (2004) Population health: what determines health? http://www.phac-aspc.gc.ca/ph-sp/phdd/determinants/determinants.html#income. Cited 15 Oct 2008

Public Safety Canada (2008) Your emergency preparedness guide: 72 hours . . . is your family prepared? http://getprepared.gc.ca/_fl/guide/national-eng.pdf. Cited 15 Mar 2009

Riedel D (2004) Human health and well-being. In: Lemmen DS, Warren FJ (eds) Climate change impacts and adaptation: a Canadian perspective. Natural Resources Canada, Ottawa

Seguin J (2008) Conclusion. In: Seguin J (ed) Human health in a changing climate: a Canadian assessment of vulnerabilities and adaptive capacity. Health Canada, Ottawa

Seguin J, Berry P (2008) Human health in a changing climate: a Canadian assessment of vulnerabilities and adaptive capacity – synthesis report. Health Canada, Ottawa

Seguin J, Clarke K (2008) Introduction. In: Seguin J (ed) Human health in a changing climate: a Canadian assessment of vulnerabilities and adaptive capacity. Health Canada, Ottawa

Sharma B, Boom D, Mahat, TJ (2008) Building resilience of mountain communities to climate change. Asia Pacific Mountain Network

Sheridan S (2006) A survey of public perception and response to heat warnings across four North American cities: an evaluation of municipal effectiveness. Int J Biometeorol 52(1):3–15

Smit B, Burton I, Klein R et al (2000) An anatomy of adaptation to climate change and variability. Clim Change 45(1):233–51

Tamburlini G (2002) Children's special vulnerability to environmental health hazards: an overview. In: Tamburlini G, Ehrenstein OV, Bertollini R (eds) Children's health and environment: a review of evidence. Environmental issues Report No. 29. Copenhagen

Tierney KJ, Lindell MK, Perry RW (2001) Facing the unexpected: disaster preparedness and response in the United States. Henry Press, Washington

Wolf J, Lorenzoni I, Few R et al (2009) Conceptual and practical barriers to adaptation: an interdisciplinary analysis of vulnerability and adaptation to heat waves in the UK. In: Adger WN, Lorenzoni I, O'Brien K (eds) Adapting to climate change: governance, values and limits. Cambridge University Press, Cambridge

Part III
Adaptation in the Industrial Sector

Chapter 15
Overview: Climate Change Adaptation in Industry

Paul Kovacs

Abstract The success of private industry has long been sensitive to weather conditions. Accordingly, companies regularly adjust their business practices with change in the weather and the climate. Accelerating climate change increases the importance for industry to manage weather risks, and it adds to the difficulty of this process. Case studies presented in this book from the electricity, construction, insurance, and forestry industries identify current barriers and constraints to adaptation, and showcase potential adaptive actions for companies seeking to manage climate risks.

Research on adaptation by industry to climate change is typically addressed within a decision-making framework where complex issues are mainstreamed into a comprehensive risk management system. The specific consideration of any one issue, like climate change, to an overall business decision may be impossible to measure, yet there is a framework for decision-making using well-established tools that can address all potential risks within the broad context that supports action.

Four industry case studies identify a number of challenges evident for private decision-makers and other interested stakeholders seeking to support better decision-making by industry with respect to adaptation to climate change. There are significant differences in climate sensitivity between industries and between companies within an industry, so support mechanisms ideally must be customized for the specific circumstance of each business. Businesses need detailed local projections for many climate variables beyond temperature and precipitation. The importance of climate change for industry decision-making is highly dependent on the planning horizon and climate sensitivity evident in each industry. Also the relative absence of research into anticipatory and reactive adaptation by private industry provides a challenge for private and public decision-makers seeking to understand industry best practices.

P. Kovacs (✉)
Institute for Catastrophic Loss Reduction, Toronto, ON, Canada
e-mail: pkovacs@pacicc.ca; pkovacs@iclr.org

J.D. Ford and L. Berrang-Ford (eds.), *Climate Change Adaptation in Developed Nations: From Theory to Practice*, Advances in Global Change Research 42, DOI 10.1007/978-94-007-0567-8_15, © Springer Science+Business Media B.V. 2011

Keywords Climate change • Industry • Anticipatory adaptation • Risk management • Decision-making • Insurance • Policy options • Economics • Best practices • Disaster • Disaster risk management

Introduction

Leading international business groups, like the World Economic Forum and the International Chamber of Commerce, consistently identify climate change as one of the most important issues facing society and private industry. This issue includes the challenge of reducing emissions and the importance of adapting to change in the climate.

Adaptation is widely expected to become a growing issue for industry as the rate of change in the climate accelerates. In particular, the growing frequency and severity of extreme weather events is expected to increase the importance of investing in adaptation. Some actions to adapt current business practices due to concerns about the climate have the potential for immediate benefits that may significantly exceed the costs incurred and provide a near-term opportunity to improve profitability. Moreover, most decisions concerning adaptation can be made in a context of greater certainty than decisions concerning actions to reduce future greenhouse gas emissions.

Recent Developments: From Theory to Practice

It is assumed in the academic literature that adaptation to climate change is viewed by private industry as a business issue. Adaptation is not free. Companies are expected to adapt by investing in new business practices, equipment, and systems when justified by the anticipated benefits. Accordingly, it is assumed that adaptation has been addressed in private industry, and will continue to be addressed, within the same decision-making processes as other business issues. This is described in the climate change literature as "mainstreaming."

Most authors assume that climate change is managed by private industry within a risk management context. Companies seek to identify potential risks and opportunities that may affect their success over a specific planning horizon. This process seeks to address a broad range of issues that may affect a company, and positions current knowledge about these risks within a comprehensive decision-making process. Climate change is one dimension of a complex risk management process, so it may not be possible to attribute the specific role of climate change in driving business decisions.

Private sector risk management processes are constantly evolving. The business literature presently recognizes "enterprise risk management" as the industry best practice for risk management, and an approach that is increasingly being adopted by private companies across a broad range of industries. Enterprise risk management differs from earlier industry risk management processes because it provides a process for the company to unify and expand upon risk management systems that had been run independently, such as investment risk and business continuity risk. Also, enterprise risk management builds on proven methods to reduce the risk of adverse impacts with the introduction of processes to identify and pursue potential opportunities for gains.

The emerging literature on the economics of adapting to climate change stresses the demonstrated effectiveness of traditional tools, like cost–benefit analysis, to support decision-making. Uncertainty concerning the regulatory environment, climate projections, and the benefits expected from changing practices increases the difficulty of making business decisions; nevertheless, this literature argues that this uncertainty can be addressed within proven and widely accepted decision-making support tools.

The disaster management literature is the largest and most mature body of research documenting the management of climate issues by the general public, governments, and private industry. This literature extends over a period of several decades and largely focuses on responses during and after extreme events including flood, drought, and tropical storms. Over time this research has increasingly balanced the study of disaster response with a growing body of analysis of options to mitigate potential future loss of life and damage to property. Recently, the climate change research community has begun to forge links with those involved in the study of disaster risk reduction, enriching the available information documenting industry management of climate risks.

Key Challenges

There are large differences in vulnerability to climate risks between industries, and between companies within the same industry. Best practices to adapt to climate change may significantly differ from company to company. Nevertheless, a common starting point involves an assessment of the potential financial and operational impact of a range of plausible climate futures. A major challenge for private industry is the acquisition of climate information, historic weather data, and climate forecasts that are relevant to their specific business operations. Further challenges are found in the development of an appropriate corporate understanding of the uncertainty involved in the use of this climate information to influence business decisions.

Industry sensitivity to climate risks often involves specific weather concerns, like farmers concerned about hail or homebuilders concerned about peak wind speed. The major climate models focus on long-term temperature and precipitation forecasts under a range of scenarios, and are often unable to provide the specific climate projections required by private industry.

The primary planning horizon differs significantly between industries, affecting the extent that climate change is considered. Some industries, like power generation and forestry, include aspects of strategic decision-making that extend over periods of several decades – periods where future change in the climate is likely to have a material impact on current decision-making. Nevertheless many industries, including most service industries, operate primarily with a focus on a period of time too short for change in the climate to be as important as natural variability in the weather.

The largest body of existing research concerning how private industry manages climate risks is found in the disaster risk management literature. This research includes studies of how companies around the world in many different industries responded to extreme weather events, including flood, drought, and severe wind. Many studies identify a window of opportunity that opens following extreme events when industry leaders may invest in adaptive measures or introduce new business practices. Much of this analysis is viewed by the climate research community as evidence of reactive adaptation.

Some emerging climate research is seeking to document business efforts to anticipate and proactively adapt to projections of future change in the climate. Much of this analysis has focused on farming practices, but some research is emerging for other industries. However, the lack of published research to identify and evaluate adaptation by industry, particularly anticipatory adaptation, presents a challenge for private and public decision-makers seeking information about best practices to help industry adapt.

Opportunities and Future Directions

The climate literature has frequently identified the potential role for industry to provide leadership in promoting adaptation to climate change; however, the case studies below show that this potential is yet to be realized. Peak load pricing by energy companies, building codes that reflect projected future climate instead of just historic weather experience, insurance companies using their damage claims expectations to influence the behavior of policyholders and building codes, and embedding climate scientists in decision support processes of forestry companies are illustrations of the potential role of industry to promote adaptation, but these actions are only emerging at this time.

The promotion of adaptation in industry will require climate projections that can be readily included in decision-making processes. The projections need to cover a period of time that fits into the planning process. For some industries, like forestry, energy, and built infrastructure, this may cover a period of 50–100 years. For insurers and most other industries, however, strategic plans address a much shorter period of time. The climate information must also address the specific local concerns of industry. This will require regional models and projections of many climate variables beyond temperature and precipitation. It is essential that industry decision-makers can articulate their specific needs to those involved in providing historic weather data and climate models.

A special opportunity is emerging in the management of risks associated with extreme events. These rare but potentially high consequence events have created a growing field of research that has the potential to directly enhance business decisions and the management of future climate risks. Industry is seeking to understand the potential change in the frequency and severity of these perils, and the best practices for managing this risk. The greatest evidence of adaptive behavior in industry is found in response to extreme weather events.

Case Studies

The book includes four case studies that describe the experience of industry in addressing adaptation to climate change. The industries – energy, construction, insurance, and forestry – were among the first within the private sector to recognize climate change as an issue that must be addressed and to take action at the company or industry level. The adaptive actions identified vary considerably across the industries, suggesting that best practices may differ considerably between industries and between companies.

The first industry case study explores adaptation options for the electricity sector. Rothstein and Parey assess the actions taken by the German energy company Energie Baden-Württemberg AG (EnBW) and the French company Electricité de France (EDF). The authors observe that climate change and volatility can have a wide range of adverse impacts that may affect electricity production, transmission, or demand. The specific impacts identified focus on recent extreme events in Europe, including severe flooding in 2002 and the heat wave in 2003.

Severe weather events have the potential to reduce peak capacity in the supply of electricity due to the absence of sufficient water for cooling; disruption as a result of flood damage; or the collapse of transmission systems due to lightning, severe wind, or ice accumulation. Observed short-term adaptation measures to address the potential disruption in electricity production or transmission focus on controlling demand through load management. Some consumers may be cut off. Rolling blackouts can be conducted. Electricity suppliers are seeking to identify controllable consumers, particularly industrial bulk consumers that will agree to reduce demand when required. Suppliers are also introducing variable tariffs, where energy prices are higher in peak load times than during other periods of the day.

These short-term adaptation measures for French and German electricity companies are likely representative of electricity companies around the world, as current production processes do not support the storage of electricity, so the industry focuses on controlling demand. Other energy industries may pursue some combination of supply and demand management.

Medium-term adaptation options identified in the case study include research into climate impacts on energy production and demand, and the identification of options for managing both the supply and demand for electricity. The study does not identify the specific actions taken, but found that in 2005 EnBW initiated a research project to assess the impact of climate change on the business operations of energy companies and to map out possible adaptation strategies. EDF has supported climate change research since the late 1980s. Following the 2003 heat wave, EDF announced adaptation measures to extreme events and committed to long-term research on climate change.

Electricity production and transmission systems are particularly vulnerable to climate change because they have a life expectancy of many decades. Rothstein and Parey observe that around 50% of all German power plants will be shut down within the next 15 years. This renewal presents a significant opportunity to invest in adaptation. The authors stress the value of longer term actions that focus on establishing a diverse range and design of power plants.

The second case study focuses on construction and maintenance of built infrastructure. Steenhof and Sparling explore the challenges and opportunities associated with using codes, standards, and related instruments to mitigate the risk of failure or underperformance of built infrastructure due to climate risks. The specific examples presented in the case study deal with Canada, yet similar codes and standards are in place in most developed nations and are emerging in developing countries, so the findings appear widely representative.

The process of creating codes and standards is ideally based on identifying and documenting consensus within the expert community. A major challenge for this process as it seeks to address climate change involves the management of uncertainty. Uncertainty can arise from the inadequacy of local weather data or from legal liability risks.

The authors find that codes and standards presently focus on historic weather experience and do not yet attempt to assess the climate expected over the life of the infrastructure. As a result, standards may fail to anticipate the conditions under which the built infrastructure will ultimately perform. The risk of error is greatest for assets with longer service lives, as climate conditions may change significantly over the next 50 or 100 years. If codes and standards can be adapted to focus on the climate conditions most likely to prevail during the life of the built infrastructure, this would introduce a powerful mechanism for society to adapt to change in the climate.

Steenhof and Sparling present several recent examples of environmental pressures that have overwhelmed the built infrastructure in Canada and resulted in extensive damage. These examples include severe weather events like the 1998 ice storm and a number of floods, and also some slow onset risks like permafrost degradation that causes infrastructure failures in Canada's North. These examples illustrate the importance of developing codes and standards that reflect local experience and knowledge, and the challenges associated with extreme events, including downbursts and other perils projected to increase in frequency and severity as a result of climate change.

Codes and standards are the primary mechanisms society uses to identify reasonable practice for the planning, design, construction, and management of built infrastructure. Insurance companies and the courts, for example, typically use codes and standards to assign liability when built infrastructure fails.

The poor quality of historic weather data is a challenge for current code and standard development. Uncertainties associated with climate models and long-term forecasts are perhaps a greater challenge for reorienting codes and standards to reflect the expected future climate. In addition, concerns about legal liability appear to have curtailed progress integrating climate change into codes and standards, as some parties appear reluctant to disclose specific infrastructure failures where climate change may have played a role, while others appear to be concerned that this process will result in higher standards of due diligence and accountability.

The third case study explores the potential for insurance and other risk transfer mechanisms to promote adaptation. The authors, Cook and Dowlatabadi, conclude that insurance companies will likely play a minor role in promoting anticipatory actions that reduce the adverse impacts of climate change on built property. They believe that proactive adaptation to climate change will have to be pursued and implemented primarily by government agencies.

Insurance is identified in the case study as the main mechanism in developed nations for risk management in built environments. The authors warn that insurance industry reliance primarily on historic loss experience serves to constrain the potential for insurance practices to promote adaptation to anticipated change in the climate. Insurance premiums that reflect past weather damage experience, not anticipated future climate conditions, will fail to provide incentives for adaptation actions like relocation of property owners away from high-risk zones and the promotion of more robust building codes.

Insurance companies have demonstrated their skill in managing large numbers of uncorrelated losses through the successful provision of life, auto, and fire insurance. The authors question the skill of insurance companies to manage the risks associated with correlated, catastrophic, and infrequent events by providing recent examples of severe weather loss events that resulted in significant and prolonged disruption of insurance markets.

The authors identify the recent introduction of climate models as a development in the insurance industry that should improve the process of setting prices and other insurance terms to reflect future climate risks rather than just historic experience. This process has begun through a greater sharing of climate risk knowledge between insurance companies and their policyholders, but it is yet to evolve further to include, for example, collaboration between insurance companies, city planners, and developers. Such collaboration is essential if insurance practices are to contribute to proactive adaptation.

The marked increase over the past three decades in severe weather damage losses around the world has lead to a reexamination of insurance practices, particularly those associated with low probability but high consequence events. This process involves insurance companies but also other stakeholders including consumers, regulators, the media, and public officials. This review has the potential to build a stronger relationship between governments and insurance companies with respect to the management of climate risks, perhaps following a similar direction to that established over many years with respect to road safety and fire prevention. Despite the potential for insurance practices to promote anticipatory adaptation, Cook and Dowlatabadi observe that this potential has largely been unrealized to date.

The fourth industry case study focuses on forest management. Johnston et al. spoke with government and industry forest managers across Canada to assess their capacity to adapt to climate change and identify barriers to action. They found that the current adaptive capacity of forest managers in Canada is high, but there is evidence that a growing gap may emerge.

The case study found that decision leaders widely recognize that climate change is a growing challenge for forest management. This view is widely supported by research identifying the increasing frequency and intensity of wildfires, increased outbreaks of insects and disease, and growing frequency of drought leading to forest dieback.

The study explores seven dimensions of adaptive capacity. In four areas, capacity was found to be high or sufficient. These include: access to technology; healthy institutions in position to promote knowledge generation and transfer; information about the impact of climate change on forests (although this was often lacking in sufficient local detail to support decision-making); and general awareness of the issue.

Three areas of concern identified in the case study include financial constraints, an aging community of experts, and limitations in the potential for risk spreading. More than 45,000 jobs have been lost in the forest sector over the past decade, and 100 mills were closed. The current industry focus on near-term survival constrains the financial capacity to identify and address actions to adapt to climate change. Moreover, forest management practices are a process that adapt over time, so severe near-term financial challenges have the potential to disrupt industry practices over an extended period of time.

Human capital in the forestry profession in Canada is high. However, an aging workforce and extensive recent layoffs will have a long-term impact. The authors also found that established professional societies, continuing education, and professional schools are not addressing climate change impacts and adaptation.

Significant climate change is expected to occur within the rotation lengths commonly used for forest management, increasing the importance of risk management and risk spreading. If an insect outbreak occurs, for example, spraying pesticides and salvage harvest may be the only options as there may not be sufficient time to consider less susceptible tree species. A more diverse portfolio of commercial tree species on a given land base may be an option for addressing this risk, but this option may be constrained by regulations which frequently stipulate that whatever species is harvested must be replanted.

The authors provide an extensive assessment of critical institutions in the forestry industry, government agencies, and nongovernmental organizations. Johnston et al. identify the concept of embedded science as an effective way for practitioners to deal with complex issues like climate change. Academic and government scientists could work closely with company managers and planners to mainstream science knowledge into the development of forest management plans.

Chapter 16
Impacts of and Adaptation to Climate Change in the Electricity Sector in Germany and France

Benno Rothstein and Sylvie Parey

Abstract The topic of this chapter is climate change adaptation options for the electricity sector in Germany and France. The impacts of weather and climate change on this sector are described first. Based on this, practical adaptation options are then specified for the better use of opportunities and minimizing the preventable risks in the energy industry. Since the impacts and possible adaptation measures need to be considered in regional and company-specific scales, the German energy company Energie Baden-Württemberg AG (EnBW) and the French company Electricité de France (EDF) are focused upon.

The options for adaptation are distinguished between short- and medium-term measures on one hand, and long-term measures on the other. One short-term option for action is load management. By recognizing the changing consumer behaviors, both quantitatively and qualitatively, loads can be adjusted to the momentarily optimized generation of energy. An important medium-term option for action is research into impacts and adaptation. Identifying the impacts of climate change on the sector is essential, because even two companies' susceptibility in the same industry can differ considerably.

A long-term measure for adaptation is to produce a diverse range and design of power plants. While this strategy should be initiated for several reasons, the risks of climate change can be dispersed better, because every type of power generation is affected by climate change in its own specific way.

B. Rothstein (✉)
University of Applied Forest Sciences, Rottenburg am Neckar, Baden-Württemberg, Germany
e-mail: rothstein@hs-rottenburg.de

S. Parey
Électricité de France (EDF) R&D, Chatou CEDEX, France
e-mail: sylvie.parey@edf.fr

J.D. Ford and L. Berrang-Ford (eds.), *Climate Change Adaptation in Developed Nations: From Theory to Practice*, Advances in Global Change Research 42, DOI 10.1007/978-94-007-0567-8_16, © Springer Science+Business Media B.V. 2011

Keywords Climate change • Adaptation • Cooling • Inland navigation • Electricity generation • Transport and distribution • Electricity sector • Consumer behavior • Load management • Power plants design

Introduction

Energy companies are affected by weather and climate. In order to adapt to the impacts of climate change, first the possible weather-caused losses and gains must be identified, and how these impacts of weather patterns can be influenced advantageously. Therefore, some impacts of climate change on different business activities of energy companies are covered in this chapter. Secondly, initial approaches for adaptation are identified. These approaches aim to take advantage of opportunities and minimize preventable risks for the energy industry.

Because the impacts and the possible adaptation measures need to be considered on both regional and company-specific scales, the German energy company Energie Baden-Württemberg AG (EnBW) and the French electricity supplier Electricité de France (EDF) are used as examples. Whereas business activities of EnBW are mainly concentrated in southwest Germany, EDF operates all over France.

The differences between climate change and weather volatility need to be discussed. Of course there are links between them, but heat waves and diminished stream flow events are not necessarily due to climate change. The long-term trends of climate change lead to one set of impacts while the volatility of the weather leads to another. The change in volatility accompanying climate change leads to a combination of both. Whether climate change or weather volatility is the motivation of adaptation is not discussed in this chapter. Furthermore, it must be taken into account that the adaptation strategies which are covered are often not only necessary because of climate change impacts, but can also be useful for strengthening the competitive advantage, regardless of changes in climate. Distinguishing between competition and climate change as driving forces for necessary changes in the activities of companies is difficult, however, and is not discussed in this chapter. Rather, this chapter discusses some risks and opportunities presented to the energy industry by climate change.

Impacts of Weather and Climate Change for the Electricity Sector

In this section, the main impacts of climate change in the areas of generation, transport, distribution, and consumption of electricity are covered briefly.

Generation of Electricity

Weather-related interferences occur rarely at thermal power plants (especially coal-fired, gas turbine, and nuclear power plants). Hence, the frequency of weather incidents occurs less from material loss, but rather due to several restrictions, such as a diminished operational availability. Some possible impacts are specified below.

Water for Cooling Power Plants

Thermal power plants need water for their cooling processes. In order to protect bodies of water, legislation regulates the water usage of every site. These water quality requirements regulate not only the supply and release but also, in case of low water, the evaporation and the warming rate of the body of water.

In Germany, basic principles for the assessment of cooling water released into bodies of water have been devised by LAWA, a German working group on water issues of the Federal States and the Federal Government represented by the Federal Environment Ministry. For example, the maximum allowed temperature increase of cooling water is 10 K, and the maximum allowed water mixture temperature is 28°C. Due to ecological considerations, a river temperature of 28°C is the limit for cooling water release. To ensure this, regular measurements are taken at every thermal power plant in 30-min intervals (EnBW 2004). In France, regulations are negotiated by the regional and national authorities for each power plant.

Climate change impacts on thermal power plants result from two possibly simultaneous reasons: legal constraints for warming and maximum temperature of the body of water, as well as constraints on the amount of water extracted during low water situations to ensure that a minimum discharge is maintained.

The cooling method has a considerable influence on the legal constraints for thermal power plants. The various restrictions are often connected and can intensify one another. In certain atmospheric conditions, problems with maximum temperature or low water situations dominate, as was the case in winter 2005–2006. Often, as in the hot summer of 2003 in central Europe, high temperatures and low water situations occur together. At the Neckar River in Germany, the hot summer of 2003 led to the restriction that only power plants with a cooling tower were allowed to operate (LfU 2004). At EDF in France, agreements have been obtained to maintain production if essential for the network equilibrium, even if the environmental constraints are exceeded, but were used very infrequently.

Furthermore, the inlet and purification of cooling water at thermal power plants may be affected by weather and climate change. A low water situation can lead to a cooling water shortage in the power plant. The inlet to the corresponding building can be dredged, but this must be requested and decided for each individual case. Therefore, a preliminary lead time has to be taken into account.

Hydropower plants need cooling water as well, but compared to thermal power plants, the amount of water withdrawn is marginal. No major impacts on the environment take place.

Water for Inland Navigation

This aspect is much more important for Germany than for France. This is because hard coal-fired power plants play a much more important role for electricity production in Germany (>20%) than in France (4%).

Coal-fired power plants can only be operated economically if their location is in close proximity to the coal district or if the supply of hard coal can be shipped cost-effectively via waterways. Three million tons (t) of coal is transported to the different power plants in Baden-Württemberg in southwest Germany over the Rhine River every year. In comparison with the Ruhr area in northwest Germany, Baden-Württemberg has a serious disadvantage in location due to the higher transport costs of coal (WM 2004). The risk of extreme water levels, which can degrade the reliability and security of the inland water navigation, adds to the high transport costs (Rothstein et al. 2008).

In order to point out the importance of coal transport, consider the coal requirements of a power plant at the Neckar River: the plant needs more than 200 t hard coal per hour, which adds up to a daily use of 3,000–5,000 t of hard coal. This equals two to four shiploads a day or two shiploads and two trainloads, respectively. A trainload consists of 20–25 wagons with altogether 1,000–1,400 t of coal. A cargo ship on the Neckar can transport around 1,200–1,600 t. Seventy percent of the coal arrives at the power plant by cargo ship and the remainder by train; in the future, however, ships will take a more dominant role. Because of high costs, truck transport is generally not suitable for coal supply (Rothstein 2007).

Water levels have always changed due to climate variability. The debate on climate change has raised the question of whether or not the current strategies used by the authorities responsible for management of inland waterways and industrial managers will be suitable to cope with future conditions.

Air Temperature

The efficiency of gas turbine power plants depends primarily on the compressor and the power that is necessary to densify the air that is sucked in. An increase in ambient temperature of 10°C lets the turbine output drop approximately 8–10% (Lepold 1984). Additionally, gas turbine power plants are particularly prone to being affected by air temperatures, i.e., they are less efficient in summer than in winter.

Gas turbine combined cycle power plants are similarly vulnerable to ambient temperatures. The gas process is more dependent upon environmental influences than the steam process. In a 250 MW block, for example, a 30 K lower ambient temperature results in an increased power of 30 MW in the gas turbine but only 3 MW in the steam process (Johnke and Mast 2002).

Water for Run-of-River Power Plants

EnBW and EDF operate several run-of-river power plants on German and French rivers as well as jointly at the Rhine River (e.g., the plants Gambsheim and Iffezheim). In general, the head (i.e., the difference of the water levels) defines the generation of electricity in a run-of-river power plant (Kaltschmitt et al. 2006).

Despite a higher head, the production of electricity lowers with a decreasing discharge. At extreme low water events, the plant may have to be shut down because the turbines work only with a minimum discharge. There is no consistent limit, however, as it depends on the particular power plant. In Baden-Württemberg, the so-called hydropower order ("*Wasserkrafterlass*") has been regulating the identification of the minimum discharge for small hydropower plants (up to 1,000 kW) since 1993 (LfU 2005).

With rising discharge, the production can be increased until the maximum limit of the turbine is reached. During high water and flood events, the generation of electricity may also be shut down and the weir may be opened (Kaltschmitt et al. 2006; BUWAL et al. 2004). Here, a consistent limit is missing as well. Moreover, run-of-river power plants work as regulators for flood waves by means of the corresponding adjustment of the weir (EnBW 2005).

Transport and Distribution of Electricity

Overhead lines are affected by atmospheric influences in several ways, such as failures by lightning, wind, additional loads such as ice or snow, low temperatures, humidity, and moisture.

First, lightning is divided into direct strokes of lightning and induced overvoltage, occurring when lightning strikes close to an overhead line. The high energy resulting from lightning strikes burns the conductor rope and demolishes the isolators, which are protective devices like overvoltage protection (EnBW 2004). Additionally, strikes in pylons, or underground cables, cause a reverse flashover.

Wind-induced failures depend on the wind speed. In addition to the direct wind loads, failures occur when branches or other foreign substances are directed to the circuit by wind. They bypass the conductor rope and cause short-circuits. Moreover, falling trees can damage pylons and conductor ropes, especially at the comparatively low overhead lines of medium voltage (EnBW 2004). Therefore, the density of wood around overhead lines is directly related to the magnitude of failure

(Martikainen 2005). Furthermore, the ropes of overhead lines often swing due to wind, which can also produce a short-circuit (Kiessling et al. 2001).

Failures caused by ice and snow loads arise mainly from ice accretion, which is classified as ice from precipitation and ice from fog or clouds. The combination of additional loads and wind is regarded as having the highest impact on overhead lines (Kiessling 2002). In the autumn of 2005, this combination caused the biggest blackout in Germany since 1945, leaving 250,000 people without electricity and causing a loss of production over 100 million €. In consequence, a guideline regarding the ice loads for pylons is currently being modified. Of course those singular events are not definitely attributable to climate change. Nevertheless, they reveal the strong impact of weather and climate on the electricity branch.

Extreme heat has also been associated with damages to underground cables. During the hot summer of 2003 in France, a number of failures with underground cables occurred, especially around the Paris area. Further to these atmospheric influences, the electricity sector is also vulnerable to droughts and floods. Droughts result in limited conduction capacities (EDF 2003), and floods can cause damage to or even break pylons (Rothstein 2007).

Consumption of Electricity

Air temperature has distinct influence on the demand for electricity, especially in summer and winter. Of course, various other reasons affect the demand as well, e.g., cloudiness, economic trends, and electricity prices.

As electricity cannot be stored, every energy utility has to orient its generation towards the demand for electricity as precisely as possible. To achieve this, the range of power plants is designed so that the demand in base, medium, and peak load can be met at the lowest costs. However, the sought after equilibrium between offered and demanded electricity can become unbalanced for energy utilities as well as for consumers.

In Germany and France, generally the highest annual consumption per month occurs in January, and the lowest occurs in August. The divergence between the off-peak times in summer and the peak times in winter is currently smaller than it was in previous decades. The difference in network utilization between the maximum in January and the minimum in August was about 10% in 2004, but was around 20% in 1974 (VDN 2005) in Germany.

In the hot summer of 2003, the demand in August in Germany rose well above the average, due to the use of air conditioning and ventilation. The trend toward glass-fronted office buildings added to this demand: on bright days, blinds have to be lowered and the lights switched on to be able to work in front of computer screens, and then air-conditioning has to be installed because the buildings heat up on summer days.

The described higher demand for electricity in periods with high temperatures coincides with limited generation capacities that resulted in summer 2003 from preassigned renovations in several power plants, as well as cooling water restrictions. The hot summer of 2003 shows clearly that extreme temperatures affect the generation and consumption of electricity, which can create an imbalance between produced and demanded electricity. Because the phenomenon is spatially wide, exchanges of electricity within the European integrated network are limited.

When estimating the long-term electricity need, the future development of the customer structure is crucial for the energy utility. Especially in France, the possible relocating of companies with high electricity consumptions would play an important role for the energy utility EDF, because the climate between the southern and northern regions is very different.

First Practical Options of Adaptation

Of course, electricity companies have always used numerous measures in order to minimize their exposure to weather and climate. Nevertheless, in recent years it has become increasingly important to deal with weather and climate risks. One reason is the numerous occurrences of extreme weather events, which have induced a higher perception for weather and climate risks. Damage analyses of these events have shown their huge financial impact. For example, the Elbe flooding of 2002 resulted in losses of more than 50 million € for the Eastern German energy industry (EnBW 2003); EnBW estimated the losses resulting from the hot summer 2003 at several million euros (Hartkopf 2004); EDF has quantified the maximum risk of production outage too.

Adaptation needs first to identify the potential impacts of climate change on each company, because the susceptibility of two companies, even of the same industry sector, can differ considerably.

EDF has conducted research projects on climate change since the end of the 1980s. However, uncertainties have limited the development of adaptation measures. Then, EDF designed a *Plan aléas climatique*, which announced adaptation measures to extreme weather events after the 2003 heat wave. Since then, a long-term research project on climate change impact has been assembled. The initial points are the analyses of changes in mean values as well as in the meteorological extremes.

In addition to this project, EDF research and development (R&D) carries out further projects related to climate change. These projects are of interest for many other EDF departments (e.g., risk management, sustainable development).

German electric companies have also dealt with the question of forecasts and impacts of numerous climate parameters on their power plants for several years. In 2005, a research project was initiated by EnBW. This project deals explicitly with the impacts of climate change on business operations of electric companies. Analogous to the research project of EDF, the objectives of this project are to define the specific impacts of climate change on the utility, as well as to map out possible adaptation strategies. Due to the collaboration of EDF and EnBW, an intensive and mutual exchange of project results takes place.

In the next section, some practical options for adaptation are presented. The differentiation between generation, transport, distribution, and demand for electricity is not applicable here, since adaptation measures are often strongly connected with each other as a functional chain. Therefore, using a structure based on time (short-, medium-, and long-term) is more suitable.

Short- and Medium-Term Measures

One essential short-term option for action (which is certainly also necessary without climate change) is load management. By recognizing the changing consumer behaviors, both quantitatively and qualitatively, loads can be adjusted according to the rise and fall of demand to optimize the generation of energy, so that power plants are working according to their capacities and higher or lower voltage levels are avoided.

The load managements comprise multiple options. One simple method is to cut off consumers. This is necessary during flood events or low voltage in the electricity network. Another option is contracts with controllable consumers (especially industrial bulk consumers, but also private electric night storage heating units). Those consumers agree by contract that they will obtain less electricity in case of need, and the energy utility can transfer loads and balance peaks; in return, they pay lower prices. Furthermore, rolling blackouts can be conducted. These are announced over the radio when necessary so that the people involved (e.g., bulk consumers, local authorities, and regions) have enough time for preparation. Calls for saving electricity, especially between 11 a.m. and 3 p.m., are another option. In the summer of 2003, EnBW and EDF addressed the population with this type of press release. A further possibility for the energy utilities to influence the demand, especially of private customers, is to use variable tariffs. This means that the energy price is higher in peak load times (e.g., noon) than during times with sufficient electricity supply (e.g., night). Although the energy utility is primarily interested in a load balance for optimized power plant efficiency, variable tariffs can help to avoid power shortages, especially during heat events. A relatively new measure related to variable tariffs and contracts in the area of Demand Side Management is called Demand Side Bidding. It gives customers or groups of customers (industrial bulk customers or private customer associations) the option of participating in electricity trading with short-term bilateral contracts. Furthermore, the provisions for electricity consumption at EDF are based on modified climatologies, in order to take climatic evolutions into account.

An important medium-term option for action is the use of research project results to improve adaptation. In EDF, one research project deals with a better understanding of climate and weather risks in a medium- and long-term view in order to guide the adaptation of the running power plants to new conditions, for example more frequent heat waves, or to guide the design of new facilities.

Long-Term Measures

A long-term measure for adaptation will be a diverse range and design of power plants. Although this strategy will be implemented for several reasons, risks of climate change can be dispersed better, because every type of power generation is affected by climate change in its own specific way.

In Germany about 40,000 MW power plant capacities will have to be replaced between 2010 and 2030 due to age. The phasing out of nuclear energy results in the need to compensate an additional 21,000 MW. Thus, by the year 2025 around 50% of all German power plants will be shut down (BMU 2005; Schilling 2004; Salzmann 2003; Festerling 2003; Hustedt 2003).

This situation is often regarded as an opportunity (e.g., Ashton 2004; Hustedt 2003) to decide on how the structure and design of the German power plants will look until 2050 or 2060. These decisions need to be made at the present (Festerling 2003). The building of a new power plant costs about 1.2 billion €. Against the background of these long-term binding investments, the plants already have to be designed according to the possible impacts of climate change. Only this will prevent the misdirection of decisive investments (Fischedick and Luhmann 2004). For example, concrete authorization standards (e.g., the ratio between cooling water extraction and warming of cooling water) will be defined at water-related regulations. Thus, the building of new power plants will open up the opportunity to adapt to the existing and expected climate changes (e.g., water temperature, discharge hydrographs) for the future generation of electricity. For example, the German research initiative "Power plants in the 21st century" (KW 21) analyzes to what extent less or almost no cooling water will be needed by power plants in the future (abayfor 2008).

EDF has carried out several studies regarding adapted power plants or power plant operation under modified conditions with a special emphasis on flow conditions and water temperatures.

Outlook

It was shown that the energy industry is affected by climate change and weather risks in different business units. Then, several options for actions that have already been developed in order to help to reduce these impacts were described.

The energy industry is facing a double challenge: to assist in developing climate change mitigation policies and to protect itself against the impacts of climate change. It plays an important role for further mitigation measures, as it is the biggest CO_2 emitter in Germany. The further development of renewable energies and more efficient measures in power plant engineering provides additional possibilities to reduce the energy-related CO_2 emissions in the future. In France, electricity production is already a low CO_2 emitter. The energy industry also has to shield itself against climate change impacts. It is, therefore, essential to find the most efficient and cost-effective combination of mitigation and adaptation.

References

Arbeitsgemeinschaft der Bayerischen Forschungsverbünde [abayfor] (2008) Forschungsinitiative Kraftwerke des 21 Jahrhunderts. http://www.abayfor.de/kw21/projekte_liste.php. Cited 22 Aug 2006 [in German]
Ashton J (2004) WWF Report: Europe feels the heat – the power sector and extreme weather. World Wildlife Fund Power Switch Campaign, WWF International, Gland
Bundesministerium für Umwelt, Naturschutz und Reaktorsicherheit [BMU] (2005) Klimaschutz und Atomausstieg kein Widerspruch. Press release 013/15 [in German]
Bundesamt für Umwelt, Wald und Landschaft [BUWAL], Bundesamt für Wasser und Geologie [BWG], Bundesamt für Meteorologie und Klimatologie [Meteo Schweiz] (eds) (2004) Auswirkungen des Hitzesommers 2003 auf die Gewässer, vol 369, Schriftenreihe Umwelt. BUWAL, Bern [in German]
Electricité de France [EDF] (2003) Plan aléas climatiques. EDF, Paris [in French, unpublished]
EnBW (2004) Ethik- und Nachhaltigkeitsbericht 2004 – Verantwortung übernehmen. EnBW, Karlsruhe [in German]
EnBW (2005) Wasser ist Energie – Wasserkraft bei der EnBW. EnBW, Karlsruhe [in German]
Energie Baden-Württemberg AG [EnBW] (2003) Jahrhundertflut – große Schäden für die Strombranche. EnBW Energie Impulse, vol 1 [in German]
Festerling M (2003) Klimaschutz ist ein Markt ungeahnter Größe. In: Frankfurter Rundschau, 08 Sept 2003 [in German]
Fischedick M, Luhmann HJ (2004) Anpassung an nicht mehr vermeidbaren Klimawandel. http://www.wupperinst.org/download/anpassung-klimawandel.pdf. Cited 5 Sept 2005 [in German]
Hartkopf, T (2004) Annual report 2003. Speech presented at the EnBW annual meeting, Karlsruhe, Germany, 29 Apr 2004
Hustedt M (2003) Vom atomaren zum solaren Zeitalter. ATW 48(2):83 [in German]
Johnke T, Mast M (2002) Leistungsbooster – Steigerung der Gasturbinenleistung bei Bedarf. Siemens power journal online. http://www.powergeneration.siemens.com/download/pool/Mast_d_4.pdf. Cited 24 Nov 2006 [in German]
Kaltschmitt M, Wiese A, Streicher W (eds) (2006) Erneuerbare Energien: Systemtechnik, Wirtschaftlichkeit, Umweltaspekte, 4th edn. Springer, Berlin [in German]
Kiessling F (2002) Grundlagen der Bemessung und. In: Freileitungsnorm in neuer Gestalt. EN 50341 (VDE 0210). Paper presented at the ETG-Fachtagung, Wuerzburg, 15–16 May 2002 [in German]
Kiessling F, Nefzger P, Kaintzyk U (2001) Freileitungen: Planung, Berechnung, Ausführung, 5th edn. Springer, Berlin [in German]
Landesanstalt für Umweltschutz Baden-Württemberg [LfU] (2004) Das Niedrigwasserjahr 2003, vol 85, Oberirdische Gewässer, Gewässerökologie. LfU, Karlsruhe [in German]
Leopold J (1984) Gasturbinen. In: Allianz Versicherungs-AG (ed) Allianz-Handbuch der Schadenverhütung, 3rd edn. VDI, Düsseldorf [in German]
LfU (2005) Mindestabflüsse in Ausleitungsstrecken: Grundlagen, Ermittlung und Beispiele, vol 97, Oberirdische Gewässer, Gewässerökologie. LfU, Karlsruhe [in German]
Martinkainen A (2005) Effects of climate change the electricity network business. In: Kirkinen J, Martikainen A, Holttinen H et al (eds) Impacts on the energy sector and adaptation of the electricity network business under a changing climate in Finland. FINADAPT Working paper 10. Finnish Environment Institute Mimeographs, vol 340. Helsinki

Rothstein B (2007) Elektrizitätswirtschaft als Betroffene des Klimawandels – Eine Identifikation
    von Betroffenheiten und Ansätze zur Anpassung an den Klimawandel dargestellt am Beispiel
    der Energieunternehmen EnBW und EDF. Postdoctoral Lecture Qualification, University of
    Wuerzburg, Wuerzburg [in German, unpublished]
Rothstein B, Scholten A, Nilson E et al (2008) Sensitivity of bulk-cargo dependent industries to
    climate change – first results of a case study from the River Rhine. In: Leal Filho W (ed)
    Interdisciplinary aspects of climate change. Lang, Frankfurt [in press]
Salzmann M (2003) Branche fürchtet keinen Gau im Stromnetz. In: Frankfurter Rundschau, 30
    Sept 2003 [in German]
Schilling HD (2004) Wie haben sich die Wirkungsgrade der Kohlekraftwerke entwickelt und was
    ist künftig zu erwarten. http://www.energie-fakten.de/html/wirkungsgrade.html. Cited 13 Sept
    2005 [in German]
Verband der Netzbetreiber [VDN] (2005) Gleichmäßigerer Stromverbrauch. http://www.vdn-
    berlin.de/akt_strom_netz_belastung_2005_06_13.asp. Cited 09 May 2006 [in German]
Wirtschaftsministerium Baden-Württemberg [WM] (2004) Energiebericht. Bertsch, Leinfelden-
    Echterdingen [in German]

Chapter 17
The Role of Codes, Standards, and Related Instruments in Facilitating Adaptation to Climate Change

Paul Steenhof and Erik Sparling

Abstract Society has developed numerous mechanisms for mitigating risks associated with the potential failure or underperformance of built infrastructure. An important subset are the codes, standards, and related instruments (CSRI) that establish tenets of reasonable practice with respect to the planning, engineering, construction, and management of built infrastructure. After introducing CSRI as fundamental risk management tools for society, this chapter identifies some of the ways in which climate change could begin to undermine these same critical functions of CSRI in the future. It investigates how CSRI and the processes involved in their development must change in order to properly account for climate change, and thereby allow CSRI to play a proactive role in facilitating adaptation on the part of planners, engineers, builders, and managers of built infrastructure. The chapter concludes by offering a number of recommendations for future work in the area of climate change and CSRI. Canada is the region of focus for this chapter, but the issues discussed are relevant for developed nations generally, since all developed nations rely upon CSRI to help set levels of safety and performance in relation to built infrastructure.

Keywords Climate change • Adaptation • Codes and standards • Risk • CSRI • Infrastructure • Built infrastructure • Risk management • Weather • Design values

P. Steenhof (✉) • E. Sparling
CSA Standards, 155 Queen Street, Suite 1300, Ottawa, ON, K1P 6L1, Canada
e-mail: paul.steenhof@csa.ca; erik.sparling@csa.ca

J.D. Ford and L. Berrang-Ford (eds.), *Climate Change Adaptation in Developed Nations:* *From Theory to Practice*, Advances in Global Change Research 42, DOI 10.1007/978-94-007-0567-8_17, © Springer Science+Business Media B.V. 2011

Introduction

Climate-related events regularly affect built infrastructure, in some cases causing disruptions to or the outright failure of services. In response, developed nations have devised numerous mechanisms for mitigating the risks associated with the potential failure or underperformance of built infrastructure. Foremost amongst these mechanisms are the codes, standards, and related instruments (CSRI) that establish tenets of reasonable practice with respect to infrastructure planning, design, construction, and management (Nash and Ehrenfeld 1997).

Climate change presents a host of risks which remain largely unaccounted for in the suite of CSRI that currently inform our planning, design, engineering, and use of built infrastructure (Auld 2008). In particular, climate and weather information contained within or referenced by most CSRI are historical in nature; to the extent that climate change leads to significant shifts in average conditions or in the frequency, intensity, or duration of extremes, CSRI may therefore fail to adequately predict the conditions in which built infrastructure must ultimately perform. This could especially be the case for assets with longer service lives since climate conditions in many locations could change considerably over a period of 50–100 years (Natural Resources Canada 2007; Auld 2008).

This chapter focuses on the role of CSRI in society with respect to built infrastructure and climate change-related risks. It investigates how CSRI and the processes involved in their development must change in order to properly account for climate change, and thereby allow CSRI to play a proactive role in facilitating adaptation on the part of planners, engineers, builders, and managers of built infrastructure. Key challenges that are investigated include the updating and derivation of climate design values, the requirement for better climate data, and potential liability issues associated with climate-sensitive CSRI. We also suggest a number of necessary actions so that CSRI can become instruments that mainstream adaptive practices within the infrastructure community.

Canada is the region of focus for this chapter, but the issues discussed are relevant for other developed nations as well, since all developed nations rely upon CSRI to help set levels of safety and performance in relation to built infrastructure (Sanchez-Silva and Rosowsky 2008).

Built Infrastructure, Climate Impacts, and Relevance for CSRI

Recent Climate Impacts

A number of recent Canadian examples help illustrate the types of impacts that climate-related events can have on built infrastructure. They also highlight a number of issues with respect to the role of CSRI in the mitigation of climate-related risks.

Freezing Rain: Electrical Infrastructure Collapse and the 1998 Ice Storm

Freezing rain events are most common in Canada's southeastern regions where they may cause either minor inconveniences and interruptions, or more significant impacts. In January 1998, a particularly severe ice storm paralyzed eastern portions of the country with over 100 mm of freezing rain in parts of Ontario, Quebec, and New Brunswick (Environment Canada 2008). As reported by Environment Canada (2008), about one million Canadian households lost telecommunications services and electricity, many for weeks on end. Downed power and telephone lines blocked major transportation routes, impeding the relocation of people and the provision of emergency services. The power failure also led to the loss of heating capacity in homes and buildings, and shortages of water where pumping stations and treatment plants lacked sufficient on-site emergency power capacity (Chang et al. 2007).

Infrastructure practitioners directly involved with the transmission and distribution of electricity in Quebec have reported that as a result of the 1998 Ice Storm, there has been a reassessment of climatic design approaches for transmission lines built in the province (personal communication with J. Carrière, Hydro-Quebec, *Oct 29*, 2009). In particular, Hydro Quebec, the main provider of electricity in that province, convened an in-house committee to reassess the potential implications of combined wind and ice loads on transmission infrastructure. This represented a shift in approach away from what had been the largely independent consideration of these important design parameters.

At the national level, the overhead design standard was changed in 2001. Instead of dividing Canada into areas susceptible to "heavy" ice loading (half-inch radial ice thickness) and areas susceptible to "medium" ice loading (quarter-inch radial ice thickness), it moved to the identification of susceptibilities in accordance with "severe" (19 mm), "heavy" (12.5 mm), "Medium B" (12.5 mm), and "Medium A" (6.5 mm) rankings. At the same time, probabilistic design methods for determining climatic loads were included for the first time. This gave designers a methodology for designing for weather loads of any return period, i.e., 50, 150, or 500 years. As explained by representatives from the Canadian Standards Association (CSA), the Standards Development Organization responsible for this standard, the methodology was further refined in the 2006 edition of the national standard (personal communication with J. O'Neil, CSA Standards, *February 11*, 2009).

Flooding: Overwhelmed Urban Storm Water Drainage Systems

Over the last number of years, a series of particularly extreme precipitation events have rendered the capacity of a number of urban storm water drainage systems in Canada inadequate, resulting in enormous damage. For example, in the summer of 2005, a single day of flooding in Toronto, Ontario resulted in over 13,000 sewer backup insurance claims at a value of $247million (Sandick 2007). Meanwhile, Hamilton, Ontario experienced consecutive flooding events in 2004 and 2005 that overwhelmed the city's storm water infrastructure, leading to the flooding of many residential basements (Gainham and Davis 2007).

The vulnerability of Canadian storm water infrastructure to extreme precipitation events is exacerbated by a number of factors. In urban core regions, storm water infrastructure is often 100 years old or older, with much of it at or exceeding its design life. Furthermore, population growth together with the process of urbanization has led to a near-constant increase in the amount of runoff and discharge with which the systems must contend. To the extent that climate change results in more frequent and intense precipitation events (Kharin and Zwiers 2000; Ohmura and Wild 2002; Bruce et al. 2003), the capacity of Canada's aging storm water systems to accommodate peak runoff events will be increasingly compromised. At the same time, much remains to be done in order to update the climatic information within or referenced by the CSRI that inform the character and configuration of storm water systems across communities and watersheds (Denault et al. 2007; Prodanovic and Solobodan 2007; Fowler et al. 2007; Auld 2008).

Permafrost Degradation: Infrastructure Failures in Canada's North

In general, Canada's northern communities have experienced significantly more warming than those in the south (Natural Resources Canada 2007). One of the many serious impacts of this rapid warming has been a pronounced degradation of permafrost in many regions of the circumpolar north, in turn impacting the infrastructure that relies on permanently frozen ground for stability or support (Nelson and Anisimov 2002; Smith and Burgess 2005; Natural Resources Canada 2007; Zhang et al. 2006; CSA 2010). Natural Resources Canada (2007) has confirmed that the continued degradation of permafrost might require remedial action or further engineering modifications in terms of existing infrastructure. Of particular concern are the waste retention ponds and water reservoirs that rely on dams constructed with "impervious" permafrost cores for the retention of hazardous materials or drinking water.

At the same time, many CSRI used in the design and engineering of northern infrastructure were originally developed with southern conditions in mind (CSA Standards 2008). Canada's National Model Building Code and associated standards, for example, have yet to provide design-specific guidance related to permafrost environments. Only recently was a national guideline developed in order to support parties assessing the potential impact of climate change on permafrost and related infrastructure design considerations (CSA 2010). Historically, engineering firms working in the north have developed their own best practices with respect to cold climate engineering, adopting these into master specifications for the infrastructure (CSA Standards 2008).

The challenges currently being experienced in the Canadian North underscore the importance of CSRI which address the assessment and management of regionally and locally specific climate change impacts.

Future Climate Change

Canada's climate is projected to continue to change over the foreseeable future, including increasing average temperatures across most regions; more intense storms and storm surges in some regions; changes in the timing, character, amount, and intensity of precipitation; and changes in ecosystem characteristics, including species composition and water availability (Natural Resources Canada 2007). These changes in climate and accompanying environmental conditions are expected to have a wide range of impacts on built infrastructure in Canada, which in many cases will drive the need for changes in CSRI (see Box 17.1).

Box 17.1 Impacts of Climate Change on Built Infrastructure

The impacts of climate change on built infrastructure have been investigated and documented in numerous sources (Auld and MacIver 2006; Infrastructure Canada 2006; Auld et al. 2007; Auld 2008). Some of the most significant impacts are summarized below.

Water infrastructure

A number of potential interacting factors associated with climate change are expected to impact water infrastructure, including higher withdrawal requirements, generally poorer water quality due to diminished supply, increased risk of flooding, as well as higher incidences of contamination due to sewer overflows.

Transportation infrastructure

Climate change impacts on transportation are expected to include the potential for heightened degradation of roads, runways, and bridges as the result of an increase in freeze–thaw cycles; damage to or loss of roads or bridges as the result of more intense or frequent flooding; and impacts to port facilities as the result of higher intensity storm events and storm surges. Meanwhile, beneficial impacts could include such factors as less snow clearing in regions with less wintertime snowfall.

Building infrastructure

Climate change could result in both beneficial and detrimental impacts on buildings. Some of the most significant beneficial impacts are expected to include a reduction in heating requirements, while major detrimental impacts could include increased snow loads, increased incidences of high intensity winds, and permafrost thaw or degradation leading to failed foundations.

(continued)

Box 17.1 (continued)

Energy infrastructure

Climate change could impact on both energy generation and transmission infrastructure. Changes in precipitation regimes could affect, for example, hydropower generation capacity due to more or less erratic precipitation and higher rates of evaporation. Impacts on energy transmission are expected to include a relative reduction in required transmission capacity for winter-peaking areas, but higher transmission capacity requirements in summer-peaking areas where higher temperatures will lead to a greater use of air-conditioning. This could be compounded by a simultaneous reduction on the lines' transmission capacity due to warmer ambient air temperatures. Meanwhile, there could be physical impacts to transmission facilities as the result of high intensity wind events, freezing rain, forest fires, or other natural hazards such as landslides and avalanches.

Infrastructure Deficit

A key issue that could well be exacerbated by the changing climate is the already poor state of infrastructure in Canada (Policy Research Initiative 2009). North America, in general, faces a daunting "infrastructure deficit" as a result of a multidecadal downward trend of investment in infrastructure relative to other societal expenditures. Consequently, billions of dollars must now be invested over the coming years in order to improve the quality of existing infrastructure and build new infrastructure (American Society of Civil Engineers 2008). The magnitude of these imminent expenditures, coupled with the prospect of climate change compromising the efficacy and sustainability of the eventual assets, provides impetus to critically examine the role CSRI might play in ensuring the adequacy of built infrastructure in a future with climate change.

CSRI and the Management of Climate-Related Risk

Societal Role of CSRI

CSRI play a fundamental role as knowledge sharing- and risk management-oriented tools, shaping the planning, design, engineering, and management of built infrastructure (Grigg 2003; Mills 2003; Liso 2006). Businesses, professionals, and trades people look to CSRI for indications of the most commonly agreed-upon tenets of reasonable practice. As such, many CSRI are referenced in regulations and other legal documents in order to set requirements for infrastructure design and performance. For infrastructure practitioners, such as engineers and those with vocations in the trades, adherence to the guidance and directives contained within CSRI is generally assumed to constitute due diligence vis-à-vis the protection of human and environmental health and safety (CSA Standards 2007; Torys LLP 2008).

CSRI are developed by bringing together key technical know-how with the perspectives of user groups, government regulators, and other societal actors. Most infrastructure-related CSRI are consensus based. They are created with the intent of tapping into well-established technical expertise, in order to establish directives for how to maintain acceptable levels of risk where infrastructure performance, failure, or loss is concerned (Sanchez-Silva and Rosowsky 2008). Hence, CSRI are key mechanisms for consolidating knowledge, as well as de facto risk management decisions for society.

CSRI and Insurance

The adequacy of the rules and guidelines detailed within CSRI with respect to infrastructure-related risks is not only important for infrastructure professionals, practitioners, and their clients, but for other key societal actors as well. Insurers in particular rely to a large extent on the accuracy of CSRI in terms of underwriting property loss or insuring professional practices (Bowker 2002; Mills 2003; Carter and Rausch 2006; Torys LLP 2008). Because site-by-site inspections would in most cases be uneconomical for insurers to carry out themselves, insurers may instead rely on CSRI processes, including appropriate application and enforcement, as proxies for good practice (Kleindorfer and Kunreuther 1999).

Where property insurance is concerned, it is considered prudential practice for contracts to consider whether or not CSRI were and are abided by in the establishment and care of the infrastructure. Where liability insurance is concerned, to the extent that infrastructure professionals, practitioners, and owners diligently apply relevant CSRI, they will generally be protected against liability for damage relating to the failure or poor performance of that infrastructure. Similarly, to the extent that the client has abided by the relevant CSRI, the insurer defending against a liability-related suite in a court of law will have a stronger legal argument in their support (Torys LLP 2008).

But what if guidance within the relevant CSRI fails to accurately reflect the true character or level of risk confronting a particular infrastructure category or process? To the extent that the shortcoming is generally unknown or has yet to be acknowledged, insurers will likely continue to accept adherence to the CSRI as a proxy for good risk management on the part of the insured, but in doing so, will be exposed to unmanaged risk. On the other hand, if knowledge of the shortcoming exists and is reasonably available, strict adherence to the instrument by the insured may no longer constitute the exercise of due diligence, nor, therefore a reliable defense in a court of law (Torys LLP 2008).

Professionals and stakeholders who rely on CSRI, including for indemnification against infrastructure-related liabilities, have a significant stake in ensuring that the instruments are rigorous and up to date, as well as widely referenced and accepted.

Challenges in Accounting for Climate-Related Risks Within CSRI

Since climate affects infrastructure in many ways, CSRI incorporate a wide variety of assumptions and directives in relation to climate and weather conditions, such as temperature averages and ranges, precipitation amounts and durations, wind pressures, and climate-related events such as flooding and freeze–thaw cycles (Auld and MacIver 2006). To enable the adaptation of built infrastructure to the impacts of climate change, CSRI guiding the specifics of infrastructure design, construction, and management must therefore accurately reflect recent and current climate conditions, as well as the likelihood and magnitude of future change. The CSRI community faces a significant challenge in this respect, and progress is only just now beginning to be made.

Reliance on Design Values Based on Long-term Historical Climate Data: The use of climate data and information in the development and application of CSRI, and hence the planning, design, construction, and management of infrastructure, takes place largely through the establishment of climate design values. These values are derived from calculated return periods for extreme weather (e.g., rain, wind, snow, cold, freezing rain) of varying intensities and durations, as well as through the consideration of average conditions (Auld and MacIver 2004). Climate design values are subsequently employed in algorithms that determine the probability and consequences of a failure or lapse in the performance of infrastructure, once climate-related impacts are coupled with other pressures to the system.

Climate design values used in and referenced by CSRI tend to be informed entirely through an analysis of long-term historical conditions over many decades. However, as a result of changing climate conditions, especially over the last 15 years, historical normals in the developed world tend to be far out of date (Holubec et al. 2009; Livezey et al. 2007). Compounding this issue is that capacity at many national-level government bodies for developing and updating design values has been reduced over the course of the last 20 years, as a result of funding cuts (Schiermeier 2006). Many long-term climate normals drawn upon in the application of CSRI were last calculated over 15 years ago.

Notwithstanding the fact that currently many CSRI in Canada are reliant upon out-of-date climate information, the climate continues to change and will do so over the foreseeable future. Thus, normals calculated even today could fail to reflect average conditions or climate extremes just 15 or 20 years from now (Auld 2008; Livezey et al. 2007). Yet many infrastructures are currently being designed, built and managed without regard for potential change in climate (Livezey et al. 2007; Myer 2008). These same infrastructures will be expected to continue delivering their services for, in many cases, much longer than 20 years.

Need for Consensus on Means for Integrating Future Climate Change: Consensus is still evolving on the means for integrating future climate change into CSRI. A number of tools and approaches exist that could become increasingly used in the derivation of forward-looking assumptions and design values. Regional climate models, for example, offer the potential to provide information detailed enough for an approximation of regional change impacts and adaptive responses, including design factors, based upon future climate conditions (Mailhot et al. 2007; Laprise 2008; Maoh et al. 2008; Tighe et al. 2008). Trends analysis is also an approach that could offer a way for practitioners to consider recent changes in climate conditions, and to then use this knowledge in order to consider how climate-related risks might change into the future (Livezey et al. 2007; Steenhof and Gough 2008; Boe et al. 2009; van Oldenborgh et al. 2009). While work has begun to establish approaches for the use and combination of such tools (Mailhot et al. 2007; Livezey et al. 2007; Holubec et al. 2009), it is still in its preliminary stages, and this area will require significant research and effort.

Legal Liability and the Response to Climate Change: Thus far, concerns relating to legal liability appear to have, to some degree, curtailed progress on factoring climate change into CSRI. The character of this obstacle is twofold. First, a significant portion of infrastructure professionals continue to see climate change-related risks as overly ambiguous and therefore difficult to manage; some of these parties may therefore resist returning to the CSRI committee table to commit to a process that could result in higher standards of due diligence and accountability (personal communication with M. Girard from CSA Standards, *October 7*, 2009).

Second, parties may be prone to withholding the disclosure of specific infrastructure failures in which climate-related factors have played a role. While these are the very sorts of lessons or forensic studies required by the CSRI community in order to strive towards the new climate change-informed instruments (personal communication with H. Auld from Environment Canada, *October 12* 2009), disclosure of such cases does not come with the assurance of indemnification (Torys 2008), opening up the potential for other parties to the failed project to sue.

Poor Availability of Data and Information: A general challenge experienced by the CSRI community in respect of climate risk is the dearth of good information at geographically and temporally appropriate scales (CSA Standards 2007, 2008; Auld 2008). In many regions of Canada's vast territory, weather stations are often too diffuse to adequately capture data related to climate change, and especially to capture changes in climate extremes. Compounding this, as discussed previously, is the fact that over the last 15 years, both meteorological monitoring and public service capacity for analyzing collected data in Canada have decreased markedly.

For certain infrastructures, data for some of the most important climate-related design parameters are insufficient with respect to the needs of infrastructure practitioners. For example, in Canada's northern regions data are required on the expected stability or degradation of warm permafrost. Publicly available data on warm permafrost are so sparse that many stakeholders in the private sector actively collect their own data (CSA Standards 2008), some of which may never make it into the public domain.

Conclusion

As the general character of climate change and associated impacts on natural and built environments becomes increasingly well understood, those with responsibilities related to the design, funding, construction, management, and insurance of physical infrastructure and related services will experience increasingly little latitude within which to ignore climate change-related risks. It is therefore in the interest of all parties with a role in the purveyance of infrastructure and associated services to ensure that CSRI appropriately support the management of climate change-related risks.

At the most fundamental level, the ability of CSRI to appropriately and accurately reflect climate change-related risks will be profoundly influenced by the quantity and quality of climate- and weather-related information. The availability of pertinent information is a function of, amongst other things, public sector commitment to meteorological and environmental monitoring; capacity for analyzing data; means for establishing consensus with respect to the development and application of new types of information (e.g., forward-looking climate design values); and reliable means for sharing the information that results. Less often noted is the importance of including on CSRI committees representatives from those regions in which the code, standard, or related instrument is expected to apply, since the inclusion of local knowledge can help to illuminate considerations that would not otherwise be taken into account.

Clearly, climate change poses a significant challenge with respect to built infrastructure and the CSRI that inform its planning, design, building, and management. The costs of renewing climate design values, developing new forward-looking methodologies for their derivation, and improving weather- and climate-related monitoring and analysis will be significant. At the same time, in many developed nations, significant investments are being made in the roads, sewers, transmission lines, and other structures that will support our economies over the years to come. A failure to appropriately factor in current, emerging and future climate-related risks during a period of infrastructure renewal could result in far greater societal costs in the future.

References

American Society of Civil Engineers [ASCE] (2008) Report card for America's infrastructure. http://www.asce.org/reportcard/2009/RC_2009_noembargo.pdf. Cited 2 Jul 2009

Auld H (2008) Adaptation by design: the impact of changing climate on infrastructure. J Public Work Infrastruct 1(3):276–288

Auld H, MacIver D (2004) Cities and communities: the changing climate and increasing vulnerability and infrastructure. In: Fenech A, MacIver D, Auld H et al (eds) Climate change: building the adaptive capacity. Meteorological Service of Canada, Environment Canada, Toronto

Auld H, MacIver D (2006) Changing weather patterns, uncertainty and infrastructure risk: emerging adaptation requirements. In: Proceedings of Engineering Institute of Canada Climate Change Technology Conference, Ottawa, May 2006. Updated version as Occasional Paper 9, Adaptation and Impacts Research Division, Environment Canada, Toronto

Auld H, Klaasen J, Comer N (2007) Weathering of building infrastructure and the changing climate: adaptation options. Adaptation and Impacts Research Division, Environment Canada, Toronto

Boe JL, Hall A, Qu X (2009) September sea-ice cover in the Arctic Ocean projected to vanish by 2100. Nat Geosci 2(5):341–343

Bowker P (2002) Making properties more resistant to floods. Proc Inst Civ Eng Munic Eng 151(3):197–205

Bruce J, Martin H, Colucci P et al (2003) Climate change impacts on boundary and transboundary water management – report submitted to Natural Resources Canada. Climate Change Impacts and Adaptation Program, Natural Resources Canada, Ottawa

Canadian Standards Association [CSA Standards] (2007) Climate change and infrastructure engineering: moving towards a new curriculum. CSA Standards, Ontario

Carter H, Rausch E (2006) Management in the fire service. National Fire Protection Association, Massachusetts

Chang SE, McDaniels TL, Mikawoz J et al (2007) Infrastructure failure interdependencies in extreme events: power outage consequences in the 1998 Ice Storm. Nat Hazard 41(2):337–358

CSA (2010) Infrastructure in Permafrost: a guideline for climate change adaptation. CSA Standards, Mississauga

CSA Standards (2008) Adapting Canada's northern infrastructure to climate change: the role of codes and standards. CSA Standards, Ontario

Denault C, Miller RJ, Lence BJ (2007) Assessment of possible impacts of climate change in an urban catchment. J Am Water Resour Assoc 42(3):685–697

Environment Canada (2008) Ice storm. Government of Canada. http://www.msc-smc.ec.gc.ca/media/icestorm98/index_e.cfm. Cited 25 Jun 2009

Fowler HJ, Slankinsop S, Tebaldi C (2007) Linking climate change modelling to impact studies: recent advances in downscaling techniques for hydrologic modelling. Int J Climatol 27(12):1547–1578

Gainham C, Davis G (2007) Wet weather woes – Hamilton's storm event response. Influents – Official publication of the Water Environment Association of Ontario. Spring 2007(2):54–56. http://www.weao.org/archive/2007_01spring/Influents_2007_Spring.pdf. Cited 25 Jun 2009

Grigg N (2003) Water, wastewater, and stormwater infrastructure management. CRC Press, Florida

Holubec I, Auld H, Fernandez S et al (2009) Climate (temperature) design criteria for permafrost regions under climate change. Environment Canada, Government of Canada, Downsview

Infrastructure Canada (2006) Adapting infrastructure to climate change in Canada's cities and communities: a literature review. Government of Canada. http://www.infc.gc.ca/research-recherche/results-resultats/rs-rr/rs-rr-2006-12_02-eng.html. Cited 25 Jun 2009

Kharin VV, Zwiers FW (2000) Changes in the extremes in an ensemble of transient climate simulations with a coupled atmosphere-oceans GCM. J Clim 13(21):3760–3788

Kleindorfer PR, Kunreuther HC (1999) Challenges facing the insurance industry in managing catastrophic risks. In: Froot KA (ed) The financing of catastrophe risk. National Bureau of Economic Research, Cambridge, MA

Laprise R (2008) Regional climate modelling. J Comput Phys 227(7):3641–3666

Liso KR (2006) Integrated approach to risk management of future climate change impacts. Build Res Inf 34(1):1–10

Livezey R, Vinnikov K, Timofeyeva M et al (2007) Estimation and extrapolation of climate normals and climatic trends. J Appl Meteorol Climatol 46(11):1759–1776

Mailhot A, Duchesne S, Caya D et al (2007) Assessment of future change in intensity-duration-frequency (IDF) curves for southern Quebec using the Canadian regional climate model (CRCM). J Hydrol 347(1–2):197–210

Maoh H, Kanaroglou P, Woudsma C (2008) Simulation model for assessing the impact of climate change on transportation and the economy in Canada. Transp Res Rec 2067:84–92

Mills E (2003) The insurance and risk management industries: new players in the delivery of energy-efficient and renewable energy products and services. Energy Policy 31(12):1257–1272

Myer M (2008) Design standards for U.S. transportation infrastructure: the implications of climate change – report submitted to Transportation Research Board. http://pubsindex.trb.org/document/view/default.asp?lbid=864525. Cited 2 Jul 2009

Nash J, Ehrenfeld J (1997) Codes of environmental management practice: assessing their potential as a tool for change. Annu Rev Energy Environ 22(1):487–535

Natural Resources Canada (2007) From impacts to adaptation: Canada in a changing climate 2007. Government of Canada, Ottawa

Nelson FE, Anisimov OA (2002) Climate change and hazard zonation in the circum-Arctic permafrost regions. Nat Hazard 26(3):203–225

Ohmura A, Wild M (2002) Is the hydrological cycle accelerating? Science 298(5597):1345–1346

Policy Research Initiative (2009) Climate change adaptation in the Canadian energy sector. Government of Canada, Ottawa

Prodanovic P, Solobodan S (2007) Development of rainfall intensity, duration frequency curves for the city of London under the changing climate – report for the city of London, Ontario, Canada. http://publish.uwo.ca/~pprodano/WGReport_rev1.pdf. Cited 2 Jul 2009

Sanchez-Silva M, Rosowsky DV (2008) Risk, reliability and sustainability in the developing world. Proc Inst Civ Eng Struct Build 161(4):189–197

Sandick D (2007) Basement flooding: lessons from Edmonton and Toronto. Risk Management, Dec 2007 edition:9–11

Schiermeier Q (2006) Arctic stations need human touch. Nature 441:133

Smith SL, Burgess MM (2005) Recent trends from Canadian permafrost thermal monitoring network sites. Wiley, Toronto

Steenhof PA, Gough WA (2008) The impact of tropical sea surface temperatures on various measures of Atlantic tropical cyclone activity. Theor Appl Climatol 92(3–4):249–255

Tighe SL, Smith J, Mills B (2008) Evaluating climate change impact an low-volume roads in southern Canada. Transp Res Rec 2053:9–16

Torys LLP (2008) Legal liability as a driver of and barrier to climate change adaptation in infrastructure projects. Torys LLP, Toronto

van Oldenborgh GJ, Drijfhout S, van Ulden A et al (2009) Western Europe is warming much faster than expected. Clim Past 5(1):1–12

Zhang Y, Chen W, Riseborough D (2006) Temporal and spatial changes of permafrost in Canada since the end of the Little Ice Age. J Geophys Res 111(D22):14

Chapter 18
Learning Adaptation: Climate-Related Risk Management in the Insurance Industry

Christina L. Cook and Hadi Dowlatabadi

Abstract Insurance is a prominent, well-established mechanism for risk transfer in developed countries. While North American governments have stalled on both mitigation of and adaptation to climate change, the insurance industry (globally and in North America) is already viewing recent catastrophic events as being partially climate change related and exploring new adaptation initiatives. In general, the intent of these initiatives is to assure the prosperity of the insurance sector, not to prevent damage to life and property. Reliance by insurers on predictive risk modeling continues to be limited, as new initiatives are prompted by extreme events rather than modeled projections of damage. As another example of reactive behavior, insurers rely on legal judgments to determine the extent of their liabilities. This pattern of learning and response has two implications. First, opportunities for anticipatory adaptation prompted by insurer initiatives are very limited, which guarantees continued large losses from extreme events into the future. Second, proactive risk mitigation will have to be pursued and implemented on behalf of public welfare by the relevant branches of government and cannot be left to market forces.

Keywords Anticipatory adaptation • Insurance • Climate change • Institutional learning • Climate-related risk management • Climate change damages • Risk management • Property risk • Catastrophic events • Risk modeling

C.L. Cook (✉)
Institute of Resources, Environment and Sustainability, University of British Columbia,
Vancouver, BC, Canada
e-mail: clcook@interchange.ubc.ca

H. Dowlatabadi
Institute of Resources, Environment and Sustainability and Liu Institute for Global Issues,
University of British Columbia, AERL, 422–2202 Main Mall, Vancouver, BC, V6T 1Z4 Canada
e-mail: hadi.d@ubc.ca

J.D. Ford and L. Berrang-Ford (eds.), *Climate Change Adaptation in Developed Nations:* 255
From Theory to Practice, Advances in Global Change Research 42,
DOI 10.1007/978-94-007-0567-8_18, © Springer Science+Business Media B.V. 2011

Introduction

In the debates on climate policy, mitigation and adaptation are often viewed as complementary actions. In other words, effective mitigation obviates the need for adaptation and vice versa. For adaptation to be effective, we need to be able to project impacts of climate change, plan risk-reducing responses, and implement them. These anticipatory adaptation measures reduce the impacts of climate change.

Insurance as a main mechanism for risk management in built environments can be an effective adaptation and risk reduction tool. Insurance premiums can prompt relocation away from high-risk zones and promote more robust building codes (Kunreuther 1996). However, as long as the industry uses "experienced risks" as their basis for risk assessment, anticipatory adaptation to climate change will not be realized. Models can be used for projecting the impacts of climate change and other trends on risk to property and persons, design of appropriate risk pools, and rate setting. In other words, continued reliance on past loss experiences, the traditional means of setting insurance rates, will not encourage proactive adaptation to climate change.

This chapter reports on the role of past experience, ambiguity in insurance coverage contracts, and the role of legal precedent to conclude that insurance companies will, in all likelihood, play a minor role in anticipatory actions that reduce impacts of climate change to built property.

Risk Management

Risk management involves first, the identification, characterization, and quantification of risk occurrence probability and consequences; and second, the development of strategies to reduce event probabilities and/or ameliorate their adverse consequences once they have occurred. Global climate change was first identified as a risk in the 1970s. The first stage – characterization and quantification of climate change risk – has taken time and involved many bodies, the most prominent being the Intergovernmental Panel on Climate Change (IPCC). Article 4.8 of the United Nations Framework Convention on Climate Change (UNFCCC) details measures for risk management and has been the subject of increasing international attention as the magnitude of climate change impacts and inadequacy of current mitigation efforts grow more evident. Effective climate change risk reduction must address both the drivers and the impacts of climate change. Drivers of climate change have been the subjects of the Conference of the Parties annual negotiations to the UNFCCC. Attempts at reducing the impacts of climate change have not enjoyed a similar concerted effort.

Trends in losses due to extreme weather over the past three decades are unmistakable (Lott and Ross 2006). While the debate over the relative roles of climate change and human factors in explaining them rages on (Changnon et al.

2000; Pielke et al. 2005), the need to manage these risks more effectively remains. In the three decades since climate change risk was first identified, advocates of climate mitigation have focused on climate-related losses starting with Cline (1992) and continuing with Stern (2006), while others have tried to develop the risk context as modified by the dynamics of climate change (e.g., Casman and Dowlatabadi 2002; Pielke and Sarewitz 2002; Reiter et al. 2003). The latter have tried to identify and rank climate-related risks within the myriad set of risks facing natural and human-mediated systems, and to focus on the need to address the multiplicity of factors that lead to vulnerability writ large (Pielke 2007).

Although systematic risk mitigation is generally rare, it could effectively address many risks (including those related to climate change). The new awareness of risks from climate change may incite the adoption of systematic risk management in arranging human activities from supply chain logistics to urban planning, which would surely be a benefit to society. Insurance has long been a risk management tool and is now recognized as a key element in public–private initiatives for spreading risks temporally, geographically, and among diverse social and commercial communities, especially in climate change-related impacts (Mills 2005; Mills et al. 2005, p. 45). In order for systematic risk mitigation initiatives to be successful, insurers and institutions within which they operate need to learn about emergent risks and develop workable strategies to use projections of future risk when assessing risk tolerance, developing rate schedules, and taking the extra step of being proactive in reducing risks.

The Industry

Insurance companies are not simply providers of an important societal function. Risk underwriting is only half of their core business, with the complement being asset management. Premiums collected for insurance coverage provide capital for asset acquisitions, the management of which provides the financial means to underwrite risk and maintain the profitability of the industry. When interest rates are low, higher risk clients are not courted. A higher risk client represents a potential loss that may make an underwriting less profitable than simply borrowing the money from the capital markets. On the other hand, when interest rates are high, the insurance industry has higher tolerance for clients that would, in a low interest market, be considered uninsurable.

The insurance industry is considered an exemplar of the science of risk management involving large numbers of uncorrelated losses (e.g., life, car, fire). However, the relatively large losses sustained by national insurance companies due to geographically concentrated weather events, such as hurricanes in Florida and Mississippi, hint at an industry that has yet to assess and manage risks from correlated, catastrophic, and infrequent events successfully. These significant correlated losses have produced feedbacks among insurance regulators, the insurance industry, and capital markets. First, rating agencies have put large insurers on notice for possible ratings downgrades (Mills et al. 2005, p. 45). Second, to limit exposure to natural catastrophe risk, insurers have begun development of financial instruments for sale into the capital markets (Catovsky 2005). While natural catastrophes present the risk of large correlated losses, climate change is expected to aggravate this situation. The challenge, as usual, extends beyond ambiguities in climate change and probability of extreme events to include significant changes in land use, building technologies, demographics, and finally economic and social interdependencies. In addition to the challenge of assessing the risk of an extreme event occurring in a given location, we need to project the impacts of such an event in that region.

Event-Based Learning

Historically, catastrophic events have been the main driver of insurance provision. Fire is one such catastrophe. The history of the provision of fire insurance is demonstrative of the evolution of insurance provision. In the early days, private fire companies provided fire services and exemplified private risk management contracts, but these grew competitive and inefficient in coverage. As a result, a public approach emerged, and fire services were municipalized (Carlson 2005). Today, the public fire service is augmented with private fire insurance in hybrid risk management. Public–private risk sharing arrangements and insurer regulations are also the norm in property theft (public policing and private insurance) and auto accident insurance (public safety infrastructure and regulations, and private insurance) among others.

In parallel with the emergence of public–private risk management regimes, climate change risk makes it imperative that insurers consider concentrations of risk in space, category, time, etc. – the risk of large correlated losses. Here, insurance providers have to balance the benefits of understanding a risk, against the challenge of being exposed to a concentration of loss events in that area.

Experience demonstrates that it is not the magnitude of loss from extreme events that determines the performance and stability of insurance companies, but rather the health of their assets. In 1992, Hurricane Andrew resulted in $22 billion (USD 2004) of insured damage: a payout that drove 11 insurance providers to file for receivership (Catovsky 2005). By contrast, although losses from the 2005 hurricane season (including Katrina and others) were nearly thrice the losses from Andrew, only one insurance firm was forced into insolvency. This more fortunate outcome has been attributed to improved asset management (not better assessment of risks from extreme weather): in 1992, industry assets were performing poorly, whereas in 2005, assets were highly profitable and largely unaffected by contemporaneous hurricanes (personal communication with D. Hoffman, Insurance Industry Expert, Washington DC, 8 May 2006).

Notwithstanding this limited loss to the industry, some insurance providers found themselves overexposed in certain markets. The 2004 and 2005 hurricane seasons finally influenced the industry to pay attention to geographically correlated risks.

In May 2006, just weeks before the start of the 2006 hurricane season, Allstate Insurance Company refused to renew coverage or issue new policies in 14 coastal counties of Texas, as well as New York City, Long Island and Westchester County in New York (Adams 2006). The company reasoned that their leadership position in homeowner underwriting had created too much correlated risk exposure on the east coast of the United States. Exposure in the more northerly latitudes of the Atlantic coast was curtailed on the basis that the region was "due" for a large hurricane event – the last such disaster having been in 1938. Allstate was not alone: Metropolitan Life Insurance Company also reduced provision of new homeowners' policies near the coast (Adams 2006).

Other examples of event-based learning can be found in fire insurance (e.g., the Great Fire of London of 1666, see Carlson 2005), earthquake coverage (e.g., the 1906 San Francisco Earthquake and Fire, see Guatteri et al. 2005), and flood damage insurance (e.g., Hurricane Betsy, 1965, precipitating the development of the National Flood Insurance Program in the United States, see Federal Emergency Management Agency 2002).

Model-Based Learning

Probabilistic risk analysis has been a major part of the engineering toolkit for half a century. It was first adopted in insurance for estimation of risks from earthquakes (Cornell 1968). Weather-related risk assessment modeling is more recent. Increasingly, probabilistic models are being used to estimate the risks from extreme events. The state governments of Florida (for hurricanes) and California (for earthquakes) have developed a process of model approval and certification. Models are then used to set rates and structure the coverage (i.e., deductible, cap, and premium) offered to potential clients.

An important part of any insurance contract is the "deductible" portion of the payout in the event of a loss. Frequent smaller loss claims can accumulate into large sums over time. To that end, insurers have begun to mandate percentage deductibles, rather than fixed-value deductibles (Mills et al. 2005, p. 45). Insurance companies can help their clients understand the risk inherent in where they have chosen to live, for example, and how to mitigate such risk. However, this is often done after a particular tract of land has been developed. What is desperately needed is a close collaboration between insurance companies, city planners, and developers. Collaboration is essential to ceasing development of high-risk regions and adopting building practices that are appropriate to local conditions – a true *ex ante* risk reduction.

Litigation-Based Learning

Modeling and experience of large losses have refined insurance providers' understandings of which risks are insurable and, consequently, their underwriting behavior. Hazards that have proven too risky for private insurance have devolved to public instruments except where governments have mandated private insurance provision. Insurance contracts enumerate covered perils or risks and can include anticoncurrent causation clauses (ACC clauses) that limit coverage by excluding damages considered indivisibly caused by an excluded and covered peril. In situations where damage is difficult to attribute, also referred to as the problem of proximity, parties may contest at court the extent of a policy's coverage. A case that arose from damage caused by Hurricane Katrina on the Gulf Coast, *Tuepker* v. *State Farm Fire & Casualty Company*, is illustrative of the problem of proximity and ACC clauses (2007 WL 3256829, US Ct of Apps, 5th Cir. 6 November 2007; 2006 WL 1442489, S.D. Miss. 24 May 2006). The plaintiff Tuepkers sought to recover from their State Farm Policy damages for flooding caused concurrently by wind and rain. The lower court and appeal court dealt directly with the issue of proximity and ACC clauses, which operate, in effect, to limit proximity. At appeal, the Fifth Circuit Court reviewed the law relating to excluded perils and ACC clauses, and held that the lower court correctly found that damage attributable to flooding, an excluded peril, was not covered. Regarding the ACC clause, the lower court found it "ambiguous and ineffective to exclude damage proximately caused by wind or rain." Given the ambiguity, the lower court, construed the policy against the insurer (as mandated by Mississippi law) and awarded flood damages to the Tuepkers, with the burden of proof to State Farm to determine and pay out the damages quantum attributable to wind, the covered peril. The appeal court overturned the lower court's decision and found instead that the ACC clause was not ambiguous. More specifically, the court found the policy clearly stated that "indivisible damage caused by both excluded perils and covered perils or other causes is not covered." Thus, the appeal court held that the Tuepkers' State Farm Policy did not cover any part of the flood damage regardless of what caused it. The parties settled the damages quantum out of court.

Settlement of this case limits the ability to learn from it. Arguably, the *Tuepker* case clarifies nothing on the extent of insurance coverage in Mississippi: the issue of proximity was resolved by interpretation of a clause in the insurance policy. Nonetheless, it does highlight that damages claimed from climate change-related perils will be fraught with problems of proximate cause. The outcome of a legal action can rarely be anticipated with accuracy and will often take considerable time to be determined. These two facts alone imply litigation should not be a primary means of learning about climate change risk liability.

Another factor that militates against recourse to litigation in order to tweak risk allocation is that insurance is a risk management tool that functions by assigning risk *ex ante*. Legal judgments can disrupt contractual risk shifting and contribute to "chronic effects that undermine the optimal functioning of liability insurance markets" (Abraham 2002). In effect, the result is an *ex post facto* expansion of coverage by the courts for unforeseen occurrences and a subsequent contraction of coverage *ex ante* of future insurance contracts by providers (Cummins 2002). Moreover, as in the *Tuepker* case, the ability to "learn" from litigation is severely constrained by the propensity of litigants to settle.

Climate change may represent the next frontier of liability litigation (for academic discussion of climate change litigation, see Grossman 2003; Healy and Tapick 2004; Hsu 2008). In the last few years, a handful of cases have been launched in the United States for climate change damages. Many of these cases allege public nuisance, a notoriously difficult cause of action and several have been dismissed. A notable success in climate change case litigation in the United States is *Massachusetts v. EPA* 549 US 497 (2007), where the US Supreme Court found that greenhouse gases are pollutants under the *Clean Air Act*[1] which the Environmental Protection Agency (EPA) has the right to regulate.

In February 2008, "the first climate change case that seeks damages from industry for the loss of property due to global warming" was filed (Native American Rights Fund 2008). Kivalina, an Alaskan Inuit village, seeks damages for loss of property resulting from greenhouse gas emissions of extraction industry (petroleum and mining) and electric power companies because the "[i]mpacts of global warming have damaged Kivalina to such a grave degree that Kivalina is becoming uninhabitable and must now relocate its entire community" (Center on Race Poverty & the Environment and Native American Rights Fund 2008). According to Hsu (2008), a case such as *Kivalina*, brought by aboriginal plaintiffs, in nuisance for property damages against defendants including electric power companies is ideal to test the limits of climate change liability.

Nonproperty Insurance

In much of this chapter, we have focused on underwriting of property risks due to climate change. Of course, the industry also offers many other types of insurance coverage including drought, crop, life, business interruption, and director and officer liability. Director and officer liability insurance has been among the less defined forms of exposure. However, it may be one of the most potent weapons in stakeholder efforts to incite a greater sense of urgency and action at large energy producing and greenhouse gas emitting companies (Mills et al. 2005, p. 45). At a workshop in 2006, the insurance industry was asked about their preparation for this eventuality, and whether modeling would be a useful approach to inform them of their potential exposure and hence needed revision to the price of and terms of coverage. A prominent insurance association representative responded that the industry would only learn the true extent of their exposure for director and officer liability in a court case. While this may be true in part, the authors believe that the industry's reluctance to examine its exposure in the area of director and officer insurance demonstrates a lack of preparedness.

---

[1]42 U.S.C., Chap. 85.

Institutional Setting

The insurance industry operates within a complex social and institutional framework. Caught between profitable lines of actuarial risks and uncertain catastrophic risks, the industry is forced to underwrite and cross-subsidize risk without full knowledge of the extent of exposure. Catastrophic events lead to institutional failures and opportunities to rewrite the rules of engagement affecting three aspects of risk management: (1) what is insurable in the private sector; (2) how much more risk mitigation will be carried out by public institutions; and (3) other provisions for covering privately uninsured/uninsurable losses. Alas, events, not models, have been leading the evolution of the industry.

While the industry has attempted to restrict coverage in some markets (e.g., in the United States, flood insurance, since 1968) and declined to renew policies it considers too risky (e.g., coastal areas that are at great risk for hurricane damage), states have mandated industry participation (e.g., the Florida Property and Casualty Joint Underwriting Association, see Florida Disaster 2008). These opposing forces have had two outcomes. First, the industry has been forced to underwrite coverage it has determined is too high risk. To mitigate this increased risk, the industry has: stipulated better building practices and investment in damage mitigation by the insured; made reinsurance more expensive; and used alternative risk transfer instruments such as industry loss warranties, catastrophe bonds, and sidecars to share underwriting risks (Jeffee and Russell 1997; Catovsky 2005; Wharton Risk Center 2007).

Second, state-run insurers have become ubiquitous and are now, in some cases, insurers of first choice rather than last resort (Insurance Information Institute 2008). The Insurance Information Institute (2008) reports that state-run property insurers (e.g., Florida Citizens Property Insurance Corporation, Louisiana Citizens Property Insurance Corporation, Mississippi's Fair Access to Insurance Requirements plans) have experienced "explosive growth . . . total exposure to loss in the plans surged from $54.7 billion (USD) in 1990 to $656.7 billion (USD) in 2006." Perhaps not surprisingly, "many [state-run insurers] operate at deficits, or from slim positions of surplus, even in years with little or no catastrophe losses" (Insurance Information Institute 2008).

It is often suggested that the government, as the insurer of last resort, is the best placed to experience internalized costs of risky activity. However, mitigation of risk through displacement of existing settlements is always unpopular (Priest 2003). Before a disaster, such interventions are perceived as unnecessary and heavy-handed, while after the disaster they are seen as insensitive and callous (e.g., the furore over discussions to not rebuild the lower Ninth Ward in New Orleans). These challenges reflect the unfortunate reality of higher risk settlements being the only possibility for the lowest income groups in some regions. Such development characteristics lead to a regressive distribution of risks from natural disasters: lower income neighborhoods are also most likely to be underinsured. So, while addressing such sociopolitical concerns is difficult, they must be approached head-on if we hope to improve disaster management.

The current situation is far from ideal because the risks of continuing climate-sensitive activities – e.g., farming where there is an ever increasing risk of drought, or building where there is high risk of storms or inundation – are not yet internalized. Appealing to market-based mechanisms will not solve this issue, because markets will soon learn the extent of their exposure and will either seek assurances against catastrophic losses or refrain from supplying insurance.

Conclusion

So far, extreme event losses have dealt only a small blow to insurance industry coffers. Increased exposures and new probabilistic models estimating the risks quantitatively are having a greater influence on underwriting decisions. Unfortunately, this has not prompted the industry to offer better-informed terms for underwriting. Rather, the industry has learned that some former underwritings are not insurable at rates that are acceptable to consumers. We expect this will eventually lead to a renegotiation of risk management through a coordination of private and public entities that will be best achieved in an atmosphere of cooperation.

In countries with a well-established insurance industry, climate concerns are forcing a careful reexamination of underwriting and risk mitigation practices. We hope this reexamination will recognize the inadvisability of property developments in hazardous areas, transfer some risks and burdens of mitigation to property owners, and engage the government and new instruments for provision of risk coverage. The recognition that many such risks are created by our own lack of foresight could entail significant public benefits. In the interim, the costs of transitioning to a more enlightened pattern of land use are likely to fall on the shoulders of the poor, who cannot afford insurance, and on the government, as an insurer of last resort, while fumbling to find the right mix of policies to mitigate risks.

Postscript

This chapter was completed before the financial crisis emerged in 2008. The events since then further emphasize the importance of asset risk management to the survival of the insurance industry. In the United States, the Federal Reserve, as insurer of first resort, has granted at least $144 billion (USD) to insurance giant American International Group Inc. While much of the discussion of climate-related losses has focused on impacts due to extreme events, these observations demonstrate the importance of balanced risk management of both assets and underwritings. The near failure of major insurers highlights the need for insurers to approach risk assessment systematically.

Acknowledgments The authors are grateful for the helpful comments of two anonymous reviewers. Thank you to Jean-Nöel Guye, Daniel Hoffman, Howard Kunreuther, Lester B Lave, Erwan Michel-Kerjan for many helpful suggestions. And thank you to Iris Grossmann for her insightful review.

This research was made possible through support from the Climate Decision Making Center (CDMC) Department of Engineering and Public Policy at Carnegie Mellon University. This Center has been created through a cooperative agreement between the National Science Foundation (SES-0345798) and Carnegie Mellon University.

References

Abraham KS (2002) The insurance effects of regulation by litigation. In: Viscusi WK (ed) Regulation through litigation. Donnelley, Harrisonburg

Adams M (2006) Strapped insurers flee coastal areas. USA Today (26 April 2006)

Carlson JA (2005) The economics of fire protection: from the great fire of London to rural/metro. Econ Aff 25(3):39–44

Casman EA, Dowlatabadi H (2002) The contextual determinants of malaria. Resources for the Future, Washington, DC

Catovsky S (2005) Financial risks of climate change. Association of British Insurers, London

Center on Race Poverty & the Environment, Native American Rights Fund (2008) Kivalina complaint. San Francisco. http://www.climatelaw.org/cases/country/us/kivalina/kivalina. Cited 22 Sep 2008

Changnon SA, Pielke RA, Changnon D et al (2000) Human factors explain the increased losses from weather and climate extremes. Bull Am Meteorol Soc 81(3):437–442

Cline WR (1992) The economics of global warming. Institute for International Economics, Washington, DC

Cornell C (1968) Engineering seismic risk analysis. Bull Seismol Soc Am 58(5):1583–1606

Cummins JD (2002) Comment: the insurance effects of regulation by litigation. In: Viscusi WK (ed) Regulation through litigation. Donnelley, Harrisonburg

Federal Emergency Management Agency (2002) National flood insurance program description. Federal Emergency Management Agency, Washington DC. http://www.fema.gov/library/viewRecord.do?id=1480. Cited 8 Sep 2008

Florida Disaster (2008) Florida Division of Emergency Management. Hurricane loss mitigation program. Tallahassee. http://www.floridadisaster.org/brm/fhlmp_section1.htm. Cited 8 Sept 2008

Grossman DA (2003) Warming up to a not-so-radical idea: tort-based climate change litigation. Columbia J Environ Law 28(1):1–61

Guatteri M, Bertogg M, Castaldi A (2005) A shake in insurance history: the 1906 San Francisco earthquake. Swiss Reinsurance Company Economic Research & Consulting, Zurich. http://www.swissre.com/pws/media%20centre/news/news%20releases%202006/swiss%20re %20publication%20recounts%20one%20of%20the%20most%20significant%20events%20in %20insurance%20history%20%20the%201906%20san%20francisco%20earthquake%20and %20fire.html. Cited 8 Sept 2008

Healy JK, Tapick JM (2004) Climate change: it's not just a policy issue for corporate counsel – it's a legal problem. Columbia J Environ Law 29(1):89–118

Hsu S-L (2008) A realistic evaluation of climate change litigation through the lens of a hypothetical lawsuit. Social Science Research Network. http://ssrn.com/paper=1014870. Cited 22 Sept 2008

Insurance Information Institute (2008) Insurance Information Institute paper analyzes growth of state-run property insurance plans: number of policyholders doubled between 1997 and 2006; stabilized in 2007–2008. http://www.iii.org/media/updates/press.788572/. Cited 19 Aug 2008

Jeffee DM, Russell T (1997) Catastrophe insurance, capital markets, and uninsurable risks. J Risk Ins 64(2):205–230

Kunreuther H (1996) Mitigating disaster losses through insurance. J Risk Uncertain 12(2):171–187

Lott N, Ross T (2006) Tracking and evaluating U.S. billion dollar weather disasters 1980–2005. NOAA's National Climatic Data Center, Ashville

Mills E (2005) Insurance in a climate of change. Science 309(5737):1040–1044

Mills E, Roth RJ Jr, Lecomte E (2005) Availability and affordability of insurance under climate change: a growing challenge for the U.S. CERES, Boston

Native American Rights Fund [NARF] (2008) NARF & Alaskan native village sues 24 oil and energy companies for destruction caused by global warming. Kivalina, AK. http://narfnews. blogspot.com/2008/02/narf-alaskan-native-village-sues-24-oil.html. Cited 22 Sept 2008

Pielke RA (2007) Future economic damage from tropical cyclones: sensitivities to societal and climate changes. Philos Trans R Soc A 365(1860):1–13

Pielke R, Sarewitz D (2002) Wanted: scientific leadership on climate. Issues Sci Technol 19:27–30

Pielke RA, Agrawala S, Bouwer LM et al (2005) Clarifying the attribution of recent disaster losses: a response to Epstein and McCarthy. Bull Am Meteorol Soc 86:1481–1483

Priest GL (2003) Government insurance versus market insurance. Geneva Pap Risk Insur Issues Pract 28:71–80

Reiter P, Lathrop S, Bunning M et al (2003) Texas lifestyle limits transmission of dengue virus. Emerg Infect Dis 9(1):86–89

Stern N (2006) The economics of climate change: the Stern review. Cambridge University Press, Cambridge/New York

Wharton Risk Center (2007) Managing large-scale risks in a new era of catastrophes: the role of the private and public sectors in insuring, mitigating and financing recovery from natural disasters in the United States. In: Doherty N, Grace M, Hartwig R et al (eds) The Wharton Risk Center extreme events project, in conjunction with Georgia State University and the Insurance Information Institute, Philadelphia

Chapter 19
Adaptive Capacity of Forest Management Systems on Publicly Owned Forest Landscapes in Canada

Mark Johnston, Tim Williamson, Harry Nelson, Laird Van Damme, Aynslie Ogden, and Hayley Hesseln

Abstract The degree to which Canadian forest management policies, institutions, and other factors either support or hinder the development of climate change adaptive capacity is discussed. The analysis is based on discussions with government and industry forest managers across Canada. Managers feel that they have the tools and the technical capability to successfully adapt. However, while these tools and abilities are available to forest managers, they are not always utilized due to policy barriers or lack of resources. Also, the adaptive capacity requirements of forest managers may be increasing as a result of global warming, as well as broader social, economic, and market trends. A model of "embedded science,"

M. Johnston (✉)
Saskatchewan Research Council, Saskatoon, SK, Canada
e-mail: johnston@src.sk.ca

T. Williamson
Natural Resources Canada/Canadian Forest Service, Edmonton, AB, Canada
e-mail: twilliam@nrcan.gc.ca

H. Nelson
Faculty of Forestry, Forest Sciences Centre, University of British Columbia, Vancouver, BC, Canada
e-mail: hnelson@forestry.ubc.ca

L. Van Damme
KBM Forestry Consultants, Thunder Bay, ON, Canada
e-mail: vandamme@kbm.on.ca

A. Ogden
Forest Management Branch, Yukon Department of Energy Mines and Resources, Whitehorse, YT, Canada
e-mail: aeogden@gov.yk.ca

H. Hesseln
Department of Bioresource Policy, Business, and Economics, College of Agriculture and Bioresources, University of Saskatchewan, Saskatoon, SK, Canada
e-mail: h.hesseln@usask.ca

J.D. Ford and L. Berrang-Ford (eds.), *Climate Change Adaptation in Developed Nations: From Theory to Practice*, Advances in Global Change Research 42, DOI 10.1007/978-94-007-0567-8_19, © Springer Science+Business Media B.V. 2011

in which scientists closely interact with forest managers in planning exercises leads to increased adaptive capacity. Some institutions, such as forest certification, have the potential for providing a framework for determining adaptation and adaptive capacity requirements. However, they will need to be modified in order to realize that potential. Forest management policy generally supports adaptation, but may limit the implementation of adaptation options in cases where the required innovation lies far outside of business-as-usual activities. Forest management policy needs to become more flexible and forward-looking, focusing on expected future outcomes under potentially different conditions, while at the same time acknowledging the uncertainty in expected outcomes. Reforming existing forest tenure arrangements and providing forest managers with more flexibility and local autonomy will allow more timely adaptation to climate change as well as other sources of change.

Keywords Forest management • Adaptation • Adaptive capacity • Forest sector • Canada • Forest policy • Forest Products Association of Canada • Climate change • Climate change impacts • Forests

Introduction

Forested regions of Canada are expected to experience greater impacts of climate change than many areas of Canada and the rest of the world (Field and Mortsch 2007). Table 19.1 provides an overview of current and potential future impacts of climate change on Canada's forests. Impacts of particular concern to forest managers include increased frequency and intensity of fires (Flannigan et al. 2005), increased outbreaks of both insects and diseases (Johnston et al. 2006), increased frequency of drought leading to forest dieback, particularly on the southern fringe of the boreal forest (Hogg and Bernier 2005), and changes to growth and amount of harvestable wood volume (Johnston and Williamson 2005). The nature and extent of these (and other) biophysical impacts is becoming clearer, although further detail is required before specific adaptation options can be identified (Johnston et al. 2006). In contrast, the effects of these impacts on forest management institutions and planning are poorly understood, and we have a limited understanding of the capacity of the forest management system to adapt to future climate change.

This chapter presents a preliminary assessment of the adaptive capacity of forest management in Canada. The discussion is based on extensive consultation with government and industry forest managers, forest-based nongovernmental organizations (NGOs) and senior executives in some of Canada's largest forest companies during 2007–2008. The findings are preliminary because the understanding of climate change impacts, adaptation, and adaptive capacity in Canada's forest sector are at an early stage.

Table 19.1 A summary of current and future climate change impacts on forests

Current	Longer growing season
	Increased forest fire activity
	Changes in phenology
	Elevation increases in tree lines
	The 2001–2003 drought
	Mountain pine beetle outbreak
	Spruce bark beetle outbreak
	Dothistroma needle blight epidemic
	Shorter winter harvest season
Future	Increase in extreme weather and climate events and increased climate variability
	Increase in frequency and intensity of severe fire years
	Increase in frequency and intensity of severe insect and disease outbreaks
	Effects on tree physiology
	Variable changes in productivity (spatially and temporally)
	Change in species composition, distribution, and structure of forest ecosystems

Summarized from Williamson et al. (2009)

Approach

Adaptive capacity is defined by Adger et al. (2007) as "... the ability or potential of a system to respond successfully to climate variability and change, and includes adjustments in both behavior and in resources and technologies." The adaptive capacity of the Canadian forest sector was assessed through semi-structured discussions with forest managers and industry executive in 50 companies, government agencies and NGOs located in nearly all provinces and territories in Canada. Commonly recognized determinants of adaptive capacity, as identified by Smit and Pilofosova (2001) and Moser et al. (2008), were used to structure the discussions. The determinants include:

- Available technological options for adaptation
- Availability of financial resources
- Institutional design
- Human and social capital
- Access to risk-spreading tools, processes, and mechanisms
- Information availability and access
- Awareness and understanding

Rather than asking specific questions about each of these factors directly, the above topics provided a general guide for a more open-ended discussion. This allowed participants maximum freedom to express their thoughts on adaptive capacity. The results of these discussions are presented below. It should be noted that the following discussion provides forest managers' perspectives on their adaptive capacity rather than an objective arms-length assessment.

Assessment of Adaptive Capacity

Technological Options

Forest managers generally have access to technology that is sufficient and appropriate under current conditions. In discussing the role additional or different technology might play in adaptation, the primary concern was cost relative to its potential return. For example, high-flotation tires on harvesting equipment may allow operations during unfrozen winter conditions in some cases. However, the concern was that this option is expensive, can require additional maintenance, and may only be required sporadically. Therefore the expense would be difficult to justify.

Another example of new technology to enhance adaptive capacity could be the use of genetic modification in producing tree varieties that are better adapted to future conditions. Cost and uncertainty about future conditions were identified as concerns, especially in the boreal region, given slow growth rates and low return on investment for this technology. Public acceptance of genetically modified organisms is still a problem in agriculture and forestry. On the other hand, there is significant research and activity among both industry and government regarding modified seed transfer zones, which stipulate the location of seed sources used for regeneration. The current approach is to restrict the use of seeds to the area from which it was collected (i.e., local is best rule). Managers are now beginning to look at seed zones relative to where suitable locations may occur in the future, so that seed is "matched" to the future climate. This is currently being done on a test basis but is rapidly evolving into a management practice (O'Neil et al. 2008).

Technology may have an important role in allowing new species to be used for forest products. Tree species ranges are expected to shift as climate change unfolds (McKenney et al. 2007; Williamson et al. 2009), resulting in the replacement of traditional commercial species with new ones (but only over very long time frames if unassisted). Companies that are able to adopt new technology to provide new products will have enhanced adaptive capacity. However, investment in the forest sector is currently low (FPAC 2007a), and it is unlikely that investment in new facilities and equipment will occur in anticipation of future species availability (particularly given the long periods of time required for species changes to occur).

Availability of Resources

Availability of resources for adaptation, especially financial resources, is extremely limited in the Canadian forest sector today. As of 2008, over 45,000 jobs had been lost over the previous 10 years and over 100 forest product mills had been closed nationally (Van Damme 2009). It was made very clear in our discussions that most forest managers in the industry are focused on day-to-day survival and do not have time or resources to consider climate change and adaptation, even in cases where they know this will be important in the future. Over and above the North American housing market collapse and global economic downturn, the forest industry in Canada has also been affected by high exchange rates, increased competition with low-cost offshore producers, and fluctuating costs of energy. These factors further reduce resources that might be available for adaptation to climate change (FPAC 2007a). Finally, most forest products facilities are long-lived and require very large capital investments. Therefore, making rapid adjustments in technologies employed in response to changes in the natural or economic–political operating environments is difficult. Adaptation through changes in forest management policies is a process that needs to be phased in over time to allow firms the time and flexibility to adapt and adjust their capital assets to new operating environments.

Critical Institutions

Our interviews indicated that institutional factors are the most important in determining adaptive capacity among forest managers. While the financial position of the forest industry is also important, the general perspective of forest management professionals is that in some respects, this is a temporary problem and part of the business cycle. On the other hand, institutional factors are seen as being long-term and a fundamentally inherent part of the forest sector's governance. In addition, significant institutional change requires political change, which may make them more difficult to implement. Canada also has a number of nongovernmental institutional options that may encourage and support adaptive capacity.

Industry

Forest industry managers indicated that industry organizations are important to them as a source of credible information. The Forest Products Association of Canada (FPAC) was identified as the first stop for FPAC members for gathering information. Working with industry associations like FPAC would be a good way of bringing information to companies from a source they trust. An example of a previous successful initiative is the large amount of work FPAC and other organizations have done in helping companies reduce greenhouse gas emissions. The forest industry in Canada has reduced greenhouse gas emissions by 44% since 1990 while increasing production by 20% (FPAC 2007b).

Nearly all jurisdictions in Canada require some type of long-term forest management plan, typically on a 20-year time horizon. Our discussions with forest managers indicate that the forest management planning function provides an excellent vehicle for considering climate change impacts and adaptations. The relatively long time horizon and the generally strategic focus of the plans means that climate change can be considered at a temporal and spatial scale consistent with the current state of understanding of climate change impacts. This provides an important example of "mainstreaming" climate change impacts and adaptation, as recommended by the recent *Canadian National Climate Change Assessment* (Lemmen et al. 2008).

Currently there are no guidelines for including climate change into long-term plans. A useful initiative, therefore, would be the development of planning guidelines that could be used across all jurisdictions, in order to provide guidance on how impacts and adaptation considerations could be integrated into forest management plans. These would necessarily be general in order to accommodate variability among jurisdictions and biophysical conditions, but could be developed in a way that would be helpful to both industry and government planners. The concept of "embedded science" (Van Damme et al. 2008) has also been shown to be an effective way for practitioners to deal with complex science-oriented issues like climate change. In this model, scientists from government or academia work closely with company managers and planners in incorporating scientific analyses into the forest management plans. This collaboration is established at the beginning of the planning cycle so that the direction and approach used by the scientists support the objectives of the plan.

An additional issue in forest management is that given that trees require a long-time period to mature, decisions made today will persist for several decades. In contrast, decisions in agriculture about what crops to plant or what management practices to follow can be modified annually. However, the outcome of forest management decision-making is difficult to predict – particularly given the high likelihood of a changing climate during the next several decades. This uncertainty reduces the adaptive capacity of forest managers, and points out the need for climate-sensitive projection models that can help managers anticipate future conditions (e.g., temperature, rainfall, soil moisture, length of winter harvest season, disturbance) and forest productivity (Spittlehouse and Stewart 2003; Spittlehouse 2005).

Regulators

In Canada, provincial governments manage Crown forestlands under legislation that prescribes responsibilities of provincial regulators and the forest industry. Rights to harvest wood are granted to forest companies through forest management agreements which stipulate harvest levels and other activities that companies must undertake: reforestation of harvested areas; submission of forest management plans; protection of fish and wildlife habitat; and others. Provincial regulators oversee forest companies by requiring 20-year and operational plans, as well as enforcing regulations under forest management legislation.

A few cases show that the process of adaptation by regulators has started. At the national level, The Council of the Federation (made up of provincial and territorial premiers) has directed forestry ministers, through the Canadian Council of Forest

Ministers, to complete a series of studies looking at adaptation of tree species, forest landscapes, and the forest sector to climate change. The Canadian Council of Forest Ministers has also recently released *A Vision for Canada's Forests: 2008 and Beyond* (see http://www.ccfm.org/pdf/Vision_EN.pdf). According to this document, climate change is considered to be one of two issues of national importance to Canada's forest sector (along with forest sector transformation). At the provincial level, forest management agencies are beginning to evaluate their adaptation requirements. For example, the British Columbia Ministry of Forests and Range is moving forward on an adaptation strategy under the umbrella of its Future Forest Ecosystem Initiative (see http://www.for.gov.bc.ca/mof/Climate_Change/). Irrespective of these early initiatives, there remain important institutional barriers to adaptation. Institutional barriers to adaptive capacity among provincial regulators are related to forest policy that is usually based on the assumption that a forest remains substantially the same over time. This is particularly a problem with climate change, given the uncertainty about future conditions. This may make acceptance of innovative ideas difficult, especially if the proposed alternative lies far outside of accepted practice and assumptions. There is also much current discussion in the forest sector regarding the need for change in the forest tenure system. Long-term agreements that are stipulated by government may reduce the adaptive capacity of both industry and provincial regulators by "locking-in" levels of harvest or other aspects of forest management, and may prevent adaptation options from being implemented (Haley and Nelson 2007). Innovative forest management practices that have both immediate and long-term benefits may be difficult to apply, given relatively inflexible tenure agreements.

Nongovernmental Organizations

A number of forest-based NGOs have developed in the past few decades that support the principles and application of Sustainable Forest Management (SFM). Forest certification bodies have developed standards for forest management that stipulate how SFM is to be achieved, and will certify a company's products as having come from a sustainably managed forest estate. As of 2008, over 147 million ha of forest land in Canada have been certified, representing over 90% of the managed forest land in Canada (FPAC 2008).

While certification standards promote SFM, neither potential climate change impacts on forests, nor adaptation and adaptive capacity, are explicitly considered. In general the standards assume a relatively unchanging forest, and usually tend to support the protection of existing species and habitats and maintenance of ecosystem processes at current locations. There is little recognition that forests may change for reasons unrelated to forest management practices, or that forest practices need to adapt to in order to mitigate impacts on forests. However, certification is seen by the forest industry as essential to continued market access, and companies will continue to seek this designation. This suggests the need for incorporating climate change considerations into forest certification standards. This would necessarily be at a fairly general level but could provide guidance to companies on how to address the critical questions about likely impacts, vulnerability, and adaptation options.

In the 1990s, The Montreal Process resulted in international agreement on a set of Criteria and Indicators (C&I) for Sustainable Forest Management (MPWG 2005). In Canada, the Canadian Council of Forest Ministers (comprising the provincial and federal ministers responsible for forestry) oversaw the Canadian Criteria and Indicators (C&I) framework (CCFM 2003). The Canadian framework recognizes six Criteria of Sustainable Forest Management: (1) Biological Diversity; (2) Ecosystem Condition and Productivity; (3) Soil and Water; (4) Role in Global Ecological Cycles; (5) Economic and Social Benefits; and (6) Society's Responsibility. Each is supported by several indicators used to measure progress toward satisfying the criterion. In its current configuration, the C&I framework is mostly a backward-looking instrument (i.e., its focus is on assessing existing forest management practices as to their impacts on biodiversity, ecosystem condition, etc). However, as is the case with certification, the C&I generally do not take account of the potential impacts of climate change with respect to our ability to achieve SFM objectives. As with certification systems, the existing C&I framework could provide an already accepted vehicle for identifying biophysical climate change impacts (Criteria 1–4) and adaptations, socioeconomic impacts and adaptive capacity (Criteria 5 and 6). Modifying the C&I framework to incorporate these changes would seem like a logical next step in its evolution.

Human and Social Capital

The forestry profession has a long history in Canada, with the first professional schools established in the early 1900s. The profession is committed to the principles of SFM. Professional societies enforce standards for education and ethical practices, and provide incentives for continuing education. Forest science and management research is active in the professional schools across Canada. At the level of forest management expertise, human capital in the forestry profession in Canada is high.

However, our discussions indicated that human capital relative to thinking about and planning for climate change impacts and adaptation is not high. The level of understanding about climate change impacts and adaptation among forest managers is relatively low. Moreover, science capacity (which is of growing importance for dealing with science-oriented issues like climate change) in companies and in regulatory agencies is sometimes low. Compounding the relatively low scientific capacity is the fact that current information on climate change impacts is generally not available at spatial and temporal scales relevant to forest management planning and operations. In addition, most forest companies today are focused on surviving the economic downturn. Even the companies that take climate change seriously and that have scientific capacity find it difficult to address this issue when day-to-day survival is the primary concern.

Social capital (networks, trust, and engagement) contributes to adaptive capacity (Williamson et al. 2007). Stronger collaboration, cooperation, sharing, and coordination across jurisdictional boundaries will contribute to enhanced capacity to adapt to climate change. Industry forest managers identified industry organizations such as FPAC as important in terms of providing both technical support to companies and a voice for the industry. Their feeling was that information available from industry organizations was generally credible and relevant to their interests. Similarly, the provincial professional societies and the Canadian Institute of Forestry have continuing education programs that could be important for communications and education about climate change impacts and adaptation.

Risk-Spreading Processes

Forest companies, like any other, engage in risk management as a part of normal business practice. However, significant climate change is expected to occur within the rotation lengths commonly used in Canadian forests. Thus, it is anticipated that climate change will contribute to increased risk to forest capital and increased requirement for risk management. Some aspects of forest management make risk management more difficult. For example, it is difficult to change the species in a forest stand once they are established. If an insect outbreak occurs, expensive treatments after the fact (e.g., spraying pesticides, salvage harvesting) are usually the only option, rather than replacing existing tree species with those that are less susceptible.

A common risk management approach in business is to diversify the portfolio of assets held. One option for forest management, therefore, might be to establish a more diverse portfolio of commercial tree species on a given land base. Forest managers, however, are constrained both by the natural environment (i.e., only certain species will grow on their land base) and by policy which usually stipulates that whatever species is harvested must be re-established. Thus, the ability to diversify the species mix on the landscape may in some cases be limited. For example, current biodiversity policy may constrain the introduction of alien (not naturally occurring) species, even where these are determined to provide an important and viable adaptation option.

Information and Knowledge About Local Impacts

Larger organizations generally have adequate knowledge-sharing mechanisms, such as in-house training sessions, newsletter, and periodic meetings. However, corporations vary in their culture of sharing information. Some leaders see knowledge as power and will only share with those who will not use it against them. Also, corporate culture regarding information and science can be an issue. If innovation is required to adapt to future conditions, but the culture does not value new information or ways of thinking, adaptive capacity will be reduced.

An important barrier to adaptation is a lack of information about future impacts at locally relevant scales. In most cases, forest managers are aware of climate change and they understand its general implications at a large scale. Climate change is a serious concern to forest managers (Williamson et al. 2005; Colombo 2006). Despite this awareness, there is limited action to "mainstream" adaptation into decision-making. Some of the reasons have been previously noted. However, an important barrier identified was a lack of information about what to expect in terms of future climate and future growing conditions at local scales. This suggests the need for an increase in climate change impacts and adaptation science.

Awareness and Understanding

The general level of understanding of climate change, and the willingness to take it seriously, is increasing in the Canadian forest sector. The question among managers is not whether climate change is real, but rather what the local impacts will be and what adaptation actions need to be taken. As mentioned above, current information on impacts is provided at temporal and spatial scales that are quite different than those required for planning and operations. Downscaling and ecosystem modeling techniques exist that could partially answer some of these questions. However, they require fairly sophisticated expertise and have large uncertainties associated with them. Finally, climate change is only one source of change affecting the forest sector. Other sources of change include demographic shifts in rural populations, the effects of global market forces on the forest industry, local and national political change, and changes in society's expectations of the values and benefits available from the forest. The integration of climate change with these other agents of change is a challenge for forest managers, given high levels of uncertainty, current economic instability, and poor market conditions.

Conclusion

In this assessment, we find that the current adaptive capacity of forest managers is relatively high. Nonetheless, the adaptive capacity requirements of forest managers under climate change is expected to grow, potentially leaving a gap between what currently exists and what will be needed. One of the more important constraints is the lack of detailed, site-specific information on potential impacts. Further development of more detailed climate models and forest ecosystem impacts scenarios will assist managers in determining adaptation options. Communication with and education of forest practitioners needs to be ongoing. Also, the concept of "embedded science" (Van Damme et al. 2008) should be included in more forest management planning exercises to increase scientific capacity and provide a closer integration of science and management. In fact, there is a need for a general increase in science capacity across the spectrum of the forest management system, including regulators and industry licensees.

Institutional barriers will probably be a more important constraint to adaptation than availability of technical forest management options. Current forest policy is unlikely to be sufficiently flexible to accommodate the novelty and surprise associated with future climate change. Forest management policy and planning needs to become more forward-looking and flexible, focusing on anticipated outcomes and outcome ranges (e.g., sustainable forest ecosystems, future forest conditions, social and economic benefits) rather than deterministic outputs (Ogden and Innes 2007). Forest managers will need to acknowledge and develop systems to deal with increased risk and uncertainty. Reforming tenure systems and providing local managers with the increased flexibility to change operations in response to (or in anticipation of) change will assist in timely adaptation.

References

Adger WN, Agrawala S, Mirza MMQ et al (2007) Assessment of adaptation practices, options, constraints and capacity. In: Parry ML, Canziani OF, Palutikof JP et al (eds) Climate change 2007: impacts, adaptation and vulnerability – contribution of Working Group II to the Fourth Assessment Report of the Intergovernmental Panel on Climate Change. Cambridge University Press, Cambridge

Canadian Council of Forest Ministers [CCFM] (2003) Defining sustainable forest management in Canada: criteria and indicators 2003. CCFM, Secretariat Office, Canadian Forest Service. http://www.ccfm.org/current/ccitf_e.php. Cited 23 Aug 2010

Colombo SJ (2006) How OMNR staff perceive risks related to climate change and forests. Climate change research information – note no.2. Applied Research and Development Branch, Ontario Ministry of Natural Resources, Ottawa

Field C, Mortsch L (2007) North America. In: Parry ML, Canziani OF, Palutikof JP et al (eds) Climate change 2007: impacts, adaptation and vulnerability – contribution of Working Group II to the Fourth Assessment Report of the Intergovernmental Panel on Climate Change. Cambridge University Press, Cambridge

Flannigan MD, Logan KA, Amiro BD et al (2005) Future area burned in Canada. Clim Change 72(1–2):1–16

Forest Products Association of Canada [FPAC] (2007a) Industry at a crossroads: choosing the path to renewal. Forest Products Industry Competitiveness Task Force, FPAC. http://www.fpac.ca/en/who_we_are/pdfs/Publications/TASK_FORCE.pdf. Cited 23 Aug 2010

FPAC (2007b) New Green Plan fails to substantially recognize early action – punishes proactive industries, says forest products industry. Press release by Avram Lazar, FPAC CEO (26 Apr 2006). http://www.fpac.ca/en/media_centre/press_releases/2007/2007-04-26_emissionsRegulations.php. Cited 23 Aug 2010

FPAC (2008) Certification status – Canada & the globe: statistics. http://www.certificationcanada.org/english/status_intentions/status.php. Cited 12 Dec 2009

Haley D, Nelson H (2007) Has the time come to rethink Canada's Crown forest tenure systems? For Chron 83(5):630–641

Hogg EH, Bernier PY (2005) Climate change impacts on drought-prone forests in Western Canada. For Chron 81:675–682

Johnston M, Williamson T (2005) Climate change implications for stand yields and soil expectation values: a northern Saskatchewan case study. For Chron 81:683–690

Johnston M, Williamson T, Price D et al (2006) Adapting forest management to the impacts of climate change in Canada. Final Report, Research Integration Program, BIOCAP Canada Foundation. http://www.biocap.ca/rif/report/Johnston_M.pdf. Cited 23 Aug 2010

Lemmen DS, Warren FJ, Lacroix J et al (2008) From impacts to adaptation: Canada in a changing climate. Government of Canada, Ottawa

McKenney D, Pedlar J, Lawrence K et al (2007) Potential impacts of climate change on the distribution of North American trees. BioSci 57(11):939–948

Montreal Process working Group [MPWG] (2005) The Montréal process. http://www.rinya.maff.go.jp/mpci/. Cited 23 Aug 2010

Moser S, Kasperson RE, Yohe G et al (2008) Adaptation to climate change in the northeast United States: opportunities, processes, constraints. Mitig Adapt Strateg Glob Change 13:643–659

O'Neill GA, Hamann A, Wang T (2008) Accounting for population variation improves estimates of the impact of climate change on species' growth and distribution. J Appl Ecol 45:1040–1049

Ogden AE, Innes JL (2007) Incorporating climate change adaptation considerations into forest management planning in the boreal forest. Intnl For Rev 9:713–733

Smit B, Pilifosova O (2001) Adaptation to climate change in the context of sustainable development and equity. In: McCarthy J, Canziani OF, Leary NF et al (eds) Climate change 2001: impacts, adaptation and vulnerability – contribution of Working Group II to the Third Assessment Report of the Intergovernmental Panel on Climate Change. Cambridge University Press, Cambridge

Spittlehouse DL (2005) Integrating climate change adaptation into forest management. For Chron 81(5):691–695

Spittlehouse DL, Stewart RB (2003) Adaptation to climate change in forest management. BC J Ecosyst Manag 4(1):1–11

Van Damme L (2009) Can the forest sector adapt to climate change? For Chron 84:633–634

Van Damme L, Duinker PN, Quintilio D (2008) Embedding science and innovation in forest management: recent experiences at Millar-Western in west-central Alberta. For Chron 84(3):301–306

Williamson TB, Parkins JR, McFarlane BL (2005) Perceptions of climate change risk to forest management and forest-based communities. For Chron 81:710–716

Williamson TB, Price DT, Beverly J et al (2007) A framework for assessing vulnerability of forest-based communities to climate change – information report NOR-X-414 Northern Forestry Center. Natural Resources Canada, Government of Canada, Ottawa

Williamson T, Colombo S, Duinker P et al (2009) Climate change and Canada's forests: from impacts to adaptation. Sustainable Forest Management Network and Natural Resources Canada. http://www.sfmnetwork.ca/docs/e/SP_ClimateChange_English.pdf. Cited 23 Aug 2010

Part IV
Adaptation in the Urban Environment

Chapter 20
Overview: Climate Change Adaptation in the Urban Environment

Thomas J. Wilbanks

Abstract This overview chapter considers five questions that cut across the four case studies in the section to follow: Why are urban environments of particular interest? What does an "urban environment" mean as a focus for adaptation actions? What do we know about climate change vulnerabilities and adaptation potentials in urban areas? What can we expect in the future with adaptation in urban areas? And what is happening with climate change adaptation in urban areas?

Keywords Climate change • Adaptation • Climate adaptation • Urban communities • IAV • Cities • Urban development • Adaptation research • Local adaptation

Introduction

After decades of inattention, adaptation to risks and impacts of climate change is now receiving long overdue attention, and it is only natural that a considerable share of this attention is focused on the places where most people live. This section considers climate change adaptation in the urban environment, defined as settings where human populations cluster – generally implying relatively large clusters, but not excluding smaller settlements that operate as coherent geopolitical and economic entities. Consistent with the topic of the book, the emphasis of this overview will be on urban environments in developed countries, but it will also draw on knowledge being developed from urban experiences across the globe.

T.J. Wilbanks (✉)
Climate Change Science Institute, Oak Ridge National Laboratory, Oak Ridge,
TN 37831-6103, USA
e-mail: wilbankstj@ornl.gov

J.D. Ford and L. Berrang-Ford (eds.), *Climate Change Adaptation in Developed Nations:*   281
From Theory to Practice, Advances in Global Change Research 42,
DOI 10.1007/978-94-007-0567-8_20, © Springer Science+Business Media B.V. 2011

Why Are Urban Environments of Particular Interest?

Simply stated, urban areas are of interest in considering climate change risks and responses because that is where most of the earth's population lives. According to the United Nations, 2008 was the first year in human history when more than half of the world's population was living in cities and towns; and the urban proportion is growing steadily (Fig. 20.1), driven mainly by urban growth in developing countries. The UN projects that the world's urban population will increase from 1.9 billion in 2000 to nearly 4 billion in 2030. Over this same period, the urban proportion in Africa and Asia will grow from 39% to 54–55%.

Population concentrations have certain advantages in responding to threats to their well-being, including both intellectual and financial resources; but they also present special risks. For instance, major climate events can be less than predictable as to their geographic locations, but whenever they overlap with urban environments they affect large numbers of people and large stocks of physical and economic capital, sometimes with catastrophic results. Moreover, cities are often the location of a large share of a nation's most disadvantaged citizens – the poor, the disabled, the disadvantaged – which further adds to potentials for catastrophe, as Paris discovered during the European heat wave of August 2003 (Wilbanks et al. 2007).

For a variety of historic reasons, in fact, populations are often concentrated in areas especially vulnerable to climate change impacts. Examples include (1) coastal areas and river valleys, reflecting both the influence of water transportation in times when urban patterns were emerging and also the attraction of scenic and other

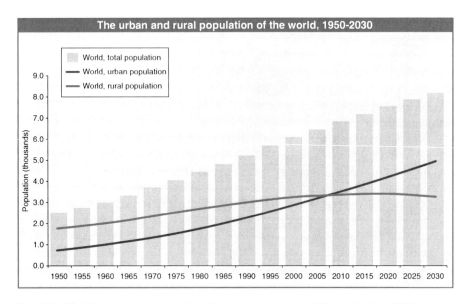

Fig. 20.1 World population trends and projections; urban vs. rural (United Nations 2005)

amenities in these types of locations; and (2) areas where the urban economic base is rooted in climate-sensitive sectors such as agriculture, forestry, and tourism.

But urban areas are connected with climate change and its impacts in more complex ways than these. In terms of both carbon emission reduction and climate change adaptation, urban areas are the locus of responses to concerns about climate change: as financial centers, as industrial centers, as innovation centers, as media centers from which information diffuses, and – in democratic societies – as political forces where one person means one vote. They are society's laboratories, where most new ideas arise and without whose engagement very few of the ideas take root.

What Does an Urban Environment Mean as a Focus for Adaptation Actions?

In considering climate change impacts and adaptation potentials, the usual practice is to organize research and discourse around "sectors": categories of systems associated with particular economic sectors and often professional expert communities, such as agriculture or water. Like "coastal areas," urban areas are not exactly a sector. At the same time, urban environments are not exactly a human system, because they are too diverse, complex, and linked with natural systems as well as human systems. They are, in a sense, a system of systems, many of which do not respect organizational or governmental boundaries.

Urban areas, however, do have meaning. They are more than an arbitrary cut across a geographic scale continuum from very small to very large. They represent a mesoscale between neighborhood and region associated with a degree of self-organization, able to mobilize collective decision-making for some purposes. In many cases, sustainability scientists have found urban environments, including their peripheries, to be the most coherent focus for place-based integration of a wide range of driving forces, sectoral activities, and socioeconomic and environmental factors.

Recent Developments: From Theory to Practice

What Do We Know About Climate Change Vulnerabilities and Adaptation Potentials in Urban Areas?

Attention to urban climate change impacts, adaptation, and vulnerability (IAV) issues can be traced to at least two ancestors. One has been the work of the Intergovernmental Panel on Climate Change (IPCC), which began considering human settlement issues in its *First Assessment Reports* in 1990 and 1992. Since that time, IPCC has devoted a chapter of its Working Group II (IAV) Report to human settlement issues in the *Second Assessment Report* (1995), the *Third Assessment Report* (2001), and part of a chapter in the *Fourth Assessment Report* (2007).

A second has been attention by the natural hazards research community to vulnerabilities of urban areas to environmental changes and events (e.g. Mitchell 1999). This literature pioneered considerations of risk and vulnerabilities to events ranging from hurricanes to earthquakes, including both risks and responses through emergency preparedness.

Within the United States, the first *National Assessment of Potential Impacts of Climate Variability and Change* (NACC), completed in 2001, included a "regional" chapter focused on an urban area: the "Metropolitan East Coast" study, concerned with the New York City metropolitan area (Rosenzweig and Solecki 2001). Soon after, a subsequent urban impact assessment was conducted in the Boston area from 1999 to 2004 (*Climate's Long-term Impacts on Metro Boston* (CLIMB)) (Kirshen et al. 2004). Internationally, London set a high standard for urban climate change impact assessments (LCC Partnership 2004), and several institutions began to focus on urban impact issues, including the International Institute for Environment and Development (IIED) and the International Human Dimensions Project (IHDP) through a new Urbanization and Global Environmental Change project.

Through this period of little more than a decade, a number of important initiatives had been undertaken to understand urban environmental roles in greenhouse gas emissions, such as ICLEI's Cities for Climate Protection Program, a number of comprehensive carbon cycle studies (such as Field and Raupach 2004), and the Urban and Regional Carbon Management Initiative of the Global Carbon Project since 2005. Toward the end of the decade, most of these initiatives were extending their scope toward cities as *targets* of climate change impacts as well as *drivers* of climate change itself. One reason was occasional reminders of urban impact vulnerabilities, from urban heat waves in Europe in 2003 and the United States in 2006, to Hurricane Katrina in the US Gulf Coast in 2005 (Wilbanks et al. 2007).

By 2007, reports on urban climate change vulnerabilities and responses were emerging in quick succession, such as: a special April 2007 issue of the journal *Environment & Urbanization* on "Reducing Risks to Cities from Disasters and Climate Change;" an Id21 report on "Climate Change and Cities," January 2008; a US Climate Change Science Program report on "Human Settlements," Chap. 3 in Synthesis and Assessment Product 4.6, 2008; an IIED book, *Adapting Cities to Climate Change*, 2009; and a World Bank guidebook on *Climate Resilient Cities*, 2009.

Meanwhile, the number of relevant research and technical assistance activities underway is growing so rapidly that no one person is aware of them all. Examples include: the US National Science Foundation's urban Long Term Ecological Research (LTER) sites, Phoenix and Baltimore; the Rockefeller Foundation's Asian Cities Climate Change Resilience Network, along with a new initiative that is expanding this focus; the Asian Development Bank's urban climate change programs; START's program initiative entitled "Cities at Risk: Developing Adaptive Capacity for Climate Change in Asia's Coastal Megacities"; and a forthcoming UN Habitat Report on *Cities and Climate Change*, the 2011 issue of the UN Global Report on Human Settlements.

Key Challenges

Some challenges have emerged for this still-young field of research. Examples include the following:

1. Far too few good case studies have been carried out to date on urban vulnerabilities to climate change impacts and possible impacts. This means that the knowledge base is generalizing from too small a sample to produce findings that can be offered with a high level of confidence. A particular concern is that a number of important types of cases are missing case studies, such as cities in arid inland areas subject to increasing water scarcity and perhaps dangers of wildfires.
2. It is very difficult to assess adaptations to climate change impacts based on evidence of results. Because the ability to attribute observed climate phenomena to climate change is just now emerging, it is generally premature to attribute benefits from actions that have been taken to reduce climate change impacts. The fact is that climate change adaptation by urban areas is still largely in the planning stages, in most cases not associated with actions, and even where actions are being taken it will be years before their results can be evaluated.
3. There are some particular areas of concern, which are high priorities for adaptation research, such as:

   (a) Large urban concentrations in especially vulnerable areas, such as Asian mega-deltas (IPCC 2007).
   (b) Implications of climate change for the intensity, frequency, and location of extreme weather events and for water scarcity, the two impact issues that are often of greatest concern (aside from temperature increases in the Arctic).
   (c) Interactions between climate change impacts and other aspects of sustainable development paths, such as poverty and infrastructure needs (see the next section). This is important as an example of complex interactions between climate change implications for urban areas and other important issues. Other examples of complex interactions include:

       (i) Interactions between climate change and other driving forces for change in urban areas: demographic change, economic restructuring and globalization, technological change, institutional change, and others (Wilbanks et al. 2007).
       (ii) Interactions between climate change adaptation and climate change mitigation. It is often important to distinguish between complementary

actions, such as improvements in the efficiency of space cooling technologies, and contradictory actions such as investments in bioenergy
production in areas facing prospects for water scarcity.

(iii) Interactions between what is happening in a particular urban area and
what is happening elsewhere. Examples include:

1. The dependence of urban centers on other areas for supplies of
food, materials, water, and other inputs and as destinations for
products, services, and wastes: how are these other areas affected
by climate change? How are movement infrastructures affected by
climate change?
2. The dependence of the economic base of an urban center on regional
comparative advantage: how are competitors and markets affected by
climate change? How are trade flows affected by climate change?

(iv) Interactions between impacts and actions in one sector and what is
happening in other sectors. For example, urban life depends on energy
services, but energy production competes for water with other sectors in
water-scarce regions. At the same time, the water sector needs energy for
water pumping, especially if water becomes scarcer. Similarly, climate
change impacts in important sectors such as health and transportation
interact with climate change impacts on the energy sector.

(v) Linkages between first-order effects and actions and second-,
third-, and nth-order effects. One characteristic of climate change impacts
in urban environments is that impacts spread in ripples like a pebble
hitting the surface of a pond. Consider, for example, temperature and
precipitation changes, along with associated changes in humidity and the
heat index. These changes can increase heat-related distress, especially
among those living and/or working without air-conditioning, and those
facing increased ozone pollution associated with a higher incidence of
respiratory distress (Wilbanks et al. 2008). But changes continue to
cascade: to challenges for food chain protection, to changes in pests and
pathogens, and even – if there are problems with drainage and flooding –
to population displacements. All of these effects can add up to a greater
total impact on public health requirements, and thus on total urban
management and fiscal resources, than the first-order impacts alone.

What Is Happening with Climate Change Adaptation in Urban Areas?

In many parts of the world, adaptation planning in practice is not waiting for
academic research results. As mentioned above, the first U.S. *National Assessment
of Potential Consequences of Climate Variability and Change* (NACC), completed

in 2001, included specific attention to one regional context defined at the scale of a metropolitan area: New York City (Rozenzweig and Solecki 2001). This study was followed by a second major study in Boston: the CLIMB study (Kirshen et al. 2004). A number of international urban impact studies are cited in Wilbanks et al. (2007) and Wilbanks et al. (2008).

In recent years, a number of localities in the United States have begun to take climate change adaptation planning seriously (Moser, Chap. 3, this volume). Since 2006, New York City has been undertaking an ambitious effort to develop a community-based adaptation plan in a broader context of an urban sustainability and growth management initiative: PlaNYC (http://www.nyc.gov/html/planyc2030/html/plan/plan.shtml). A collaboration between Seattle/King County, WA, Local Governments for Sustainability (ICLEI), the Climate Impacts Group at the University of Washington, and others have produced both an adaptation plan and a handbook to assist other communities with adaptation planning (see Chap. 24, this volume). The first of five pilot efforts to reach completion is Keene, NH, a city of 23,000 (NRC 2010).

What these early efforts have found, whether or not they began with attention to larger frames of reference such as sustainability, is that it is extremely difficult to separate climate change adaptations from participative discussions of where the community wants to go with its development paths. In most cases, the adaptation actions likely to be favored are those that offer "co-benefits": benefiting both climate change risk management and efforts to reduce other development stress points (Chaps. 21–24, this volume). It can be argued, in fact, that attention to climate change implications and responses can be a catalyst within an urban environment for more comprehensive discussions of development goals and challenges (Wilbanks 2003).

Opportunities and Future Directions

What Can We Expect in the Future with Adaptation in Urban Areas?

It appears that two forces are emerging relatively quickly that are likely to combine to make climate change adaptation in urban areas of developed countries an active area of discussion and action in coming years: (1) a growing awareness at local levels of risks to community well-being from climate change impacts; and (2) a growing level of attention to such issues from national and regional governments, including assistance with vulnerability assessments and, in some cases, sources of funding for local adaptation actions.

At the same time, two other developments will also be co-evolving: adaptation *research* and adaptation *practice*. Adaptation planning and decision-making will not wait for a decade or more for the knowledge base to be transformed, and adaptation research is limited in its ability to transform the knowledge base very quickly. Instead, the need is for each to proceed – not separately but in a rich and productive partnership: research contributing insights and lending legitimacy to local actions where possible (including advice about how to address uncertainties), while practice contributes research questions and a growing body of evidence about what works.

In both connections, we will see an immense variety of experimentation, reflecting differences in local contexts and leaderships, including locally-focused experiments with new linkages between components of communities that have not always worked together closely – for example, the public and private sectors, institutions of teaching and learning and their local communities, disadvantaged populations and neighborhoods and city leaders. Moreover, we can hope to see a new spirit of partnership across scales of governments and other institutions, from local to national, in ways that contribute to a fuller social consensus about how to respond to risks of climate change and its impacts.

References

Field C, Raupach M (eds) (2004) The global carbon cycle: integrating humans, climate, and the natural world. Island, Washington, DC

IPCC (2007) Summary for policymakers. In: Parry ML, Canziani OF, Palutikof JP et al (eds) Climate change 2007: impacts, adaptation and vulnerability – contribution of Working Group II to the Fourth Assessment Report of the Intergovernmental Panel on Climate Change. Cambridge University Press, Cambridge

Kirshen PH, Anderson WP, Ruth M (2004) Climate's long-term impacts on metro Boston – final report to the US Environmental Protection Agency [EPA], Office of Research and Development. EPA, Washington, DC

LCC Partnership (2004) London's warming: a climate change impacts in London evaluation study. London

Mitchell JK (ed) (1999) Crucibles of hazard: megacities and disasters in transition. United Nations University Press, New York

NRC (2010) Adapting to the impacts of climate change. US National Research Council, Washington, DC

Rosenzweig C, Solecki W (2001) Climate change and a global city: the potential consequences of climate variability and change – metro east coast. Report for the US Global Change Research Program, National assessment of the potential consequences of climate variability and change for the United States. Columbia Earth Institute, New York

United Nations (2005) World urbanization prospects: the 2005 revision. Department of Economic and Social Affairs, Population Division, United Nations, New York

Wilbanks TJ (2003) Integrating climate change and sustainable development in a place-based context. Clim Policy 3(S1):147–154

Wilbanks TJ, Lankao PR, Bao M et al (2007) Industry, settlement, and society. In: Parry ML, Canziani OF, Palutikof JP et al (eds) Climate change 2007: impacts, adaptation and vulnerability – contribution of Working Group II to the Fourth Assessment Report of the Intergovernmental Panel on Climate Change. Cambridge University Press, Cambridge

Wilbanks TJ, Kirshen P, Quattrochi D et al (2008) Effects of global change on human settlements. In: Gamble J (ed) Analyses of the effects of global change on human health and welfare and human systems. US Environmental Protection Agency, Washington, DC

Chapter 21
Integrated Assessment of Climate Change Impacts on Urban Settlements: Lessons from Five Australian Cases

Geraldine Li and Stephen Dovers

Abstract As evidence of climate change has increased, attention has shifted from "is it happening?" toward "what should be the response?" There is increasing focus on assessing climate change impacts at the local scale, and on developing practical responses. This chapter summarizes lessons for assessment methodologies and practical adaptive responses for small- to medium-sized urban settlements, drawing on integrated assessment (IA) of five Australia settlements. The chapter describes the IA methodology developed, the nature of detailed investigations into specific impacts, and insights regarding local assessment and adaptation.

Keywords Climate change • Climate impacts • Integrated assessment • Adaptation • Vulnerability • Urban • Australia • Local adaptation • Systems thinking • Integrated Assessment of Climate Impacts on Urban Settlements • IACCIUS

Introduction

There have been two recent shifts in research and policy attention to climate change (Dovers and Hezri 2010). First, for two decades, attention has focused on the scientific case of whether, and to what extent, human-induced climate change was occurring. Especially, since the *Fourth Assessment Report* of the Intergovernmental Panel on Climate Change (IPCC) in 2007, there is a majority acceptance in scientific and policy circles that climate change is happening, and will happen, and attention has shifted to possible policy and management responses.

G. Li • S. Dovers (✉)
The Fenner School of Environment and Society, The Australian National University, Canberra, ACT, Australia
e-mail: geraldine.li@anu.edu.au; stephen.dovers@anu.edu.au

J.D. Ford and L. Berrang-Ford (eds.), *Climate Change Adaptation in Developed Nations: From Theory to Practice*, Advances in Global Change Research 42, DOI 10.1007/978-94-007-0567-8_21, © Springer Science+Business Media B.V. 2011

Second, as attention has focused on responses, interest has shifted from global and regional (sub-national) scales of climate modeling and impact assessment, to national and local scales, where policy and management responses will be constructed. The focus in this respect is on the local, settlement scale where decisions of, for example, urban planning, water management, emergency management, maintenance of open spaces and service provision are made. These decisions are made by state or provincial governments and local municipal councils, rather than through national and international policy and policy processes. Such a focus demands fine-scale data and understanding, identification of the interactions within local economies, societies and environments, and connecting climate change to the local, practical level.

Thus far, attention to climate impact and adaptation studies, and in particular integrated assessments, has been on broader scales and nonurban, northern hemisphere settings (e.g., Edmonds and Rosenberg 2005; Holman et al. 2005; Kirshen et al. 2008; Lange 2008). This chapter focuses on smaller urban settlements, how climate impacts can be assessed in an integrated manner, and identifies possible adaptation strategies. We report findings from a multi-settlement assessment of five Australian localities, in an applied research project underpinned by: (1) a commitment to participatory research; (2) the aim to develop operational assessment processes; and (3) an integrated, systems-oriented methodology. It was also informed by the proposition that climate change impacts may not be as novel as some think; that they are not too dissimilar to the challenges of managing climate and other variability which agencies and communities are already facing (Dovers 2009). The proposition invites a "normalization" of climate adaptation by connecting to existing local agendas, empowering local agencies, and drawing on local expertise, as opposed to an unfamiliar and externally imposed threat.

The Impact Assessment Project

The work drawn on here was one of several projects funded under the Australian Government Department of Climate Change and Energy Efficiency's (DC-CEE's) Integrated Assessment of Human Settlements Sub-programme ("the Sub-programme"). This work was aimed at developing approaches to, and exploring implications of, climate impacts on urban areas. The Integrated Assessment of Climate Impacts on Urban Settlements (IACCIUS) project reported here, through engagement with five case study settlements, sought to: (1) develop and test a methodology for integrated assessment, and (2) investigate specific priority issues in each jurisdiction. Figure 21.1 provides a locality map of the case study settlements.

In each settlement, an overview risk and vulnerability assessment was undertaken using the participatory process developed by IACCIUS (discussed below). This was conducted in partnership with local, state and territory governments and other stakeholders, and was followed by a detailed investigation of priority issues

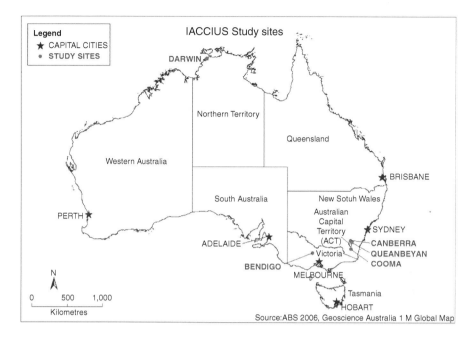

Fig. 21.1 Location of IACCIUS case study settlements (Note: this figure is derived from ABS 2006; GA 2010)

nominated by the research partners. Detailed local climate histories were also developed. These were used alongside coarser-scale scenarios to make climate change more meaningful at the local scale, and to reflect local differences (Hutchinson et al. 2010). Scenarios and local climate histories were supplemented by plain language descriptions of predicted changes to extreme events (e.g., an increase from two to five heat wave days per annum). The settlements and the specific issues were:

- Bendigo, Victoria, population 90,000: a rural center built on a gold-mining past, now servicing a rich agricultural district, with farming, educational, tourism and regional service industries. Issue: water consumption, and management of open space and recreation facilities in the face of drought.
- Canberra, population 330,000: the young, purpose-built and highly planned national capital of Australia. Issue: residential energy and water consumption, and the effectiveness of demand management measures.
- Cooma, New South Wales, population 8,000: a rural town servicing the Snowy Mountains Hydro-electric Scheme and pastoral surrounds, and gateway to Australia's alpine tourism areas. Issue: economic impact on the town of decreasing snow cover and increased wild fire occurrence. Analysis used proxy data that quantified visitation rates to the nearby alpine national parks.
- Darwin, population 110,000: the capital of the Northern Territory, with a high Indigenous population, situated closer to Asia than to other Australian cities

in a cyclone-prone tropical location. Issue: differential vulnerability to climate extremes in different areas across the city, and water and energy consumption issues.

• Queanbeyan, population 35,000: a rural city in New South Wales adjacent to, but separately governed from, the national capital of Canberra, and with a dormitory suburb role and diverse urban form. Issue: change to urban land cover (pervious versus impervious surfaces) and stormwater management, which is looked at in the context of current and likely increases in drought occurrence and more frequent storm events.

The selection of these settlements provides a range of different urban characteristics, socioeconomic attributes, and climatic zones and impacts. Unlike other projects in the Sub-programme, IACCIUS did not comprehensively analyze one place. Instead it used five different settlements as reference points against which to develop rigorous and practical approaches to assessing climate impacts, while also investigating priority issues in each place. IACCIUS was one of five integrated assessment (IA) exercises funded through the Sub-programme. The IACCIUS research team organized two cross-project meetings to allow comparison and learning. Most of the lessons reported were evident in the other projects.

Research and policy agendas of sustainability require integrated approaches in order to address the interdependence of human and natural systems (Dovers 2005a, p. 3). Integration in research incorporates knowledge across disciplines, whereas integration in policy connects agencies and sectors, and thus *informs* policy, and/or more directly, *formulates* policy. IACCIUS aimed to inform policy through: (1) providing information to local agencies for their use; (2) feeding assessments into decision-making via actively engaged officials; and (3) developing ways of undertaking IA elsewhere. It was agreed, however, that the formulation of policy responses was the role of the responsible authorities, not of the external assessment team.

The IACCIUS Methodology

The IACCIUS project addressed the gaps left by earlier IA studies by developing an initial approach (see Li 2007) and elaborating upon it through case studies. IACCIUS developed its methodology in the context of smaller Australian towns and cities, treating the urban settlements as complex adaptive human–environment systems, and emphasizing local realities, information needs, and decision-making. Settlements are not static, independent entities, but parts of the world dominated by the interactions of humans and their surroundings, and are complex and constantly adapting to internal and external pressures. This requires an approach that is integrative, flexible, critical and adaptive, and suited to the urban context.

To address these requirements, the IACCIUS methodology draws on a range of theoretical and methodological fields, including the following (for detail, see Li 2010; Li et al. 2010):

- Climate science
- Critical realism and adaptive theory
- Participatory research methods
- Systems thinking and tools
- Urban studies
- Public policy and institutional design
- Risk management and analysis of uncertainty
- Vulnerability analysis

Informed by these fields, and tested through diverse applications across five settlements, a methodological framework termed the Integrative Systems Risk and Vulnerability Assessment (ISRaVA) process was developed. The process is summarized in Box 21.1 (from Li et al. 2010). Integrated assessment is a process and intent rather than a method, and utilizes diverse methods and tools in different contexts. Box 21.2 evidences the diversity of suitable methods depending on local context and issues. The ISRaVA methodology is based on appropriate roots in relevant disciplines. It balances rigor with flexibility in varying practical contexts, and was found to be operational and accessible to local decision-makers.

The following two sections draw out more practical insights that have been derived from the development and application of this methodology. The aim is to indicate the challenges of local-scale climate impact and adaptation assessment, and the kinds of adaptation possibilities that arise.

Box 21.1 Integrative Systems Risk and Vulnerability Assessment (ISRaVA) Process[1]

Establishing the context

1. Accept the problem of local urban-scale climate change impacts as a complex adaptive human–environment system problem, requiring whole-of-system and whole-of-government engagement and response
2. Study, to understand within uncertainty limits, past, present, and future local climate change and variability

Identifying and analyzing what's at risk/vulnerabilities

3. Conduct a participatory urban system risk analysis process that:

   - "Teaches" participants basic systems concepts and associated tools
   - Enables the agreement of where to place system boundaries
   - Identifies urban system parts at risk of climate change impacts
   - Identifies key constraining relationships between system parts

(continued)

Box 21.1 (continued)

- Collectively constructs influence diagrams of the urban system
- Identifies systems and subsystems of interest
- Identifies priority subsystems for further in-depth investigation

4. Develop a communication and feedback system between researcher and stakeholder as the research proceeds, especially as may be related to concurrent policy processes
5. Identify relevant policy history and ongoing policy processes that may be relevant to the research process
6. Assess vulnerability for the priority subsystems of interest by using appropriate methods to collect and analyze primary or secondary data that:

- Enable a better understanding of the parts of the system at risk (exposure units)
- Enable a qualitative or quantitative description of system sensitivities
- Enable an exploration of past and present adaptation actions taken, future possible strategies, and, where possible, adaptive capacities
- Enable a more in-depth systems analysis taking into account vulnerability findings (this may appear as reworked influence diagrams, or a more detailed systems model)

7. In the assessment of vulnerability, use participatory processes with stakeholders to further identify future possible adaptation strategies, and current and future capacities to adapt, including possible policy recommendations

Evaluating and reviewing the process

8. Identify gaps in the analysis that may require:

- Further research or investigation
- The collection of further primary or secondary data to fill data gaps

9. Finalize and write up analyses, disseminate and communicate findings to all stakeholders

Treating the risks/vulnerabilities

10. Stakeholders responsible for adaptation strategies and policy-making at the city wide level to:

- Take into account findings of the integrated assessment and commence or continue management and the policy-making and implementation process
- Iterate through any of the above steps as required

---

[1] See also AGO (2006) for a comparative risk assessment based approach.

Box 21.2 Methods to Support Integrated Assessment

Goal	Method
Social and environmental assessment	Risk and vulnerability assessment (e.g., Adger 2006); social impact assessment (e.g., Henk and Vanclay 2003); environmental impact assessment (e.g., Morgan 1998); strategic environmental assessment (e.g., Dovers and Marsden 2002); sustainability assessment; ecological footprint analysis (e.g., Chambers et al. 2000); material flows analysis; life cycle analysis (e.g., Guinée 2002); state of environment reporting (e.g., DEHWA 2009)
Systems analysis	Systems thinking tools (e.g., Sterman 2000); influence/causal loop diagrams; reference modes (time series analysis); stock and flow modeling; group model building
Decision analysis	Game theory (e.g., Hanley and Folmer 1998), Bayesian decision networks (e.g., Jensen 1996; Cain 2001); meta-models, agent-based, expert systems; coupled-component; multi-criteria analysis
Spatial analysis	Demographic mapping; hazard mapping; remote sensing; geostatistics, GIS (e.g., Chang 2008)
Policy analysis	Policy monitoring and evaluation (e.g., Dovers 2005b); historical policy analysis; scenario planning; comparative policy analysis; comparative policy instrument choice analysis
Economic analysis	Triple-bottom-line (e.g., DEH 2003), cost–benefit analysis; nonmarket valuation; natural resource accounting
Historical analysis	Diachronic and synchronic historical policy analysis; oral history environmental history (e.g., Dovers 2000; Proust 2004); ethnographies
Discourse and document analysis	Government, private, archived, internet, newspaper, academic literature, local knowledge (e.g., Kitchin and Tate 2000)
Participatory and deliberative methods	Workshops; surveys; interviews (e.g., Hay 2005); participatory modeling; scenario building; local information gathering; deliberative techniques; focus groups, joint problem-framing; citizens juries (e.g., James 2003); consensus conferences; team-work approaches

Methodological Lessons

The applications of the IACCIUS methodology indicated strong potential of a systems-oriented, participatory approach to identifying risks and vulnerabilities at the local settlement scale. The benefits accrued at both a general, overview scale, and through more detailed assessment of priority issues. The participatory approach allowed joint learning to occur, increasing agency awareness of climate change impacts, and for local imperatives to be identified and pursued. The following methodological lessons suggest ways to render IA more practical through design and scoping of assessments.

Time, Data, and Skills

Climate change is a new concern for most local decision-makers, and an unfamiliar issue within existing administrative, decision-making, human resource ,and informational capacities. Increasing numbers of local agencies are investigating the implications of climate change. A wide communication of methods and applications would assist in driving a more rapid, rigorous and efficient assessment of risks and vulnerabilities, and development of response and adaptation options. In particular, local agencies require understanding not only of ways in which to investigate climate change implications, but the likely investment of time and resources needed. IA is demanding of skills, and precise needs are difficult to predict until the overview assessments have identified priority issues. When this is done, it may be clearer whether skills such as GIS, demographic, climate science, economic, hydrological, and/or engineering are needed.

IACCIUS recorded the number of work days required to conduct the IA of the case study settlements in order to indicate the level of commitment more generally. In total, the number was in the order of 200–250 work days per case study settlement. This included the overview assessment, the localizing of climate science, and investigations of four to five priority issues.

These are significant commitments for local agencies, and in addition to the time required to complete the tasks, there are challenges related to dealing with data and the availability of skills. As is evident from the range of methods potentially required for an IA (Box 21.2), there is a matching demand for acquisition and management of diverse and often complex datasets. These may include energy and water consumption, local climate records, demographic data, remotely sensed topographic and other land surface data, and economic data; these datasets also usually need spatial representation and manipulation. Three issues arise. First, many existing datasets will be unsuitable for the required use because of spatial resolution, limited time series, coverage or data quality. Second, large and diverse datasets will require a sophisticated data management system, the provision and servicing of which will need to be included in the project design and funding. Third, difficulties will likely arise in accessing data, through cost of purchase, time constraints of data custodians, or the evocation of commercial-in-confidence or privacy provisions. Often, key data custodians will not be included in the original project design but will be identified later, and custodians may not be committed to, or accept the importance of, the assessment.

The range of data and analytical demands is matched by the range of skills required. Investigation of specific issues may require, for example, skills in statistics, demography, local economic profiling, policy analysis, climate science, hydrology, engineering and asset appraisal, or risk assessment, as well as generic skills in facilitation, communication, and project management. Assessment teams and local agencies will be hard pressed to provide all these skills.

Relevance to Local Decision-Making

Of critical importance in local-scale assessment of the implications of climate change is an established connection to local decision-making and cycles of planning and policy. Participatory IA exercises, such as IACCIUS, seek this to some extent by including local decision-makers (e.g., local government planners) in the assessment process. However, it is also important to develop an understanding of local policy processes and to match the IA to these. A local government may undertake strategic planning once every 5 years, meaning that an IA process at a particular time may be either very timely, or incapable of being incorporated into strategic planning for another few years. IAs will often be performed by contracted research or consultant teams as a discrete exercise, and so the process of developing cooperative, positive relationships and an understanding of local imperatives may take time to achieve.

It is also necessary to include wide representation from local agencies, especially in the overview assessment phase, to ensure that all impacts, risks, vulnerabilities, and interests are identified, and that interactions are mapped. Buy-in at this stage will increase the likelihood of collaboration in subsequent investigation of specific issues, and will require support by senior or central people in an agency, for example, a general manager or mayor of a local council, or the central planning or similar agency at state or territory level.

Practical Lessons

The above methodological lessons identified practical issues generic to local integrated assessment (IA). The following illustrates the kind of practical adaptation options that emerged, using examples from the five IACCIUS case studies. Note that implementation of specific strategies is a matter for local and other organizations beyond the life of the IACCIUS project (completed September 2008).

Bendigo

In the rural city of Bendigo, IACCIUS investigated the impacts of current and likely drought and water shortages under changed climatic conditions. This priority emerged from the local council and stakeholders, with outdoor sport and recreation and facilities amenities being important for local lifestyles, public health and cultural identity, and to the economy by being the location of major sporting events. While on a wider scale, the survival of sporting activities may not feature in climate debates, at a local level such priorities are central to regional settlements and are the responsibility of local governments.

Although the local council and stakeholders were already familiar with drought and active management of water shortages, new insights emerged from the analysis. This included the issue of equity of access to water for water-dependent sports, which was identified from socioeconomic and spatial analysis of facilities across the community. It also revealed the equity issue related to the possible under-resourcing of nonwater-dependent activities (e.g., indoor sports) at a time when investments may be skewed toward maintenance of water-dependent ones. Also identified were differential impacts on sports and recreational activities tied to their relative water- (or grass cover-) dependence, and thus varied opportunities to decouple use from water availability.

Canberra

In Canberra, the intent was to investigate fine-scale water and energy consumption of households, to explore the impact and potential of appliance rebate and other demand management interventions as strategies for climate adaptation. Existing datasets are not organized at a useful scale for this, being averaged over settlement and suburb level, and historically organized for billing rather than consumption analysis. Although previously made available in Canberra and other cities in Australia, these data were not made available to the project on privacy grounds, and the analysis could not proceed. This evidences the importance of data suitability and availability noted above, and of gaining collaboration of data custodians in a timely fashion.

Cooma

In Cooma, local stakeholders identified climate change-driven threats to the tourism income of the area as a priority, with specific reference to declining snow cover and impact of wild fires. This required confirmation of past events and likely future event scenarios using local climate history and climate projections, and revealed the difficulty of projecting future visitation numbers in a useful fashion. Instead, IACCIUS examined the impact of past poor snow and major fire seasons, which revealed: (1) some inadequacies in past visitation data, and (2) highly differentiated impacts of poor seasons. Through interviews with local tourist-dependent businesses, it was revealed that some had already adapted to existing climate variability in ways that prepared them for future increased variability.

It was concluded that many options for ensuring resilience in the local economy to climate change were similar to familiar recommendations for local economic development. That is, better knowledge of local economic trends, better visitation data and understanding of the visitor market, coordinated land use planning, economic diversification, and improved marketing and communication efforts. Also revealed was a finer-scale understanding of roles and responsibilities for local climate assessment and adaptation across local businesses and communities, local government, wider industry associations, and state and national levels of government. This may inform coordination and efficiencies in data provision and adaptive strategies.

Darwin

In tropical Darwin, the threat of heat stress and increased cyclone intensity dominated the study, and from this emerged the need to better understand differential vulnerability across the city. This was done using fine-scale spatial, biophysical and socioeconomic data, water and energy consumption data and population data. This approach allowed identification of areas within the city where a combination of these factors indicated varying levels of vulnerability and adaptive capacity to different climate impacts. This approach also allowed analysis of the adequacy of some adaptation measures, such as the distribution of cyclone shelters, when analyzed against population distribution and mobility. Areas of potential improvement were identified in data availability to describe important variables such as fine-scale water and energy demand, population turnover and socioeconomic status, threats of storm surge, and building stock characteristics. Again, existing data and information pathways will be tested by the new questions posed by local climate impact assessments. Furthermore, the IA process may identify improvements useful for other purposes such as public health, emergency management, or building stock assessment and improvement.

Queanbeyan

In Queanbeyan, drought and rapid urban development had produced a changing urban surface, intensified by shifts in household landscape preferences and loss of grass cover. The latter included replacement of landscaping products in the face of drought and water restrictions. Urban water runoff is accelerated and stormwater management complicated by such changes. Beneath an assumed impervious surface proportion of approximately 40% for suburban landscapes, innovative fine-scale analysis of remotely sensed data revealed a range in the presence of impervious surfaces, between 17% and over 80%, across the settlement, depending on urban form and housing stock age. High imperviousness in newer, low-density developments was an unexpected finding. Such insights may inform future subdivision layout, landscaping requirements, and stormwater provision.

Discussion

Generic determinants of vulnerability and adaptive capacity are commonplace when the concern relates to a need to compare across regions or nations (see for example Brooks et al. (2005)). However, at the local scale, investigation into the context-specific determinants of vulnerability and adaptive capacity may be more relevant. In this integrated assessment study, issues of political and other sensitivities arose at the local scale. Broader-scale climate policy debates identify more sensitive regions, sectors and subsets of populations affected by direct climate impacts, indirect physical or economic impacts, and implications of policy responses. Locally, though, impacts and implications identified at individual property- or entity-scale are defined. This personalizes the issue of climate change in a highly political and sensitive manner, raising issues of business viability, individual household vulnerability, or property prices. This cannot be avoided if we are to assess and respond to climate change at the local scale where many planning and other decisions are made. Assessment processes must at least recognize this sensitivity and plan the participatory assessment process and communication strategy accordingly.

Recommendations

In answer to the above issues, the IACCIUS project recommended that IA, reflecting the ISRaVA process (Box 21.1), would be most effectively and efficiently undertaken in three stages, and in the following sequence:

- A systems-oriented overview assessment of climate impacts, risks, and vulnerabilities. The last part of this would be a stakeholder-led prioritization of key issues requiring more detailed, time- and data-intensive investigation, allowing definition of the problem and hence, the required methods, resources, data, and skills.
- Project redesign for detailed analysis of priority issues.
- Re-integration of findings of detailed investigations against the overview assessment to capture cross-sectoral interactions; for example, impact of water demand management on public health, or emergency management evacuation plans on public transport.

A final generic issue is the fate of insights developed during what is normally a "one-off" assessment process. Even with connection to local policy-making systems, there is the need to integrate climate change considerations into ongoing information streams and decision-making processes. This issue of climate policy integration requires further research and policy development.

Conclusion

This chapter has summarized the methodology developed and tested in five diverse, small- to medium-sized Australian settlements, and identification of some of the fine-scale practical impacts and response options. The ISRaVA process, or similar integrated assessment approach using multiple methods, have significant potential to inform local discussions and decision around climate change adaptation. As with other applications of IA, more methodological development and testing is required, but sufficient knowledge now exists to progress with further practical applications.

References

Adger WN (2006) Vulnerability. Glob Environ Change 16(3):268–281

Australian Bureau of Statistics [ABS] (2006) Census data. Australian Government. http://www.abs.gov.au/websitedbs/d3310114.nsf/home/census+data. Cited 30 Aug 2010

Australian Greenhouse Office [AGO] (2006) Climate change impacts and risk management: a guide for business and government. Australian Government. http://www.greenhouse.gov.au/impacts/publications/risk-management.html. Cited 19 Jan 2009

Brooks N, Adger WN, Kelly PM (2005) The determinants of vulnerability and adaptive capacity at the national level and the implications for adaptation. Glob Environ Change 15(2):151–163

Cain JD (2001) Planning improvements in natural resources management: guidelines for using Bayesian networks to support the planning and management of development programmes in the water sector and beyond. Centre for Ecology and Hydrology, Wallingford

Chambers N, Simmons C, Wackernagel M (2000) Sharing nature's interest: ecological footprints as an indicator of sustainability. Earthscan, London

Chang K (2008) Introduction to geographic information systems, 4th edn. McGraw-Hill, Boston

Department of the Environment and Heritage [DEH] (2003) Triple bottom line reporting in Australia: a guide to reporting against environmental indicators. Australian Government. http://www.environment.gov.au/settlements/industry/publications/indicators/pubs/indicators.pdf. Cited 19 Jan 2009

Department of the Environment, Water, Heritage and the Arts [DEWHA] (2009) State of the environment (SoE) reporting. Australian Government. http://www.environment.gov.au/soe/index.html. Cited 19 Jan 2009

Dovers S (ed) (2000) Environmental, history and policy: still settling Australia. Oxford University Press, Melbourne

Dovers S (2005a) Clarifying the imperative of integration research for sustainable environmental management. J Res Pract 1(2): Article M1

Dovers S (2005b) Environment and sustainability policy: creation, implementation, evaluation. Federation, Sydney

Dovers S (2009) Editorial: normalizing adaptation. Glob Environ Change 19(1):4–6

Dovers SR, Hezri AA (2010) Institutions and policy processes: the means to the end of adaptation. Wiley Interdiscip Rev Clim Change 1:212–231

Dovers S, Marsden S (eds) (2002) Strategic environmental assessment in Australasia. Federation, Annandale

Edmond JA, Rosenberg NJ (2005) Climate change impacts for the conterminous USA: an integrated assessment summary. Clim Change 69(1):151–162

Geoscience Australia [GA] (2010) Digital topographic data: global map Australia 1 M. Australian Government. http://www.ga.gov.au/topographic-mapping/digital-topographic-data/index.jsp. Cited 30 Aug 2010

Guinée JB (ed) (2002) Handbook on life cycle assessment: operational guide to the ISO standards. Kluwer, Dordrecht

Hanley N, Folmer H (eds) (1998) Game theory and the environment. Edward Elgar, Cheltenham

Hay I (ed) (2005) Qualitative research methods in human geography. Oxford University Press, South Melbourne

Henk AB, Vanclay F (eds) (2003) The International handbook of social impact assessment: conceptual and methodological advances. Edward Elgar, Cheltenham

Holman IP, Nicholls RJ, Berry PM et al (2005) A regional, multi-sectoral and integrated assessment of the impacts of climate and socio-economic change in the UK. Clim Change 71(1–2):43–73

Hutchinson MF, Campbell-Wilson AM, Davis C (2010) The Integrated Assessment of Climate Change Impacts on Urban Settlements (IACCIUS) project: report on local climate variability and change in Bendigo, Canberra & Queanbeyan, Cooma and Darwin. Fenner School of Environment and Society, The Australian National University. http://fennerschool-research.anu.edu.au/iaccius/publications/. Cited 30 Aug 2010

James RF (2003) Citizens' juries and deliberative valuation: evaluating their potential use in participatory environmental management. PhD thesis, Australian National University, Canberra

Jensen FV (1996) An introduction to Bayesian networks. Springer, New York

Kirshen P, Ruth M, Anderson W (2008) Interdependencies of urban climate change impacts and adaptation strategies: a case study of Metropolitan Boston USA. Clim Change 86(1–2):105–122

Kitchin R, Tate NJ (2000) Conducting research in human geography: theory, methodology and practice. Prentice Hall, Harlow

Lange MA (2008) Assessing climate change impacts in the European north. Clim Change 87(1–2):7–34

Li GM (2007) Approaching integrated assessment of climate change impacts on urban settlements. State of Australian Cities Conference Proceedings, Adelaide, 28–30 Nov 2007. http://www.unisa.edu.au/SOAC2007/program/papers/0102.PDF. Cited 22 Apr 2008

Li GM (2010) A Methodology for Integrated Assessment of Climate Change Impacts on Urban Settlements (IACCIUS) in Australia. Fenner School of Environment and Society, The Australian National University. http://fennerschool-research.anu.edu.au/iaccius/publications/. Cited 30 Aug 2010

Li GM, Dovers S, Sutton Pet al (2010) IACCIUS: Synthesis Report on Integrated Assessment of Climate Change Impacts on Urban Settlements. Fenner School of Environment and Society, The Australian National University. http://fennerschool-research.anu.edu.au/iaccius/publications/. Cited 30 Aug 2010

Morgan RK (1998) Environmental impact assessment: a methodological perspective. Kluwer, Dordrecht

Proust KM (2004) Learning from the past for sustainability: towards an integrated approach. PhD thesis. Australian National University, Acton. http://thesis.anu.edu.au/public/adt-ANU20050706.140605/index.html. Cited 7 Oct 2006

Sterman JD (2000) Business dynamics: systems thinking and modeling for a complex world. Irwin/McGraw-Hill, Boston

Chapter 22
The Role of Local Government in Adapting to Climate Change: Lessons from New Zealand

Andy Reisinger, David Wratt, Sylvia Allan, and Howard Larsen

Abstract Local government plays an important role in facilitating adaptation to climate change at the community and regional level. Experiences and lessons from New Zealand suggest that the following elements together contribute to an enabling environment for local government: (1) raising community awareness of climate change; (2) engaging and developing the local expertise of professionals and decision-makers by presenting climate change science, scenarios and their uncertainties in locally relevant contexts and through interactive workshops; (3) adoption of a sequential approach to assess risks and identify vulnerabilities in the context of other socioeconomic and location-specific pressures; and (4) support from central government through regulation and guidance material. We outline the measures undertaken in each of those areas and discuss remaining barriers and uncertainties.

Keywords Local government • Vulnerability • Risk assessment • Regulatory environment • Guidance material • Mainstreaming • Bottom-up • Top-down • New Zealand • Risk management

A. Reisinger (✉)
Climate Change Research Institute, Victoria University, Wellington, New Zealand
e-mail: andy.reisinger@vuw.ac.nz

D. Wratt
National Institute of Water and Atmospheric Research (NIWA), Wellington, New Zealand
e-mail: d.wratt@niwa.co.nz

S. Allan
MWH NZ Ltd, Wellington, New Zealand
e-mail: sylvia.allan@ihug.co.nz

H. Larsen
Ministry for the Environment, Wellington, New Zealand
e-mail: howard.larsen@mfe.govt.nz

Introduction: New Zealand Context and the Role of Local Government

Local government agencies at city, district, and regional levels play a key role in responding to the risks posed by climate change. This chapter outlines experiences from New Zealand on the role of local government in societal adaptation to climate change, including lessons that may be applicable in other countries.

This introduction provides an overview of the New Zealand climate, economy, and local governance arrangements. Subsequent sections discuss: past experiences and barriers to adaptation until the early 2000s; the role of science communication and challenges created by scientific uncertainty for decision-making processes involving diverse stakeholders; the benefits of a shift toward a vulnerability-based approach in risk assessment; and finally the importance of central government regulation and guidance material in facilitating adaptation at the local level. The concluding section summarizes our key findings.

New Zealand consists of two large islands extending from about 33 to 46°S in the southwest Pacific and numerous smaller islands in the subtropics and sub-Antarctic region. Its climate is maritime, dominated by the moderating influence of the Southern ocean that leads to moderate summers and warm and wet winters, and by westerly winds that lead to strong orographic rainfall differences between western and eastern regions. High intensity rainfall and steep terrain lead to dynamic flow regimes in many rivers. El Niño-Southern Oscillation (ENSO) and the Interdecadal Pacific Oscillation contribute to inter-annual climate variations (Hennessy et al. 2007).

The New Zealand economy is open and trade-oriented, with a strong focus on tourism, agriculture and forestry, manufacturing and services. Economic and population growth have led to increased coastal development and land-based production and growing demand for energy and water (irrigation and hydropower). Many aspects of these sectors are sensitive to climate variability and extremes (Stroombergen et al. 2006), and recent extreme events such as floods and droughts have had demonstrable impacts on GDP and rural activity (Buckle et al. 2002; Tait et al. 2005). Climate-related changes in ecosystems, including invasive species, pests and fire risk have further potential to affect New Zealand's natural assets (Hennessy et al. 2007).

The management of natural resources and hazards in New Zealand is almost entirely devolved to local government, not only in the urban environment but also at regional (watershed, county, or state) scales. The functions and responsibilities of local government, set out in the Resource Management 1991 and Local Government 2002 Acts (NZ 2008a, b), include managing development and hazards in the natural and built environment through policies, plans and consents, providing essential services such as water supply, civil defense and waste management, and managing natural resources.

A critical challenge for local government in New Zealand and elsewhere is the need to assess climate change risks and prioritize adaptation responses in the local context of many other diverse functions and responsibilities. Councils frequently find themselves at the interface of competing public and private interests and need to weigh up economic, social and environmental aspects, with both near- and long-term perspectives, often with limited resources and in direct interaction with the community. Forward-looking adaptation decisions that may constrain near-term development opportunities (such as coastal set-back zones that take long-term sea level rise into account) in particular need to be understood and supported by the local community (Adger et al. 2007; Yohe et al. 2007).

Past Experiences and Barriers to Local Government Adaptation

Community awareness of climate change issues emerged during the early 1990s. In New Zealand, as in most other developed countries, this was driven partly by central government engaging in international negotiations, and partly by increasing scientific research and publications on climate change and its potential impacts (RSNZ 1988, 1990, 1992; MfE 1990; Wratt et al. 1991). However, material responses from local government were limited, as the issue had only begun to emerge.

In 1993, the New Zealand Ministry for the Environment published a booklet aimed at raising climate change awareness among local government elected representatives, senior managers, policy advisers, and planners (Allan 1993). The publication represented the first effort by central government to directly engage local government in actions to address both the impacts of climate change and to limit greenhouse gas emissions. It advised local government that it had an important role in supporting central government, but provided no guidance on concrete decisions or approaches ("the government is not in a position, nor would it necessarily wish to dictate to local authorities on how they should respond to climate change" (Allan 1993)). This approach was seen as consistent with general devolution of decision-making to local authorities, the diverse nature of local communities in New Zealand, and the interplay of climate change impacts with other socioeconomic and resource pressures at the local scale.

A 1993 review found very few examples of local government practice directed explicitly at climate change under any of the relevant areas of responsibility. These areas had been identified as regional and district policy- and plan-making, regional energy policy, urban and rural planning including management of urban form, subdivision and building design, transport planning and management, waste management, avoiding and mitigating natural hazards, and community education and advocacy (Allan 1993).

On a world scale, the New Zealand experience is not unusual. Throughout much of the 1990s and early 2000s, local governments faced significant barriers to action, including the coarse resolution of climate scenarios and impacts, and the perception in some councils and sections of the community that many impacts and even climate change itself were highly speculative. Such perceptions created concern that planning decisions based on climate change scenarios could compromise other, apparently more pressing local priorities. At the same time, central governments mostly focused on the development of policies to limit greenhouse gas emissions on a national basis, but provided little further guidance on the roles that local government could play in responding to climate change. Even for the period 2001–2007, only three out of 39 Cabinet papers on climate change focus on adaptation policies and related measures by central government (MfE 2007a; released under Official Information Act).

An additional barrier possibly more specific to New Zealand was a largely market-led approach to decision-making. In a period focused on a free market economy, monetarist economic policy and privatization of state-owned industries, interventions that impinged on private property rights were subject to rigorous examination and challenge under the Resource Management Act. Under this Act, councils wishing to impose controls on land use or specific activities have to demonstrate the negative consequences of a proposed activity, rather than proponents of the activity having to demonstrate its sustainability. Thus both the political climate and the regulatory environment limited the ability of local authorities to address either the causes or effects of climate change.

The turn of the millennium saw a growing awareness at the level of local government of the reality of climate change, as well as the obligations arising from the Kyoto Protocol. In New Zealand this was reflected in increasing efforts by regional planners and resource scientists to incorporate climate change effects into local government policies, plans and decisions, and at central government level, a Parliamentary Select Committee Inquiry into the role of local government in contributing to domestic climate change objectives (NZ 2001).

Workshops in 2001–2002 organized by the Ministry for the Environment highlighted that significant barriers to more effective adaptation actions by local government remained. These included: limited effectiveness of science communication and limited community awareness; the difficulty of integrating climate change projections and their uncertainties into local government decisions while balancing the interests of various stakeholder groups; and the lack of regulatory certainty and guidance regarding the mandate, priorities and options to respond to climate change impacts (Reisinger 2002). The following sections outline experiences gained from efforts in New Zealand to overcome these barriers.

Communicating Science and Uncertainty to Facilitate Adaptation

Communicating climate change science, projected impacts and uncertainties to elected representatives, council staff and the community pose major challenges but are vital for the implementation of effective adaptation measures. Councils in New Zealand generally operate very closely within their communities. They are

Table 22.1 New Zealand attitudes to climate change, based on polls considered representative of the New Zealand population (ShapeNZ 2007a, b, 2008)

	April 4–7, 2007	June 14–26, 2007	September10–11, 2008[a]
An urgent problem	35%	35%	25%
A problem now	42%	39%	51%
A problem for later	13%	15%	9%
Not a problem	8%	10%	13%
Don't know	2%	1%	2%

[a]Note: New Zealand Parliament passed emissions trading legislation on September 10, 2008

thus very responsive to views on climate change held by key stakeholders (including industry and business organizations, environmental and social advocacy groups, and the Indigenous population) and sentiments held in the community at large. Adaptation policies and measures by councils are thus largely shaped by priorities expressed by the local community, while in turn council programs can facilitate actions by individuals and other local organizations.

One major challenge in New Zealand has been the mixture of views in the community regarding the reality and/or seriousness of climate change, which directly limits the ability of councils to proactively address climate change-related risks. Although the profile of climate change has increased, significant pockets of the community remain skeptical about climate change. Community surveys in 2001 revealed that climate change was ranked well behind other environmental and sustainability concerns (MfE 2001). By 2007–2008, around 75% of New Zealanders regarded climate change as an urgent and/or present problem, but some 20–25% considered it to be not a problem, or only a problem for later (Table 22.1). In particular, some farmers and councils with a strong rural base have remained rather skeptical about climate change science (Fulton 2008). This skepticism may in part be motivated by concerns about policies to reduce agricultural greenhouse gas emissions (Brenmuhl 2008).

We found that such political barriers to adaptation can often be reduced by separating discussions of climate change impacts, and the necessity of adaptation at the local scale, from discussions about mitigation policies, which often involve different stakeholders and/or different levels of decision-making (Kenny and Fisher 2004; Kenny 2005; Klein et al. 2007).

A second challenge is that alternative emissions scenarios and differences between climate models often generate a broad range of potential future changes at local scales. Councils that wish to revise existing policies, plans, and hazard zones on the basis of risk assessments using such broad scenarios are open to challenge by members of the community who may be concerned about the short-term opportunity cost of adaptation measures. Council staff therefore often seek "single-number" projections for future climate changes at specific time horizons that could be added to existing planning provisions (e.g., flood or coastal hazard zones), especially where such measures could be challenged by well-funded business interests (e.g., in high-value coastal housing developments). In such situations, uncertainty in climate projections can act as a barrier to proactive adaptation as councils seek to avoid lengthy and expensive court cases.

Improved science communication alone cannot resolve such conflicts, but it can help communities address them more effectively by presenting climate change projections and their uncertainties within their relevant local context. Past experiences with extreme events, assets at risk, benefits and risks of infrastructure developments, socioeconomic pressures and community values regarding short-term development and sustainability all carry their own uncertainties, but nonetheless collectively shape a community's views on acceptable risks and appropriate responses. This suggests that uncertainties related to climate change can be interpreted more meaningfully if they are integrated ("mainstreamed") into such broader community decision-making processes (Kelly and Adger 2000; Burton et al. 2002; Smit and Wandel 2006; van Aalst et al. 2008).

Council staff, professional bodies, and local interest groups have considerable knowledge about local vulnerabilities to climate extremes and the management of such risks within competing interests from various parts of the community. Recognizing and valuing such local expertise through interactive workshops that include experts as well as key local stakeholders appears crucial to turn scientific "top-down" information on climate change into practical measures that are likely to be supported by the local community.

Successful examples of such integration in New Zealand are dune restoration programs that create multiple benefits (MfE 2007b), and upgrading of stormwater systems in response to recent flood events (MfE 2007c; see also Box 22.1). Dedicated workshops have also been held for farming communities in eastern regions vulnerable to drought (Kenny and Fisher 2004; Kenny 2005) and for the indigenous Māori population. Māori are regarded as vulnerable to climate change due to greater reliance on natural resources, high exposure of traditional coastal and river settlements, generally lower socioeconomic status, and threats to some traditional value systems (Packman et al. 2001; Cottrell et al. 2004; NIWA 2006).

Box 22.1 Examples of Bottom-Vulnerability Assessments and Adaptation Responses

Case Study: Stormwater infrastructure – Tauranga City (MfE 2007c)
Tauranga is a coastal city in the Bay of Plenty, on the northeast of the North Island of New Zealand. In 2005, the western Bay of Plenty area was hit by storms that severely overloaded the stormwater system in Tauranga (and caused widespread damage elsewhere).

In response, the Tauranga City Council undertook a study of the historical rainfall and the possible changes due to climate change. Council staff used the methodology set out in MfE (Mullan et al. 2004) to determine first that there was a potential risk from climate change, and then to develop projections for heavy rainfall events. The study found that more intense rainfall events are likely in the future, despite an expected decrease in average annual rainfall.

New rainfall design charts, incorporating the projected climate changes, are now used in designing all new stormwater systems and during the upgrade of existing systems. In partnership with the regional government (the Western Bay of Plenty District Council and Environment Bay of Plenty), the projected increase in heavy rainfall has been incorporated into the planning blueprint for growth and development in the region.

Case Study: Industrial Rezoning – Invercargill City
Invercargill is New Zealand's most southern city, lying close to the southern coast of the South Island. The City Council identified the need for more industrial land adjacent to the city. Following a site-selection exercise, some 600 ha of land were purchased by the Council, a concept plan was developed, and a change of zoning was notified for public submissions. During this process submitters, including the Southland Regional Council, questioned the potential effects of climate change, given the site's coastal location and natural drainage pattern, including the adequacy of provisions for future stormwater management.

As a result, the City Council delayed the processes of rezoning while additional science and engineering advice was obtained, particularly relating to sea level rise and the effects of storms on coastal processes in the area, and on likely future rainfall intensities. The development of a range of scenarios based on local conditions, and the application of a screening study, followed by more detailed risk assessment, determined that most of the site would be appropriate for industrial use and development, but that a more extensive coastal buffer should be retained and buildings and infrastructure should be set back further from flood prone areas. The concept plan, which had already included provisions for significant natural drainage areas and a coastal setback, was modified by the City Council prior to the submissions being heard. As a result of the changes, the Regional Council withdrew its opposition to part of the rezoning.

The process demonstrates both the need for best possible information at the time of major land use change, the effectiveness of a risk-based approach, and the desirability of consultation with key stakeholders early in planning processes.

Adaptation by councils also relies on consultants (e.g., engineers and planners) being able to incorporate relevant climate change projections and risks assessments into their reports. Workshops and training courses for practicing engineers have therefore also received increased attention in New Zealand to enhance the transfer of climate change knowledge to councils. In these workshops, climate scientists cover the underpinning science, and engineers present and discuss case studies and translate climate projections into their practical implications. Such workshops have the double benefit that council staff become sensitized to the implications of climate change, and external engineers and other consultants become better equipped to include climate change issues in their offers of service and professional communications to councils (Gluckman and Boyle 2003).

Overall, these experiences indicate that "bottom-up" interactions between climate specialists, consultants, local government planners and engineers, elected representatives, residents and local pressure groups are important components in communicating climate science for adaptation. Documents, reports, and websites are necessary but not sufficient to communicate climate science: many of these stakeholders also like to hear from knowledgeable persons, ideally colleagues, at workshops or public presentations where they can ask challenging questions and raise issues in a peer-to-peer environment (Kenny and Fisher 2004; Wratt et al. 2006). In turn, such interactions can help organizations involved in producing and disseminating scientific knowledge to tailor their communications to stakeholder needs (Tribbia and Moser 2008).

Toward a Vulnerability-Based Approach

Consistent with these science communication perspectives, a critical step to improving the integration of climate change information into local government decision-making in New Zealand was to move away from predominantly providing top-down information on climate change and its impacts toward more emphasis on bottom-up assessments of local vulnerability and risks. This approach helps councils determine the relevance and relative priority of climate change within the context of other local functions and responsibilities, and to identify adaptation responses that are consistent with other environmental or socioeconomic pressures and development plans (Burton et al. 2002; UKCIP 2003, 2007; Adger 2006; O'Brien et al. 2007; van Aalst et al. 2008).

The specific steps involved in a sequential assessment of climate change-related risks from a local government perspective were promulgated in guidance material issued by the Ministry for the Environment (Mullan et al. 2004, 2008) and disseminated in workshops, case studies and publications in planning and engineering journals.

Sequential risk assessment encourages councils to undertake a series of steps that are characterized by increasing complexity and data requirements (Fig. 22.1). The first step is to consider whether climate qualitatively plays a significant role in the

Fig. 22.1 Flow-chart indicating sequential decisions of increasing complexity and resource requirements in assessing climate change risks (adapted from MfE (Mullan et al. 2008), © New Zealand Ministry for the Environment)

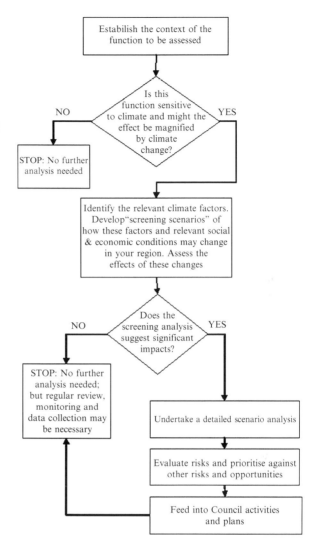

council's ability to carry out or meet any particular function or responsibility. If it does, the next step is to undertake a preliminary "screening" assessment using broad quantitative ranges of projected future climate change and relevant socioeconomic changes over comparable time frames. If the screening assessment indicates a significant potential for negative consequences (or opportunities), councils are advised to undertake a full risk assessment using more complex and location-specific data, projections, and impact models (Mullan et al. 2008).

Such a sequential approach encourages consideration of climate change risks even where resources are limited, and aims to help integration of climate change risk assessments into standard council processes. The bottom-up approach is also intended to help councils prioritize risks and potential adaptation actions based on the presence of other locally relevant stresses that could exacerbate vulnerability to climate change (Yohe et al. 2007). Relevant additional stresses include vulnerability to current climate variability and extremes, absence of safety margins or techno-logical solutions, number of people or size of resource affected, and the potential for lock-in of long-lived infrastructure or developments in areas of increasing risk (e.g., in areas exposed to gradually increasing river or coastal flood risk). These stresses can be used to prioritize risks and are consistent with criteria used by the Intergovernmental Panel on Climate Change (IPCC) to identify so-called key vulnerabilities (Schneider et al. 2007). Key risk areas identified in this process are usually those with high vulnerability even to current climate variability and extremes, such as infrastructure and water resources and areas exposed to coastal erosion and flooding.

The sequential risk-management approach outlined here does not reduce the inevitable uncertainty about quantitative climate change projections and their impacts at the local scale, nor the divergence of views in the community about appropriate responses. Its main advantage is that it allows consideration of scientific uncertainties and alternative scenarios in a more relevant context that can be managed by councils themselves. Councils have well-developed mechanisms to deal with uncertainties as part of risk management processes and to decide whether to focus on avoidance of catastrophic risks, the exceeding of certain thresholds, or minimization of net costs to the community. Such decisions are inevitably, and appropriately, influenced by other non-climate considerations, including com-munity perceptions, value judgments, and legislative frameworks. Integration of climate change decisions into wider nonclimate contexts ensures that appropriate community consultation mechanisms contribute to decisions, which increases the chances of support for appropriate adaptation even if shorter-term development opportunities may be reduced or curtailed. In contrast, separating the consideration of climate change from established standard processes increases the costs of climate risk assessments, and makes it harder to identify and build on synergies between adaptation options and other plans for community development and sustainable resource management (Adger et al. 2007; Klein et al. 2007).

Regulation and Guidance Material to Support Adaptation

As noted earlier, under the Resource Management Act the burden of proof generally falls on councils that wish to impose controls on activities that may be unsustainable in the face of climate change, unless there are relevant additional central government regulations or direct interventions. Prior to 2004, the only additional legal obligation for local government in New Zealand to consider climate change existed through the New Zealand Coastal Policy Statement produced under the Resource Management Act (Minister of Conservation 1994). This requires council policy and decision-makers to consider the "possibility of sea level rise" but provides no guidance on the magnitude or time frame that should be considered. Other provisions of the Resource Management Act encourage a precautionary and sustainability-focused approach but do not refer to climate change or sea level rise explicitly. As a result, some councils have been challenged by private individuals and land developers in the courts, arguing that climate change impacts were either too uncertain to factor into decisions, did not justify high near-term opportunity costs from foregone development, or were outside the mandate of councils to consider.

The lack of regulatory certainty was addressed in 2004 through an amendment to the Resource Management Act. Section 7(i) of the Act now requires all persons exercising functions and powers under the Act to have "particular regard [...] to the effects of climate change" (NZ 2008b). Personal communications from council planners and recent reviews of regional policy documents (Willis 2007; Environment Bay of Plenty 2008b) suggest that this legislative change is leading to a more explicit and comprehensive focus on the potential effects of climate change in the development of regional policies and planning provisions. However, there is as yet insufficient evidence whether any improvements in local government policy and practice can be attributed to this legislative change or might equally be due to greater community awareness and general acceptance of climate change.

Intensive effort has also been devoted to the production and dissemination of suitable guidance material for local government to ensure that climate change considerations become a regular component of relevant council functions and are recognized by staff, elected representatives, and regional industry and community groups. In May 2004, the Ministry for the Environment coordinated and published a detailed guidance manual, *Climate Change Effects and Impacts Assessment* (Mullan et al. 2004). This manual provides guidance on the sequential risk assessment approach outlined above, including extensive regional information on climate change projections based on downscaled scenarios. This general guidance was supplemented with a specific guidance document on *Coastal Hazards and Climate Change* (Ramsay and Bell 2004)

A nontechnical summary of this guidance was produced to introduce local government professionals such as planners, engineers, asset managers, and hazard managers to the key issues and sequential decision-making steps. This 30-page booklet, *Preparing for Climate Change: a guide for local government in New Zealand* (MfE 2004), proved popular not only with local government but also with engineers, planners, and others in private practice.

In addition, a number of case studies have been conducted to provide more specific examples of risk assessments and possible adaptation actions, e.g., on urban stormwater design, transport networks, and coastal flooding and erosion (MfE 2008c). Climate change workshops (over 40 in 2007–2008) have been run with various professional groups. Feedback from engineers has been that while they started by using the summary publication, they quickly found themselves referring to the detailed technical manual. Planners, lawyers, and insurance industry professionals, on the other hand, have indicated they found the high-level conceptual summary most useful.

Following the publication of the *IPCC Fourth Assessment Report*, and in light of feedback from users, the guidance material has recently been updated and extended using downscaled scenarios from the *IPCC Fourth Assessment Report* (MfE 2008a; Mullan et al. 2008; Ramsay and Bell 2008). Additional guidance is provided on integrating climate change into Long-term Council Community Plans that are mandated under the Local Government Act 2002 (MfE 2008b).

The optimal extent and prescriptive power of regulations regarding methods and scenarios for climate risk analysis continues to be a subject of debate in New Zealand. Some councils argue that guidance on scenarios and methods for climate risk assessment should be nationally binding and include specific figures, especially for sea level rise, because it would avoid delays, costs and uncertainties resulting from challenges against the particular choices that councils otherwise have to make as they draw from the wider scientific literature (LGNZ 2006). A current review of the New Zealand Coastal Policy Statement, which is required at 10-year intervals, provides a further potential platform for the provision of more specific and binding guidance (Department of Conservation 2008; Environment Waikato 2008a).

The benefits and potential details of a more prescriptive approach will need further evaluation, since they are tied to human and financial resource constraints, particularly in small councils (LGNZ 2007, 2008). In addition, the current inability of climate science to provide an upper bound for sea level rise projections (IPCC 2007) suggests that a managed adaptive approach that does not rely on a single upper bound for sea level rise may be more appropriate in some circumstances (e.g., as used for the United Kingdom Thames estuary; see Defra 2006; Ramsbottom and Reeder 2008). However, effective implementation of such an approach would require central government support to ensure that short-term special interests do not override long-term community sustainability at the local level in New Zealand.

The current regulatory framework and planning guidance also leaves open the question of how managed retreat, one key adaptation to long-term sea level rise, would be implemented in practice. Managed retreat in developed areas challenges councils who have to reconcile community cohesion, compensation for affected individuals, and personal responsibilities with their own liabilities. An appropriate regulatory framework that addresses this key adaptation option remains to be formulated (Environment Bay of Plenty 2008a; Environment Waikato 2008b; Waitaki District Council 2008).

Conclusions: Lessons, Remaining Barriers, and Links with Sustainable Development

Local government agencies play a key role in promoting and facilitating community-level adaptation to climate change. Their actions can be supported by raising community awareness and improving science communication, integrating climate change projections and their uncertainties into standard local government processes, and providing an enabling environment through central government regulation and guidance. The main steps taken in New Zealand have been:

- Encouraging a sequential approach to assessing risks and identifying vulnerabilities. This can reduce barriers related to the cost and complexity of risk assessments and promotes consideration of climate change scenarios within their appropriate regulatory and societal context.
- Providing a clear regulatory framework that defines the overall responsibilities of councils in addressing climate change. Additional guidance from central government on the use of scenarios, methods, and models aims to increase consistency and to assist in developing a shared body of case studies.
- Improving the communication of climate change science to enhance bottom-up interactions that draw on and develop local expertise and potential solutions, in addition to more traditional top-down provision of reports and model results. Raising community awareness at all levels appears critical to support local government actions on climate change.

While there appears to be an increasingly broad engagement by local government in climate change adaptation, several barriers to more effective adaptation actions remain. These include: limited availability of baseline data (in particular, short-time records for rainfall and flood risk, and limited understanding of coastal dynamics); the large range of local-scale projections from different climate models; and limited availability of probabilistic climate change projections to support quantitative risk assessments. In addition, the uneven levels of awareness of climate change make it difficult for some communities and councils to reach informed decisions about adaptation measures and priorities. This suggests the need for more detailed and prescriptive central government guidance that would help balance long-term community perspectives against powerful special interests. However, the value and specific form of such additional guidance continues to be debated. Overcoming such barriers may require additional support from both science and central government agencies.

Some of these remaining barriers are not unique to climate change. They are likely to surface wherever local governments need to manage long-term community wellbeing against pressures from short-term special and often highly localized interests, and where a range of different scientific disciplines and scenarios form the basis of decisions. Improving the general ability of councils to deal with such challenging situations, including the use of biophysical and socioeconomic scenarios for long-term risk assessment and adaptive management, would not only build adaptive capacity, but also foster achievement of broader sustainable development goals.

Acknowledgments   The authors are grateful for helpful comments and inputs from local council staff, consultants and central government policy advisers during the development of this chapter, in particular Judy Lawrence, Stephen Lamb, Blair Dickie, Frances Sullivan, Stephen Swabey, Jacqui Yeates, Helen Plume, Julie King, and Warren Gray. The comments from two anonymous reviewers and additional suggestions by the editors of this book have also helped improve and strengthen this chapter.

References

Adger WN (2006) Vulnerability. Glob Environ Change 16(3):268–281

Adger N, Agrawala S, Mirza MMQ et al (2007) Assessment of adaptation practices, options, constraints and capacity. In: Parry M, Canziani O, Palutikof J et al (eds) Climate change 2007: impacts, adaptation and vulnerability – contribution of Working Group II to the Fourth Assessment Report of the Intergovernmental Panel on Climate Change. Cambridge University Press, Cambridge

Allan S (1993) Information for the guidance of local authorities in addressing climate change. New Zealand Climate Change Programme, Ministry for the Environment, Wellington

Brenmuhl F (2008) Opinion: emissions trading and agriculture. Straight Furrow (4 Sep 2008). http://straightfurrow.farmonline.co.nz/news/nationalrural/agribusiness-and-general/general/opinion-emissions-trading-and-agriculture/1262947.aspx. Cited 16 Jan 2009

Buckle R, Kim K, Kirkham H et al (2002) A structural VAR model of the New Zealand business cycle. Working Paper 02/26, Treasury, Wellington

Burton I, Huq S, Lim B et al (2002) From impacts assessment to adaptation priorities: the shaping of adaptation policy. Clim Policy 2(2–3):145–159

Cottrell B, Insley C, Meade R et al (2004) Report of the climate change Maori issues group. New Zealand Climate Change Office, Ministry for the Environment. http://www.climatechange.govt.nz/resources/. Cited 16 Jan 2009

Department of Conservation (2008) Summary of submissions: New Zealand coastal policy statement. New Zealand Government. http://www.doc.govt.nz/getting-involved/consultations/current/new-zealand-coastal-policy-statement-2008/summary-of-submissions/. Cited 16 Jan 2009

Department of Environment, Food and Rural Affairs [Defra] (2006) Flood and coastal defence appraisal guidance: FCDPAG3 economic appraisal – supplementary note to operating authorities – climate change impacts. Her Majesty's Government, London

Environment Bay of Plenty (2008a) Submission on the proposed coastal policy statement. Department of Conservation, New Zealand Government. http://www.doc.govt.nz/getting-involved/consultations/current/new-zealand-coastal-policy-statement-2008/submissions/. Cited 16 Jan 2009

Environment Bay of Plenty (2008b) Regional policy statement review and preparation of the next regional policy statement. Environment Bay of Plenty, New Zealand Government, Whakatane

Environment Waikato (2008a) Submission on the proposed coastal policy statement. Department of Conservation, New Zealand Government. http://www.doc.govt.nz/getting-involved/consultations/current/new-zealand-coastal-policy-statement-2008/submissions/. Cited 16 Jan 2009

Environment Waikato (2008b) Submission to the board of inquiry on the proposed coastal policy statement. Department of Conservation, New Zealand Government. http://www.doc.govt.nz/getting-involved/consultations/current/new-zealand-coastal-policy-statement-2008/submissions/. Cited 16 Jan 2009

Fulton T (2008) Doubts on human-induced climate change. The New Zealand Farmers' Weekly (5 May 2008). http://www.farmersweekly.co.nz/article/7384.html. Cited 16 Jan 2009

Gluckman JDN, Boyle CA (2003) Impacts of climate change in New Zealand and the required response: an engineering focus on coastal margins, human settlements, infrastructure and water resources. IPENZ engineering treNz. http://www.ipenz.co.nz/IPENZ/Forms/pdfs/TreNz3.pdf. Cited 16 Jan 2009

Hennessy K, Fitzharris B, Bates B et al (2007) Australia and New Zealand. In: Parry M, Canziani O, Palutikof J et al (eds) Climate change 2007: impacts, adaptation and vulnerability – contribution of Working Group II to the Fourth Assessment Report of the Intergovernmental Panel on Climate Change. Cambridge University Press, Cambridge

IPCC (2007) Climate change 2007: synthesis report – contribution of Working Group I, II and III to the Fourth Assessment Report of the Intergovernmental Pannel on Climate Change [IPCC]. IPCC, Geneva

Kelly PM, Adger WN (2000) Theory and practice in assessing vulnerability to climate change and facilitating adaptation. Clim Change 47(4):325–352

Kenny G (2005) Adapting to climate change in eastern New Zealand: a resource kit for farmers in eastern New Zealand. Earthwise Consulting Ltd, Hastings

Kenny G, Fisher M (2004) The view from the ground: a farmer perspective on climate change and adaptation. Earthwise Consulting Ltd/Hawke's Bay Climate Change Adaptation Group, Hastings

Klein RJT, Huq S, Denton F et al (2007) Inter-relationships between adaptation and mitigation. In: Parry M, Canziani O, Palutikof J et al (eds) Climate change 2007: impacts, adaptation and vulnerability – contribution of Working Group II to the Fourth Assessment Report of the Intergovernmental Panel on Climate Change. Cambridge University Press, Cambridge

LGNZ (2007) National policy statement on flood and stormwater risk management: a position statement from Local Government. www.lgnz.co.nz/projects/EnvironmentalSustainability/FloodandCatchment/LGPositionStatmentonNPS.pdf. Cited 16 Jan 2009

LGNZ (2008) Submission on the proposed coastal policy statement. Department of Conservation, New Zealand Government. http://www.doc.govt.nz/getting-involved/consultations/current/new-zealand-coastal-policy-statement-2008/submissions/. Cited 16 Jan 2009

Local Government New Zealand [LGNZ] (2006) Report for the Ministry for the Environment on the adaptation to climate change workshops held from 23–30 May 2006.   http://www.lgnz.co.nz/projects/EnvironmentalSustainability/ClimateChange/ReporttoMFEonAdaptingtoClimateChangeWorkshops.pdf. Cited 16 Jan 2009

MfE (2001) NZers awareness of climate change poor – survey. Press release by the New Zealand Government. http://www.scoop.co.nz/stories/PA0107/S00143.htm. Cited 16 Jan 2009

MfE (2004) Preparing for climate change: a guide for local government in New Zealand – report ME534. New Zealand Climate Change Office, New Zealand Government, Wellington

MfE (2007a) Climate change – cabinet papers. New Zealand Government. http://www.mfe.govt.nz/issues/climate/resources/cabinet-papers/. Cited 16 Jan 2009

MfE (2007b) Coast care Bay of Plenty dune restoration – INFO 267. New Zealand Government, Wellington

MfE (2007c) Tauranga City Council prepares for more intense rainfall – INFO 268. New Zealand Government, Wellington

MfE (2008a) Preparing for climate change: a guide for local government in New Zealand – report ME891. New Zealand Government, Wellington

MfE (2008b) Climate change and long-term council community planning – report ME911. New Zealand Government, Wellington

MfE (2008c) Climate change publications. New Zealand Government. http://www.mfe.govt.nz/publications/climate/#case. Cited 16 Jan 2009

Minister of Conservation (1994) New Zealand coastal policy statement. New Zealand Government. http://www.doc.govt.nz/templates/MultiPageDocumentTOC.aspx?id=45205. Cited 16 Jan 2009

Ministry of the Environment [MfE] (1990) Climatic change: impacts on New Zealand. New Zealand Government, Wellington

Mullan B, Wratt D, Allan S et al (2004) Climate change effects and impacts assessment: a guidance manual for Local Government in New Zealand – report ME503. Ministry for the Environment, New Zealand Government, Wellington

Mullan B, Wratt D, Dean S et al (2008) Climate change effects and impacts assessment: a guidance manual for Local Government in New Zealand (2nd edn) – report ME870. Ministry for the Environment, New Zealand Government, Wellington

National Institute of Water and Atmospheric Research [NIWA] (2006) Proceedings of the second Maori climate forum Hongoeka Marae, Plimmerton, 24 May 2006. http://www.niwascience.co.nz/ncc/maori/2006-05/. Cited 16 Jan 2009

New Zealand House of Representatives [NZ] (2001) Inquiry into the role of local government in meeting New Zealand's climate change target: report of the Local Government and Environment committee. New Zealand Government. http://web.archive.org/web/20040705045703/www.clerk.parliament.govt.nz/content/537/i9ctext.pdf. Cited 16 Jan 2009

NZ (2008a) Local Government Act 2002 (no.84). New Zealand Government. http://www.legislation.govt.nz/act/public/2002/0084/latest/DLM170873.html. Cited 16 Jan 2009

NZ (2008b) Resource Management Act 1991 (no.69). New Zealand Government. http://www.legislation.govt.nz/act/public/1991/0069/latest/DLM230265.html. Cited 16 Jan 2009

O'Brien K, Eriksen S, Nygaard LP et al (2007) Why different interpretations of vulnerability matter in climate change discourses. Clim Policy 7(1):73–88

Packman D, Ponter D, Tutua-Nathan T (2001) Maori issues – climate change working paper. New Zealand Climate Change Office, Ministry for the Environment, New Zealand Government. http://www.climatechange.govt.nz/resources/. Cited 16 Jan 2009

Ramsay D, Bell R (2004) Coastal hazards and climate change: a guidance manual for Local Government in New Zealand – report ME502. Ministry for the Environment, New Zealand Government, Wellington

Ramsay D, Bell R (eds) (2008) Coastal hazards and climate change: a guidance manual for Local Government in New Zealand (2nd edn) – report ME892. Ministry for the Environment, New Zealand Government, Wellington

Ramsbottom D, Reeder T (2008) Adapting flood risk management for an uncertain future: flood management planning on the Thames Estuary – paper 04a-3. Paper presented at the flood and coastal management conference, Manchester University, Manchester, 1–3 July 2008

Reisinger A (2002) Adapting to the impacts of climate change: what role for local government? - report for the New Zealand climate change programme. LGNZ. http://www.lgnz.co.nz/library/files/store_001/impacts-of-climate-change.pdf. Cited 16 Jan 2009

Royal Society of New Zealand [RSNZ] (1988) Climate change in New Zealand. RSNZ, Wellington

RSNZ (1990) New Zealand climate report 1990: miscellaneous series 18. RSNZ, Wellington

RSNZ (1992) 1992 supplement to the New Zealand climate report of 1990: bulletin 32. RSNZ, Wellington

Schneider SH, Semenov S, Patwardhan A et al (2007) Assessing key vulnerabilities and the risk from climate change. In: Parry M, Canziani O, Palutikof J et al (eds) Climate change 2007: impacts, adaptation and vulnerability – contribution of Working Group II to the Fourth Assessment Report of the Intergovernmental Panel on Climate Change. Cambridge University Press, Cambridge

ShapeNZ (2007a) New Zealanders' views on climate change and related policy options. Report for the NZ Business Council for Sustainable Development. http://www.nzbcsd.org.nz/_attachments/ShapeNZ_Climate_Change_survey_interim_result_April_4-7_2007.pdf. Cited 16 Jan 2009

ShapeNZ (2007b) Climate change II survey June 2007. Report for the NZ Business Council for Sustainable Development. http://www.nzbcsd.org.nz/_attachments/Climate_Change_II_survey_June_2007.pdf. Cited 16 Jan 2009

ShapeNZ (2008) Climate change and emissions trading – views of business decision makers. Report for the NZ Business Council for Sustainable Development. http://www.nzbcsd.org.nz/_attachments/Shape_ETS_support_REPORT_business_decision_makers_filter_2204_11_Sept_08.doc. Cited 16 Jan 2009

Smit B, Wandel J (2006) Adaptation, adaptive capacity and vulnerability. Glob Environ Change 16(3):282–292

Stroombergen A, Tait A, Renwick J et al (2006) The relationship between New Zealand's climate, energy, and the economy in 2025. Kotuitui NZ J Soc Sci Online 1:139–160

Tait A, Renwick J, Stroombergen A (2005) The economic implications of climate-induced variations on milk production. NZ J Agr Res 48(2):213–225

Tribbia J, Moser SC (2008) More than information: what coastal managers need to plan for climate change. Environ Sci Policy 11(4):315–328

UK Climate Impacts Programme [UKCIP] (2003) Climate change and local communities – how prepared are you? UKCIP, Oxford

UKCIP (2007) Local climate impacts profile [LCLIP]: a local climate impacts profile. UKCIP, Oxford

van Aalst MK, Cannon T, Burton I (2008) Community level adaptation to climate change: the potential role of participatory community risk assessment. Glob Environ Change 18(1): 165–179

Waitaki District Council (2008) Submission on the proposed coastal policy statement. Department of Conservation, New Zealand Government. http://www.doc.govt.nz/getting-involved/consultations/current/new-zealand-coastal-policy-statement-2008/submissions/. Cited 16 Jan 2009

Willis G (2007) Evaluation of the Waikato regional policy statement. Enfocus, Auckland

Wratt D, Mullan B, Clarkson T et al (1991) Climate change: the consensus and the debate. Ministry for the Environment, New Zealand Government, Wellington

Wratt D, Mullan B, Kenny G et al (2006) New Zealand climate change: water and adaptation. In: Chapman R, Boston J, Schwass M (eds) Confronting climate change: critical issues for New Zealand. Victoria University Press, Wellington

Yohe G, Lasco R, Ahmad QK et al (2007) Perspectives on climate change and sustainability. In: Parry M, Canziani O, Palutikof J et al (eds) Climate change 2007: impacts, adaptation and vulnerability – contribution of Working Group II to the Fourth Assessment Report of the Intergovernmental Panel on Climate Change. Cambridge University Press, Cambridge

Chapter 23
Perceptions of Risk and Limits to Climate Change Adaptation: Case Studies of Two Swedish Urban Regions

Louise Simonsson, Åsa Gerger Swartling, Karin André, Oskar Wallgren, and Richard J.T. Klein

Abstract This study analyzes processes of adaptation to climate change through participatory research in Sweden's two largest cities, Stockholm and Gothenburg. Perceptions of climate risks and constraints to adaptation are discussed. Practitioners from the public and private sector have identified stakeholders who are, and who should be, giving attention to adaptation, including the risks and threats facing the regions and how and which factors hinder the implementation of adaptation. In this study, it is found that those issues where adaptation is considered most difficult are mainly related to response capacity.

Keywords Risk perception • Stakeholder analysis • Participatory methods • Adaptation constraints • Adaptive capacity • Urban regions • Sweden • Stockholm • Gothenburg • Climate risk • Climate change vulnerability • Adaptation

Introduction

As centers of population and economic activity, cities are key in both mitigation and adaptation to climate change. More than half of the world's population now live in cities, and urban lifestyles tend to lead to greater greenhouse gas emissions than rural ones. Collectively, cities are responsible for 80% of total global greenhouse gas emissions. At the same time, urban dwellers and urban infrastructure are often

L. Simonsson (✉) • K. André
Centre for Climate Science and Policy Research, Linköping University, Norrköping, Sweden
e-mail: louise.simonsson@tema.liu.se; karin.andr@tema.liu.se

Å.G. Swartling • O. Wallgren • R.J.T. Klein
Stockholm Environment Institute, Stockholm, Sweden
e-mail: asa.swartling@sei.se; oskar.wallgren@sei.se; richard.klein@sei.se

J.D. Ford and L. Berrang-Ford (eds.), *Climate Change Adaptation in Developed Nations: From Theory to Practice*, Advances in Global Change Research 42, DOI 10.1007/978-94-007-0567-8_23, © Springer Science+Business Media B.V. 2011

highly exposed to climate risks, as was markedly illustrated by the impact of the 2003 heat wave on Paris, and that of Hurricane Katrina on New Orleans in 2005. Yet most of the literature on adaptation to climate change pays only scant attention to the vulnerability of cities and their inhabitants. This oversight is due largely to the fact that adaptation research and policy is typically organized by economic sector: agriculture, water resources, transport, health, tourism, and so on. In cities, all these and other sectors interact in a relatively small area, made up of heterogeneous units, to create a system that cannot be understood and managed by analyzing and addressing sector processes alone.

While most literature has identified resource constraints as being the most significant determinant of adaptation, empirical research on adaptation has so far largely neglected psychological factors, such as risk perception and perceived adaptive capacity, in determining adaptation (Grothmann and Patt 2005). Risk perceptions based on knowledge and rationality, as well as on experience of climate-related crises, are important to initiating and developing policies, encouraging decision-making, building institutional capacity, and implementing practical adaptation measures. In Sweden, the 2005 storm Gudrun had severe consequences for society and individuals, with the forestry and insurance sectors particularly affected. The aftermath of Gudrun led to the initiation of the Swedish Commission on Climate and Vulnerability (the "Commission"). Together with other national and international events and publications on climate change and climate risk (e.g., the *Fourth Assessment Report* by the Intergovernmental Panel on Climate Change (IPCC), the *Stern Review*), the Commission introduced and emphasized the need for adaptation to occur alongside mitigation, the latter of which hitherto had been the main focus for Sweden's climate debate. The Commission consulted with a wide range of different actors within Swedish society to produce the comprehensive report *Sweden facing climate change – threats and opportunities* (SOU 2007:60). Taken together with other newly launched political documents, such as the proposals of the Swedish Climate Commission (SOU 2008:24) and the Government's Climate Bill (Prop. 2008/09:162), it is fair to say that all levels of Swedish society have become included in the management of climate change. The main conclusion from the Commission was that "It is necessary to make a start on adapting to climate changes in Sweden. The principal features of the climate scenarios, despite uncertainties, are sufficiently robust to be used as a basis" (SOU 2007:60, p. 11). The Commission placed particular emphasis on the increased risks of flooding and landslides with severe implications for the built environment and infrastructure. The Commission also suggested that municipalities and County Administrative Boards take on additional responsibilities. Accordingly the County Administrative Boards, the supervisory authority at the regional level, are now responsible for coordination and reporting on the progress of climate adaptation within each county, both at the regional and local level (Prop. 2008/09:162).

This chapter reports on the first research activities that take a systemic approach to assessing adaptation needs and processes in the two largest cities in Sweden, Stockholm and Gothenburg. The aim is to advance the understanding of how climate change is perceived by stakeholders, and how these perceptions shape practical limits to, and opportunities for, adaptation in urban settings. The chapter concludes with a discussion of the specific factors that constitute adaptive capacity as identified by the stakeholders in the regions.

Climate Change Adaptation in Swedish Policy and Legislation

Sweden's total population is approximately 9.4 million and the Stockholm region has a population of about two million. The city is located at the main outlet of Lake Mälaren where the lake connects to the Baltic Sea and the islands in the archipelago. The municipality of Gothenburg, on the western coast, has a population of about 500,000 inhabitants. River Göta Älv runs through the city, connecting the largest lake of Sweden, Lake Vänern, and the North Sea. The soils along the river are especially prone to landslides due to the composition of clays, and several very severe incidents have occurred over the last decades (Hågeryd et al. 2007).

The legislation relevant to adaptation among actors in Swedish urban regions primarily concerns spatial planning and the municipal responsibilities for analyzing risks and vulnerabilities and preparing for accidents and extreme events. A number of national authorities have supporting functions for adaptation. The orientation of the work by national agencies in relation to adaptation is determined by the instructions issued by the government. Generally speaking, few (if any) of the agencies had at the time of this study in 2008 had their instructions changed as a result of the growing understanding of the implications of climate change. In 2007, the Commission on Climate and Vulnerability (SOU 2007:60) proposed changes in the instructions for a large number of national sector agencies in order to spell out their specific responsibilities in relation to a changing climate. Agencies include the Swedish Environmental Protection Agency, the Swedish Meteorological and Hydrological Institute (SMHI), and The National Board of Housing, Building and Planning. Most importantly, the Commission proposed that the County Administrative Boards be given specific responsibilities for: performing regional climate change impact assessments; initiating, supporting and evaluating adaptation work among municipalities; providing support for business and government agencies active in the county; and collecting, synthesizing and disseminating information. There is no single authority where climate adaptation naturally fits and the various governmental authorities have different functions and instructions depending on their working areas and specific responsibilities (SOU 2007:60, p. 626).

The previous Regional Development Plan for Stockholm dates back to 2001, when the implications of climate change were still not well known or considered enough to merit inclusion in the plan (RTK 2002). The plan should serve as a guiding framework for the structural planning taking place in the municipalities. A new plan was passed by the County Council in May 2010. The new plan explicitly states the need to adapt to a changing climate. Particular mention is made of physical planning; water provision and wastewater treatment systems; transportation infrastructure; and electricity distribution and telecommunications networks. It also refers to ecosystem changes and deterioration of water quality as factors that may affect the region in the future (SLL 2010).

The Gothenburg region did not have a regional development plan as such. In an ongoing consultative process, The Gothenburg Region Association of Local Authorities has developed a crude map in which overarching principles for spatial development – with respect to, for example, settlements, transportation infrastructure and green space – are stated. In the final version of the document, climate change is identified as a key factor, particularly in relation to the risks associated with River Göta Älv. No tangible recommendations or guidelines are presented, which reflects the nature of the document as a general policy for key areas of cooperation, expressing the lowest common denominators among the 13 signatories.

Beyond the requirement that local and regional authorities have plans in place for dealing with accidents and extreme events, there is only weak legislative support for these actors to carry out adaptation measures. The Commission on Climate and Vulnerability proposed a number of legislative changes, and a few of these changes have already been made, but the long-term effects of the Commission's work and the recent changes in the national authorities' instructions in terms of legislation and action remain to be seen. However, while explicit mention of climate change is missing in the supporting legislation, there is nothing to stop local and regional actors from exploring and implementing adaptation measures. In the absence of both national legislation and guiding regional policies, such action has until recently primarily been based on local initiatives and the locally perceived need for adapting to a changing climate.

Stakeholder Involvement in Adaptation Processes

Engaging stakeholders has become increasingly common in adaptation research (e.g., Few et al. 2007), policy processes (e.g., the Arctic Climate Impact Assessment, ACIA; the Swedish Commission on Climate and Vulnerability), and in knowledge exchange (e.g., United Kingdom Climate Impacts Programme). If we wish to better understand the factors that determine climate responses, top-down knowledge of the climate system needs to be integrated with bottom-up knowledge of social vulnerability (see Ch. 19 in Parry et al. 2007). Local actors' perceptions of climate risk determine the course of action chosen in response to risk (Willows and Connell 2003). However, available adaptation options are perceived differently by different stakeholders (Arnell and Charlton 2009).

Adaptation will in many cases require cooperation and networks among stakeholders at different levels of society (Conde and Lonsdale 2005), and these stakeholders might have both interrelated and contrasting views on climate risks and the need for adaptation. Moreover, climate change occurs together with other changes in society (Adger, et al. 2005). Hence, if the interactions between climate change and societal change are to be understood, it is important to gain the perspectives of those people who are experiencing such change (ACIA et al. 2005; Keskitalo 2008). Vulnerability to climate change is distributed and perceived differently between different groups in society, and also by those who are inside and outside the system (ACIA et al. 2005). The definition of who is a stakeholder thus depends upon the purpose of the activity (see for example European Commission 2003; Willows and Conell 2003; Conde and Lonsdale 2005). The question of which stakeholders actually have a stake varies from issue to issue, and also influences practical adaptation processes. Therefore this question must be carefully analyzed for each case.

The Commission on Climate and Vulnerability (SOU 2007:60) has identified a number of regional and national organizations and authorities that have a direct role in facilitating national adaptation processes, as well as those actors that indirectly will have to integrate climate considerations in daily decisions. It is clear that the division of responsibilities is distributed between individual citizens, business, municipalities, and the state (p. 617). Municipalities have a role at the local level owing to their responsibilities for physical planning, which according to the Commission should be adapted to climate change (p. 16). The role of the state is to provide information and prepare basic data for decision-making. At the regional level, the County Administrative Boards have a central role through their coordinating and reporting function.

In an urban setting, there are several stakeholders involved that may, or may not, have clear areas of responsibilities, organized forms of cooperation, and formalized roles and networks. Urban regions are characterized by heterogeneity within and between municipalities (e.g. demographically, socio-economically, politically, environmentally, etc.). This fact, combined with the different objectives and methods of the public and private sectors, can create complicated and uneven power relationships that could hinder adaptation where cooperation is crucial. Furthermore, because adaptation to climate change is a rather new phenomenon, in several cases the relationships between key actors in an urban region are not obvious.

This study mainly used participatory methods where focus groups were central. The focus group participants were selected to represent local and regional stakeholders (public and private) responsible for the implementation of adaptation measures. The groups consisted of practitioners and experts within municipal and regional administration and private sectors that work with planning, technical, and insurance issues. In total, about 40 persons were met in their respective groups at three separate occasions. The "snowball sampling method" (see Kossoudji 2001) was used to map a net of organizations, key actors, and individuals from which the selections of participants were made (see André and Simonsson 2008 for details). The stakeholders' notions of who is, or should be, involved and active in regional adaptation to climate change were then analyzed to make the schema more complete and to allow for comparisons between perspectives as shown in Fig. 23.1.

Figure 23.1 shows a general overview of important stakeholders and group of actors as identified by both researchers and focus group members in the Stockholm

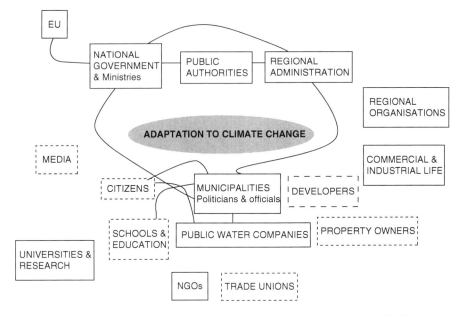

Fig. 23.1 General overview of actors important to climate change adaptation in Stockholm and Gothenburg regions as identified by researchers and stakeholders through participatory processes. See main text for detailed explanations

and Gothenburg regions. The closer the stakeholder (represented by a box) is to the centre of the image, the more important the actor is considered to be for work on adaptation. One result from the research process is that the participatory methods allowed for new knowledge to be generated as the participants contributed to a more complete understanding of the stakeholder landscape. To elucidate this in Fig. 23.1, actors in bold boxes indicate those stakeholders that were identified by both researchers and the focus groups, whereas boxes with broken lines indicate actors that were identified solely by the focus groups. The lines between some of the actors represent connections of formalized responsibility between these actors. Even though the illustration is simplified, it shows that the relationship between importance and responsibility do not follow formal hierarchical structures. Those factors that are not connected with lines might indeed influence each other greatly, but without formal responsibilities and obligations. For example, actors with no formal responsibility for societal climate adaptation – such as media, individual citizens, developers and property owners – are important actors owing to their different roles as knowledge and information disseminators, their ability to put pressure on decision-makers, and their access to resources, etc. Consequently, while studying the process of adaptation from decision-making through to implementation, it is important to consider the whole chain of actors perceived and identified by both stakeholders and researchers.

Scientific Knowledge and Stakeholder Perceptions on Climate Risk

The main physical risks and systems vulnerable to climate change in the Stockholm and Gothenburg regions, as identified in the Commission report (SOU 2007:60), result from changes in precipitation and temperature. It is anticipated there will be increased risks of flooding; landslides; sea level rise; pollution; and impacts on human health, ecological systems and biodiversity. These risks have implications for spatial planning infrastructure, transportation, nature conservation, forestry, and agriculture.

The risk and vulnerability reports from 2008 by the County Administrative Boards for the regions of Stockholm and Gothenburg (Länsstyrelsen i Stockholms län 2008; Länsstyrelsen i Västra Götalands län 2008) identify the following activities, elements and processes as posing threats and risks to the urban regions: petrochemical industries; companies at risk; transport of hazardous waste (on land, lakes, rivers, and sea); tunnels for technical infrastructure; shipping; airports; terrorist attacks/sabotage; organized crime; disease and viruses; watercourses; flooding; and landslides.

We argue that scientific and general assessments of risk need to be complemented with stakeholders' perceptions of risks at a local scale. Since their perceptions are based on both professional experience and knowledge of the region within which they are acting, they will have a deeper understanding of the factors that generate vulnerability, and it is these factors that must form the basis of adaptation discussions. Consequently, we applied participatory research methods among stakeholders to assess and identify: (1) perceptions of risks and threats related to their respective work or professional roles, and to the regions at present and in the future (around 2025); and (2) views on adaptation options and constraints. The results from the stakeholder group dialogues in the Stockholm and Gothenburg regions are shown in Table 23.1.

There were striking similarities between the general perception of climate risk and adaptation needs among stakeholders in the two regions. It was widely recognized that urban regions: involve many actors; constitute sensitive human and natural systems; have high levels of economic activity; and undergo a high rate of change and development in combination with inflexible infrastructure and buildings. This complex web of activities and multiple systems and actors in a relatively limited area creates a high demand for coordination between the various entities and sectors that together constitute the urban landscape. Political will, prioritization, and institutional set-up are thus critical for adaptation to succeed and be both sustainable and effective.

Another perceived problem is the differences and imbalances in power relationships and resources across and within the Stockholm and Gothenburg regions. Some examples include: the Stockholm region holds both Sweden's largest and smallest municipality; the proximity to the national government affects the political landscape within the Stockholm region; and large differences in socioeconomic

Table 23.1 Risk factors as identified by stakeholders in Stockholm and Gothenburg classified according to a vulnerability framework after Füssel (2005)

Socioeconomic	The built and natural environment
Internal "Response capacity"	*Sensitivity*
Lack of resources for maintenance and new investments	Coastline
Low political priority	Clay soils along River Göta Älv prone to slides (Gothenburg only)
Buildings on risky areas (e.g., near waterfronts where floodings may occur and/or increase)	Hard ground surfaces (man-made) that hinder infiltration and increase runoff
Low ability to take action due to the global nature of climate change	
Difficulties to act now for perceived distant future negative impacts of climate change	
Low interest for allocation of funds and resources due to long circulation time	
Decreased interest in, and prioritization of, climate issues	
Continuity and long-term work	
Continuity in crisis management and preparedness	
Unclear roles and responsibilities	
Lack of formalized and effective coordination	
Population changes	
Social tensions between citizen groups	
"External social stressors"	*Exposure*
Terrorism and sabotage	Floods
Social tensions between citizen groups	Severe storms (Gothenburg only)
Criminality	Land slides
Decreased trust in the society and public institutions	Increased sea level
Migration	Salt water intrusion
Unemployment	Pollution
Economy	
Population growth	

resources exist within both regions. In addition, stakeholders identified clearly defined areas of responsibility and tools and legal instruments as prerequisites for effective adaptation action.

Perceived Barriers and Constraints to Adaptation

The *Fourth Assessment Report* from IPCC's Working Group II (Parry et al. 2007) points out that there are significant barriers to implementing adaptation that include both the inability of natural systems to adapt to the rate and magnitude of climate change, as well as environmental, economic, informational, social, attitudinal, and behavioral constraints. Also, perceptions of barriers to adaptation do, in fact, limit adaptive actions, even when capacities and resources to adapt are available (ibid, Ch. 17). These perceptions vary among individuals and groups within populations. Thus, decision-makers need to be aware of these barriers and provide structural support to overcome them.

In Stockholm, adaptation to the risk of flooding is the primary focus in policy documents and risk-reducing activities. Based on the results from a survey with responsible urban planners in all of the Stockholm region's municipalities, it appears that adaptation to the increased risk of mainly flooding (but also increased rainwater discharge and landslides due to climate change) is present in daily discussions within many municipalities, and will be addressed in the new comprehensive plans under way in a handful of them during the coming 2 years (RTK 2009). However, it can be argued that adaptation to flood risk is to a large extent a reactive adaptation to existing climate variability. In our study of Stockholm, it was apparent that although the stakeholders recognized and emphasized the risk of floods they also strongly argued that it is possible to adapt to and reduce that risk.

Other issues that are related to adaptive capacity are more complicated. The risks identified by the focus group participants as being related to response capacity (see Table 23.1) were summarized into ten categories. The stakeholders were then asked to discuss and classify the obstacles according to the potential to deal with, and overcome the problems. The four categories presented in Fig. 23.2 indicate how stakeholders perceived their own ability, and the general capability, to affect the specific issues. The categories represent the following: *Barrier* here means that the problem is impossible to overcome by the stakeholders' own capacity; *Hinder* refers to that the problem is great but could possibly be overcome with substantial efforts (this implies that the stakeholders have some possibility to affect the issue themselves); *Of great significance* means that there is clearly a problem but that it should not hinder adaptation since it is possible to manage and it has previously been acknowledged in discussions; *Significant* is a joint category for those issues where adaptation is clearly possible or practical work is ongoing, or under development to remove the obstacle (it could however remain a risk if it emerges alongside other severe incidents).

Those five factors that are considered as barriers to successful adaptation are both external (e.g., issues of economy and population that are beyond control of the organizations) and internal in character. *Changes in population* is something most stakeholders identify as an important external factor that is hard to plan for, predict and affect, as it involves both international, national and regional migration as well as population growth (or decline). The *economy* is however something that the stakeholders feel they can have some effect on at the regional level, but the global and national economy is naturally beyond their influence.

The *uncertainty* of climate change is a widely recognized barrier to effective adaptation responses. The uncertainty regarding the effectiveness of adaptation efforts is likely to result in inaction in decision-making, as other policy areas that are easier to predict may be more appealing to target (Adger et al. 2005). In our study, stakeholders expressed that many formal institutions cannot deal with

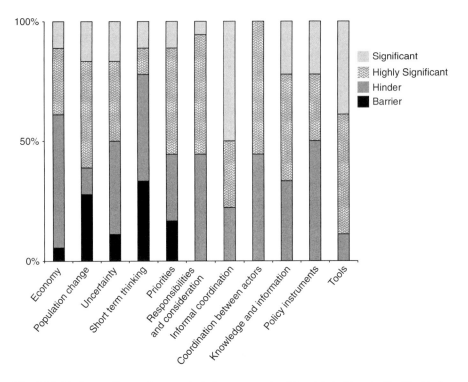

Fig. 23.2 Limits to climate change adaptation as perceived by stakeholders in the Stockholm region. The graph shows how the stakeholders categorized the respective obstacles to adaptation, based on their assessment of how difficult the constraints are to overcome, and their relative importance. See the main text for detailed explanations of the categories

uncertainty because it cannot be budgeted for if facts are absent or too vague to support and legitimize costs. The knowledge of mitigation and the relationship between emissions of greenhouse gases and global climate change are greater than that of regional and local impacts, vulnerability and adaptation. This could partly explain the greater emphasis on mitigation rather than adaptation efforts. However, as mentioned by the IPCC (Ch. 17 in Parry et al. 2007), evidence shows that there are reservations in assuming that providing individuals with scientifically sound information will result in information assimilation, increased knowledge, action and support for policies based on this information (Eden 1998; Sturgis and Allum 2004; Lorenzoni et al. 2005).

Priorities between competing concerns will always occur, and between long-term and short-term problems that need to be solved. Concerns about other risks (such as the current economic crisis) may overshadow considerations about the impacts of climate change and adaptation. It is often pointed out that developing nations need to address their current severe problems and thus cannot afford to prioritize future anticipated difficulties. But climate change is not a pressing personal priority for most people in developed nations either, despite apparent widespread public concern about climate change (Lorenzoni and Pidgeon 2006). Urban regions in developed countries also face the problem of contrasting objectives. Demands for further economic development compete with the aspiration for legal commitments to sustainable development and environmental protection to ensure today's and coming generations' well-being. This conflict is also apparent in our study, where we found that stakeholders in the regions perceive that adaptation efforts are impeded because marketing and economic growth receive a higher priority than environmental and safety concerns.

Short-term thinking is viewed as the greatest obstacle to effective and sustainable adaptation. It includes and is related to: the economy; the political system where election periods set the time frame and political agenda; and investment and planning periods for private companies. In the discussions it was also stressed that limited resources often mean that other issues are prioritized over the environment.

It is interesting to note that the lack of *coordination between actors* is considered to be highly significant and even a hindrance to adaptation. In political-administrative terms, the regional level in Sweden is relatively weak compared with other parts of Europe. The Swedish political system could be described as an hourglass where the regional level is the waist, with the least power (Petersson 1998). The limitations in regional decision-making have been pointed out repeatedly (e.g., Bryniellson 2003; SOU 2000:85) and are of particular concern with regard to the environmental problems that the regional level must help to solve. In Stockholm these needs are particularly significant. The division of responsibility for planning and implementation among actors at different levels means that it is difficult to achieve common and joint action, and the region continues to lack efficient institutions for coordination and common decisions (RTK 2002). The stakeholder mapping shown in Fig. 23.1 further underlines the importance and urgency of coordination on crosscutting issues like climate change and adaptation.

Conclusions

Sweden is a highly developed country, with a high degree of social capital and adaptive capacity (education, organizational structure, developed infrastructure, technologically advanced system, enforced laws, etc.) that can serve as foundations for successful adaptation responses. However, this study indicates that, of all the factors that appear to be obstacles to effective adaptation, the most difficult to solve are issues of: organization (e.g., too little coordination within organizations and between actors); conflicting interests; and lack of will and opportunity to prioritize adaptation measures. In contrast, it seems it is possible to solve problems around technology, information and finance if there is a will to do so. Thus, in the context of a developed nation, there are a range of difficulties that need to be highlighted if adaptation is to succeed and vulnerability to multiple stresses reduced. These difficulties are especially prominent in areas, such as cities, where cross-sector approaches are crucial (cf. O'Brien et al. 2006).

The fact that legislation and division of responsibilities was considered weak and unclear, or not implemented, was perceived as another obstacle to adaptation. At the time of the study in 2008, actors had to a large extent rely on individuals' personal commitments and perceptions of the severity of risks and the need for adaptation. The stakeholder mapping and the identification of risks, barriers, and obstacles performed in this study underline the complexity of the climate change adaptation arena. In the face of this complexity, it seems that better coordination on adaptation issues within and between municipalities, between local, regional, national and international levels and concerned private companies is essential.

Stakeholder and public support to adaptation strategies is vital, since adaptation actions need be taken at all scales of society. This study shows that there are several perceived risks and problems that shape the institutional, local and regional adaptive capacity. How, and what types of, adaptation that then comes about is a result of what is recognized as being possible to deal with and what is considered being beyond practitioners' and their respective organizations' domains to manage. There is, however, a need for greater understanding of the specifics of the parameters that shape adaptive capacity, as well as the barriers to and opportunities for adaptation. Our research aimed to pinpoint the driving forces behind vulnerability and adaptation processes to improve implementation of practical adaptation measures. This study shows that risks perceived by practitioners include multiple factors: biogeophysical (e.g., landslides, storms, floods, etc.); social (e.g., population changes, social tensions, lack of trust in society, etc.); economic; political (e.g., contrasting interests and prioritization); institutional (e.g., unclear roles and division of responsibilities, etc.); technological (e.g., maintenance and design); and environmental (e.g., pollution). Although the risk of flooding is identified as serious and urgent – and has been stressed in official documents and scientific studies – stakeholders perceive it as a risk that is possible to reduce through adaptation measures. Those risk issues where adaptation is considered most difficult are mainly related to response capacity. Our results indicate that limits to climate change adaptation in urban regions in developed countries are related to population changes, economy, uncertainty regarding the effectiveness of adaptation efforts, public and private prioritization, short-term thinking and planning, and lack of coordination. However, since stakeholders are a heterogeneous group with diverging and sometimes contrasting interests, perceptions and capabilities, we see a need for further studies of possible mismatches and gaps between stakeholder groups to deepen the understanding of adaptation processes.

Acknowledgments The authors would like to express their gratitude to all stakeholders who participated in the study. Their help has made a valuable contribution to a better understanding of the process of adaptation to climate change. Thanks are also due to Peter Rudberg and Tom Gill from Stockholm Environment Institute and the two anonymous reviewers for valuable contributions to the text. The research is funded by The Swedish Research Council for

Environment, Agricultural Sciences and Spatial Planning (2006-4871-7662-55), and the Foundation for Strategic Environmental Research (Mistra) through the Mistra-SWECIA programme (www.mistra-swecia.se).

References

Adger WN, Arnell N, Thompkins E (2005) Successful adaptation to climate change across scales. Glob Environ Change 15(2):77–86

André K, Simonsson L (2008) Who has influence on adaptation to climate change and climate risk reduction? Identification and analysis of the stakeholder landscape for the Stockholm region – report for the Mistra-Swecia research programme. Centre for Climate Science and Policy Research, Linköping University, Norrköping

Arctic Climate Impact Assessemt [ACIA], Arctic Monitoring and Assessment Programme [AMAP], International Arctic Science Committee [IASC] (2005) Arctic climate impact assessment. Cambridge University Press, Cambridge

Arnell NW, Charlton BM (2009) Adapting to the effects of climate change on water supply reliability. In: Adger NW, Lorenzoni I, O'Brien KL (eds) Adapting to climate change: thresholds, values, governance. Cambridge University Press, Cambridge

Bryniellson H (ed) (2003) På jakt efter en ny regional samhällsordning! Svenska Kommunförbundets FoU-råd, Stockholm [in Swedish]

Climate bill (2008 Prop. /09:162). En sammanhållen klimat- och energipolitik. Klimat. Miljödepartementet, Stockholm [in Swedish]

Conde C, Lonsdale K (2005) Engaging stakeholders in the adaptation process. In: Lim B, Spanger-Siegfried E, Burton I et al (eds) Adaptation policy frameworks for climate change. Cambridge University Press, Cambridge

Eden S (1998) Environmental issues: knowledge, uncertainty and the environment. Prog Hum Geogr 22(3):425–432

European Commission (2003) Common implementation strategy for the water framework directive (2000/60/CE) – guidance document no.8. http://ec.europa.eu/environment/water/water-framework/index_en.html. Cited 19 Jan 2008

Few R, Brown K, Tompkins E (2007) Public participation and climate change adaptation: avoiding the illusion of inclusion. Clim Policy 7(1):46–59

Füssel H-M (2005) Vulnerability in climate change research: a comprehensive conceptual framework – paper 6. University of California International and Area Studies, Berkeley

Grothmann T, Patt A (2005) Adaptive capacity and human cognition: the process of individual adaptation to climate change. Glob Environ Change 15(3):199–213

Hågeryd A-C, Viberg L, Lind B (2007) Frekvens av skred i Sverige. Varia 583. Swedish Geotechnical Institute, Linköping [in Swedish]

Keskitalo C (2008) Vulnerability and adaptive capacity in forestry in northern Europe: a Swedish case study. Clim Change 87(1–2):219–234

Koussidjou SA (2001) Strategies of stakeholder analysis to improve participation and project performance: concepts, fields and techniques. In: Interests groups and organization as stakeholders social development papers – paper no.35. http:www-wds.worldbank.org. Cited 15 Oct 2007

Länsstyrelsen i Stockholms län (2008) Regional risk- och sårbarhetsanalys 2007. Länsstyrelsen i Stockholms län, Räddnings- och säkerhetsavdelningen, Stockholm [in Swedish]

Länsstyrelsen i Västra Götalands län, enheten för skydd och säkerhet (2008) Risk- och sårbarhetsanalys 2007. Rapportnr: 2008:22. Länsstyrelsen i Västra Götalands län, enheten för skydd och säkerhet, Göteborg [in Swedish]

Lorenzoni I, Pidgeon NF (2006) Public views on climate change: European and USA perspectives. Clim Change 77(1–2):73–95

Lorenzoni I, Pidgeon NF, O'Connor RE (2005) Dangerous climate change: the role for risk research. Risk Anal 25(6):1387–1397

O'Brien K, Eriksen S, Sygna L et al (2006) Questioning complacency: climate change impacts, vulnerability, and adaptation in Norway. AMBIO J Hum Environ 35(2):50–56

Parry ML, Canziani OF, Palutikof JP et al (eds) (2007) Climate change 2007: impacts adaptation and vulnerability – contribution of Working Group II to the Fourth Assessment Report of the Intergovernmental Panel on Climate Change. Cambridge University Press, Cambridge

Petersson O (1998) Statsbyggnad. Den offentliga maktens organisation. SNS förlag, Stockholm [in Swedish]

Regionplane- och trafikkontoret [RTK] (2002) Regional utvecklingsplan 2001 för Stockholmsregionen. Program och förslag 2002. Regionplane- och trafikkontoret, Stockholm [in Swedish]

RTK (2009) Klimatförändringar: dags att anpassa sig? En rapport om anpassning till effekterna av klimatförändringar i Stockholmsregionen. Rapport 4:2009. Regionplane- och trafikkontoret, Stockholms Läns Landsting, Stockholm [in Swedish]

SOU 2000:85 (2000) Regionalt folkstyre och statlig länsförvaltning. Finansdepartementet, Den parlamentariska regionkommittén, Stockholm [in Swedish]

SOU 2007:60 (2007) Sweden facing climate change – threats and opportunities. Swedish Government Official Reports, Final report from the Swedish Commission on Climate and Vulnerability, Stockholm

SOU 2008:24 (2008) Svensk klimatpolitik (*Swedish Climate Policy*). Miljödepartementet, Stockholm [in Swedish]

Stockholms läns landsting [SLL] (2010) Regional utvecklingsplan för Stockholmsregionen – RUFS 2010. Stockholm [in Swedish]

Sturgis P, Allum N (2004) Science in society: re-evaluating the deficit model of public attitudes. Public Underst Sci 13(1):55–74

Willows RI, Connell RK (eds) (2003) Climate adaptation: risk, uncertainty and decision-making. UK Climate Impacts Programme [UKCIP] Technical Report. UKCIP, Oxford

Chapter 24
Asking the Climate Question: Climate Change Adaptation in King County, Washington

Elizabeth Willmott and Jennifer Penney

Abstract King County, Washington, is a North American leader in the development and implementation of climate change adaptation strategies. This chapter describes the early experiences of King County in developing and implementing a climate change adaptation program under the leadership of King County Executive Ron Sims, during the years 2005–2009. Basic review of the King County example suggests that the county can serve as a useful model for other governments in launching and maintaining a successful climate change adaptation program. The King County experience also provides insights about types of external support a local government would need when establishing such a program.

Keywords Center for Clean Air Policy • Climate Impacts Group • ICLEI – Local Governments for Sustainability • King County • Washington • Local governments • North America • Ron Sims • Urban adaptation • Urban Leaders Adaptation Initiative • United States federal government

Introduction

In 1988, King County, Washington, council members Ron Sims and Bruce Laing jointly proposed an "office of global warming" to study the effects of climate change on the region. This proposal was rejected by the county council and mocked in the media (Seattle Times 1988). The idea lay dormant for 17 years.

E. Willmott (✉)
Willmott Sherman L.L.C., Seattle, WA, USA
e-mail: elizabethwillmott1977@gmail.com; Mlle.willmott@gmail.com

J. Penney
Clean Air Partnership, Toronto, ON, Canada
e-mail: jpenney@cleanairpartnership.org

J.D. Ford and L. Berrang-Ford (eds.), *Climate Change Adaptation in Developed Nations: From Theory to Practice*, Advances in Global Change Research 42,
DOI 10.1007/978-94-007-0567-8_24, © Springer Science+Business Media B.V. 2011

In 2005, Sims, who had by that time become King County's Executive, read an editorial by a University of Washington paleontologist that reflected on "the catastrophic consequences that have resulted from seemingly small changes to […] climate" in the Earth's history (Snover et al. 2007). The article motivated him to ask King County's environmental department to hold a public conference about the impacts of climate change on the region's primary economic sectors and the County government's main areas of responsibility. The time to prepare for climate change had come (Box 24.1).

Box 24.1 About King County, Washington

Since 2005, King County has received national recognition in the United States for its work on climate change adaptation (Schulte 2006; EPA 2007; The Heinz Center 2007; Swope 2007). Located on Puget Sound in western Washington and covering over 2,100 square miles, King County is the most populous county in Washington and home to more than 1.8 million people. There are 39 cities within King County, including Seattle, and numerous high-profile organizations, including Microsoft, Amazon, Starbucks and the Gates Foundation. King County government provides regional services – such as public transit, wastewater treatment and flood management to all residents of the county – and local services such as roads, parks and land use planning to unincorporated areas. The region's economy is based on information technology, clean technology, life sciences, international trade and aerospace manufacturing, as well as fisheries, forestry and recreational sports. In 2009, King County Executive Ron Sims was nominated to become Deputy Secretary of the United States Department of Housing and Urban Development.

Early Days of the King County Adaptation Effort

King County needed help for this work. Although the County had a number of scientists on staff, including experts in hydrology and ecology, the government required expert analysis about climate change and its impacts in order to develop substantive background materials for the conference. The Climate Impacts Group (CIG) at the University of Washington was positioned to play this critical role.

Since the mid-1990s, scientists in the CIG had been evaluating impacts of climate change and variability in the US Pacific Northwest, advising water suppliers on seasonal stream flow projections, and holding workshops to raise awareness about the local effects of climate change. In 2005, the CIG was a Regional Integrated Science and Assessment (RISA) program funded by the National Oceanic and Atmospheric Administration (NOAA), to "support research that addresses complex climate sensitive issues of concern to decision-makers and policy planners at a regional level" (NOAA Climate Program Office 2009).

The CIG scientists recognized that a major area of concern in King County was the 50-year decline in snowpack, which serves as a water reservoir for the region and has provided a reliable stream flow in the late spring and summer for fish and wildlife needs, urban watersheds, irrigation and hydropower production (Mote et al. 2005). This analysis was reinforced by a 2005 winter drought that had left the mountains relatively devoid of snow.

The CIG also participated in a regional water planning process initiated by King County in 2005, which included numerous public and private water suppliers in Pierce, Snohomish and King Counties. In particular, the CIG provided analysis for a technical committee charged with investigating climate change impacts on the region's water resources. The committee developed a model that projected climate changes to 2075, the results of which showed increases in future winter stream flows, decreases in summer stream flows and impacts on the region's water supply. The projections have been integrated into hydrologic models for the County's major watersheds and into the planning of three major regional utilities. Overall, the regional water planning process helped to create consensus among government and utility leaders about impacts the region would likely face.

CIG's activities provided the scientific underpinnings for "The Future Ain't What It Used to Be – Planning for Climate Disruption," the October 2005 conference co-hosted by King County and CIG. More than 600 people attended the conference – almost two-thirds of them staff from local, regional and state governments and agencies. From presentations and background materials prepared for the event by CIG, participants learned that the Puget Sound region is facing warmer temperatures, reduced snowpack, changing streamflow patterns and more intense and frequent flooding – changes which will likely affect hydropower, water supply, forestry, fisheries and agriculture (Casola et al. 2005). They also explored potential flooding impacts to stormwater and wastewater systems. The event challenged participants to "ask the climate question" for each government operation or service: "How is climate change likely to affect our plans, programs and future investments?"

"Asking the Climate Question"

Following on the heels of "The Future Ain't What It Used to Be," the King County Executive Office formed an Executive Action Group on Climate Change – an interdepartmental team led by the County's Deputy Chief of Staff. The team was charged with integrated climate planning to reduce greenhouse gas emissions and prepare for climate change impacts.

Staff members from the County's Executive Office worked with King County's Department of Natural Resources and Parks (DNRP) to survey County departments and assess their awareness of climate changes that could affect their facilities and operations. This questionnaire helped the Executive Office decide which departments to include on the interdepartmental team.

The resulting team included approximately 10 senior decision-makers and advisors from the Executive Office; the Department of Development and Environmental Services (responsible for land use planning and building permits); the DNRP (responsible for stormwater management and wastewater, among many other areas); Seattle-King County Public Health; and the Department of Transportation. Members of the team were selected based on their decision-making authority in sectors that contribute substantially to the County's greenhouse gas emissions or have responsibility for plans, programs and investments that climate change is likely to impact. From the outset, the team included technical experts and policy advisors. Coordination of the group was the responsibility of a climate change program coordinator, who worked in the Executive Office. Not all desired departments were able to participate fully, due in part to staffing constraints. However, the broad focus of the initial scoping survey and the diverse membership of the team helped to raise awareness about climate change considerations throughout the County's departments.

Planning, Implementing, and Learning

In February 2007, the County released its inaugural Climate Plan ("the Plan"), after approximately 6 months of collective drafting by the interdepartmental team (King County Government 2007). The Plan was intended as a dynamic document that would guide King County's actions to reduce greenhouse gas emissions and prepare for climate change. It would need to be updated regularly, based on evolving information about climate change impacts. Half of the plan was devoted to mitigation, and half to adaptation.

In the first version of the Plan, the team identified six "strategic focus areas" of the County's adaptation efforts: climate science; public health and emergency preparedness; land use and infrastructure; biodiversity; water quality and water supply; and economic impacts. The Plan included an inventory of relevant existing activities as well as a series of next steps and emerging issues. In that way, the Plan was both a snapshot of the preparedness measures the County already had in place and a vision of what it would need to do to become more resilient in the future.

As of 2009, some of the highlights of the County's adaptation program were:

- Development of reclaimed water infrastructure, as part of a major expansion of the County's wastewater treatment system. The reclaimed wastewater is suitable for irrigation and industrial uses and is intended to help reduce future demand on the region's water supply.
- Creation of a Flood Control Zone District that uses countywide taxing authority to fund the rebuilding of old levees and development of new flood management

projects. The District united several previously fragmented flood districts to respond more effectively to the likelihood of increased flooding in industrial and high-density residential areas. An economic analysis of flooding impacts commissioned by the County showed that flood levees across King County protected tens of thousands of jobs and tens of millions of dollars of economic activity (ECONorthwest 2007).

- Incorporation of new climate information into capital planning. The County's Wastewater Treatment Division used a recent CIG report on sea level rise to anticipate which treatment facilities, if any, are likely to be flooded in the future (Mote et al. 2008; King County Water Treatment Division 2008). The division also explored how other aspects of treatment and conveyance could be affected, and the work was turned into a geographic information tool that can be downloaded by other governments online. The County's Roads Division likewise incorporated flooding projections into its capital projects, resulting in modified design of bridges and culverts to accommodate higher river flows.

King County's progress in these areas has been dependent on scientific analysis of climate change impacts in the region, conducted by the CIG and others, as well as on the scientific literacy of leading public officials. By 2005, the County had long been a robust internal scientific and technical capacity. Decision-makers not only drew on the expertise of staff members able to translate climate scenarios into understandable concepts and actionable language; they too were scientifically fluent.

In spite of these resources, the King County team recognized that there were gaps in understanding of how climate change would affect the Puget Sound region. Information is still limited in some areas, including health impacts of climate change on vulnerable and disadvantaged populations. In early 2009, however, a CIG study on updated projections for regional climate change impacts included new analysis on heat and health in the region, which was intended to provide more guidance for public health authorities across the state.

Even in areas that are better understood, uncertainty still exists. It is broadly expected, for example, that fall and winter flooding will become more frequent and intense, especially in basins that experience a mix of rain and snow, as temperatures increase (Mantua et al. 2009). Future precipitation changes will also affect that risk but are still not completely predictable (Palmer et al. 2006).

Moreover, although the Pacific Northwest is known for a strong environmentalist ethic, building the case for climate change adaptation has not been entirely easy. Until recently, many environmental leaders across the world and in the King County region argued that investing resources in adapting to climate change would divert energy from the urgent work of reducing greenhouse gas emissions (Pielke et al. 2007). However, with the most recent Intergovernmental Panel on Climate Change reports, consensus has grown that some climate change impacts are unavoidable and must be addressed.

Throughout this time, guided by practicality, County departments worked toward systematic integration of broad climate change concerns into facility planning. For one, the King County Wastewater Treatment Division developed plans to apply scientific projections to their understanding of the system hydraulics and related energy use of wastewater treatment and conveyance. The conclusions of this proposed work could help wastewater treatment officials to determine whether modifications to existing operations would reduce future vulnerability and avoid costly future responses. The bottom line of this effort was for the County to be informed sufficiently to take relatively low-cost adaptive steps today that would save taxpayers from higher costs of future impacts.

As of 2009, the Executive Action Group on Climate Change continued to meet biweekly, supported by a number of subgroups responsible for implementing the Climate Plan. The Executive Action Group also worked with the County's Office of Management and Budget to incorporate climate change mitigation and adaptation projects into its annual budget.

Sharing Knowledge with Other Local Governments

When King County began its adaptation efforts, only a few other local governments had begun to "ask the climate question," as Sims put it. The county thus sought to share lessons with other governments about how to take practical steps toward climate change adaptation by co-authoring *Preparing for Climate Change: A Guidebook for Local, Regional and State Governments* with the CIG (Snover et al. 2007). In the years after its publication, the guidebook was distributed widely through the international networks of CIG and ICLEI – Local Governments for Sustainability, and informed the development of ICLEI's Climate Resilient Communities Program[TM].

Around that time, a growing number of cities and urban regions were also beginning to work with climate researchers to understand their climate change vulnerabilities (Hunt and Watkiss 2007), enabling King County and others to establish networks or "communities of practice." In December 2006, for example, the county joined with the US Center for Clean Air Policy (CCAP) to launch the Urban Leaders Adaptation Initiative, a national consortium of nine partner cities and counties, all committed to sharing lessons from their early experiences in climate change adaptation. With the understanding that "local governments are the first responders" to climate change impacts, the initiative partners have intended "to serve as a resource for local governments as they face important infrastructure and land-use decisions that affect local adaptation efforts and empower local communities to develop and implement climate resilient strategies" (CCAP 2009) (Box 24.2).

Box 24.2   Other Early Adaptation Leaders

In 2007, Penney and Wieditz reviewed the early climate change adaptation efforts of six urban regions in the United Kingdom, United States and Canada. Among the early leaders identified by this study were King County, New York, London and Halifax. In the nearly 2 years since this research was completed, adaptation efforts have evolved considerably, but the report identified characteristics of adaptation leaders that remain relevant: commitment by political and/or staff leaders; creation of a dedicated adaptation team; comprehensive assessment of impacts; investigation of adaptation options; incorporation of adaptation into policies, plans and budgets; monitoring of ongoing climate changes and evaluation of adaptation responses. In the last few years many other urban centers have seriously taken up the challenge of adapting to climate change. In addition to King County, the Urban Leaders Adaptation Initiative partner governments now include Chicago, Miami-Dade, Toronto, San Francisco, Los Angeles, Nassau County, Phoenix and Milwaukee.

Some Lessons from King County

The experience of King County offers several lessons about what it takes for a municipality to launch a climate change adaptation effort and how much support governments will need from outside institutions.

Municipal government officials must signal a firm commitment to adaptation and invest financial resources for implementation. King County Executive Ron Sims showed a visible national and regional presence in arguing for both mitigation and adaptation, supporting the work with two advisors from his office and involving more than 10 senior decision-makers and staff on a biweekly basis in the Executive Action Group on Climate Change.

Sound scientific and technical advice is critical to all stages of adaptation – from raising awareness in the early stages of a program to policy development and continued improvement. King County efforts were scientifically supported by the University of Washington's CIG, which offered county decision-makers access to dozens of scientists committed to providing practical information and analysis to support adaptation decision-making. King County also commissioned climate change studies on a project-by-project basis, such as economic analysis for the Flood Control Zone District. Municipalities everywhere need access to this kind of scientific and technical information, as well as flexible funding to commission analysis that is tailored to their specific needs.

Climate change adaptation requires a dedicated commitment of human resources from across disciplines, government functions and levels of management. King County's Executive Action Group on Climate Change consisted of members from a range of departments and disciplines, as well as levels of government, with both subject matter experts and senior decision-makers. This team was charged with strategic planning both to reduce emissions and to prepare for climate change impacts. It was coordinated from the Executive Office by the Climate Change Program Coordinator and from key departments by dedicated project managers.

Climate change concerns should be incorporated into long-range planning as well as existing operations and infrastructure. The County systematically worked to incorporate climate change into major planning and decision-making. King County's 2008 Comprehensive Plan update reflected this approach, as did a range of county sub-plans, including the Flood Hazard Management Plan, and capital plans such as the Roads Strategic Plan and the Reclaimed Water Comprehensive Plan. However, it is also important to recognize and build on the existing work that may already be improving resilience to climate change. King County's reclaimed water efforts, for example, were already ongoing prior to the launch of the County's climate change efforts.

Third parties, including the federal government, can learn from municipal adaptation experiences. Although King County is still in the early stages of its work, there have been some efforts to evaluate and learn from its activities. For example, Harvard Kennedy School has completed several case studies about the County's adaptation activities, for use in master's program coursework. The US Government Accountability Office (GAO) also interviewed the County's staff for lessons on adaptation to be incorporated into a report for the House Select Committee on Global Warming (GAO 2009).

Systematic, dedicated federal resources are necessary for long-term municipal government adaptation. The King County case highlights some resources that are needed by local governments. For example, although small amounts of federal funding have been available for specific adaptation projects, as of early 2009 there has been no consistent, overarching commitment at the federal level to provide adaptation outreach or financial support to local governments. This shows a contrast with the UK government, which has underwritten the efforts of the United Kingdom Climate Impacts Programme, an organization that has provided direct support for local governments involved in adaptation planning since 1998. This gap compelled King County to promote "asking the climate question" at the federal level, by making recommendations on climate change adaptation, climate science and related services in several venues, including America's Climate Choices (a National Academy of Sciences study committee) and CCAP efforts on national adaptation legislation.

This relative lack of focus on adaptation at a national level may change, as the US federal government has discussed a number of possible models that build on NOAA's RISA experiences in providing municipalities with actionable climate information and other adaptation resources.

Conclusion

By taking early action on climate change adaptation, King County uncovered some of the steps that municipalities can take in preparing for the threats of climate change. But more than anything, the King County case study reflects how much support local leaders and their communities need now and in the future from outside sources, as climate change impacts continue to hit regions across the world. The US federal government and non-governmental organizations have both an opportunity and an imperative to join together in support of local communities adapting to climate change.

References

Casola JH, Kay JE, Snover AK et al (2005) Climate impacts on Washington's hydropower, water supply, forests, fish, and agriculture. University of Washington, Seattle. http://cses.washington.edu/db/pdf/kc05whitepaper459.pdf. Cited 10 Oct 2008

Center for Clean Air Policy (2009) Urban leaders adaptation initiative: building resiliency to climate change impacts through community action. http://www.ccap.org/index.php?component=programs&id=6. Cited 10 Oct 2008

ECONorthwest (2007) Economic connections between the King County floodplains and the Greater King County economy. Prepared for King County Water and Land Resources Division, King County Government, Seattle. http://www.metrokc.gov/dnrp/wlr/flood/flood-control-zone-district/pdf/floodplain-economic-connections.pdf. Cited 10 Oct 2008

Hunt A, Watkiss P (2007) The economics of climate change impacts and policy benefits at city scale: a conceptual framework. OECD, Paris

King County Government (2007) 2007 King County climate plan. http://www.metrokc.gov/exec/news/2007/pdf/ClimatePlan.pdf. Cited 10 Oct 2008

King County Wastewater Treatment Division (2008) Vulnerability of major wastewater facilities to flooding from sea-level rise. King County Governemnt, Seattle. http://dnr.metrokc.gov/wtd/csi/csi-docs/0807_SLR_VF_TM.pdf. Cited 4 Dec 2008

Mantua NJ, Tohver I, Hamlet AF (2009) Chapter 6: Impacts of climate change on key aspects of freshwater salmon habitat in Washington State. In: The Washington Climate Change Impacts Assessment: Evaluating Washington's Future in a Changing Climate, Climate Impacts Group, University of Washington, Seattle

Mote PW, Hamlet AF, Clark M et al (2005) Declining mountain snowpack in western North America. Bull Am Meteorol Soc 86(1):39–49

Mote PW, Petersen A, Reeder S et al (2008) Sea level rise in the coastal waters of Washington State. University of Washington Climate Impacts Group and the Washington Department of Ecology, Lacey. http://cses.washington.edu/db/pdf/moteetalslr579.pdf. Cited 4 Dec 2008

NOAA Climate Program Office (2009) Climate Program Office website. National Oceanic and Atmospheric Administration [NOAA], United States Department of Commerce. http://www.climate.noaa.gov/cpo_pa/risa/. Cited 10 Apr 2009

Palmer RN, Wiley MW, Polebitski A et al (2006) Climate change building blocks: a report prepared by the Climate Change Technical Subcommittee of the Regional Water Supply Planning Process, Seattle. http://cses.washington.edu/db/pdf/palmeretalbuilding546.pdf. Cited 4 Dec 2008

Penney J, Wieditz I (2007) Cities preparing for climate change: a study of six urban regions. Clean Air Partnership, Toronto

Pielke R Jr, Prins G, Rayner S et al (2007) Climate change 2007: lifting the taboo on adaptation. Nature 445:597–598

Schulte B (2006) Temperature rising. US News and World Report. 140(21):36–40. http://www. usnews.com/usnews/news/articles/060605/5warming.htm. Cited 12 Dec 2008

Seattle Times (1988) Hot air glut – county doesn't need a 'greenhouse office. Seattle Times (7 Sep 1988), p. A-8. Washington State Newsstand database. (Document ID: 52417049). Cited 9 Apr 2009

Snover AK, Whitely Binder L, Lopez J et al (2007) Preparing for climate change: a guidebook for local, regional, and state governments. International Council for Local Environmental Intitiatives [ICLEI] – Local Governments for Sustainability, Oakland

Swope C (2007) Local warming: it's too late to stop climate change – what we can do is plan for it. Governing Magazine. http://www.governing.com/articles/0712warm.htm. Cited 12 Aug 2008

The Heinz Center (2007) A survey of climate change adaptation planning. http://www.heinzctr. org/publications/PDF/Adaptation_Report_October_10_2007.pdf. Cited 12 Aug 2008

US Environmental Protection Agency [EPA] (2007) 2007 Climate award winners. http://www.epa. gov/cppd/awards/2007winners.html. Cited 20 Nov 2008

US Government Accountability Office [GAO] (2009) Strategic federal planning could help Government officials make more informed decisions. http://www.gao.gov/new.items/d10113. pdf. Cited 19 Sept 2010

Part V
Adaptation in the Agricultural Sector

Chapter 25
Overview: Climate Change Adaptation in the Agricultural Sector

John M. Reilly

Abstract Agriculture is vulnerable to climate change but has considerable adaptation potential. The sector will be significantly transformed over the next few decades as it incorporates new technology and management and organization, that are in part a response to changing demands from growing population, changing incomes, other competition for land, resource and environmental protection, and the effects of environmental change, including climate change, on agriculture. Adaptation research to date has largely been about educating the sector about potential risks and possible adaptation strategies. As we move to more "client-based" research, these recommendations will need to be more precise and tailored to individual decision-makers. Given the likely changes in the sector, much adaptation information may be delivered to farmers through recommendations on crops, pest management, and other inputs and products delivered by input manufactures and suppliers, and so these organizations will be as, if not more important clients for adaptation research. Research also needs to focus on where markets may fail in being adequately prepared, such as from the risk of abrupt climate change. The challenge ahead for adaptation research is how to come up with concrete, tested, and robust strategies that are responsive to the decision time frames in agriculture (10–20 years), given the significant variability and noise in climate.

Keywords Climate change • Agricultural sector • Research and policy approaches • Adaptive strategies • Technological adaptation • Institutional adaptation • Sustainable adaptation • Agriculture • Diversification • Climate change opportunities

J.M. Reilly (✉)
Center for Environmental Policy Research, Joint Program on Global Change, Massachusetts Institute of Technology Sloan School of Management, Cambridge, MA, USA
e-mail: jreilly@mit.edu

J.D. Ford and L. Berrang-Ford (eds.), *Climate Change Adaptation in Developed Nations: From Theory to Practice*, Advances in Global Change Research 42, DOI 10.1007/978-94-007-0567-8_25, © Springer Science+Business Media B.V. 2011

Introduction

Climatic conditions essentially define major agricultural regions of the world, such as large wheat growing regions in Canada, Australia, Russia, and the United States; the so-called corn-belt of the United States where maize and soybeans are grown extensively; rice paddy areas in Monsoon Asia; and cattle grazing and ranching in the Pampas of South America and the Great Plains of the United States. Many crops and agricultural activities span a wide range of climatic zones and regimes. There are limits: yields of the major grain staples fall off with high temperatures that affect grain formation; extreme conditions such as frost or extreme heat and drought can cause crop failure; and flooding, wind and hail storms can cause severe damage to most crops. Livestock productivity (meat and milk production) also falls off under extreme conditions (for further discussion, see for example, Reilly 1999).

Agriculture, through management, has also adapted to climatic conditions, extending crops into climatic regimes in which they otherwise would not thrive. Irrigation is one of the most effective adaptations to climate: on the order of 18–20% of crop land worldwide is irrigated but that land accounts for 30–40% of all crop production (Wiebe and Gollehon 2003). Livestock also benefit from shelters against the cold, and ventilation and shading can reduce heat stress. Production of vegetables under glass provides nearly complete control of climate (but the cost is high, and so this technique is not practical for the production of bulk commodities). Costs and resource availability are limits to what can be done or makes economic sense to do. The limits for irrigation are defined in part by the availability of freshwater resources, which will also be affected by climate change. That said, irrigation is generally highly inefficient and there are examples of arid regions, like countries in the Middle East, that make much better use of water through drip irrigation and other technologies than most other areas.

Technological change can overcome some climatic limits. In the United States, for example, development of shorter maturing maize varieties has allowed expansion of the crop into cooler climates (Reilly et al. 2003). Further advance in technology may change substantially the relationship of agricultural production and its sensitivity to climate. For example, Carolan et al. (2007) and Laser et al. (2009) describe bioprocessing refineries that could produce multiple products including feed grain equivalent livestock feed, ethanol, and other biochemicals from a general biomass stock. Such a process would greatly increase the economic productivity of land by producing valuable feed grain equivalent products, using the entire plant rather than just the grain. It could use a wide range of biomass feed stocks with different climatic tolerances, thus dramatically expanding the climate range suitable for providing bioprocessing feed stocks. The tropics might become major "feed grain product" (or livestock product) and fuel suppliers, with corn and soybean production in the United States disappearing and that land used instead to produce grains and other crops for direct human consumption.

Even if climate change were not a factor in the future, the agricultural system will be under significant pressure to change. In general, forces for change include:

meeting the world's growing and changing demand for food as population and income increase; adapting to other environmental issues such as tropospheric ozone, aerosol haze, and soil degradation (Reilly et al. 2007); other competition for land and water to meet energy, recreation, and urbanization needs (Gurgel et al. 2007; Antoine et al. 2008); meeting a variety of environmental constraints such as reduced soil erosion and run off of nutrients into streams, lakes, and coastal areas; continued competitive pressures from increasingly globalized agricultural markets; changes in technology; and the "industrialization" of agriculture (see Reilly 2002 for a report on the interaction with agricultural stakeholders who identified many of these forces).

Industrialization of agriculture refers to the institutional structure in which farming is practiced and how risks and rewards are shared. Rather than individual farmers growing products and taking them to market, and selling for whatever price they fetch, industrialization refers to a large central firm with a global reach contracting with or hiring people to manage production, and likely increasing specialization of production in regions best suited for particular crops. More of the decisions on what and how to produce are made by the industrial enterprise. Drivers of this trend are the demand for uniform product and assured continuous supply through the year, risk pooling against crop failure in some regions, and bringing in investment and financing at levels needed, reshaping how risks (and rewards) are shared between the land holder and the industrial enterprise. It has extended fairly far in the United States, especially in fruit, vegetable, and livestock markets; less so in grain production, likely because United States commodity programs have had some effect on farm size. What implications does the industrialization of agriculture have for climate change adaptation? Such enterprises may be able to devote more resources to optimizing production practices, taking into account climate change and may be less constrained with regard to access to capital. On the other hand, centralized decisions could be less cognizant of local conditions, or one could imagine rapid abandonment of areas if climate deteriorated leaving the local landowners out in the cold. The trend toward industrialization likely changes how climate information and inputs adapted to climate conditions are and will be delivered in the future. The traditional approach – local agricultural research and extension provided by governments – may remain important for small farmers but is less likely to be the dominant force for the bulk of global food production. Land holder and ownership traditions in some countries may slow this trend, but the urbanization of the population and the percentage of population employed directly in agriculture is likely to decrease as countries develop, as was observed in the Europe, the United States and other developed areas, and this will necessarily mean larger farms.

Given the picture of agriculture described above, the vulnerability of the sector and potential for adaptation cuts two ways. On the one hand, agriculture is vulnerable to climate change as production is fully exposed to the elements. On the other, the success of agriculture in a wide range of climates suggests its high potential for adaptation. The evidence is clear that agriculture, through technology and management, has extended production activities into what would otherwise be difficult or inhospitable climates. The fact that the sector has and will continue to change rapidly can also be seen in different ways. Climate change may be an additional stress on a sector that is facing many challenges. However, the rate of change in the sector due to technology advance and competitive pressures may exceed the rate needed to adapt to climate change, and as long as climate considerations are integral to those processes, climate change may present no additional adjustment, but just adjustment in a somewhat different direction. If because of technological change or competitive issues, farmers need to adopt a new crop variety, crop, or management practice, they can adopt one that is also suited to the changing climate. And, then will trends in the sector lead to greater or less vulnerability to climate? As discussed above, the answer is unclear. Industrialized agriculture may have greater access to financial resources and achieve economies of scale in evaluating and testing adaptive strategies, but further specialization may increase vulnerability of supply of crops if regions supplying an ever larger proportion of the crop face crop failure. Specialization may also increase risks to pest and diseases that could spread rapidly and affect a large proportion of the crop.

While recognizing the important role of adaptation, it seems clear that the goal cannot be the preservation of the status quo at the local and regional level, as even without climate change the status quo will change. Changing climate and higher CO_2 levels may be benign or even advantageous for some areas if agricultural systems adapt to take advantage – but that adaptation and change may be painful for some who fail to adapt and who succumb to competitive pressures, and the crops and landscapes' residents are familiar with may change. Even improving agricultural conditions can force some into bankruptcy because it may mean lower commodity prices, and the slow to adapt can lose even if productivity is rising (this is a process similar to the "technology treadmill" of advancing technology: technology lowers production costs, but with lower cost production, commodity prices fall and farmers need to keep adopting the new technology or face losses). Other areas are likely to become less agriculturally productive, perhaps becoming marginal agricultural areas. Attempts to prop up these areas even as climate conditions degrade could increase costs and lead to further resource misallocations and environmental problems, whether that is competition over dwindling surface water, mining of groundwater, or the increasing threat of wind or water erosion if conditions become arid or rainfall comes in more infrequent but larger storms.

Some final considerations are the roles of markets, policy, and information on adaptation. Markets can only work well with proper incentives for resource protection, and cannot foresee the unexpected. Agricultural policy continues to have strong effects in most regions and likely will continue to shape production decisions and trade patterns. These broad policy measures could retard adaptation by responding to constituencies that want to preserve the status quo (e.g. Lewandrowski and Brazee 1993). Agricultural research and extension could help provide climate change information relevant to local and regional agricultural activities, evaluate different crops, livestock, and practices suitable to the new climate, and develop new technologies that facilitate adaptation (e.g. Reilly and Schimmelpfennig 1999). While studies of the agricultural sector and climate adaptation often focus on production systems and farmers, the ultimate goal of a production sector is to provide things people need and want. Food is a basic need where poor consumers, faced with high prices and lack of availability, end up malnourished or worse. To some extent, growing incomes around the world and the availability of inexpensive products high in fat and sugar is leading to problems of obesity rather than lack of food, but famine and hunger still are a problem for millions of people, and climate change could aggravate this problem (Rosenzweig and Parry 1994). The adequacy of food and potential for malnutrition has been taken up more regularly under the "health and climate change" umbrella than under the agricultural sector (see, McMichael et al. 2003; Confalonieri et al. 2007). However, the agricultural community needs to continue to be involved in this aspect of the problem, as many food programs in the world are run by agriculture ministries (even though these programs are often run more as a mechanism for disposing of surplus commodities resulting from programs to support farmer's incomes). And the problem is not a lack of ability to produce food, but the lack among the poor to control resources that get them access to food. Income support and direct food programs are needed, as simply seeing the problem as a need for more production is not very effective in delivering food to the poor. Certainly expensive food makes the problem worse, but even if food could be produced for nothing, it would still cost something to get it to those who need it.

Recent Developments: From Theory to Practice

Research on climate impacts, adaptation, and vulnerability, in general, and in agriculture in particular extends back more than two decades. Among the early efforts were those of Waggoner (1983) and Schelling (1983). Early efforts to quantify impacts on a national (Adams et al. 1990) and global level (Rosenzweig and Parry 1994) found that impacts depended strongly on the climate scenario and effectiveness of adaptation. The work showed a general result that has appeared to hold up over time – that the warmer tropics and subtropics were more likely to be detrimentally affected by climate change whereas as one moved pole-ward, climate change, especially considering the effects of CO_2 fertilization on plants, could be beneficial. The work has also showed that differential ability to adapt could provide further advantage to developed country agriculture in temperate regions while further worsening agricultural competitiveness in tropical regions. Early work on global impacts showed that major exporting regions could actually gain economically if climate change led to a general deterioration of growing conditions globally because commodity price increases would increase their revenues more than they were hurt by reduced yields, while consuming regions might be worse off even if their agricultural prospects improved (Kane et al. 1992; Reilly et al. 1994).

Many empirical and theoretical issues related to societal adaptation to climate change were brought together in *Climatic Change* (Vol. 45) introduced by Kane and Yohe (2000). With respect to agriculture, Bryant et al. (2000) provide an excellent summary of theory and experience for Canada that identifies issues applicable to agriculture in many areas. Additional efforts have been made to generalize impact and vulnerability assessment into a handbook or regularized process (e.g., Parry 1990; Lim and Spanger-Siegfried 2005). The concept of vulnerability assessment arose out of frustration at the inability to predict climate and impacts with any accuracy. That has led to one of those half empty–half full debates on whether to focus on "vulnerability" or "adaptive capacity" where vulnerability focuses on what is at risk, and adaptive capacity on how much of that potential risk can be avoided. A general issue with these concepts is that operational definitions that would actually help identify adaptations or the vulnerable are not possible absent solid predictions of future climate or the range of future climates through formal uncertainty analysis (see e.g. Reilly and Schimmelpfennig 1999). It is possible to imagine all types of climate events to which the agricultural sector in a region would be vulnerable, and consider many things that could be done to lessen the impact to these events. But if these are nothing more than scenarios without an estimate of the chance of them occurring, then it is not possible to calculate the value of "real options" (i.e. adaptation measures) that might lessen the impact. Another adaptation concept that has been identified is that of "robust" adaptation. Clearly, the concept of a robust response that works across the range of expected climates rather than a narrowly inflexible response geared to a single prediction is an appropriate way to think about the problem of adaptation in the face of uncertainty. For some areas of impact, like those associated with sea level rise, the direction of impact is clear. Strategies to limit development in low-lying coastal areas are "robust," albeit even here one is balancing the restriction against the cost of denying people their preference of living directly by the seaside, and making judgments about how far from the coast development should be limited that depend on a how much sea level rise is really possible, and in what time frame.

On the other hand, the nature of agricultural impacts – dependent on highly unpredictable changes in precipitation as well as on how global impacts affects world prices for commodities – means that we often are not sure of even the direction of impact. Will it get wetter or drier, will commodity prices rise or fall? This makes identification of "robust" strategies that would not involve significant investment more difficult. In economics, investment to be prepared for uncertainty is referred to as real options and their value depends on the range of climate and likelihood of different extremes occurring, as well as future prices. An additional issue related to uncertainty is the time frame of climate change and that of the economics of decision-making. Many decisions in agriculture are relatively short-lived – what to plant this year – and even for long-lived investments, when one takes discounting into consideration, most of the payback for an investment is expected within 10 or 20 years. In that time horizon, any climate trend due to long-term climate change can be swamped by natural climate variability. And, it is possible to mal-adapt, reading short-term variability as consistent with a long-term trend. In this regard, Smit (1993) found some farmers switched to longer maturing maize hybrids on recent warmer seasons, only to have cooler and shorter seasons return and suffer frost damage that might have been avoided with shorter season varieties.

In the end, it is not clear that a "theory" of climate change and adaptation (or vulnerability) has developed. In fact, the community has struggled to agree on a common language with competing efforts to offer definitions for mal-adaptation, autonomous adaptation, and other concepts (e.g., Reilly and Schimmelpfennig 2000; Smit et al. 2000).

What eventually emerged from some of the academic research and writing was a focus on stakeholder involvement in impacts and adaptation research, as in the US National Assessment (National Assessment Synthesis Team 2001). This was a positive direction in that it brought potential users of scientific information and the science community together to better understand decisions at risk and how climate information could be incorporated. As initially developed, stakeholder involvement was a very general exercise with many different clients for the research brought together in a few meetings over the course of the research exercise. From that initial start, what has developed is what I would call "client-based research." Client-based research is stakeholder-driven. The stakeholders (clients) invest their own resources in the research exercise and expect tailored advice in return. In that way, they have a vested interest in the outcomes.

Key Challenges

The key challenge remaining is to bring the consideration of climate change into regular decision processes. This involves moving from analysis and evaluation of *response to climate change* to an analysis of policy formulation on resource management, and decisions people and organizations make, *where changing climate is one element factored into the analysis*. Farmers face a decision of what to plant this year recognizing that there are new seed varieties available, new equipment, and changed relative prices among crops. As part of that decision, they will need to evaluate seasonal projections for weather and climate such as El Niño-Southern Oscillation (ENSO) phenomena, and they will need to assess whether climate conditions have trended sufficiently in some direction to warrant doing something differently. This consideration on climate change may be embedded in the recommendations of seed producers on what to plant, equipment suppliers, and pest management consultants. Policymakers face decisions on how to formulate commodity policy, crop insurance, drought and disaster assistance, food stamp and food assistance programs. Climate change is just one consideration in these policy decisions. Similarly, water managers face decisions of how to manage reservoirs and whether improvements or expansion is needed. Increased population and other factors governing demand may be major drives behind these decisions, and one factor to be included is the potential for climate to change availability of water and demand for it.

A main barrier to incorporating climate change in decision processes has been the failure of decision-makers at various levels to recognize that climate is actually changing. There now appears to be more of that recognition. The next barrier is to convince these decision-makers that information on climate that is useful for their decisions and is sufficiently predictive is available. Much of the published literature and major studies on climate impacts have looked at climate change 30, 50, or 100 years in the future. Those timescales have been interesting for analysts because they have been more confident that the signal of climate change has risen from the noise, but those timescales are nearly irrelevant for most major decisions. Even large dam projects, for example, while they may have lifetimes of 100 years, must have a positive cost-benefit which, because of discounting, is mostly determined over the next 20 years. Thus, the climate projections over that period are much more important even for long-lived projects. Predictive information in this case is not an absolute prediction, but a prediction of the spread of what is likely and possible, for example more variability which then may have implications for strategies that increase flexibility. The big challenge, then, is what climate scientists and the adaptation research community can say about climate over 10 or 20 years, whether that is from model predictions or study of local trends.

Opportunities and Future Directions

Once the usefulness and predictive power of climate information is demonstrated, then it will be in the interest of decision-makers to seek out this information and to pay for it. We are now nearing that point – or at least there appears to be enough recognition and interest among some of stakeholders in agriculture to begin looking for expert help and working through how the information might be used. Again, the community needs to recognize that the main channel of adaptation information may not be directly to farmers. Climate adaptive strategies for farmers may be delivered through the inputs and practices that are recommended to them by private vendors or the more traditional agriculture extension services. As we transition to client-based research, the role of the academic research community that has debated and discussed relative adaptability of systems, and tried to model and predict impacts and adaptation, should move to assessing the relative success of different management approaches. Which approaches have been more successful, which less at incorporating climate change? Can we measure avoided damage in some way that can quantify the effectiveness of different approaches to adaptation? Has climate change affected the structure of agriculture and the delivery of services to the sector? Where are the vulnerabilities not being addressed by individuals and market participants? This latter question reflects a division that many economists would make. Much of adaptation is a matter that informed private agents will undertake – they will create a market for adaptation information by becoming adaptation research clients.

Much adaptation work so far has focused on trying to figure out what individual agents (farmers) might do differently, funded under government research grants. By doing so, this research has been trying to make the case that climate matters – and by communicating this research through assessment activities and stakeholder engagement hoping to create "informed" private agents. Once that job is done, the research community can evolve to other roles. Future research can address the lack of work that has occurred to date regarding potential events that may not be solved by markets – the large-scale and abrupt climate change and the problems they would create. Even as the community continues to hammer away at the first problem – convincing potential clients there is value in climate information – the community needs to investigate where markets could fail and what measures can correct those failures or provide ready responses should events lead to failure.

Case-Studies

The case studies in this section address some of the key issues identified above. Werners et al. apply portfolio theory to evaluate diversification of agricultural activities as a response to climate risks. Here the hypothesis is that less specialization will lead to great resiliency to climate variability. They find the Tisza River area is relatively well adapted to the current climate. This study shows that portfolio analysis is an appropriate tool for addressing resiliency in the face of climate change, and lays the groundwork for future work that can make this approach more useful for decision-makers. For example, future work could build on this study by using some measure of profit rather than gross revenue, and could incorporate radically different strategies, other than those historically adopted, into the analysis. To find where change is needed, future work could examine the margins of the region – either that belt of land where flooding is really changing or on more finely defined adaptive responses – and incorporate climate projections. This future work will be subject to the challenge of developing near-term predictive information relevant to decisions today. Will the next 10 or 20 years require significant changes in the amount of diversification?

Chapters by Helfrich and Prasad and by MacGregor and Cowan, both investigate national efforts to use agricultural policy to achieve or encourage adaptation. The former examines wheat areas, and contrasts Australian and Canadian responses with those by Defra to assure the sustainability of agriculture in the United Kingdom. Helfrich and Prasad note a difference between Canada's top down and Australia's bottom up approaches. Their chapter is a first step at assessing the effectiveness of different strategies, and thus is in line with where future research needs to go. The major issue facing their work is the lack of measurable success. How can we evaluate a strategy's effects on income, risk, or output, compared to a case where no measure was taken, to determine whether the strategy was efficacious and cost-effective? This is especially difficult for adaptation research, as 1 year or even a decade is likely too short a time period for evaluation, given how variable climate can be. The MacGregor and Cowan chapter begins with a review of four broad principles that, while abstract, are sensible for guiding adaptation. They come up with eight proposed sets of relative concrete measures that meet these general criteria. They then address what influences farmer behavior and the levers available to affect behavior, the latter of which is often missing in adaptation research. This is a key consideration in moving from the position of simply identifying things that might work, to actually changing what is done on the ground.

The three chapters together are headed in the right direction. The shortfalls identified above should not be taken as criticisms of the work per se, but indicate some of the many challenges ahead for adaptation research. In short, how do we come up with concrete, tested, and robust strategies that are responsive to the decision time frames in agriculture?

References

Adams RM, Rosenzweig C, Ritchie J et al (1990) Global climate change and U.S. agriculture. Nature 345:219–224

Antoine B, Gurgel A, Reilly JM (2008) Will recreation demand for land limit biofuels production? J Agr Food Ind Organ 6(2):Article 5

Bryant CR, Smit B, Brklacich M et al (2000) Adaptation in Canadian agriculture to climatic variability and change. Clim Change 45(1):181–201

Carolan JE, Joshi SV, Dale BE (2007) Technical and financial feasibility analysis of distributed bioprocessing using regional biomass pre-processing centers. J Agr Food Ind Organ 5(Article 10):1–27

Confalonieri U, Menne B, Akhtar R et al (2007) Human health. In: Parry ML, Canziani OF, Palutikof JP et al (eds) Climate change 2007: impacts adaptation and vulnerability – contribution of Working Group II to the Fourth Assessment Report of the Intergovernmental Panel on Climate Change. Cambridge University Press, Cambridge, pp 391–431

Gurgel A, Reilly JM, Paltsev S (2007) Potential land use implications of a global biofuels industry. J Agr Food Ind Organ 5(2):Article 9

Kane S, Yohe G (2000) Societal adaptation to climate variability and change: an introduction. Clim Change 45(1):1–4

Kane S, Reilly J, Tobey J (1992) An empirical study of the economic effects of climate change on world agriculture. Clim Change 21(1):17–35

Laser M, Jin H, Jayawardhana K et al (2009) Projected mature technology scenarios for conversion of cellulosic biomass to ethanol with coproduction of thermochemical fuels, power, and/or animal feed protein. Biofuels Bioprod Biorefining 3(2):231–246

Lewandrowski J, Brazee R (1993) Farm programs and climate change. Clim Change 23(1):1–20

Lim B, Spanger-Siegfried E (eds) (2005) Adaptation policy frameworks for climate change: developing strategies, policies, and measures. Cambridge University Press, Cambridge

McMichael AJ, Campbell-Lendrum DH, Corvalan CF et al (eds) (2003) Climate change and human health: risks and responses. World Health Organization, Geneva

National Assessment Synthesis Team (2001) Climate change impacts on the United States: the potential consequences of climate variability and change. Report for the US Global Change Research Program. Cambridge University Press, Cambridge

Parry M (1990) Methods of assessing impacts of climate change. In: Parry M (ed) Climate change and world agriculture. Earthscan, London, pp 24–36

Reilly J (1999) Climate change and agriculture: the state of scientific knowledge. Clim Change 43(3):650–659

Reilly JM (ed) (2002) Agriculture: the potential consequences of climate variability and change. Cambridge University Press, Cambridge

Reilly J, Schimmelpfennig D (1999) Agricultural impact assessment, vulnerability, and the scope for adaptation. Clim Change 43(4):745–788

Reilly J, Schimmelpfennig D (2000) Irreversibility, uncertainty, and learning: portraits of adaptation to long-term climate change. Clim Change 45(1):253–278

Reilly J, Hohmann N, Kane S (1994) Climate change and agricultural trade. Glob Environ Change 4(1):24–36

Reilly J, Tubiello F, McCarl B et al (2003) U.S. agriculture and climate change: new results. Clim Change 57:43 69

Reilly J, Paltsev S, Felzer B et al (2007) Global economic effects of changes in crops, pasture, and forests due to changing climate, carbon dioxide, and ozone. Energy Policy 35:5370–5383

Rosenzweig C, Parry ML (1994) Potential impact of climate change on world food supply. Nature 367:133–138

Schelling TC (1983) Climate change: implications for welfare and policy. In: Changing climate: Report of the Carbon Dioxide Assessment Committee. National Academy Press, Washington DC, pp 449–482

Smit B (ed) (1993) Adaptation to climatic variability and change – occasional paper no.19. University of Guelph, Guelph

Smit B, Burton I, Klein RJT et al (2000) An anatomy of adaptation to climate change and variability. Clim Change 45(1):223–251

Waggoner PE (1983) Agriculture and a climate changed by carbon dioxide. In: Changing climate: Report of the Carbon Dioxide Assessment Committee. National Academy Press, Washington DC, pp 383–418

Wiebe K, Gollehon N (2003) Irrigation economics, global. In: Stewart BA, Howel TA (eds) Encyclopedia of water science. Marcel Dekkar, Basel, pp 459–461

Chapter 26
Climate Change and Adaptation of Wheat Producing Nations: Selected Case Studies from Canada and Australia

Monique Helfrich and Vivek Prasad

Abstract The production of grain is common to many developed countries. This industry therefore provides numerous opportunities to compare the responses of developed countries, in adapting to the impacts of climate change. This chapter focuses on Canada and Australia. These countries are representative of developed countries because of their geographical locations, and wheat is a major industry in each. This chapter identifies (1) research efforts in each country that are focused on developing new adaptive technologies, strategies, and plans; and (2) current practices that encourage the implementation of adaptive strategies. This chapter compares the adaptation research and policy approaches taken by Canada and Australia, and identifies strengths, weaknesses, and best practices that can be shared among developed countries.

Keywords Climate change • Agricultural sector • Wheat producers • Research and policy approaches • Adaptive strategies • Technological adaptation • Institutional adaptation • Best practices • Holistic approach • Integration • Networks

Introduction

Globally, wheat production is second only to corn (USDA 2003). Wheat covers more of the earth's surface than any other food crop (IPNI n.d.). There is wide diversity among agricultural sectors throughout developed countries, ranging from

M. Helfrich (✉)
School of Public Policy, George Mason University, 1600 N Oak St, Unit 1830, Arlington, VA 22209, USA
e-mail: mhelfric@gmu.edu

V. Prasad
Department of Environmental Science and Public Policy, George Mason University, Fairfax, VA, USA
e-mail: vprasad1@gmu.edu

J.D. Ford and L. Berrang-Ford (eds.), *Climate Change Adaptation in Developed Nations: From Theory to Practice*, Advances in Global Change Research 42, DOI 10.1007/978-94-007-0567-8_26, © Springer Science+Business Media B.V. 2011

agricultural production on a large-scale to small-scale operations focused on niche markets. Grain is produced in many developed countries, including major producers like the United States, France, Canada, Germany, Australia, and the United Kingdom (IPNI n.d.). This global cultivation is partly due to wheat's tolerance to a wide range of moisture, temperature, and soil conditions (Atwell 2001). For these reasons, wheat production provides an opportunity to compare the responses of developed countries in adapting to the impacts of climate change.

Climate change could have many implications for agriculture. First, climatic zones could undergo latitudinal shifts to the poles, such that temperate climates conducive to agriculture are shifted north (UNFCCC 2002). Second, higher temperatures and changing precipitation patterns, including increased variability of precipitation, could alter food production patterns. Third, increased concentrations of atmospheric CO_2 could lead to increased agricultural productivity in some zones (UNFCCC 2002). Fourth, climate change could increase insect pest, weed, and disease pressure through higher temperatures, shifts in rainfall patterns and cloudiness, increased prevalence of extreme events, and elevated CO_2 levels (Burdon et al. 2006; Ziska and George 2004). Therefore, productivity could be affected indirectly. All aspects of wheat production, including planting, fertilizing, irrigating, and harvesting are susceptible to these impacts of climate change, suggesting that wheat yields could change significantly (Atwell 2001).

This chapter focuses on Canada and Australia. These countries are representative of developed wheat producers because of their geographical locations, and wheat is a major industry in each. In addition, each country spans a large latitudinal range, and, therefore, their wheat production will likely be significantly affected by climate change. This chapter briefly summarizes each country's current approach to adaptation in its wheat production sector. This chapter also identifies (1) research efforts focused on developing new adaptive technologies, strategies, and plans; and (2) current practices that encourage the implementation of adaptive strategies. This chapter compares the research and policy approaches to adaptation taken by Canada and Australia, and identifies strengths, weaknesses, and best practices that can be shared among developed countries.

Canada

Wheat production in Canada dates to the 1800s, and today, wheat is a major crop grown for both domestic consumption and export. As shown in Fig. 26.1 below (USDA 2004), production of wheat is primarily based in the western Prairie region of Canada (including Saskatchewan, Alberta, and Manitoba), although there is a smaller production capability in Ontario and Quebec.

There are many varieties of wheat grown throughout developed countries. Spring wheat comprises the majority (approximately 75%) of wheat produced in Canada, with durum wheat accounting for much of the rest of Canadian production (approximately 20% of the remainder). Current production of durum wheat is

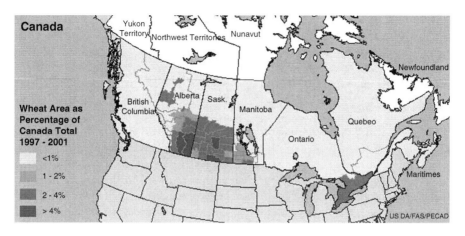

Fig. 26.1 Canadian wheat-growing areas

limited to 5% of total production due to the harsh winter climate of Canada's agricultural regions. The Prairie region is the major producer of spring wheat, while southwestern Ontario is the major producer of winter wheat. Both industry and government have invested heavily in the development of varieties of wheat grown throughout Canada (AAFCPMC 2005).

Total annual production can be significantly reduced by weather conditions, including extreme heat waves, late frosts, and, in particular, drought. Pests, weeds, and diseases can also lead to significant reductions in production (AAFCPMC 2005). Canadian agriculture has already begun to experience losses due in part to increasing variability in climate and weather conditions. According to Statistics Canada, droughts in 2001 and 2002 cost the industry $6.14 billion, while droughts in 2003 contributed to a net farm income of negative $13 million. Within individual provinces, crop insurance claims have been significant. In Ontario, insurance payments totalled approximately $1 billion from 1966 to 2000; yet, they increased by about $640 million between 2000 and 2004. These economic effects have contributed to an increased interest in proactively addressing the negative impacts associated with climate and weathers conditions (C-CIARN Agriculture 2004). It should be noted, however, that current projections of climate change impacts predict that agricultural countries in high latitudes in the northern hemisphere, such as Canada, may benefit from the effects of climate change. Canada may be able to take advantage of these changes by shifting its grain industry into the northern regions of the Prairies that were previously unsuited for production (Samson 2006).

The adaptation of Canadian agriculture, and its wheat industry in particular, are well documented in the literature. For example, Natural Resources Canada provides a comprehensive summary of current policies and their implementation from a national perspective (2007a, b, c, d, e). In this report, the authors indicate that "increases in water scarcity represent the most serious climate risk," particularly in the Prairies, where "future projections include lower summer stream flows, falling lake levels, retreating glaciers, and increasing soil- and surface-water deficits" (NRC 2007d). The authors also note that "[h]igher ... crop productivity from increased heat and atmospheric [CO_2] could be limited by available soil moisture, and dry soil is more susceptible to degradation" (NRC 2007d). Other impacts identified by the authors include the loss of "some advantages of a cold winter" such as natural limitations on pests and diseases, as well as increased frequency of flooding and drought (NRC 2007d).

The authors acknowledge that regions in Canada such as the Prairies have already exhibited considerable flexibility in responding to both drought and flooding, particularly with localized innovations, as they have developed their agricultural industry (NRC 2007e). The authors also note, however, that "[a]lthough the adaptive capacity of agriculture producers appears relatively high ... coping thresholds will be exceeded by departures from normal conditions that are outside the historical experience" and that "[g]lobalization is shifting the responsibility for adaptation to agri-business, national policy makers and the international level" (NRC 2007e). Therefore, they assert that "[p]lanning for changing future environmental conditions is a relatively new policy and management paradigm" (NRC 2007e) which will require more formal and structured approaches, including changes to practices, policies, and infrastructure, that will focus on strengthening the capability to respond to climate change and change it from a reactive mode to a proactive one (NRC 2007e).

In addition, the authors note that at the present time, adaptive capacity of the agricultural industry is not well understood, nor is it understood how and where adaptation will be used to address the potential impacts of climate change. As part of these findings, the authors identify the need for research to develop an improved understanding of adaptation and the need for institutions, both public and private, to implement policies to encourage the "mobilization" of existing and future adaptive capacity (NRC 2007d).

In response to this recognized vulnerability, the goal of the Climate Change Impacts & Adaptation Division (CCIAD), which is part of Natural Resources Canada, is to increase "Canada's resilience to climate change by raising awareness and understanding, as well as by facilitating climate change adaptation decision-making across the country" (NRC 2007a). In 2004, CCIAD produced a report that summarized the Canadian perspective on climate change impacts and adaptation and acknowledged the fact that adaptation will be influenced by policy and market conditions, as well as climate change (NRC 2007b). The authors of the report identified four categories of adaptation options that can be implemented by a variety of actors, both public and private, and which should be assessed for inclusion in an overall strategy: technological developments; government programs and insurance; farm production practices; and farm financial management (NRC 2007b). Given the fact that various actors often have competing priorities, the authors stress that it will be necessary to assess proposed adaptive strategies with respect to their appropriateness, as well as their "effectiveness, economic feasibility, flexibility, and institutional compatibility" (NRC 2007b), and to ensure that they are compatible with the overall strategy of addressing climate change impacts on agriculture.

In addition to highlighting the importance of conducting research on the potential impacts of climate change on agriculture, the authors stress the importance of conducting research to fill knowledge gaps related to: (1) understanding the processes of adaptation; (2) identifying thresholds beyond which adaptive strategies will no longer be viable; (3) identifying barriers that exist in the agricultural sector to implementing adaptive strategies; and (4) developing approaches to mitigate these barriers (NRC 2007c).

Australia

Australia also has a long history of wheat production. Today, Australia is one of the major producers and exporters of wheat in the world. Australia's major markets are the North African countries, China, and West Asia (Koo and Taylor 2008). Winter wheat (similar to Hard Red Spring wheat in terms of quality and characteristics) is the major wheat crop grown in Australia. Wheat production is concentrated in the eastern Australian states of New South Wales, Victoria, and Western Australia (see Fig. 26.2 below from ABARE 2008, p. 4). The annual average Australian

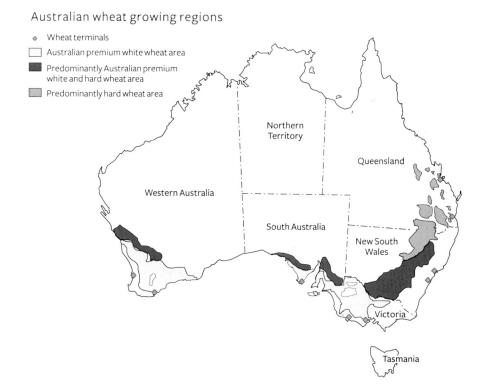

Fig. 26.2 Australian wheat type and growing areas

wheat production was 19.3 million tons from 2003 to 2007. However, in 2006 and 2007, the Australia production rate was lower – 10.5 and 13.0 million metric tons, respectively (Koo and Taylor 2008).

In the last 50 years, there has been a steady increase in the mean annual temperature in Australia, with 2005 being the warmest year in recent recorded history. During 2007 and 2008, rainfall during the winter wheat growing period was below the long-term average (ABARE 2008). This drop in rainfall was reflected in the lower production rate seen in 2007. Abrupt changes – such as the relatively recent drop in rainfall in southwest Western Australia, rise of temperature above natural variability, and unpredictability of storm patterns – indicates that climate change is having a measurable impact on Australia's agriculture (Samson 2006).

Crop models and empirical studies predict that increases in CO_2 concentrations can have a positive impact on wheat yields, but these positive effects can be offset by decreased precipitation and increased temperatures. Specifically, in the short term, climate change might benefit Australian wheat production due to increased temperatures, but in the long term, production could suffer more serious impacts than other major wheat-producing regions of the world because of decreased precipitation (Clark et al. 2006). Changes in production will be dependent on the interactions between temperature and precipitation, as well as the ability to counteract the negative effects of increases in pests, diseases, and weeds (Preston and Jones 2006; Australian Natural Resource Management Ministerial Council 2006). At the present time, changes in CO_2 concentrations and precipitation patterns have already led to significant reductions in wheat production at both the regional and national levels (Howden and Jones 2004).

According to Howden and Jones (2001), pragmatic adaptation approaches on the national level could be "worth $100M annually and also decrease the likelihood of production falling below current levels. Additional adaptations through development of residual management, seasonal climate forecasting and improved varieties could further improve these responses." Recognizing the importance of wheat farming to the national economy and international trade, the Australian government has taken a variety of proactive actions. These actions include policy formulation, research and development, creation and dissemination of knowledge and technology, and related decision-making support. In addition, there have been a number of regional and local research initiatives that have focused on climate change impacts and the development of adaptive capacity. The following discussion is an overview of the national initiative and its relationship to representative examples of regional and local initiatives.

The Australian government is currently implementing a $130 million Australia's Farming Future (AFF) initiative. Although this program is focused on the agricultural in general, the wheat industry is an integral part of this effort. The AFF initiative targets the following elements:

(a) Research, development, and demonstration (RDD) to develop commercially viable tools and management techniques for farmers, foresters, and fishers to manage emissions and adapt to climate change. The Expression of Interest

(EOI) is a major well-documented project that promotes small-scale research, which is funded under the RDD. The EOI enables farmers to be actively engaged in identifying research that is applicable to their needs.

(b) Information services to encourage producers to seek information and training on climate change and to encourage behavioral change.

(c) Targeted training activities to improve producers' capacity to understand the commercial implications of measures to address climate change and to increase their self-reliance and preparedness to adapt to climate change.

(d) Community capacity building activities to involve indigenous people, women, and young adults (18–25 years of age) in primary industries to help these industries adapt to climate change.

(e) Adjustment assistance, including professional advice and training, re-establishment assistance, and transitional income support (DAFF n.d.).

Additionally, the AFF includes research programs and institutes, such as The National Climate Change Adaptation Framework. Contributors to this Framework include: the Griffith Research Facility; The Commonwealth Scientific and Industrial Research Organisation (CSIRO) National Adaptation Flagship; Bureau of Meteorology/CSIRO Centre for Australian Weather and Climate Research; The Australian Government's Caring for Our Country initiative; and The Australian Government's Water for the Future initiative (DAFF n.d.). These programs are focused on both the local and national levels. More importantly, most of the research and development efforts are collaborative and recognize regional and local variations.

Managing Climate Variability (MCV) is another example of these research programs. This program is a partnership of multiple agencies and organizations that focus on agricultural issues. Based on research and field trials, they have developed Climate Management Tools (CMT), which is a software package that can be used by wheat farmers in their decision-making (MCVRDP n.d.). The CMT attempts to incorporate behavior of these farmers into its decision-making algorithm, as well as the physical aspects of different climatic scenarios and their impacts on crop yields. It provides farmers with a tool to assist in quick responses to actual weather conditions. Other resources include: (1) Yield Prophet®, another online model that estimates crop yields under different climate scenarios; (2) the booklet *Masters of the Climate Revisited: Innovative Farmers Coming Through Drought*, which provides insights from farmers who are successfully managing climate variability and profitability (MCVRDP n.d.); and (3) routine yield forecasts from the Australian Bureau of Agricultural and Resource Economics, based on adaptations of a simple water balance model (Potgieter et al. 2006).

Wheat simulation modeling is another research approach being carried out in different parts of wheat-growing areas. There are two basic approaches to wheat crop simulation. First, the regional sensitivity approach examines the sensitivity of a wheat-growing enterprise to a range of temperature, rainfall, and CO_2 changes. This approach uses the APSIM-N (Agricultural Production Systems Simulator) wheat cropping systems modeling framework. Second, the effectiveness of on-farm adaptation approach make a series of distinct climate projections ("low," "medium," and "high") which are then linked to the APSIM modeling framework to assess the likely impacts of climate change on a range of crops and the effectiveness of adaptation options. These approaches acknowledge the uncertainty associated with regional projections of climate change as well as the limited scope for validating the effectiveness of adaptation (Crimp et al. 2007).

Adaptation policies and related research and development have started to produce significant benefits. For example, according to Crimp et al. (2008), changing the strategy for wheat farming to take advantage of temperature increases and frost-free periods resulted in yield increase of up to 36% across ten sites, when compared with simulated yields with no change in current practices. Specifically, by switching from faster maturing to slower maturing varieties of wheat, farmers were able to take advantage of increased temperature and rainfall.

In Western Australia, the wheat production zone is affected by irregularities in rainfall. However, wheat yield has increased during recent periods of decreasing rainfall. This happened because of changes in technology and management that had been adopted. One example of management change was the adoption of no-till agriculture, which produces the same or increased yields under conditions of decreasing rainfall (Samson 2006).

Best Practices and Recommendations

Canada and Australia represent a major share of the wheat production industries in developed countries. They also represent the two ends of the spectrum in terms of both impact of climate change on developed countries and policy approaches to these impacts. Therefore, they each provide best practices and lessons learned that can be used by other developed countries that are also large-scale producers of wheat and other agricultural products for the global market.

The examples of the Canadian approach to climate change discussed in this chapter demonstrate the importance of an overarching national strategy, while the Australian examples reflect the importance of localized initiatives. Based on these examples, it appears that an effective approach to adaptation for large-scale industries that are national rather than local in scope requires a top-down approach to policy formulation and implementation that effectively incorporates input and needs from all relevant stakeholders. One of the most important stakeholders are the wheat farmers who are ultimately responsible for implementing the solutions and will either benefit or suffer from their impacts. As seen in the Australian case studies, the overall adaptive strategy needs to be holistic and deal with the wheat industry's response to multiple drivers of change, including climate change, from a countrywide perspective, which then flows down to local regions.

This holistic approach to policy should also be reflected in research and development to support adaptation. Canadian Climate Impacts and Adaptation Research Network (C-CIARN) Agriculture, which was created in 2001, was an example of such an approach, as it "worked toward the goal of building a network of representatives from industry, research, and policy communities to promote and facilitate research on climate impacts and adaptation for Canadian agriculture." Unfortunately, this organization was closed in 2007 (C-CIARN Agriculture n.d.); however, it does provide an example of a national approach to integrating research and development efforts. In contrast, although Australia is the source of many individual research initiatives that could be potentially applied to other developed countries, it is clear that Australia's climate research community must continue to move towards more integrative, collaborative approaches and away from the somewhat seemingly fragmented and competitive mode that has sometimes typified climate research in the past (Samson 2006).

However, the Australian examples do provide lessons and practices on engaging stakeholders in the development of policy and technology to respond to the impacts of climate change. For example, the Australian EOI initiative enables the wheat farmers, who are the end users of technological solutions that are developed by researchers, to be actively engaged in the process of identifying research needs, as well as to provide input on the feasibility of proposed technological solutions. In other words, the EOI initiative promotes the concept of developing "place-based" adaptation strategies that are responsive to local needs and practices and can be implemented by local wheat farmers.

Another lesson that can be learned from these examples is the importance of improving education and increasing awareness of potential future impacts and responses, in order to improve the adaptive capacity of the wheat industry. Such an emphasis on education can be used to reinforce and expand the existing social capital for adaptation already created by rural networks at local levels that have enabled farmers to adapt to past changes in the climate and ecology (NRC 2007e).

References

Agriculture and Agri-Food Canada Pest Management Centre (2005) Crop profile for wheat in Canada. Agriculture and Agri-Food Canada. Available via the Depository Services Program E-collection.   http://dsp-psd.pwgsc.gc.ca/collection_2009/agr/A118-10-16-2005E.pdf. Cited 30 Oct 2008

Atwell WA (2001) An overview of wheat development, cultivation, and production. Cereal Foods World 46(2):59–62

Australian Bureau of Agricultural and Resource Economics [ABARE] (2008) ABARE. Australian crop report no.147. http://www.abareconomics.com/publications_html/cr/cr_08/cr08_Sept.pdf. Cited 30 Oct 2008

Australian Natural Resource Management Ministerial Council (2006) National agriculture & climate change action plan 2006–2009. Australian Government, Department of Agriculture, Fisheries and Forestry. http://www.daff.gov.au/__data/assets/pdf_file/0006/33981/nat_ag_clim_chang_action_plan2006.pdf. Cited 30 Oct 2008

Burdon JJ, Thrall PH, Ericson L (2006) The current and future dynamics of disease in plant communities. Annu Rev Phytopathol 44:19–39

C-CIARN Agriculture (2004) Climate change adaptation: a producer perspective. University of Guelph, Guelph. http://www.c-ciarn.uoguelph.ca/documents/Meeting_2004.pdf. Cited 30 Oct 2008

Canadian Climate Impacts and Adaptation Research Network for Agriculture [C-CIARN Agriculture] (n.d.) Home. http://www.c-ciarn.uoguelph.ca/index.html. Cited 30 Oct 2008

Clark A, Barratt D, Munro B et al (2006) Science for decision makers: climate change – adaptation in agriculture. Australian Government, Department of Agriculture, Fisheries and Forestry. http://affashop.gov.au/PdfFiles/sfdm_climatechange10.pdf. Cited 30 Oct 2008

Crimp S, Gaydon D, DeVoil P et al (2007). On-farm management in a changing climate: a participatory approach to adaptation. Birchip Cropping Group. http://www.bcg.org.au/cb_pages/images/Adapting%20to%20Climate%20Change%20article.pdf. Cited 30 Oct 2008

Crimp S, Howden M, Power B et al (2008) Global climate change impacts on Australia's wheat crops. Garnaut Climate Change Review. http://www.garnautreview.org.au/CA25734E0016A131/WebObj/01-BWheat/$File/01-B%20Wheat.pdf. Cited 30 Oct 2008

Department of Agriculture, Fisheries and Forestry [DAFF] (n.d.) Climate change research program: research, development and demonstration – application guidelines. Australian Government. Available via the Murrumbidgee Landcare Incorporated website. http://www.murrumbidgeelandcare.asn.au/files/rdd-guidelines.pdf. Cited 30 Oct 2008

Howden M, Jones R (2001) Costs and benefits of CO_2 increase and climate change on the Australian wheat industry. Australian Greenhouse Office. Available via the Australian Government's Department of Climate Change and Energy Efficiency website. http://www.climatechange.gov.au/science/wheat/index.html. Cited 30 Oct 2008

Howden M, Jones RN (2004) Risk assessment of climate change impacts on Australia's wheat industry. 4th international crop science congress, Brisbane, Sep-Oct 2004. http://www.cropscience.org.au/icsc2004/symposia/6/2/1848_howdensm.htm. Cited 30 Oct 2008

International Plant Nutrition Institute [IPNI] (n.d.) Wheat types and allocations in the world. http://www.ipni.net/ppiweb/nchina.nsf/$webindex/B8EAD8A9458117EC482573A80031C330. Cited 30 Oct 2008

Koo WW, Taylor RD (2008) Agribusiness & applied economics report no.628: 2008 outlook of the U.S. and world wheat industries, 2007–2017. Center for Agricultural Policy and Trade Studies, North Dakota State University, Fargo. http://www.ext.nodak.edu/~aedept/aemisc/AAE628.pdf. Cited 30 Oct 2008

Managing Climate Variability Research and Development Program [MCVRDP] (n.d.) Managing climate change in Australia: farmer's stories. http://www.managingclimate.gov.au/library/scripts/objectifyMedia.aspx?file=pdf/89/35.pdf. Cited 30 Oct 2008

NRC (2007a) Climate change impacts and adaptation division. http://www.adaptation.nrcan.gc.ca/index_e.php. Cited 30 Oct 2008

NRC (2007b) Climate change impacts and adaptation: a Canadian perspective – agricultural adaptation to climate change. http://www.adaptation.nrcan.gc.ca/perspective/agri_4_e.php. Cited 30 Oct 2008

NRC (2007c) Climate change impacts and adaptation: a Canadian perspective – knowledge gaps and research needs. http://adaptation.nrcan.gc.ca/perspective/agri_5_e.php. Cited 30 Oct 2008

NRC (2007d) From impacts to adaptation: Canada in a changing climate 2007, chapter 7. http://adaptation.rncan.gc.ca/assess/2007/ch7/index_e.php. Cited 30 Oct 2008

NRC (2007e) From impacts to adaptation: Canada in a changing climate 2007, chapter 7 – 5. adaptation and adaptive capacity. http://adaptation.rncan.gc.ca/assess/2007/ch7/5_e.php. Cited 30 Oct 2008

Potgieter A, Hammer GL, Doherty A (2006) Oz-wheat – a regional-scale crop yield simulation model for Australian wheat. Queensland Department of Primary Industries and Fisheries. www2.dpi.qld.gov.au/extra/pdf/climate/Oz-Wheat/Oz-WheatReport-Abstract&Intro.pdf. Cited 30 Oct 2008

Preston BL, Jones RN (2006) Climate change impacts on Australia and the benefits of early action to reduce global greenhouse gas emissions. Commonwealth Scientific and Industrial Research Organization [CSIRO]. http://www.csiro.au/files/files/p6fy.pdf. Cited 30 Oct 2008

Samson C (2006) Farming profitably in a changing climate: a risk management approach bureau of rural sciences. Australian Bureau of Agricultural and Resource

Economics   [ABARE].   http://www.abareconomics.com/interactive/outlook06/outlook/ speeches/papers/Samson,C-ClimateVariation.doc. Cited 30 Oct 2008

United Nations Framework Convention on Climate Change [UNFCCC] (2002) Climate change information kit. http://unfccc.int/essential_background/background_publications_htmlpdf/climate_change_information_kit/items/305.php. Cited 30 Oct 2008

United States Department of Agriculture [USDA] (2003) Crop production tables. http://www.fas.usda.gov/wap/circular/2003/03-02/tables.html. Cited 30 Oct 2008

USDA (2004) Canada: wheat. http://www.fas.usda.gov/remote/Canada/can_wha.htm. Cited 30 Oct 2008

Ziska LH, George K (2004) Rising carbon dioxide and invasive, noxious plants: potential threats and consequences. World Res Rev 16(4):427–44

Chapter 27
Use of Modern Portfolio Theory to Evaluate Diversification of Agricultural Land Use as an Adaptation to Climate Risks in the Tisza River Basin

Saskia E. Werners, Éva Erdélyi, and Iwan Supit

Abstract Adaptation is gaining attention as an inevitable answer to the challenges posed by climate change. The increasingly uncertain climatic conditions to which actors are exposed are becoming a constraint for their well-being. This chapter looks at diversification of agricultural land use as a key factor in reducing risk and as a means of coping with an uncertain climate. Borrowing from economic theory, this chapter illustrates how cropping patterns influence the expected revenue and risk. The standard deviation of the land use revenue is used as a proxy for climate risk.

Agricultural land use is associated with two competing land use and water management strategies in the Hungarian Tisza River Basin: intensive agriculture protected by flood levees, and water retention areas with extensive cattle breeding and orchards. To cope with flood risk, the Hungarian government supports water retention and land use change to replace or complement the prevailing intensive agriculture dependent on flood levees and drainage. Our analysis shows that revenues from agriculture are well adjusted to the current climate variability. Considering recent revenues, a shift from intensive agriculture to extensive cattle breeding and orchards increases both the expected revenue from agriculture and the risk.

Keywords Diversification • Risk • Climate impacts • Adaptation • Crop revenue • Portfolio theory • Agriculture • Hungary • Climate change • Cropping patterns • Tisza River Basin • Flood risk

S.E. Werners (✉) • I. Supit
Wageningen University and Research Centre, Wageningen, The Netherlands
e-mail: werners@mungo.nl; iwan.supit@wur.nl

É. Erdélyi
Corvinus University of Budapest, Budapest, Hungary
e-mail: eva.erdelyi@uni-corvinus.hu

J.D. Ford and L. Berrang-Ford (eds.), *Climate Change Adaptation in Developed Nations:* *From Theory to Practice*, Advances in Global Change Research 42,
DOI 10.1007/978-94-007-0567-8_27, © Springer Science+Business Media B.V. 2011

Introduction

Future climate change adds to the uncertainties that resource management has to cope with. This research enhances our understanding of how to adapt land use and water management to climate change. More specifically, it shows how diversification of agricultural land use combines adaptation to climate-related risks with a financially attractive agro-economy. Agriculture is explored as a large land user that is at risk from climate change. While land use change is perceived as an effective adaptation strategy to cope with climate-related risks, it is not practiced yet on a large scale (Footitt and McKenzie Hedger 2007).

Agricultural and ecological research has shown that diverse systems are more robust and better able to cope with future risk. For example, research on diversified cropping systems has illustrated the importance of diversity for reducing crop failure (Vandermeer 1989; Altieri 1994). It has been shown that minimizing risk while investing in crop production can be achieved by developing a portfolio of crop types that have low covariance with respect to the risk to which they are subjected. For example, Tonhasca Jr and Byrne (1994) investigated the effect of crop diversification on mitigating pests. Others have assessed how the diversity of farming systems and agricultural landscape structures influences the vulnerability of yields and biodiversity (Sendzimir and Flachner 2007; Reidsma and Ewert 2008). Furthermore, crop diversity has been used in a range of African settings as an indicator of both ecosystem resilience and a strategy for food security (Unruh 1994; Blocka and Webb 2001). Fraser et al. (2005) elaborate on the importance of portfolio management and diversification for reducing vulnerability in agro-environmental systems.

In the area of financial services, planning, and investing under uncertainty is common. An approach that is widely used to determine investment strategies under uncertainty is Modern Portfolio Theory (MPT) (Markowitz 1952). It shows how different investments can be combined in a portfolio with a lower risk than that of the individual investments. Although diversification is commonly studied in agricultural and economic research to meet demand fluctuations (Isik and Devadoss 2006), fewer attempts are made to quantify the benefits of diversification in relation to coping with climate change (Werners et al. 2007). Diversification is fundamentally different from crop rotation or substitution in that it seeks to combine crops within a growing season.

This chapter uses MPT to evaluate diversification of agricultural land use as an adaptation to climate risk in the Hungarian Tisza River Basin. The main climate-related risk in the Tisza is the frequent occurrence of floods and droughts (Láng 2006).

Two agricultural production systems have dominated land use in the Tisza River Basin. Until the eighteenth century, socioeconomic activities were mainly organized around the operation of a system of creeks and channels regulating the water flow between the main river bed and the floodplain (Balogh 2002). The inundation frequency determined land use. Mosaic floodplain production systems combined plough land, forest, floodplain orchards, meadows, fishing, and cattle (Andrásfalvy

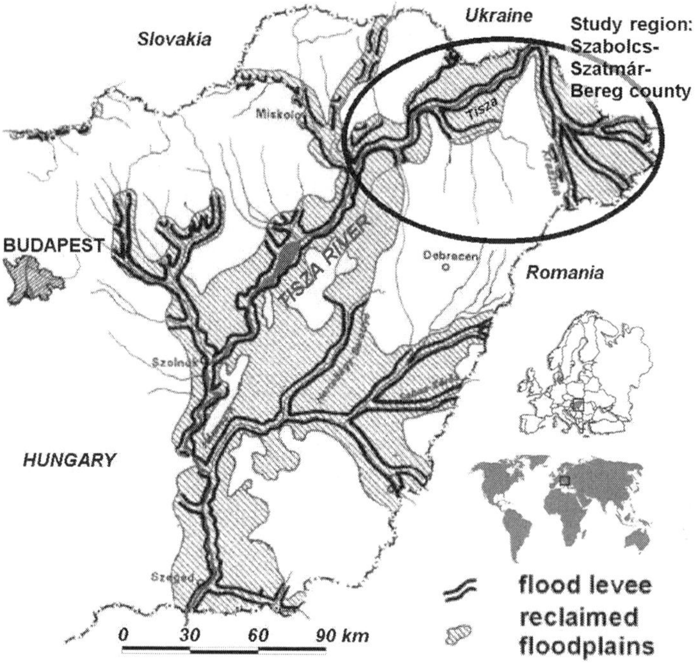

Fig. 27.1 Tisza River Basin (Tóth and Nágy 2006)

1973; Bellon 2004). Since then, agricultural production has shifted to tillage. To cater to large-scale intensive agriculture and river transport, the river was canalized and straightened and the floodplains drained (Fig. 27.1).

Following major floods in 1998 and 2000, a new water management plan was issued to cope with flood risks. The plan marked a substantial shift in land use and water management, as it recognized rural development and nature conservation as important objectives along with flood protection (VITUKI 2004). With this plan, the Hungarian government supported water retention and land use change to replace or complement the prevailing intensive agriculture dependent on flood levees and drainage (Werners et al. 2009). It is against this background that we analyze the diversification of crops and agricultural production systems. This chapter has three main components. Firstly, the methods are introduced to assess diversification as an adaptation strategy to reduce climate-related risks. Secondly, the results are discussed for selected land use patterns together with possible extensions of the approach. Thirdly, the conclusions are presented about the use of portfolio theory to assess revenue and risk of individual crops and agricultural production systems in the Hungarian Tisza River Basin. By focusing on climate risk, this chapter offers decision-makers a tool to take climate adaptation into account along with the many non-climatic factors that influence decisions on land use and water management.

Methods

Selection of Agricultural Land Use, Yield Data, and Calculation of Revenue

In response to climate change, the agricultural sector has to cope with (1) the impact of climate change and variability on crops, and (2) the shift in water management from flood levees to retention areas. For the purpose of this analysis, we consider two types of agricultural production systems. Systems that rely on flood levees are referred to as "intensive agriculture." Those relying on retention areas are referred to as "floodplain agriculture." The main difference between these agricultural systems is their potential to cope with temporal water cover. On one hand, intensively produced crops are sensitive to water cover. On the other, extensive cattle breeding and floodplain orchards can withstand water cover for a certain amount of time, and even benefit from annual flooding. This distinction does not mean to imply that these agricultural systems exist only under dry or wet conditions, respectively. Some intensive agriculture is currently practiced at locations suffering from water-logging, and floodplain agriculture exists outside the active floodplain in areas behind the flood levees. Yet, distinguishing between the two agricultural systems facilitates the analysis and interpretation of the results.

Annual crop yield data at the county level were used from the period 1998–2007. Criteria for crop selection include predominance in the chosen region and on data availability. The crops considered in this analysis are summarized in Table 27.1 (note that although cattle is included, we refer to "crops" throughout). The study region is the Szabolcs-Szatmár-Bereg county (see Fig. 27.1). The data were obtained from the Hungarian Central Statistical Office (available from http://portal.ksh.hu). We use a 10-year time slice because the results of the years before bear a strong imprint of the agricultural reforms after the fall of the communist regime (Erdélyi 2008). In addition, the data covers a period with both floods and droughts in the Tisza region.

To compare crops, gross crop revenue (hereafter, "revenue") was calculated by multiplying annual crop yield by the annual procurement price of the crop. For reference, Table 27.1 lists average procurement prices. By excluding variable production costs, opportunity costs, and subsidies, gross revenue allows the analysis to focus on climate impacts on crop yield and to therefore inform stakeholders about this particular aspect of agricultural decision-making.

Assessing Revenue, Correlation, and Risks Associated with Land Use Strategies

The assessment builds on MPT to identify land use and water management strategies with low climate-related risks. MPT quantifies how different activities can be

Table 27.1 Characteristics of the assessed agricultural land use

	Land use	Characteristics	Average yield [kg ha^{-1}]^a	Average fraction of total area [ha]/revenue [%]	Average price^b [€ kg^{-1}]
Intensive agriculture	Wheat	Well adapted to different soil types and the regional climate. Predominantly winter wheat. Grown rain fed. The growth phases most sensitive to drought are shooting, flowering, caryopsis filling. Sensitive to water-logging.	3,700	23/20	0.11
	Maize	Water is important limiting factor. In Tisza region grown rain fed.	5,000	20/20	0.10
	Sunflower	Heat tolerant without extra maintenance. Rain fed.	2,000	10/10	0.23
Floodplain agriculture	Fruit	Fruit orchards. Fruit trees can withstand temporary inundation depending on time of the year and fruit species. For example, nut trees benefit from shallow flooding and high soil moisture.	5,800^c	5/3	0.14^d
	Cattle	Extensive cattle breeding mostly for meat production. Traditionally, cattle were held on land subject to water logging or temporary flooding.	520^e	28/25	1.03

^{a}Ten-year average (1998–2007) of the annual yield
^{b}Average procurement price (1998–2007)
^{c}Average total fruit production (1998–2007)
^{d}Weighted average of procurement price of apples, pears, sour cherries, plums, apricots, peaches
^{e}Average of annual number of cattle per hectare grassland multiplied by cattle weight [kg/ha]

combined into a portfolio that has a lower risk than the activities individually. MPT can be applied when four conditions are met: (1) there is more than one possible investment; (2) these investments are subject to risk; (3) there is information about the historical and/or expected revenue of the investments; and (4) the same external conditions do not affect all investments equally (Elton and Gruber 1995; Fraser et al. 2005). These conditions apply to crop selection.

Following MPT (Harvey 1995), investments (in this case, different agricultural land uses) are characterized by mean revenue and risk. They are combined into cropping patterns. Risk is calculated as the standard deviation of revenue over a 10-year period (1998–2007). It is assumed that farmers aim to maximize land use revenue and minimize the probability of low revenue (risk). They prefer land uses with higher expected revenues and/or lower risks. The expected mean revenue of a cropping pattern, $\overline{R}_{cropping\_pattern}$, with n different crops each grown at a share X_i (as a fraction of the total area or the total revenue) is estimated:

$$\overline{R}_{cropping\_pattern} = \sum_{i=1}^{n} \overline{R}_i * X_i \qquad (27.1)$$

Where, $\overline{R}_i$ is the average expected revenue per hectare over the period 1998–2007 for a particular crop or land use i.

The risk $SD_{cropping\_pattern}$ of a cropping pattern is represented by the standard deviation SD_i of its revenues, which can be estimated with the formula:

$$SD_{cropping\_pattern} = \sqrt{\sum_{i=1}^{n} X_i^2 SD_i^2 + 2\sum_{i=1}^{n}\sum_{j=i+1}^{n} X_i X_j \rho_{ij} SD_i SD_j} \qquad (27.2)$$

With X_i being the share of the individual crops in the cropping pattern; SD_i the standard deviation of the revenue of land use i over the assessment period; and ρ_{ij} the correlation of the revenue of the crops i and j over the assessment period.

A negative correlation ρ_{ij} between two crops i and j indicates that when the revenue of crop i turns out to be above its expected value, then the revenue of crop j is likely to be below its expected value, and vice versa. A positive correlation suggests that when i's revenue is above (below) its expected value, then j's will also be above (below) its expected value.

Equation 27.2 indicates that the standard deviation (risk) of a cropping pattern is less than the weighted sum of the standard deviations of the individual crops when the correlation between the crops is less than one. In other words, diversification is beneficial as long as there is less than perfect positive correlation between crop revenues.

Diversification for two crops or land use patterns is illustrated in Fig. 27.2. The figure plots the expected revenue and standard risk of a hypothetical cropping pattern. The curved lines were obtained by changing the crop share in the cropping

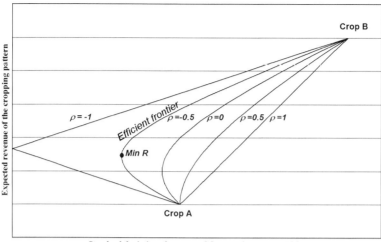

Fig. 27.2 Expected revenue of cropping patterns of two crops for different correlations between the crops revenues

pattern from growing only A (point A) to only B (point B). The different curves correspond to different values for the correlations ρ between the two crops.

The lower the correlation between the two crops, the greater the bend of the curve, indicating that the same revenue can be earned at lower risk. The point MinR (minimum risk) on each of these curves represents the minimum risk cropping pattern. The share of the crops in the minimum risk cropping pattern is determined by the standard deviation *and* correlation of the crops. The backward bending always occurs if $\rho \leq 0$, but may or may not occur if $\rho > 0$. No farmer wants to grow crops with expected revenue below the minimum risk cropping pattern. Therefore, the so-called efficient frontier representing all efficient cropping patterns for a given correlation lies between MinR and B.

Results and Discussion

Yields, Revenue, Risk, and the Diversification of Land Uses

Table 27.2 lists the average crop yield, crop revenue, and risk (standard deviation) in the assessment period (1998–2007). Recall that "revenue" refers to gross revenue. Table 27.3 lists the yield correlations of the different crops in the assessed period.

Crops differ in their development over consecutive growth phases from planting to harvesting. Thus the climate affects crops differently, determining the correlation

Table 27.2 Average crop yield, revenue, and risk in the Szabolcs-Szatmár-Bereg county (1998–2007)

Crop	Yield [kg ha^{-1}]	Risk [kg ha^{-1}]	Revenue [€ ha^{-1}]	Risk [€ ha^{-1}]
Wheat	3,730	690	404	82
Maize	4,966	1,359	500	119
Sunflower	1,839	300	426	81
Fruit	12,527	3,859	1,765	410
Cattle	470	25	504	112

Table 27.3 Correlation of crop revenue in the Szabolcs-Szatmár-Bereg county

Correlation	Wheat	Maize	Sunflower	Fruit	Cattle
Wheat	1	–	–	–	–
Maize	0.92	1	–	–	–
Sunflower	0.76	0.84	1	–	–
Fruit	−0.04	−0.20	−0.60	1	–
Cattle	−0.57	−0.44	0.06	0	1

of crop growths. The low correlation between sunflower and wheat, for example, can be explained by their different growth seasons. Wheat is sown in autumn or early spring, whereas sunflower is a summer crop. Thus, events like a wet period early in spring or a summer drought will affect these crops differently.

A correlation smaller than unity suggests a benefit from diversification, as the risk associated with the crop combinations is less than the weighted sum of the risk of the individual crops (see Eq. 27.2). This is examined in Fig. 27.3 for the main crops used in intensive agriculture (wheat, maize, and sunflower) and in floodplain agriculture (cattle and fruit). Following the example in Fig. 27.2, Fig. 27.3 illustrates the trade-offs between average revenue and risk for these crops. First, average wheat revenue is lower than that of sunflower and its risk higher. Although wheat may seem like a less attractive crop, its medium correlation with sunflower makes it an attractive crop for risk diversification. Second, maize is a high revenue crop grown in intensive agriculture that comes at a high risk. That is, the variation in revenue between consecutive years is larger than for the other two crops. Third, the revenue of sunflower is lower, yet the risk is also lower. Over the last 10 years, wheat, maize, and sunflower were grown on average at a ratio 1:3:1 of the total area for intensive agriculture. The revenues and risks associated with the present crop mix in the study region are shown in Fig. 27.3 and prove to be on the efficient frontier. This suggests some risk reduction is traded in for revenue. The same holds true for floodplain agriculture. Cattle production yields less risks and lower revenues compared to fruit production. Yet, over the last 10 years, cattle and fruit were produced at a ratio 2:1 of the total area for floodplain agriculture.

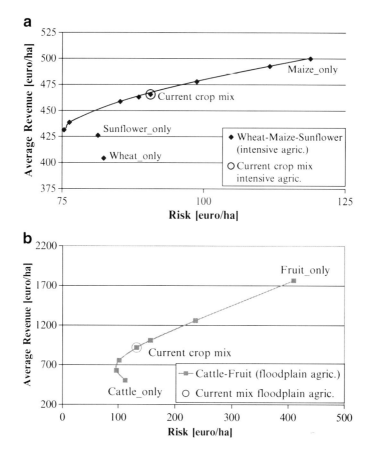

Fig. 27.3 Average revenue and risk of combinations of main crops in intensive agriculture (**a**) and floodplain agriculture (**b**) in the Szabolcs-Szatmár-Bereg county

Diversification of Intensive Agriculture and Floodplain Agriculture

The previous section presented results for crop diversification within the two main agricultural systems: intensive agriculture protected by flood levees, and floodplain agriculture with the possibility of water retention. This section investigates whether a combination of these two strategies reduces risks for the agricultural sector. To this end, the current crop mix in respectively the intensive agriculture and the floodplain agriculture are treated as two investment options in a portfolio. Figure 27.4 illustrates the revenue and risk of a portfolio with a mix of intensive agriculture and floodplain agriculture. It shows that the present mix of these production systems (60% intensive and 40% floodplain agriculture) has the lowest risk, illustrating the benefit from diversification. A lower risk can only be achieved by changing the

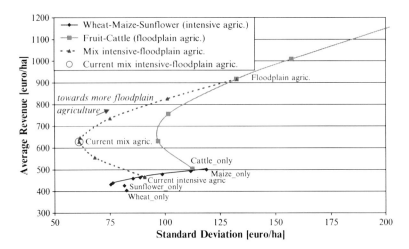

Fig. 27.4 Average revenue and risk of combinations of intensive agriculture and floodplain agriculture in the Szabolcs-Szatmár-Bereg county

fraction of wheat, maize, sunflower, fruit, or cattle within the current production systems or by altering the way these crops are produced. For example, cattle breeding may be made more efficient or irrigation may be considered. The analysis illustrates the benefit of diversification given the existing production systems in the region. It also illustrates how people in the region may perceive a shift from one production system to another – as supported by the Hungarian government's new water management plan – based on their present experience and risk perception.

As summarized in Fig. 27.4, results suggest that shifting from intensive agriculture to floodplain agriculture in the study region will presently not reduce risks in the region, unless the practice within one agricultural production system is changed. At the same time the figure shows that a shift toward more floodplain agriculture can increase the average revenues from agriculture, at the cost of accepting higher risks.

Table 27.4 lists strengths and limitations of the approach presented in this chapter.

The mechanism of diversification can be used to adapt to changes in (the correlation of) crop yields under climate variability and change. The presented approach intends to support decision-makers by estimating the benefits of diversification as a climate adaptation strategy. As a result, climate risks can be included as one component in decision-making about agricultural systems, together with information about subsidies and regulation, for example. The approach allows the inclusion of non-climate factors, like variable production costs, by replacing historic gross revenues with net revenues or (farm) net value added. Likewise, expected net revenues and price speculation can be included, for example, by using results from whole-farm design models that explore the effect of a variety of farmer's decisions and objectives at the farm system level (Sterk et al. 2006). Costs and benefits however must be carefully assigned to specific actor groups. In the Hungarian case,

Table 27.4 Strengths and limitations of the presented approach

Strengths	Limitations
+ Explicitly designed to deal with risk	− Difficult to include social and environmental considerations in revenue
+ Well-documented methods and metrics to calculate risks and revenue along portfolio composition	− Land use and other activities are treated as homogeneous and non-location specific
+ Shows there are limits for reducing risks through diversification	− Assumes that revenues impacted by climate variability and extremes are normally distributed and well-represented by their average value, risk, and covariance
+ Allows combination of data from different sources	
+ Allows comparison between spatial and temporal scales, as well as future scenarios	− Presently focuses on gross revenue and climate risk, ignoring farm-level decision factors such as costs (labor, seed, irrigation, fertilizer, opportunity costs), subsidies, risk perception, personal experience, and preferences
+ Both net revenues and gross revenues can be used in assessment	

the government offers to compensate farmers for land use change, opportunity costs or flood damages.

Another extension of this work is the addition of crops and crop management options that enable adaptation to climate change, such as irrigation techniques and changes in the timing of planting and crop handling. Stakeholders in the region have expressed interest in the benefits of diversification with non-agricultural activities such as tourism or alternative agricultural production systems like vini- and horticulture. Whereas, the analysis in this chapter can be broadened to include other activities and non-climate factors, it ultimately remains up to individual farmers and regional decision-makers to handle trade-offs between risks and revenue.

Conclusion

Whenever people can chose between different activities that are subject to risk, portfolio theory can be used to develop robust sets of activities. These conditions hold for land use planning that aims to cope with climate change, and for land use and water management in particular. The Tisza River Basin example shows that agricultural land uses can be combined to reduce climate-related risks. It shows how MPT can help in determining cropping patterns that generate the highest revenue under an acceptable risk. The standard deviation of revenue is used as a proxy for risk.

Although relatively simple, this example confirms that MPT allows the relationship between the revenue and risk of individual crops and agricultural production systems to be discussed systematically. It also shows the importance of understanding the correlation of the revenues of different crops. Given the land uses in a region and their dependence on climatic conditions, there is an upper limit to diversification beyond which risks cannot be "diversified away." Current land use minimizes regional risk. Land use change toward floodplain agriculture raises the revenue from land use, but also increases the probability of low revenue. The regional water management plan suggests land use change to reduce flood damages locally and downstream. To encourage people to change land use, compensation for additional risks associated with land use change is recommended.

Acknowledgments The work has been supported by a grant from the European Commission through the EU research project ADAM (Project no. 018476-GOCE), the Dutch Science Foundation NWO and the Dutch Ministry of Agriculture, Nature Conservation and Food Quality. We are grateful for the valuable comments and suggestions of two anonymous reviewers and our research partners, especially Peter Balogh, Paulina Hetman, Stefan Hochrainer, Joanne Linnerooth-Bayer, Fulco Ludwig, Jan Sendzimir, and Rik Leemans.

References

Altieri MA (1994) Biodiversity and pest management in agroecosystems. Haworth, New York

Andrásfalvy B (1973) Ancient floodplain and water management at Sarkoz and the surrounding area before the rived regulations. Vízü Tör Füz 6 [in Hungarian]

Balogh P (2002) Basics and method of floodplain management on Middle-Tisza valley. VATI Kht, Budapest

Bellon T (2004) Living together with nature: farming on the river flats in the valley of the Tisza. Acta Ethnogr Hung 49(3):243–256

Blocka S, Webb P (2001) The dynamics of livelihood diversification in post-famine Ethiopia. Food Policy 26:333–350

Elton EJ, Gruber MJ (1995) Modern portfolio theory and investment analysis. Wiley, New York

Erdélyi É (2008) Uncertainty and risk in winter wheat production of Hungary. In: Mihailovic D, Miloradov MV (eds) Environmental, health and humanity issues in the Down Danubian Region, multidisciplinary approaches. Proceedings of the 9th international symposium on interdisciplinary regional research, Novi Sad, 21–22 June 2007. World Scientific, Singapore

Footitt A, McKenzie Hedger M (2007) Climate change and water adaptation issues, EEA technical report no.2/2007. European Environment Agency, Copenhagen

Fraser EDG, Mabee W, Figge F (2005) A framework for assessing the vulnerability of food systems to future shocks. Futures 37(6):465–479

Harvey CR (1995) Optimal portfolio control. www.duke.edu/~charvey/Classes/ba350/control/opc. htm. Cited May 2008

Isik M, Devadoss S (2006) An analysis of the impact of climate change on crop yields and yield variability. Appl Econ 38(7):835–844

Láng I (2006) The project "VAHAVA", executive summary. Hungarian Ministry for the Environment and Water Management [KvVM] and the Hungarian Academy of Sciences [MTA], Budapest

Markowitz H (1952) Portfolio selection. J Finance 17:77–91

Reidsma P, Ewert F (2008) Regional farm diversity can reduce vulnerability of food production to climate change. Ecol Soc 13(1):Article 38

Sendzimir J, Flachner Z (2007) Exploiting ecological disturbance. In: Scherr SJ, McNeely JA (eds) Farming with nature: the science and practice of ecoagriculture. Island, Washington, DC

Sterk B, van Ittersum MK, Leeuwis C et al (2006) Finding niches for whole-farm design models – contradictio in terminis? Agr Syst 87(2):211–228

Tonhasca A Jr, Byrne DN (1994) The effects of crop diversification on herbivorous insects: a meta-analysis approach. Ecol Entomol 19(3):239–244

Tóth S, Nágy L (2006) Dyke failure in Hungary of the past 220 years. In: Marsalek J, Stancalie G, Balint G (eds) Transboundary floods: reducing risks through flood management, Nato Science Series: IV: Earth and Environmental Sciences. Springer, Dordrecht

Unruh J (1994) The dilemma of African agrobiodiversity: Ethiopia and the role of food insecurity in conservation. Indiana University, Bloomington

Vandermeer J (1989) The ecology of intercropping. Cambridge University Press, Cambridge

VITUKI (2004) Rebirth of the River Tisza – The new Vásárhelyi Plan. In: Vásárhelyi Plan Intersectorial Committee Bulletin. National Environment, Conservation and Water Authority; Water Resources Research Institute (VITUKI), Budapest. http://www.vizugy.hu/vtt/altalanos_english.pdf. Cited June 2008

Werners SE, Incerti F, Bindi M et al (2007) Diversification of agricultural crops to adapt to climate change in the Guadiana River Basin. In: Proceedings from the international conference on climate change, Hong Kong, 29–31 May 2007

Werners SE, Matczak P, Flachner Z (2009) The introduction of floodplain rehabilitation and rural development into the water policy for the Tisza River in Hungary. In: Huitema D, Meijerink S (eds) Water policy entrepreneurs: a research companion to water transitions around the globe. Edward Elgar, Camberley

Chapter 28
Government Action to Promote Sustainable Adaptation by the Agriculture and Land Management Sector in England

Nicholas A. Macgregor and Caroline E. Cowan

Abstract Agricultural land covers over 70% of England and provides a wide range of important benefits to the society. These benefits are vulnerable to both the direct and indirect effects of climate change. Successful adaptation by the agriculture and land management sector is therefore vital, and this adaptation must be sustainable.

This chapter introduces the concept of sustainable adaptation and discusses the United Kingdom government's approach to adaptation by the agriculture and land management sectors in England. We explain the importance of agriculture in England and the multiple benefits provided by agricultural systems, and briefly outline the pressures from climate change. We then outline a set of principles and a framework for decision making we have developed to help achieve sustainable adaptation. We describe how this framework is being applied to the development of adaptation policy for agriculture, and summarize some initial adaptation priorities that have been identified.

We discuss how sustainable adaptation by farmers and land managers can be encouraged, focusing on the role government can play, including the likely contribution that existing agricultural policies will make to adaptation. Significant challenges exist, not least uncertainty about the future. Nevertheless, the adaptation measures that we have identified as initial priorities correspond closely to existing agricultural and environmental good practice, providing a clear starting point for action.

Agriculture policy and adaptation policy are largely devolved to the national assemblies of the constituent countries of the United Kingdom. Therefore this chapter will largely focus on England. However, in some cases policy or research (for example, the Climate Change Risk Assessment required by the Climate Change Act) is undertaken on a UK-wide basis, some statistics are collected for the UK as a whole, and many of the issues discussed are applicable across all administrations.

N.A. Macgregor (✉) • C.E. Cowan
Natural England, Hercules House, Hercules Road, London SE1 7DU, UK
e-mail: nicholas.macgregor@naturalengland.org.uk; caroline.cowan@naturalengland.org.uk

J.D. Ford and L. Berrang-Ford (eds.), *Climate Change Adaptation in Developed Nations: From Theory to Practice*, Advances in Global Change Research 42, DOI 10.1007/978-94-007-0567-8_28, © Springer Science+Business Media B.V. 2011

Keywords Sustainable adaptation • Agriculture • Ecosystems • Resilience • Vulnerability • Biodiversity • Defra • Climate Change Act • Farming Futures • Environmental Stewardship

Introduction

Covering over 70% of English land, agriculture is an essential part of England's environmental, social and economic capital, and agricultural systems (a term we use in this chapter to encompass both agricultural production and the natural ecosystems on which it depends) produce a wide range of benefits to the society.

As well as producing a large proportion of the food the United Kingdom consumes (Defra 2008a), agricultural systems provide vital ecosystem services such as regulation of water supply and quality, alleviation of flooding, and storage of significant amounts of carbon (Millennium Ecosystem Assessment 2005; FAO 2007; Thompson 2008). These systems also support significant biodiversity – more than half of England's Sites of Special Scientific Interest are on farmland (Defra 2008a). In addition, agricultural land represents the dominant image of nature for English people, providing significant non-material benefits such as land for recreation and valued landscapes (Matless 1998). More than three-quarters of England's National Park area is agricultural land (National Parks are landscape rather than wildlife designations in the UK) (Defra 2008a). The benefits provided by agricultural land are likely to be affected directly or indirectly by climate change, and agriculture has a central role to play in the overall adaptation effort. Policymakers therefore need to develop integrated adaptation solutions for the sector that addresses multiple environmental, social, and economic objectives.

The Risks Posed by Climate Change

The services provided by agricultural land are likely to be affected significantly by climate change. Agricultural systems are vulnerable to changes in climate, and this vulnerability is exacerbated by existing non-climate stresses.

Climate projections for the United Kingdom indicate warmer, wetter winters and hotter, drier summers, as well as serious impacts from sea level rise and increased storminess, with a large measure of regional variation (Hulme et al. 2002; Murphy et al. 2009). This is likely to present some potential benefits for agricultural systems, such as longer growing seasons, increased geographical ranges for some crops and the opportunity to introduce new crops and livestock breeds. However, with increasing climate change, these benefits are likely to be offset by a range of negative consequences, including greater variability in yields; pests and diseases; increased damage from floods and storms; water pollution, water scarcity, and drought; loss of soil moisture; changes to habitats and ecosystems; and loss of biodiversity (e.g.,

Hopkins et al. 2004, 2007; Jaggard et al. 2007; Mitchell et al. 2007; Evans et al. 2008; Hughes et al. 2008; Twining et al. 2008; Broadmeadow et al. 2009a, b; Moran et al. 2009; Semenov 2009).

Human responses to these direct impacts will further affect agricultural ecosystems. Land managers can be expected to change their practices (what they produce, where and how) in response to climatic changes. These responses could affect biodiversity, water management, soil quality, food availability, and landscape character. Given the great influence that farmers have on natural resources (FAO 2007; Pretty 2008; Chesterton 2009), these indirect effects of climate change could be much more significant than the direct effects (e.g. Mitchell et al. 2007). Furthermore, indirect effects could be caused not just by climatic changes in the United Kingdom, but also by changes to the climate in other parts of the world (Parry et al. 2004).

Effects of climate change are also likely to interact with existing non-climate pressures (Parry et al. 2007). For example, climate change combined with nutrient pollution could exacerbate eutrophication of shallow lakes (Feuchtmayr et al. 2009); in turn eutrophication increases the sensitivity of aquatic ecosystems to higher temperatures and reduced water flow (Natural England 2008a).

At the same time, the ecosystem services provided by agricultural land, such as flood regulation, will be vital to our efforts to adapt to climate change (Millennium Ecosystem Assessment 2005; Parry et al. 2007); good land management practices will be needed to maintain these services (FAO 2007; Pole et al. 2008).

Given these factors – namely, the importance of agricultural systems to society, their vulnerability to the direct and indirect effects of climate change and their position at the intersection of a range of climate-related and other environmental issues – successful adaptation for the agriculture sector is vital. Adaptation for agricultural businesses and adaptation for the natural environment on agricultural land cannot be considered in isolation from each other (e.g. Berry et al. 2006). Integrated adaptation solutions are needed to address multiple environmental, economic, and social objectives.

A Framework for Sustainable Adaptation

A range of different approaches to adaptation are being developed (e.g., Adger et al. 2005; Berkhout 2005; Smit and Wandel 2006; Parry et al. 2007; McGray et al. 2007; Eakin et al. 2009). Adaptation policy is also developing at a rapid pace in the United Kingdom, with the introduction of a statutory framework, local government requirements, and a national adaptation strategy (Defra 2008b; UK Parliament 2008; Communities and Local Government 2008). There is, however, a risk that action in one area may prevent adaptation in another. For example, increased irrigation by the agriculture sector could have negative impacts on supply for human consumption, or threaten freshwater ecosystems. We feel that a key consideration, not always sufficiently emphasized in the adaptation literature, is the need to assess adaptation options for their sustainability and to recognize the multiple values of systems;

overlooking this risks the promotion of sector-based solutions that potentially limit adaptation in other areas.

We are developing a framework for decision making that we hope will address some of these issues and promote a more sustainable approach to adaptation. It is based on the following principles which apply to adaptation in general but are particularly relevant to agriculture:

1. *Adaptation should aim to maintain or enhance the environmental, social, and economic benefits provided by a system, while accepting and accommodating inevitable changes to it.* It is important for adaptation to be based on a clear set of objectives that frame the problem in terms of "what are we adapting for?" (i.e., focusing on the benefits we want to obtain).

2. *Adaptation should not solve one problem while creating or worsening others. We should prioritize action that has multiple benefits and avoid creating negative effects for other people, places, and sectors.* Many adaptation responses to address socioeconomic factors will have wider consequences for natural systems and vice versa. Taking an ecosystem-based approach (Millennium Ecosystem Assessment 2005), and applying principles of sustainable development (Defra 2005) across sectors will be necessary to identify integrated, sustainable adaptation solutions.

3. *Adaptation should seek to increase resilience to a wide range of future risks and address all aspects of vulnerability, rather than focusing solely on specific projected climate impacts.* Actions must be taken in the face of uncertainty in relation to future climate changes, socioeconomic change, and the interaction between them. Therefore it is important to build resilience to cope with a range of plausible futures (Dessai and Hulme 2007; Dessai et al. 2009; Eakin et al. 2009).

4. *Approaches to adaptation must be flexible and not limit future action.* Increasing the resilience of what we have is a good initial strategy but, in the longer term and under more extreme climate change, transformative approaches will increasingly be needed (Horrocks and Harvey 2009; Howden et al. 2010). The application of "adaptive management" approaches (Holling 1978) can help ensure flexibility. This allows early adaptation action while reducing the risk of over-committing or acting inappropriately (UKCIP n.d.). New approaches to adaptation will need to be tested and monitored at the appropriate scale (Smithers et al. 2008) so that we can learn from the experience and revise our approaches accordingly.

The principles have been used to inform a decision-making framework, drawing on existing environmental frameworks (EPA 1998; Willows and Connell 2003). In this framework, the first step is to identify the *benefits* provided by the whole system. *Vulnerability* of those benefits can then be assessed and specific risks identified. Finally, possible adaptation *responses* are identified and evaluated against aspects of sustainability discussed above.

Applying the Framework for Sustainable Adaptation to Agriculture

Agricultural adaptation is being addressed by the UK Department of Environment, Food and Rural Affairs (Defra) through a project in partnership with Natural England, the Environment Agency and the Forestry Commission. The aim is to identify actions and policy responses that will help to increase the resilience to climate change of both agricultural production and the natural ecosystems that underpin farming, in order to maintain a sustainable supply of food and to provide other important ecosystem services and to enable wild species to adapt and so enhance biodiversity.

The project has been applying the framework for sustainable adaptation outlined above, to identify and deliver integrated solutions that will maximize benefits for the whole agricultural system. First, the essential services and benefits provided by agricultural systems were identified, and from these, a set of 15 desired environmental, social, and economic outcomes for the sector were compiled, encompassing agricultural production, animal welfare, rural communities, biodiversity, and a range of ecosystem services. The project group then reviewed the direct climate risks and potential adaptation options from the available literature, including a substantial body of government-commissioned research (summary available from the authors on request). This identified more than 100 possible adaptation measures that could be taken at farm level, covering a similarly wide range of activities to those identified by other reviews (e.g. Howden et al. 2007; Paterson 2009). These can be grouped into the following categories, which give an indication of the wide range of activities that land managers collectively will need to carry out:

1. *Risk assessment, and business and contingency planning*, including planning for change and loss, as well as being able to exploit new opportunities.
2. *Changing or diversifying crop types, livestock breeds or species, and/or farm type*, to replace old varieties that cannot grow in new conditions; spread risks from variable weather; or take advantage of new opportunities from longer growing seasons and changing crop ranges.
3. *Land management to create "green infrastructure"* including trees, buffer strips, wetlands, and sustainable drainage, to address a wide range of climate risks, such as effects on biodiversity, water shortages and floods, heat stress of livestock, and increased erosion and runoff.
4. *Altering infrastructure and introducing new technology*, such as improved seed and crop storage facilities, in order to deal with projected changes in moisture, temperature and pest infestation, and increasing water storage capacity.
5. *Water management*, especially to improve efficiency and reduce wastage to cope better with shortages, and to cope with flooding.
6. *Fertilizer and pesticide management*, to optimize the quantity applied and timing of application to avoid pollution, the risk of which is likely to increase with projected increases in heavy rainfall.

7. *Livestock management*, to adjust management practices and their timing to changed conditions, for example, to reduce soil erosion and maintain animal welfare standards.
8. *Crop management*, to adjust management practices and to changed conditions, for example, changing the timing of sowing and harvest, and taking measures to reduce erosion.

To evaluate the sustainability of different adaptation measures, we evaluated the likely positive and negative effects of each measure on each desired outcome. This allowed us to identify an initial list of priority adaptation measures that would address a wide range of climate risks and have multiple benefits against outcomes, thereby contributing strongly to sustainable adaptation. It should be noted that adaptation is a very time- and place-specific activity and will depend on local conditions and the specific decisions that need to be made. Also, the wider positive and negative effects of different actions can vary substantially depending on where and how they are implemented (Paterson 2009). Therefore there is a limit to the extent to which one can prioritize or rank adaptation measures at a national level. However, our analysis has created a very useful shortlist of promising options that can be tailored to specific locations in partnership with farmers and other members of local communities. It has also helped to identify key areas for further policy analysis.

Identifying Effective Levers to Encourage Appropriate Action by Farmers

Having identified a range of sustainable adaptation measures, it will be essential for the government to identify effective levers to encourage their uptake. Successful adaptation for the agriculture sector is likely to involve a mixture of measures to reduce sensitivity and exposure to risks, increase adaptive capacity and resilience to change and losses, and take advantage of opportunities (Street 2008; see also Smit and Skinner 2002 for a typology of different adaptation strategies). The industry is dynamic and responsive to changing environmental conditions and will act autonomously in some areas. However, to ensure sustainable adaptation, the government will likely have to intervene to influence the practices of farmers and land managers to encourage and reinforce forward-looking and climate-sensitive behavior and avoid undesirable outcomes (Berkhout 2005). The choice of appropriate policy instruments must therefore be based on an understanding of the factors that determine farmers' behavior and the potential barriers that might need to be overcome.

What Influences Farmers' Behavior?

A behavior-orientated approach putting the farmer, as decision maker, at the center of the policy process has been developed by Defra's Agriculture Change and Environment Observatory (Pike 2008). This identifies four factors that influence both the short- and long-term behaviors of farmers: their attitudes; their habits; social factors, including the attitudes of neighbors, farm advisers, etc.; and external factors, like costs, market conditions, and policy interventions. The relative influence on the behavior of these different factors varies greatly among farmers; the farming sector is, to a great extent, a collection of individual decision makers (Pike 2008).

These findings have important implications for identifying appropriate policy interventions. Typically, government interventions have focused on external factors, such as environmental regulations or financial incentives. Clearly, external factors can be an important barrier to or incentive for adaptation action. However, "internal" characteristics (attitudes, habits, social factors) can hinder or facilitate adaptation action equally strongly; they might, for example, prevent some farmers from exploiting new opportunities from climate change even if there is an economic incentive to do so. Therefore, if we are to encourage sustainable adaptation in the agriculture sector, interventions should be designed with an understanding of those to be targeted. Interventions should include a range of options that allow for the uptake of different levels of adaptation, recognizing both business diversity and varying internal objectives and motivations. In many ways, adaptation for agriculture lends itself to the utilization of interventions that act on internal barriers, given that action will be essential from a business perspective as well as to address political and wider societal goals.

Existing Levers for Action

An important part of Defra's work on adaptation for agriculture between 2005 and early 2011 was the Rural Climate Change Forum. This was the government's principal advisory body on policy, research, and communication on climate change issues (both adaptation and mitigation) relating to the agriculture, forestry, and land management sectors. It brought together the main organizations with an interest in this field, including government bodies, non-governmental organizations and industry groups, and helped to generate awareness and discussion of the issues.

One of the Rural Climate Change Forum's most important achievements was the development of a collective approach to providing information for farmers and land managers. Stimulating attitudinal change and changing social norms, as discussed above, will be essential to make adaptation a more mainstream activity. To achieve this will require real engagement with farmers, not just one-way communication, although appropriate advice delivered by a trusted source is a key element of successful engagement (Dwyer et al. 2007; Pike 2008). To foster this engagement, the Rural Climate Change Forum worked closely with Farming Futures, a communications initiative that provides information, through fact sheets, case studies and workshops, to farmers and land managers on the threats, opportunities, and responsibilities presented by climate change (Farming Futures 2009).

These efforts appear to be succeeding. In a survey of English farmers by Farming Futures in 2009, 50% of the farmers surveyed said they were already affected by climate change, and 63% expected to be affected in the next 10 years. Almost half reported that they believed that climate change presented more threats than opportunities to their businesses – unpredictable weather, flooding, and droughts were the most common concerns. Despite this, at least half of the respondents thought climate change would bring some opportunities. Around 30% of the farmers surveyed were taking action to adapt to the impacts of climate change (Farming Futures 2009). The results of the Defra Farm Practices Survey 2008 suggested a similarly high level of awareness among farmers in England, and indicated that some farmers are, or are considering, taking action to adapt to climate-related threats and opportunities (Defra 2008d).

We are already seeing some innovative farmers starting to take advantage of the changing climate in the United Kingdom by diversifying their businesses to include crops and produce not typically associated with northwest Europe. This is particularly evident in southwest England (the warmest part of the country), with tea being produced in Cornwall (BBC 2005) and a "climate change farm" in Devon that has England's only olive groves, along with pecans, almonds, apricots, persimmons, and other warm climate fruit (Farming Futures 2009).

Adaptation for agriculture is, of course, not taking place in a policy vacuum. The Climate Change Act (UK Parliament 2008) has provided an overall framework for adaptation, including: the requirement of a statutory risk assessment to be undertaken by the government; a statutory national adaptation program; a power for the Secretary of State to require any public body to report on the risks they face from climate change; and the establishment of an Adaptation Sub-Committee of the Committee on Climate Change to provide advice and scrutiny. Further to these, the Climate Change Act sets out a long-term framework for emissions reduction. The importance of adaptation is also increasingly being recognized at European level, with the recent publication of a White Paper on adaptation and an accompanying discussion paper on adaptation by agriculture (European Commission 2009a, b).

The economic and environmental context of farming in the United Kingdom is also heavily influenced by the European Union's Common Agricultural Policy (CAP) and a considerable body of legislation related to natural environment protection, much of it derived from the European Commission. There is a wide range of environmental regulations, a number of schemes that provide grants and other financial incentives for further environmental management, and a range of sources of information and advice for farmers (e.g., Defra 2003; 2008c, e; RDS 2005; Forestry Commission 2008; Natural England 2008b; Natural England, Environment Agency, Defra 2008; Rural Payments Agency and Defra 2008).

In many cases, adaptation will not require new institutional arrangements, as much will be delivered through existing mechanisms (Dovers 2009). It is likely that existing agricultural policies will play an important role in helping the sector adapt, by reducing other sources of pressure on ecosystems and encouraging actions that will help retain or restore the environment's capacity to adapt. For example, an important contribution will be made by Environmental Stewardship, the most significant of the agri-environment schemes that now cover almost 70% of English agricultural land (Natural England 2010). Environmental Stewardship supports the adaptation of the natural environment to climate change by funding management aimed at increasing the size, heterogeneity, and connectivity of habitats and species populations at the wider landscape scale. Natural England, which delivers Environmental Stewardship, is currently evaluating how to further enhance the scheme's contribution to both adaptation and mitigation. However, to our knowledge there has not yet been a full evaluation of how well the full suite of existing agricultural policies encourages desired adaptation measures. This will be addressed by the Defra project discussed above.

Challenges

There are risks for sustainable adaptation if the options chosen by farmers, such as intensification of production, or the widespread planting of biofuels (see Gallagher 2008), have negative impacts on ecosystems and the natural environment. Current biodiversity legislation can protect existing protected sites, and there is some control over other natural resources. However, there is still limited provision to ensure widespread ecosystem protection outside protected areas. This is an important area for government intervention to educate and create incentives and support for farmers in making sustainable decisions. That we are still failing to meet some of our environmental targets (in particular halting the decline of biodiversity and improving water quality; Natural England 2008a) suggests that the institutional arrangements even to meet existing needs might be insufficient, and therefore that there are likely to be challenges in integrating adaptation into policy and delivery.

A particular challenge for government action will come from the range of important non-climate drivers that influence agricultural land use in a globalized world (see e.g. Rounsevell et al. 2006), and the difficulty of predicting their effects. For example, the combination of some drivers – higher prices for agricultural commodities, the drive for increased supply of bioenergy crops, and fears linked to food security – could lead to further intensification in the agricultural sector. This intensification has knock-on effects on the natural environment and potential implications for the role and effectiveness of agri-environment policies. The government has limited influence over such drivers; so needs to ensure that the policy can be responsive to such changes and provide alternative and appropriate incentives to encourage continued long-term planning by the agricultural sector. This is especially important in areas, such as farm woodland, that require long-term decision making (e.g. Read et al. 2009). Further work is needed to consider possible future scenarios of land use and the effect they would have on provision of ecosystem services by agricultural systems, and to identify where potential conflicts and trade-offs might occur. Responses to climate change will need to be integrated with consideration of other pressures (Howden et al. 2007, 2010)

Conclusion

Agriculture forms a vital part of the environmental, social, and economic capital of England. Ensuring it is able to cope with the effects of climate change is thus a high priority for the government. Given the complex interactions between economic production, ecosystems, and cultural values, the sector provides an important case study for testing integrated approaches based on the principles for sustainable adaptation outlined above. Defra and its partners have identified the key types of adaptation that will be necessary to support the wide variety of policy objectives and legislative requirements relating to agriculture and the natural environment, and have begun to identify a number of existing levers and mechanisms that can be used to influence the behavior of farmers and land managers and help them make appropriate decisions. A key achievement is the recognition of the variety of factors that influence behavior, beyond the traditional "external" drivers, and Defra and its partners are working to address those factors.

The work is still at an early stage, and policymakers face a number of challenges to the implementation of effective adaptation. Some of these derive from adaptation's status as a relatively new area of policy. At this stage, there are few clear examples of successful action on adaptation; combined with uncertainty about the future, this can cause decision makers to struggle to identify the best course of action. Nevertheless, application of the principles and framework for sustainable adaptation outlined above has identified an initial list of priority adaptation measures for agriculture. It is striking how many of these correspond to existing good environmental and agricultural management practices, which points to a clear starting point for action (see also Dovers 2009). With effects of climate change on natural and human systems already starting to become apparent (Parry et al. 2007; Rosenzweig et al. 2008), and at least 30 years of some level of further climate change now inevitable (Murphy et al. 2009), it is vital that uncertainty does not act as an excuse for inaction.

Acknowledgments We are grateful to Tony Pike for providing information about his work on factors influencing farmers' behavior and discussing its relevance to adaptation policy. We would also like to thank Humphrey Crick, Jessica Tipton, and two anonymous reviewers for their very helpful comments on drafts of the chapter, and other colleagues who provided or checked information for us. The work described in the chapter to identify priority adaptation measures for agriculture is a project of the Department for Environment, Food and Rural Affairs in partnership with Natural England, the Environment Agency and the Forestry Commission; we are very grateful to the colleagues who have worked on various aspects of this project and helped us to develop our thinking in this complex area of adaptation research and policy, particularly Jamie Letts, Mark Broadmeadow, David Thompson, Emmanuelle Bensaude, Ingrid Doves, Ian Pickard, Jessica Tipton, and Kate Sugden.

References

(*Note*: Reports for all Defra projects cited here can be searched for by their project number at the following website: http://randd.defra.gov.uk/).

Adger WN, Arnell NW, Tompkins EL (2005) Successful adaptation to climate change across scales. Global Environ Change 15(2):77–86

Berkhout F (2005) Rationales for adaptation in EU climate change policies. Clim Policy 5(3): 377–391

Berry PM, Rounsevell MDA, Harrison PA et al (2006) Assessing the vulnerability of agricultural land use and species to climate change and the role of policy in facilitating adaptation. Environ SciPolicy 9:189–204

British Broadcasting Corporation [BBC] (2005) Company plans Cornish tea centre. BBC News (20 Jun 2005). http://news.bbc.co.uk/1/hi/england/devon/4112272.stm. Cited 3 Aug 2009

Broadmeadow MSJ, Morecroft MD, Morison JIL (2009a) Observed impacts of climate change on UK forests to date. In: Read DJ, Freer-Smith PH, Morison JIL et al (eds) Combating climate change – a role for UK forests: an assessment of the potential of the UK's trees and woodlands to mitigate and adapt to climate change. TSO, Edinburgh

Broadmeadow MSJ, Webber JF, Ray D et al (2009b) An assessment of likely future impacts of climate change on UK forests. In: Read DJ, Freer-Smith PH, Morison JIL et al (eds) Combating climate change – a role for UK forests: an assessment of the potential of the UK's trees and woodlands to mitigate and adapt to climate change. TSO, Edinburgh

Chesterton C (2009) Natural England research report NERR030: environmental impacts of land management. Natural England, Sheffield

Commission F (2008) English woodland grant scheme: general guide to EWGS. Her Majesty's Government, Cambridge

Communities and Local Government (2008) National indicators for local authorities and local authority partnerships: handbook of definitions. Her Majesty's Government, London

Defra (2005) Securing the future: UK Government sustainable development strategy. Her Majesty's Government. http://www.defra.gov.uk/sustainable/government/publications/uk-strategy/documents/SecFut_complete.pdf. Cited 3 Aug 2009

Defra (2008a) Agriculture in the United Kingdom 2008. Her Majesty's Government. https://statistics.defra.gov.uk/esg/publications/auk/2008/AUK2008.pdf. Cited 3 Aug 2009

Defra (2008b). Adapting to climate change in England: a framework for action. Her Majesty's Government. http://www.defra.gov.uk/environment/climate/documents/adapting-to-climate-change.pdf. Cited 3 Aug 2009

Defra (2008c) Environmental permitting core guidance for the Environmental Permitting (England and Wales) Regulations 2007. Her Majesty's Government, London

Defra (2008d) Farm practices survey 2008 – England. Her Majesty's Government. https://statistics.defra.gov.uk/esg/publications/fps/FPS2008.pdf. Cited 3 Aug 2009

Defra (2008e) Sustainable farming and food strategy – indicator data sheet. Her Majesty's Government. http://statistics.defra.gov.uk/esg/indicators/pdf/h6a.pdf. Cited 3 Aug 2009

Department for Environment Food and Rural Affairs [Defra] (2003) Sites of special scientific interest: encouraging positive partnerships – code of guidance. Her Majesty's Government. http://www.defra.gov.uk/rural/documents/protected/sssi-code.pdf. Cited 3 Aug 2009

Dessai S, Hulme M (2007) Assessing the robustness of adaptation decisions to climate change uncertainties: a case study on water resources management in the East of England. Global Environ Change 17(1):59–72

Dessai S, Hulme M, Lempert R et al (2009) Climate prediction: a limit to adaptation? In: Adger WN, Lorenzoni I, O'Brien KL (eds) Adapting to climate change: thresholds, values, governance. Cambridge University Press, Cambridge

Dovers S (2009) Normalizing adaptation. Global Environ Change 19(1):4–6

Dwyer J, Mills J, Ingram J et al (2007) Understanding and influencing positive behaviour change in farmers and land managers: a project for Defra. Defra, Majesty's Governemnt. http://randd.defra.gov.uk/Document.aspx?Document=WU0104_6750_FRP.doc. Cited 3 Aug 2009

Eakin H, Tompkins EL, Nelson DR et al (2009) Hidden costs and disparate uncertainties: trade-offs in approaches to climate policy. In: Adger WN, Lorenzoni I, O'Brien KL (eds) Adapting to climate change: thresholds, values, governance. Cambridge University Press, Cambridge

European Commission (2009a) Adapting to climate change: towards a European framework for action. European Commission, Brussels

European Commission (2009b) Adapting to climate change: the challenge for European agriculture and rural areas. European Commission, Brussels

Evans N, Baierla A, Fitt BDL et al (2008) Range and severity of plant disease increased by global warming. J R Soc Interface 5:525–531

Farming Futures (2009) Farming Futures website. http://www.farmingfutures.org.uk/. Cited 3 Aug 2009

Feuchtmayr H, Moran R, Hatton K et al (2009) Global warming and eutrophication: effects on water chemistry and autotrophic communities in experimental hypertrophic shallow lake mesocosms. J Appl Ecol 46(3):713–723

Food and Agriculture Organization of the United Nations [FAO] (2007) The state of food and agriculture 2007: paying farmers for environmental services. FAO, Rome

Gallagher E (2008) The Gallagher review of the indirect effects of biofuels production. Renewable Fuels Agency, Her Majesty's Government, London

Holling CS (ed) (1978) Adaptive environmental assessment and management. Wiley, New York

Hopkins A, Richter GM, Coleman K et al (2004) Impacts of climate change on the agricultural industry: a review of research outputs from Defra's CC03 and related research programmes – Defra project CC0366. Defra, Her Majesty's Government, London

Hopkins JJ, Allison HM, Walmsley CA et al (2007) Conserving biodiversity in a changing climate: guidance on building capacity to adapt. Defra, Her Majesty's Government, London

Horrocks L, Harvey A (2009) The implications of $4+°C$ warming for adaptation strategies in the UK: time to change? Paper presented at '4 degrees and beyond' international climate conference, Oxford, 28–30 Sept 2009

Howden SM, Soussana J-F, Tubiello FN et al (2007) Adapting agriculture to climate change. PNAS 104:19691–19696

Howden SM, Crimp SJ, Nelson RN (2010) Australian agriculture in a climate of change. In: Jubb I, Holper P, Cai W (eds) Managing climate change: papers from the GREENHOUSE 2009 conference. CSIRO Publishing, Melbourne, pp 101–111

Hughes G, Wilkinson M, Boothby D et al (2008) Changes to agricultural management under extreme events: likelihood of effects and opportunities nationally – Defra project CC0361. Defra, Her Majesty's Government, London

Hulme M, Jenkins GJ, Lu X et al (2002) Climate change scenarios for the United Kingdom: the UKCIP02 scientific report. Tyndall Centre for Climate Change Research, Norwich

Jaggard KW, Qi A, Semenov MA (2007) The impact of climate change on sugar beet yield in the UK: 1976–2004. J Agric Sci 145:367–375

Matless D (1998) Landscape and englishness. Reaktion, London

McGray H, Hammill A, Bradley R et al (2007) Weathering the storm: options for framing adaptation and development. World Resources Institute, Washington, DC

Millennium Ecosystem Assessment (2005) Ecosystems and human well being: opportunities and challenges for business and industry. World Resource Institute, Washington, DC

Mitchell RJ, Morecroft MD, Acreman M et al (2007) England biodiversity strategy: towards adaptation to climate change – Defra project WC02019. Defra, Her Majesty's Government, London

Moran D, Topp K, Wall E et al (2009) Climate change impacts on the livestock sector – Defra project AC0307. Defra, Her Majesty's Government, London

Murphy J, Sexton D, Jenkins G et al (2009) Climate change projections. Met Office Hadley Centre, Her Majesty's Government, Exeter

Natural England (2008a) State of the natural environment 2008. Natural England, Sheffield

Natural England (2008b) Entry level stewardship handbook, 2nd edn. Natural England, Sheffield

Natural England (2010) Environmental Stewardship update 11, Sept 2010. http://www.naturalengland.org.uk/Images/ES%20Update%2011%20Final_tcm6-22131.pdf. Cited 19 Oct 2010

Natural England, Environment Agency, Defra (2008) Catchment sensitive farming delivery initiative: the first phase – a compendium of advice activity examples. Defra, Her Majesty's Government, London. http://www.defra.gov.uk/farm/environment/water/csf/pdf/ecsfdi-compendium.pdf. Cited 3 Aug 2009

Parry ML, Rosenzweig C, Iglesiasic M et al (2004) Effects of climate change on global food production under SRES emissions and socio-economic scenarios. Global Environ Change 14(1):53–67

Parry ML, Canziani OF, Palutikof JP et al (eds) (2007) Climate change 2007: impacts adaptation and vulnerability – contribution of Working Group II to the Fourth Assessment Report of the Intergovernmental Panel on Climate Change. Cambridge University Press, Cambridge

Paterson JS (2009) Agriculture. In: Berry PM (ed) Biodiversity in the balance – mitigation and adaptation conflicts and synergies. Pensoft Publishing, Sofia, pp 27–88

Pike T (2008) Understanding behaviours in a farming context. Defra, Her Majesty's Government. https://statistics.defra.gov.uk/esg/ace/research/pdf/ACEO%20Behaviours%20Discussion%20Paper.pdf. Cited 3 Aug 2009

Pole J, Collier R, Lillywhite R et al (2008) Ecosystem services for climate change adaptation in agricultural land management – Defra project AC0308. Defra, Her Majesty's Government, London

Pretty J (2008) Agricultural sustainability: concepts, principles and evidence. Phil Trans R Soc Lond B 363:447–465

Read DJ, Freer-Smith PH, Morison JIL et al (eds) (2009) Combating climate change – a role for UK forests: an assessment of the potential of the UK's trees and woodlands to mitigate and adapt to climate change. TSO, Edinburgh

Rosenzweig C, Karoly D, Vicarelli M et al (2008) Attributing physical and biological impacts to anthropogenic climate change. Nature 453:353–357

Rounsevell MDA, Reginster I, Araújo MB (2006) A coherent set of future land use change scenarios for Europe. Agric Ecosyst Environ 114:57–68

Rural Development Service [RDS] (2005) Higher level stewardship handbook: terms and conditions and how to apply. Her Majesty's Government, Cheltenham

Rural Payments Agency, Defra (2008) Single payment scheme: the guide to cross-compliance in England. Defra, Her Majesty's Government, London

Semenov MA (2009) Extreme impacts of climate change on wheat in England and Wales. J R Soc Interface 6(33):343–350

Smit B, Skinner MW (2002) Adaptation options in agriculture to climate change: a typology. Mitig Adapt Strat Glob Change 7:85–114

Smit B, Wandel J (2006) Adaptation, adaptive capacity and vulnerability. Global Environ Change 16(3):282–292

Smithers RJ, Cowan C, Harley M et al (2008) England biodiversity strategy: climate change adaptation principles. Defra, Her Majesty's Government, London

Street RB (2008) Climate change projections for the UK: a farming perspective. Paper presented at the HGCA conference 'arable cropping in a changing climate', Lincolnshire, 23–24 Jan 2008

Thompson D (2008) Carbon management by land and marine managers: Natural England research report NERR026. Natural England, Sheffield

Twining S, Procter C, Wilson L et al (2008) Impacts of 2007 summer floods on agriculture. Defra, Her Majesty's Government, London

UK Climate Impacts Programme [UKCIP] (n.d.) Identifying adaptation options. http://www.ukcip. org.uk/images/stories/Tools_pdfs/ID_Adapt_options.pdf. Cited 3 Aug 2009

UK Parliament (2008) Climate Change Act 2008 (c.27). Her Majesty's Stationary Office [HMSO]. http://www.opsi.gov.uk/acts/acts2008/ukpga_20080027_en_1. Cited 3 Aug 2009

US Environmental Protection Agency [EPA] (1998) Guidelines for ecological risk assessment. EPA, Washington, DC

Willows R, Connell R (eds) (2003) Climate adaptation: risk, uncertainty and decision-making – UKCIP Technical Report. UK Climate Impacts Programme [UKCIP], Oxford

Part VI
Adaptation in Rural
and Resource-Dependent Communities

Chapter 29
Overview: Climate Change Adaptation in Rural and Resource-Dependent Communities

Stewart J. Cohen

Abstract Anthropogenic climate change will create a new decision environment for governments, both national and regional. In rural regions and small towns, which are already facing a wide range of social, economic, and environmental pressures, this challenge will be especially difficult because of lower capacities to adapt. There are, however, encouraging examples of rural regions and small communities that are finding innovative ways to create shared learning partnerships that could empower them to create their own response paths. While researchers offer new knowledge on future climate change impacts, local-based practitioners and knowledge holders provide insights on local systems and the importance of attachment to place. As rural regions and small towns begin to take on leadership roles in adaptation planning, researchers and higher levels of government can use their capacities to enable local adaptation as part of long-term development planning. Examples are provided from the literature, followed by five cases contributed to this volume.

Keywords Anticipatory adaptation • Climate change • Mainstreaming • Adaptation • Rural communities • Resource-based communities • Community-based research • Local governance • Capacity building

Introduction

Throughout human history, a majority of the world's peoples have lived in rural or small town settings. However, it was projected that the world's population would be equally urban and rural by 2010 (United Nations 2006), a clear signal of the global

S.J. Cohen (✉)
Adaptation & Impacts Research Division (AIRD), Environment Canada, Department of Forest Resources Management, University of British Columbia, 4617-2424 Main Mall, Vancouver, BC V6T 1Z4, Canada
e-mail: stewart.cohen@ec.gc.ca; scohen@forestry.ubc.ca

J.D. Ford and L. Berrang-Ford (eds.), *Climate Change Adaptation in Developed Nations: From Theory to Practice*, Advances in Global Change Research 42, DOI 10.1007/978-94-007-0567-8_29, © Springer Science+Business Media B.V. 2011

urbanization trend that has been underway throughout the last century. Urbanization has occurred more rapidly in developed countries, with the United States and Canada, for example, recording urban majorities beginning in the 1920s. Currently, 80% of United States inhabitants live in urban areas (United Nations 2006), which occupy only 2% of the country's land area, so urban population density is much higher than rural density. However, not all regions within these countries have followed this trend. In Canada, for example, Newfoundland, Northwest Territories, and Nunavut continue to have primarily rural populations (Bollman and Clemenson 2008).

As climate change evolves as an environmental phenomenon, and is superimposed on a socioeconomic development trend of urbanization, what impacts are projected for rural and small town communities? Is urbanization affecting rural regions' capacity to adapt to external stresses? And, if these communities are to adapt to climate change impacts in a proactive way, what challenges do they face that may be different from their urban cousins?

Projected impacts of climate change on ecosystems, water resources, agriculture, forestry, coastal zones, and health may be felt more acutely in rural regions than in urban centres. Rural residents are much closer to resource production activities, the farms, the timber supply areas, sawmills, the fish-bearing streams, parks, conservation areas, and traditional Aboriginal hunting lands. Infrastructure is likely to be relatively small in size and capacity, such as water towers and wells instead of secondary treatment facilities, unpaved and winter roads instead of paved roads, nursing stations instead of hospitals, and volunteer emergency services with distant support instead of local full-time professional providers. In a small town, a hazardous event can be locally disastrous simply because it is likely to affect a relatively high proportion of the population (Sauchyn and Kulshreshtha 2008).

Recent reports of the Intergovernmental Panel on Climate Change (IPCC) have summarized impacts on physical and biological systems, and have highlighted specific issues for countries, watersheds, coastal zones, ecosystems, food systems, and built environments (Parry et al. 2007). Within the developed countries, a key research priority is determining the risk that projected rates of climate change might overwhelm existing engineered and social infrastructure. Within the developing countries, there is concern about the potential for climate change to disrupt livelihoods and prevent countries from achieving sustainable development (Wilbanks et al. 2007).

In rural areas, services are likely to be more dispersed and located farther away from residents. On the other hand, rural and small town residents are likely to have a more intimate knowledge of local governments and service utilities, and the people who work for them, including elected officials. How do these attributes affect rural adaptive capacity? And, how would new information on climate change be used by rural interests in long-term planning, as they seek to balance a wide range of management objectives for the benefit of their constituents and neighbors?

This overview explores some recent literature on climate change adaptation in rural and small town contexts, and sets the stage for the collection of case studies included in this section of the book.

Recent Developments: From Theory to Practice

A warmer climate can be described in terms of changing means, deviations and probabilities of extremes. However, the translation of new climate statistics into impacts requires knowledge of relationships between climate indicators and attributes of interest, such as water supply, corn yield, and wildfire or flood damage. These relationships might be based on an observation period in which the strength of such relationships could be calculated. In recent years, however, there has been increasing recognition of the need to account for concurrent changes in both climate statistics and "facts on the ground." Technological and policy changes have led to changes in agricultural practice, infrastructure construction designs, fire policy, and early warning systems. Attitudes about protection from extreme events are also changing, but this is context-specific. For example, in post-Katrina Louisiana and other US Gulf Coast states, new levees have been constructed to provide greater protection from storm surges. This was seen as the preferred choice for urban areas within this coastal zone (Karl et al. 2009). Meanwhile, in Delaware, Massachusetts, South Carolina, and several other states, regulations do not permit such structures unless they are setback from the beachfront (Rubinoff et al. 2008), so that long-term development incorporates the reality of rising sea level by allowing coastal ecosystems to adjust unimpeded by engineered structures. Various regions within a single developed country are exhibiting different attitudes about a similar biophysical risk from climate change.

Local attitudes about development are reflections of local opinions about vulnerability, resilience, response capacity, and attachment to place. The latter is an especially important part of the social context of adaptation that is very difficult to incorporate into quantitative assessments of future climate change risks. This would also affect projections of effectiveness of adaptation measures, either as individual acts or as parts of portfolios of measures integrated into local development plans. The idea that development pathways can influence climate change vulnerabilities is not a new one, given the lengthy experiences from natural hazards research (e.g. Burton et al. 1993), and the well-known concept of "double exposure" in which globalization, interacting with climate change, is seen as creating new winners and losers (O'Brien and Leichenko 2000; Leichenko and O'Brien 2008). For example, cold regions would benefit from longer growing seasons enabling greater agricultural production, as long as cropland is protected from urban sprawl, while low-lying tropical coastal zones and small islands would lose due to sea level rise and flooding making many settlements uninhabitable.

Attachment to place and corresponding property development may partially explain increasing property losses from weather-related extreme events (Pielke et al. 2005). Market demand to invest in risky areas continues to grow because of the positive attributes these areas already have (beautiful ocean views, urban–forest interface, etc.). However, this kind of response is also a sign that adaptation is limited by societal values, perceptions, and power structures. Adger et al. (2009) argue that decision-making processes themselves will be more important for adaptation than having more precise knowledge of future climate change, and that undervaluing places and cultures may limit the range of adaptation actions. Urban-based notions of adaptation can encounter resistance in rural areas if these clash with entrenched visions of rural lifestyles as pillars of local economic development. Many coastal and urban–forest interface developments will therefore be protected in place, not through retreat or relocation. How does climate change adaptation adapt to this?

While social development factors are important, it must also be recognized that anthropogenic climate change may be leading to changes in climate statistics that are outside of current design standards for resource systems and engineered structures. Past observations lead to future expectations, and these form the basis of building codes, land use zoning, insurance rates, infrastructure design, and emergency response plans (Peterson et al. 2008). Weather-related property damages have increased at a rate faster than population or inflation (Mills 2005). Changing weather statistics may be influencing this, along with changing construction practices, reactive adaptation, and in some cases, lack of enforcement of building codes. There are also differences among agencies in calculating and reporting damage costs (Peterson et al. 2008).

Recognition of the need for a structured approach for conducting risk and vulnerability assessments that could enable planning and decision-making on adaptation led United Nations Development Program (UNDP) to publish a guidance document (Lim et al. 2004) primarily for developing countries which were seen as having greater vulnerabilities due, in part, to their primarily rural-based economies. This is one of a number of adaptation guides being prepared for various audiences, in anticipation of a growing demand for well-described methods that could facilitate regions and communities to initiate and conduct their own assessments to support long-term adaptation planning.

For developed countries, Perkins et al. (2007) reviewed adaptation guidebooks from the United States, Canada, United Kingdom, Australia, New Zealand, as well as adaptation planning cases from various large urban centers and some small towns. Increased awareness of climate change information by planners and municipal leaders has led them to seek collaborative opportunities with researchers to jointly construct step-by-step assessment processes that can be carried out at the local level. An example is the King County guidebook (Snover et al. 2007), which lays out a roadmap for building teams of local experts and interests, linking them with climate change researchers, and together, creating shared narratives on projected risks. This leads to dialogue on setting priorities for adaptation. Dialogue is used as a form of soft knowledge transfer, as few examples of quantitative assessments of adaptation measures are available (Adger et al. 2007). In the absence of quantitative scenario-based information specific to the local context, planners and decision-makers are advised on methods for ranking impacts, to enable decision-makers to incorporate adaptation into ongoing planning processes. This would provide a basis for future quantitative analyses that would support changes in design, operation and management of climate-sensitive infrastructure and resource systems. Specific language on adaptation could then be included in Official Community Plans, Water Management Plans, and other similar planning and governance documents.

Many early adopters of the idea of mainstreaming climate change into official plans are large cities, such as New York (City of New York 2008), London (Land Use Consultants et al. 2006) and Toronto (Toronto Environment Office 2008). However, small towns are also becoming active. One small town example is Keene New Hampshire, with a population of around 23,000. Keene, supported by ICLEI, produced an adaptation plan (City of Keene and ICLEI-Local Governments for Sustainability 2007) that contained goals and targets for building codes, green development, infill development, stormwater management, and protection of energy distribution systems, wetlands, local food security and public health. There is even a goal for creating support services for people in cold-climate industries (e.g. maple syrup production) who may need assistance to adapt their businesses to a warmer climate. This is part of a larger discussion within the plan on promoting a "Local Climate Appropriate Economy."

Another example is the Columbia Basin Trust, which has a program called "Communities Adapting to Climate Change." In 2008–2009, the Trust supported adaptation planning in two small communities in southeast British Columbia, Kimberley, and Elkford, with populations of 7,500 and 2,500, respectively. Both communities now intend to incorporate adaptation into their official community plans (Columbia Basin Trust 2009), and the program has expanded in 2009–2010 to include Rossland, Castlegar and Kaslo/Regional District of Central Kootenay Area "D".

Key Challenges

The nuts and bolts of adaptation are place-specific, including the planning and implementation of engineering-based or rules-based measures. Although there can and should be standards supported and enforced by higher scales of authority (government, professional body (for example, Canadian Council of Professional Engineers 2009), each place will want to chart its own path for designing and carrying out adaptation. This means that climate change impacts will have to be translated across scales, disciplines, and cultures, so that each rural region and small town will know their own specific "damage reports".

Climate change is often perceived as a long-term global "green" issue, rather than a local development issue with links to global climate change, but still relevant to existing local planning horizons. A key challenge is to overcome the perception that there is no immediate local role. Connected to this is the skeptical view of anthropogenic climate change that is still held by many people. Some skeptics do not believe the results of global climate models. Others do not accept that warming could create problems for ecosystems and damage to society. Skeptics argue that acting on climate change would be riskier than any potential impacts of climate change (Dyson 2005). Regardless of the underlying motives of skeptics, their presence can influence public involvement in climate change response, and also professional and government commitment to adaptation measures. At the same time, other actors in the community will be already convinced of the nature of the problem and the need to act. An effective engagement process is one that incorporates different opinions, backgrounds, and agendas among participants (Gardner et al. 2009).

Another challenge emerges from the rural–urban contrast in adaptive capacity. It is generally assumed that rural regions tend to have lower adaptive capacities because of lower levels of investment in infrastructure, weaker information systems, and weaker institutions. But how does one assess adaptive capacity? Although there is no standard approach for measuring this capacity, some examples are available. One is for Canada's Prairie Provinces (Swanson et al. 2009), describing the contrast in adaptive capacity between urban and rural regions on the basis of differences in economic resources, technology, infrastructure, information availability, and skills. The adaptive capacity index in this study focuses on agriculture, and identifies regions with lower rankings to be associated with lower off-farm earnings, less diversity of employment opportunities, less use of computers in farm management, lower density of transportation networks, and less proximity to specialized education institutions.

Choices of indicators in the Prairie agriculture case study reflect the reality of rural-based livelihoods being dependent on smaller networks of people and income streams that may be more climate-sensitive compared to urban regions. But adaptation to external stresses is not an unknown challenge, regardless of the nature of the stress, environmental or socioeconomic. Adaptation portfolios will likely consist of familiar technologies and governance measures, perhaps applied in new ways. Agrawal and Perrin (2008) classify rural adaptation responses into four types: mobility, storage, diversification, and communal pooling. Mobility contributes to reducing risk across space. Storage reduces risk across time. Diversification reduces risks across assets. Pooling reduces risk among individuals. The ability of rural and small town residents and businesses to use these responses will depend on availability of information, support by investors and other levels of government, and local leadership. A fifth class of response, exchange, can be used instead of the first four, if there is ready access to markets.

So, if there are familiar adaptation tools that are available, are resources (financial, human) and uncertainties (scientific) the only obstacles to implementing climate change adaptation plans? Even among rural communities whose leaders want to plan for climate change, there may be another challenge to face: how to effectively start the dialogue on adaptation within an environment of transparency and trust. Existing planning and governance mechanisms already offer opportunities to review past extreme events and performance of local emergency services (e.g. Walkerton inquiry (O'Connor 2002), Firestorm 2003 (Filmon 2004), Hurricane Katrina (Townsend 2006), and the 2003 Europe heat wave (Lagadec 2004; Parry et al. 2007)). These kinds of dialogue processes are meant to offer pathologies of observed extreme weather/climate episodes, in order to understand what went wrong, so that improvements can be made to enhance future performance of emergency preparedness services and infrastructure. A dialogue on adapting to future climate change, however, is different because there may be different visions of adaptation, even if there is consensus on the projected biophysical impacts (Cohen and Waddell 2009). How would rural lifestyles change within an adaptation scenario? For example, how does one reconcile the desirable location of housing within the urban–forest interface with the projected increase in wildfire risk? Would adaptation mean densification of small town residential areas and an end to interface housing? Some might see this adaptation scenario as contrary to the "non-urban" lifestyle that rural regions want to promote for their economic development. If rural communities don't like "urban" adaptation pathways, what would a rural adaptation pathway look like? Also, in the absence of quantitative assessment of the effectiveness of adaptation portfolios within future scenarios of climate change, on what basis can adaptation choices be made?

Opportunities and Future Directions

Planned adaptation to climate change is an opportunity for shared learning. Impacts and adaptation assessments require dialogue with researchers, practitioners (engineers, planners, foresters, water managers, etc.), and decision-makers (e.g., local governments, investors). Defining the problem can be facilitated by quantitative analysis and dialogue among researchers and stakeholders. But climate change adaptation has to be part of ongoing planning and policy processes, or "mainstreamed" into best practices. Limitations in access to relevant information, and the lack of tools to facilitate integration of existing knowledge into decision-making, are barriers to knowledge transfer. Informing key audiences and engaging them in a proactive way would likely lead to an expansion of adaptation planning and implementation (Burton 2008).

For example, group-based model building can be used to combine research and practitioner knowledge for desktop exercises on assessing effectiveness of adaptation measures. In a study of the Okanagan Basin, a decision model was constructed based on local knowledge of the Okanagan system of reservoirs and operations for irrigation, domestic uses and in-stream flows for fisheries, combined with research on climate change impacts on water supply and demand (Cohen and Neale 2006; Langsdale et al. 2007, 2009). Some of the authors of the Okanagan Basin Water Board's proposed Sustainable Water Management Strategy (Okanagan Water Stewardship Council 2008) had also been participants in the decision model-building exercise.

Visualization of future scenarios offers a forward-looking exercise in shared learning. One example is the use of computer graphic visualizations of sea level rise and possible adaptation strategies in a case study of Delta, British Columbia (Shaw et al. 2009). The images shown to local residents and Delta's planning staff were based on previous sea level rise research for this area, combined with local knowledge on coastal protection.

Given the growing recognition of the importance of stakeholder engagement on climate change adaptation, how might this engagement process be carried out

Table 29.1 Stakeholder engagement on climate change adaptation (Adapted from Gardner et al. 2009)

Stage of process	Adaptation pathway	Drivers	Barriers
Initial	Clear understanding of climate change	Information about climate change	Misinformation, uncertainty, skepticism
	Understanding of own climate change vulnerability	Assessment of local capacity and potential impacts	Negative emotional reactions of fear, hopelessness
	Sense of responsibility for developing a solution	Group values, culture, social influence	Expectations that a solution will be provided by an external agency
	Willingness to engage in adaptation planning	Capacity for strategic planning	Lack of resources
End	Adaptation planning		

more effectively? Gardner et al. (2009) offer a best practice process of engagement, summarized in Table 29.1. A key element is that at the initial stage of the engagement process, projected impacts and adaptation need to be framed within the local development context, recognizing the diversity of values that drives local decisions. Specific local problems will attract more attention than a more general topic. Skepticism and emotional reactions need to be addressed directly through open discussion on available knowledge and uncertainties, and identification of tangible actions that can be taken to adapt. The process needs to create a sense of trust among participants, and a sense of responsibility for finding solutions.

In the absence of quantitative information on projected effectiveness of adaptation measures (Adger et al. 2007), shared learning partnerships will play a crucial role as knowledge brokers for decision-makers and other interested parties. Joint construction of impact and adaptation narratives can enable the dialogue to progress from problem identification to policy debate. The key element is that this exercise is about shaping future adaptive development visions, complementing rather than repeating other processes that already perform pathologies of historic extreme events. As illustrated by the city of Keene, the Columbia Basin Trust, Okanagan Basin Water Board, and others, it is possible for rural areas and small towns to join larger urban centers in using scenario-based approaches to plan to adapt to climate change.

Case Studies

This section on rural regions and small towns includes five case studies. Three are from Canada, and one each from Australia and the Tibetan Plateau, China.

Nunavut, a territory in Arctic Canada, is a sparsely populated region containing small Inuit settlements. Boyle and Dowlatabadi describe the anatomy of the region's vulnerabilities to development pressures, environmental and social changes, and ask how climate change adaptation can be incorporated into current planning and governance. Their analysis of specific attributes of decision-making and Nunavut's current capabilities leads to the conclusion that available adaptive capacity is not sufficient. Although there has been considerable research on climate change, little quantitative information is available on future scenarios of local impacts and effectiveness of possible response measures, and there has yet to be an effective way of combining Western and Inuit knowledge. This barrier to shared learning may be hampering efforts to incorporate climate change adaptation into Nunavut's planning and governance programs. Indeed, 5 years after completion of the *Arctic Climate Impact Assessment* (ACIA), which brought widespread attention to both biophysical and human dimensions aspects of climate change in the Arctic, research has yielded only moderate practical benefits for local planners and decision-makers (Sommerkorn and Hamilton 2008).

Change Islands, Newfoundland, and Alert Bay, British Columbia, are two coastal communities in Canada, both of which originated as fishing ports. Alert Bay is also home to the "Namgis First Nation." As is the case with Nunavut, McLeman et al. describe local vulnerabilities stemming from their dependence on a small number of economic activities, their small size and isolated locations. Lack of local resources and lack of local-scale climate change impacts information are important constraints to adaptation planning. Local planners and community leaders are interested in incorporating adaptation into long-term community planning, and would welcome greater collaboration with researchers.

Rural areas of southern Ontario are located close to Canada's major urban centers, yet local planning and governance capacities are similarly challenged when faced with external environmental pressures, as has been noted in the other case studies from coastal and Arctic Canada. A particular concern is water resources, and de Löe provides some background on the region's recent failure to protect water quality, as illustrated by the well-known contamination of the Town of Walkerton's water supply in 2000 following extreme rain events. This experience, as well as the availability of regional impacts information from climate change researchers, has led provincial authorities to initiate efforts at incorporating adaptation into water source protection planning. This case offers an assessment of an adaptation planning instrument, known as the Clean Water Act, and de Löe concludes that the Ontario government has partially mainstreamed climate change into this instrument. Local technical capacity continues to be a barrier to more fully implementing adaptation into resource management and planning. Nevertheless, some progress has been made in linking global climate science and regional-scale decision-making.

Another encouraging case study is a scenario thinking exercise described by Smith et al. Hamilton is a rural farming community in Victoria, Australia, and experienced devastating bushfires in February 2009. Prior to this event, in 2008, a social learning project was conducted, in which local residents and decision-makers were asked to create future stories on adaptation challenges and opportunities, as seen through the eyes of individuals in their future professional or social roles in the community. These stories were not quantitative assessments of futures, nor were they vulnerability assessments of the past and present. They were narratives that resulted from participants sharing their perspectives (farming, aboriginal, immigrant, etc.), creating several different vignettes that could be the basis for expanding the dialogue on incorporating climate change adaptation into long-term planning.

The case study from the Tibetan Plateau concerns pastoralism and the capacity of local herders to adapt to extreme climatic events, particularly drought and snowstorms. Klein et al. describe a case study in Nagchu, which experienced a severe snowstorm in the late 1990s. Local- and national-scale response to this event, as well as the occurrence of drought, has increased herders' vulnerability, rather than reduce it. Traditional coping mechanisms, such as temporary migration of herds, has been replaced by reduced access to some grazing lands, infrastructural improvement, increased information, increased grain storage and more food aid. There is concern that these new policies could change social networks and reduce local resilience to future extremes. The authors draw a parallel to the Arctic, where adaptation choices may inadvertently create new vulnerabilities and new dependencies on external parties.

These five cases illustrate the importance of shared learning, in which local (practitioner) and external (research) perspectives, when shared, can lead to shared ownership of narratives on future conditions, or at least, an increased desire to achieve this for policy purposes. As we seek ways for moving the discourse beyond the climate change damage report, it is encouraging to see examples where local and regional champions can generate support for incorporating climate change adaptation into local development planning. The challenge is to sustain the effort to support local adaptation initiatives, thereby encouraging other local champions to emerge in rural regions and small towns around the world.

References

Adger WN, Agrawala S, Mirza MMQ et al (2007) Assessment of adaptation practices, options, constraints and capacity. In: Parry ML, Canziani OF, Palutikof JP et al (eds) Climate change 2007: impacts, adaptation and vulnerability – contribution of Working Group II to the Fourth Assessment Report of the Intergovernmental Panel on Climate Change. Cambridge University Press, Cambridge

Adger WN, Dessai S, Goulden M et al (2009) Are there social limits to adaptation to climate change? Clim Change 93(3–4):335–354

Agrawal A, Perrin N (2008) Climate adaptation, local institutions, and rural livelihoods. IFRI working paper no.W08I-6, School of Natural Resources and Environment, University of Michigan, Ann Arbor

Bollman RD, Clemenson HA (2008) Structure and change in Canada's rural demography: an update to 2006. Rural and Small Town Canada Analysis Bulletin 7(7): 1–27. Statistics Canada catalogue no. 21-006-X, Minister of Industry, Ottawa

Burton I (2008) Moving forward on adaptation. In: Lemmen DS, Warren FJ, Lacroix J et al (eds) From impacts to adaptation: Canada in a changing climate 2007. Government of Canada, Ottawa

Burton I, Kates RW, White GF (1993) The environment as hazard. Guilford Press, New York

Canadian Council of Professional Engineers (2009) PIEVC engineering protocol for climate change infrastructure vulnerability assessment – Part I. Version 9, Apr 2009. Engineers Canada, Ottawa

City of Keene, ICLEI-Local Governments for Sustainability (2007) Keene, New Hampshire – adapting to climate change: planning a climate resilient community. International Council for Local Environmental Initiatives [ICLEI] – Local Governments for sustainability, Oakland

City of New York (2008) Assessment and action plan report 1: a report based on the ongoing work of the DEP Climate Change Task Force. Department of Environmental Protection [DEP], New York

Cohen S, Neale T (eds) (2006) Participatory integrated assessment of water management and climate change in the Okanagan Basin, British Columbia. Final report, Project A846. Submitted to Natural Resources Canada, Ottawa. Environment Canada and University of British Columbia, Vancouver

Cohen SJ, Waddell MW (2009) Climate change in the 21st century. McGill-Queen's University Press, Montreal

Columbia Basin Trust (2009) Communities adapting to climate change. http://www.cbt.org/newsroom/?view&vars=1&content=Publication&WebDynID=900. Cited 24 Sept 2010

Dyson T (2005) On development, demography and climate change: the end of the world as we know it? Popul Environ 27(2):117–149

Filmon G, Provincial Review Team (2004) Firestorm 2003 provincial review. Government of British Columbia. http://www.2003firestorm.gov.bc.ca. Cited 30 Apr 2010

Gardner J, Dowd AM, Mason C et al (2009) A framework for stakeholder engagement on climate adaptation. CSIRO climate adaptation flagship working paper no. 3, Commonwealth Scientific and Industrial Research Organization [CSIRO]. http://www.csiro.au/resources/CAF-working-papers.html. Cited 28 Jul 2009

Karl TK, Melillo MJ, Peterson TC (eds) (2009) Global climate change impacts in the United States. Cambridge University Press, New York

Lagadec P (2004) Understanding the French 2003 heat wave experience: beyond the heat, a multi-layered challenge. J Conting Crisis Manag 12(4):160–169

Land Use Consultants, Oxford Brookes University, CAG Consultants et al (2006) Adapting to climate change impacts – a good practice guide for sustainable communities. Department for Environment, Food and Rural Affairs, Her Majesty's Government. http://www.london.gov.uk/lccp/index.jsp. Cited 05 Aug 2009

Langsdale SM, Beall A, Carmichael J et al (2007) An exploration of water resources futures under climate change using system dynamics modelling. Integr Assess J 7(1):57–79

Langsdale SM, Beall A, Carmichael J et al (2009) Exploring the implications of climate change on water resources through participatory modelling: case study of the Okanagan Basin, British Columbia. J Water Resour Plan Manag 13(5):373–381

Leichenko RM, O'Brien KL (2008) Environmental change and globalization: double exposures. Oxford University Press, Oxford

Lim B, Spanger-Siegfried E, Burton I et al (eds) (2004) Adaptation policy frameworks for climate change: developing strategies, policies and measures. United Nations Development Programme, Cambridge University Press, Cambridge

Mills E (2005) Insurance in a climate of change. Science 309(5737):1040–1044

O'Brien KL, Leichenko RM (2000) Double exposure: assessing the impacts of climate change within the context of economic globalization. Global Environ Change 10:221–232

O'Connor DR (2002) Part 1 – a summary, report of the Walkerton inquiry: the events of May 2000 and related issues. Ontario Ministry of the Attorney General, Toronto

Okanagan Water Stewardship Council (2008) Okanagan sustainable water strategy Action Plan 1.0. Okanagan Basin Water Board, Coldstream

Parry ML, Canziani OF, Palutikof JP et al (eds) (2007) Climate change 2007: impacts adaptation and vulnerability – contribution of Working Group II to the Fourth Assessment Report of the Intergovernmental Panel on Climate Change. Cambridge University Press, Cambridge

Perkins B, Ojima D, Corell R (2007) A survey of climate change adaptation planning. The H. John Heinz III Center for Science, Economics and the Environment, Washington

Peterson TC, Anderson DM, Cohen SJ et al (2008) Why weather and climate extremes matter. In: Karl TR, Meehl GA, Miller CD et al (eds) Weather and climate extremes in a changing climate – regions of focus: North America, Hawaii, Caribbean, and U.S. Pacific Islands. A report by the U.S. Climate Change Science Program and the Subcommittee on Global Change Research, Washington, DC

Pielke RA Jr, Agrawala S, Bouwer LM et al (2005) Clarifying the attribution of recent disaster losses: a response to Epstein and McCarthy. Bull Am Meteorol Soc 86(10):1481–1483

Rubinoff P, Vinhateiro ND, Piecuch C (2008) Summary of coastal program initiatives that address sea level rise as a result of global climate change. Rhode Island Sea Grant, Coastal Resources Center, University of Rhode Island, Narragansett, RI

Sauchyn D, Kulshreshtha S (2008) Prairies. In: Lemmen DS, Warren FJ, Lacroix J et al (eds) From impacts to adaptation: Canada in a changing climate 2007. Government of Canada, Ottawa

Shaw A, Sheppard S, Burch S et al (2009) Making local futures tangible: synthesizing, down-scaling, and visualizing climate change scenarios for participatory capacity building. Global Environ Change 19(4):447–463

Snover AK, Whitely Binder L, Lopez J et al (2007) Preparing for climate change: a guidebook for local, regional, and state governments. In association with and published by ICLEI – Local Governments for Sustainability, Oakland

Sommerkorn M, Hamilton N (eds) (2008) Arctic climate impact science: an update since the ACIA. World Wildlife Fund International Arctic Program, Oslo

Swanson DA, Hiley JC, Venema HD and Grosshans R (2009) Indicators of adaptive capacity to climate change for agriculture in the Prairie region of Canada: comparison with field observations. Working paper for the Prairie Climate Resilience Project. International Institute for Sustainable Development, Winnipeg. http://www.iisd.org/pdf/2009/pcr_adaptive_cap_ag.pdf. Cited 24 Aug 2010

Toronto Environment Office (2008) Ahead of the storm: preparing Toronto for climate change. City of Toronto. http://www.toronto.ca/teo/adaptation.htm. Cited 15 May 2008

Townsend FF, Katrina Lessons Learned Review Group (2006) The federal response to Hurricane Katrina: lessons learned. The White House, Washington, DC

United Nations (2006) World urbanization prospects: the 2005 revision. Department of Economic and Social Affairs, United Nations, New York

Wilbanks TJ, Romero Lankao P, Bao M et al (2007) Industry, settlement and society. In: Parry ML, Canzioni OF, Palutikof JP et al (eds) Climate change 2007: impacts adaptation and vulnerability – contribution of Working Group II to the Fourth Assessment Report of the Intergovernmental Panel on Climate Change. Cambridge University Press, Cambridge

Chapter 30
Scenarios for Engaging a Rural Australian Community in Climate Change Adaptation Work

Jodi-Anne Michelle Smith, Martin Mulligan, and Yaso Nadarajah

Abstract The Hamilton region in Victoria, Australia is a rural farming community consisting of several small towns and the regional center of Hamilton. The region is already experiencing climate change, with a steady decline in annual rainfall and available groundwater, and increased frequency of droughts. A prolonged drought has necessitated ongoing water restrictions and forced farmers to alter cropping and stocking practices. Rainfall patterns are predicted to shift further toward the dry, which will affect farm viability, as will increased transport costs due to rising oil prices. The challenges the community face have led to a high local interest in understanding and responding to climate change. When the authors organized a public meeting in April 2007, over 70 people attended. They wanted to take immediate action on climate change, not wait for new national policies.

A scenario thinking workshop was held in February 2008. Forty-one representatives of different sectors within the community participated. They developed four different stories of the future and undertook an initial analysis to identify implications and adaptation strategies. This revealed that climate change could have far more complex impacts on the region than first imagined. Possible impacts included higher levels of financial pressures, stress, mental illness, and addictive behaviors, affecting community cohesion and quality of life, plus possible farm closures, high unemployment, and associated population losses, affecting the viability of small towns. Strategies identified to reduce the vulnerability of the region included altering farming practices, ensuring water security, building social cohesion, attracting new residents, and diversifying employment opportunities. The

J.-A.M. Smith (✉) • Y. Nadarajah
Global Cities Institute, RMIT University, GPO Box 2476, Melbourne, VIC 3001, Australia
e-mail: jodi-anne.smith@rmit.edu.au

M. Mulligan • Y. Nadarajah
Globalism Research Centre, RMIT University, GPO Box 2476, Melbourne, VIC 3001, Australia
e-mail: martin.mulligan@rmit.edu.au; yaso.nadarajah@rmit.edu.au

J.D. Ford and L. Berrang-Ford (eds.), *Climate Change Adaptation in Developed Nations:* 413
From Theory to Practice, Advances in Global Change Research 42,
DOI 10.1007/978-94-007-0567-8_30, © Springer Science+Business Media B.V. 2011

local shire council, regional health service, and others have used the workshop outcomes to rethink their strategic plans.

Keywords Scenario thinking • Community-engaged research • Climate change adaptation • Australia • Farming community • Agriculture • Rural community • Drought • Water availability • Local climate change policy

Introduction

Situated 300 km from the metropolitan center of Melbourne, Hamilton is a significant regional center in the relatively prosperous farming districts of western Victoria. The area is part of the larger Glenelg Hopkins catchment of Victoria. Hamilton's development as a regional center over the past 150 years has been founded on a steady expansion of an agricultural sector based predominantly on wheat, sheep, and cattle grazing.

A prolonged drought in the region, which in part reflects the onset of climate change, has meant tighter water restrictions, and has compelled farmers to alter cropping and stocking practices. Nationally, the Australian Bureau of Agricultural and Resource Economics (ABARE) (Gunasekera et al. 2007) has identified a range of climate change adaptation options for farmers to use in order to diversify their farm production. Locally, the Victorian Department of Primary Industries (DPI) (2008) is working with farmers to implement climate change adaptation options. In Hamilton, the demarcation between areas of farming land suitable for cropping and that for stocking is changing. The Commonwealth Scientific and Industrial Research Organisation (CSIRO) and Bureau of Meteorology (2007) predict a warming, drying climate with a consequent reduction in groundwater volumes and increased frequency and severity of droughts. Rainfall patterns are expected to become even less predictable, and the combination of drier and hotter conditions increases the risk of dangerous grassfires and bushfires. Since these predictions were made, Victoria has experienced record high summer temperatures, and the hot and dry conditions resulted in the devastating bushfires of February 7, 2009, which claimed over 200 lives and more than 2,000 homes.

In an area where the economy is so dependent on farming, water supply is already a critical issue with a current "supply deficiency of 1000ML per annum based on the last 10 years water yield from the Catchment" (Wannon Water 2007, p. 1). Wannon Water (2007) has investigated options to ensure water supply. It is now deciding between two options to bring water via a pipeline or outlet channel, both from over 40 km at an estimated cost of $29 million each. The lack of water has required changes not only to farming practices, but also to water use within homes and gardens in Hamilton itself.

Local-Global Project

Researchers from Royal Melbourne Institute of Technology (RMIT) University's Community Sustainability Program within the Global Cities Institute have been conducting a "Local-Global Project" in the Hamilton region since 2004. The project aims to explore local responses to issues of global scope and character in consultation with a Community Reference Group. Social learning opportunities are created for the community to undertake dialogue and shared research and decide on and respond to issues of growing concern across the region (see Mulligan and Nadarajah 2008 for further details). Climate change became an area of concern raised by the Community Reference Group in 2006. In response to these concerns, the researchers organized a range of guest speakers to present at a climate change public forum in April 2007.

More than 70 people from across the Hamilton region attended the forum. Different views were expressed regarding the likely impacts of climate change in the region, but the prevailing sentiment was that the time for sitting on the fence had passed. The organizers were asked to look for ways to build on the momentum of this gathering. It was decided that a scenario thinking workshop on the impacts of climate change and the future of the region would be a useful next step.

Scenario Thinking

Scenario thinking has been used for the development of military and business strategy (Shell International 2003), complex global problem solving (NI 2007), the development of the nonprofit sector (Scearce and Fulton 2004), the exploration of countries' futures (le Roux 1992; Kahane 1998; Institute of Economic Affairs and Society for International Development 2000), the development of the government policy arena (Ringland 2002), and regional planning (Robertson et al. 2007; Wang et al. 2007; Meadowlark Institute 2009). The value of using scenario approaches for exploring climate change impacts and adaptation options has been recognized by both the United Nations Development Programme Global Environment Facility (2003) and the United Nations Framework Convention on Climate Change (UN-FCCC) Secretariat (2005). Both have provided guidance on how to use the tool in this adaptation context. Despite the recommendations, the scenario process has mostly been used for predicting climate change and its impacts (Dessai et al 2005).

It is only relatively recently that the approach has been used as a way to explore adaptation to climate change. Turnpenny et al. (2005) and Jordan et al. (2001) applied scenario thinking to climate change adaptation. However, they used a "top-down" approach to scenarios development. In essence, this involved their team of "experts" creating the scenarios and then seeking input, feedback, and discussion of the scenario implications with people in the targeted regions.

In contrast, a "bottom-up" approach was used in the current study. A group of diverse community members from the Hamilton region participated in the workshop and created the scenarios themselves. They discussed the implications of the scenarios and brainstormed next steps for action. The authors did not expect the community members to develop "scientifically or technically sound answers" on how the community should adapt to climate change. The scenario thinking workshop was undertaken as a process to (1) find out what community members currently knew about the climate change adaptation issue; (2) engage them in dialogue on the complexity of the issues involved; (3) gather their adaptation ideas; and (4) build their awareness that many different possible futures may unfold based on actions they do or do not take, as well as changes that occur to the climate.

Through undertaking this process, it was hoped that participants would continue the discussions outside of the workshop, leading to ongoing planning and action on climate change adaptation within the region. The scenario stories, once written up, would be a tool for furthering these discussions with those that had not participated in the workshop. The scenario stories would also be useful for government educators and policy makers who could use them to identify any misunderstandings or gaps in awareness of climate change issues within the community. Targeted information and capacity building programs to address such gaps could then be produced.

Scenario Thinking Workshop

The Hamilton scenario thinking workshop was held on February 4–5, 2008 using an inductive scenario approach. Forty-one people participated (18 female, 23 male), including several farmers, a retired school principal, a church minister, an Aboriginal community leader, a publican (i.e., hotel operator), local shire councillors, a Country Fire Authority representative, artists, business personnel, and two new migrants to the area. Participants were split into four small groups and taken through a process for identifying and discussing the many factors likely to affect their community's development in the future. They explored what they felt were the most critical factors facing their community – those which would have a high level of impact and where the outcome of that impact was highly uncertain. The groups identified their five most critical factors. These were shared in plenary to produce a top 20 list of critical factors, which were then voted on by the individual participants to identify the four overall factors seen as most critical. Each group was then allocated one of these top four critical factors to use as the starting point for generating their scenario. Throughout the 2 days, a number of plenary sessions were held where each small group reported back on their developing scenario and received input from the other participants. This ensured differentiation between the scenarios.

The scenarios produced did not focus on disputes about what level of climate change would occur or on the effectiveness of different adaptation strategies. They focused on the adaptation challenges facing the region and the likely impact of these.

By lunchtime on day 2, four broad scenarios were mapped out. In plenary each group told their scenario, and then a discussion was held about the implications of that scenario for the future of the region. After hearing all four scenarios, a list of possible strategies and desired next steps were generated.

After the workshop, the four broad scenarios and all the associated workshop materials were given to two local writers who were allocated the task of turning them into more detailed scenario stories. The writers worked in consultation with a range of workshop participants to develop four plausible stories set well into the future. Due to space restrictions, only synopses of the stories can be included here. For copies of the complete scenario stories, see Nadarajah et al. (2009).

Hamilton Future Stories in Synopsis

Lake Condah Sustainable Development Project, 2030

After many years of persistent drought in the Hamilton region, many farmers have gone out of business and many enterprises associated with farming are struggling to survive. There has been a significant dieback of trees and exposed grasslands and a noticeable reduction in biodiversity. Early mornings that were once filled with birdsong have become eerily quiet. However, the wetland system surrounding the old Aboriginal mission at Lake Condah, which was restored before the long dry began, has become a refuge for birds. The Aboriginal community is benefiting from the success of the Lake Condah Indigenous Discovery Centre and the Lake Condah Bush Foods factory and outlet. Local farmer and bird enthusiast "Old Jack" Murphy had thought the reflooding of the Lake Condah wetlands a waste of time and money when it took place; but 20 years later, on a visit to the Lake Condah Indigenous Discovery Centre in 2030, he is blown away by the refuge that has been created, and especially by the presence of so many species of birds that are no longer sighted anywhere else in the region. He has to admit to Aboriginal community leader Billy Lovatt, whom he had coached in junior football, that his earlier criticism of the project had been well wide of the mark. "It only took you whitefellas about 200 years to realize that you might have something to learn from us about how to look after this country around here," Lovatt told his old coach.

Danny's Story

Danny Brown did it tough. He grew up with struggling parents and he found himself alone, with his beloved dog Beetle, when both parents died soon after he managed to find a job drawing cartoons for the *Weekly Times* newspaper. Danny's talent for drawing had been noticed by one of his teachers at high school and she helped him to get into an appropriate course at technical college and then into the job at the *Weekly Times*. Danny had always been a survivor and, in particular, he managed to accumulate some personal assets by selling rare plastic bottles that he had collected as a child. Because he knew what it was like in his childhood to struggle with so little support and resources, Danny made a special effort to support young people living in the camp for Vietnamese climate change refugees that was set up at the refloodings in 2030. He became a regular visitor to the tent city. However, others in the town were much less hospitable to the refugees who had been assigned to the region. Violent conflicts broke out between rival youth gangs from the camp and the town in 2035, resulting in seven deaths. Hostility toward the refugees intensified when it was learnt that hungry camp residents had killed and eaten some "stray dogs" who had come into their camp; when Danny discovered that his beloved Beetle had gone missing he went to the camp to find him. When Danny confronted some people he knew about what might have happened to his dog, a scuffle broke out and Danny died when his head struck a rock after he was knocked to the ground. Danny's death provoked some strong reactions in town and in the refugee camp. Danny's tragic death helped Hamilton's community leaders to see that they must take action to prevent a further escalation of dangerous social conflicts, and it helped some of the young people he worked with in the refugee camp to understand that there are people in the wider community who could help them break out of their isolation and anger.

Damian McCrae and Georgia D'Ambrosia

When Damian McCrae took over the management of the historic family farm in the district of Cavendish from his father Donald McCrae, the effects of climate change were already causing major adjustments in farming strategies. Damian had just returned from agricultural college with new ideas about how to survive the crisis – ideas that his father found difficult to accept. Soon after taking responsibility for the farm, Damian met and married regional health worker Georgia D'Ambrosia. Georgia had grown up in Melbourne in a family that most farmers would disparagingly label as "urban greenies." Georgia had not been adequately prepared for the role of rural health promotion officer at a time when new and acute health problems were emerging as a result of severe heat stress and the spread of new diseases related to climate change. To make matters worse, a lack of transport options and deteriorating roads made it expensive and difficult to cover the district, and she felt she could never keep up with the demands of her job. Georgia had chosen a rural life because she wanted to be "closer to nature," but nature was making her life difficult and her husband Damian was equally stressed. Unexpectedly, Georgia found solace in talking to Damian's grandfather, Fergus, who was still living on the farm, and she came to understand that his deep local knowledge could be more of an asset in the new conditions than Damian had ever thought.

Nguyen Pham's Campaign Speech in 2050

Nguyen Pham arrived in Hamilton as a climate change refugee from Vietnam when she was just 14. She lived with her parents, a brother and a sister in the tent city established at the Hamilton showgrounds in 2030, and the whole family found the bitter winter hard to take. As dysentery and tuberculosis claimed the lives of other children in the camp, Nguyen and her siblings were cheered by the generous donation of clothes and toys from the local Red Cross and Combined Churches group, and the whole family benefited from English lessons run by some retired teachers in the Ram Sale shed at the showgrounds. Sadly, Nguyen's older brother was killed in the gang wars of 2035 that rocked the Hamilton. The son of the eminent Hamilton heart surgeon Charles Cameron, Brett Cameron had been the nominal leader of the gang responsible for Nguyen's brother's death, and although Brett was not present at the time, Charles decided to make a special effort to support the bereaved family. Many years later Nguyen Pham and Brett Cameron met by chance in the United States, and a shared love of music brought them into a relationship that resulted in marriage. Nguyen and Brett returned to live and work in the Hamilton region, and in the year 2050, Nguyen is launching her campaign to become mayor. Her opening campaign speech recounts the story of what has happened in the district since the tragic death of her brother in 2035, and explains why she thinks that things are looking up for the district in 2050.

Analysis and Implications Arising from Scenarios

As a result of creating the scenarios, participants in the workshop could see a range of implications that they as a community needed to consider and for which they needed to plan responses. They saw a need to find ways to influence individuals to take action to reduce their carbon emissions and adapt their homes and lifestyles sooner rather than later. They saw the opportunity to learn from other cultures and those who have already taken action. They recognized that their current mental health and community welfare services were not sufficient to cope with the increased demand predicted in many of the scenarios. They identified a need for preventative programs and early intervention programs for farmers and families at risk. The idea of creating a National Centre for Farmer Health was raised. Nutrition and healthy lifestyle education programs were also suggested as a way to prevent health problems. Diversifying sources of farm income by moving away from monocultures to a range of different crops and other sources of income were seen as ways to minimize the vulnerability of farmers to climate change.

The participants saw a need to undertake activities to build community cohesion and support each other through tough times. They saw a need to develop disaster response plans to cater for a possible influx of climate refugees from southern Asia and the Pacific; they wanted to start cultural exchange initiatives and awareness raising initiatives now, so that it would not be such a shock to the current population if an influx of migrants did occur. Participants also highlighted a need to consider ways to attract people and businesses to the region. Expanding educational and employment opportunities were seen as key in this regard. This was also seen as a strategy to help stop the trend of youth leaving the region. The participants saw an opportunity for their educational organizations to specialize in environmental education and teach Aboriginal and Asian perspectives of environmental management.

The participants also identified a range of further research required. This included research into ways to farm effectively under drought conditions, minimize water use, and ensure water security. It also included research on the implications of a future shortage of oil and oil-based products on farming practices, transport of produce, and the functioning of other aspects of the community life. Research into the implications of "corporate farms" run by large multinational companies and increased mechanization of farming practices on employment levels in the region were also recommended.

Other Outcomes

Toward the end of the workshop, participants were asked what they would like to see as the next steps of this project. The participants identified a range of strategies including: generating a report on the scenarios that could be widely distributed throughout the community; having secondary school students comment on the scenario stories, possibly to make movies about them; and having the scenarios narrated on Radio National. Of course, a starting point would be for participants to share insights they had gained at the workshop with family and friends in order to inject more urgency into community discussions about future lifestyles in the region, which, in turn, might encourage people to think more urgently about what they can do now to reduce greenhouse gas emissions. Some participants raised the difficulties involved in getting local organizations and agencies to work together, and so an emphasis was placed on a need to build more effective partnerships. There was enthusiasm for completing the local future stories so that the challenges they raise could be discussed widely across the community. It was noted that the shire council and other government organizations needed to review their strategic plans for the future.

One participant in the scenario thinking workshop – regional health worker Rosie Rowe – was sufficiently stimulated by the workshop to go back into the Western District Regional Health Service and work on a policy for how health services might respond to a wide range of health challenges, and this report has been circulated across Victoria (Rowe and Thomas 2008).

Conclusions

The authors are encouraged by the outcomes of the scenario thinking workshop. A plethora of ideas and suggestions for action were generated. To move from ideas into action will require engaging the wider community with the stories generated. Stories based on a combination of expert knowledge and local experiences have the ability to mobilize hearts as well as minds. As a result, rather abstract predictions about climate change impacts become more real to local sensitivities and good stories can have a rather "visceral" impact on people who may have thought that the problems of climate change can be safely left to scientists and politicians.

This use of scenario thinking techniques to provoke deeper community engagement with the challenges of climate change adaptation is the first step of many to be taken. Ongoing initiatives will be required to ensure that the scenario stories are widely distributed and that discussions, planning, and adaptation actions occur within the Hamilton region.

References

Commonwealth Scientific and Industrial Research Organization [CSIRO], Australian Bureau of Meteorology (2007) Climate change in Australia: technical report 2007. CSIRO, Canberra

Department of Primary Industries [DPI] (2008) Farming systems scenario development. Australian Government. http://www.dpi.vic.gov.au/dpi/vro/vrosite.nsf/pages/climate_vccap_scenario_development. Cited 15 Sept 2008

Dessai S, Lu X, Risbey JS (2005) On the role of climate scenarios for adaptation planning. Global Environ Chang Part A 15(2):87–97

Gunasekera D, Ford M, Tulloh C (2007) Climate change: issues and challenges for Australian agriculture and forestry. Aust Commod 14(3):493

Institute of Economic Affairs [IEA-Kenya], Society for International Development [SID] (2000) Kenya at the crossroads: scenarios for our future. IEA-Kenya and SID, Nairobi

Jordan A, Lorenzoni I, Hulme M et al (2001) Co-evolutionary approach to climate change impact assessment: a scenario-based study in the UK. CSERGE Working paper 2000-01, Norwich, UK

Kahane A (1998) Destino Colombia: a scenario-planning process for the new millennium. Deeper News 9(1):1–32

le Roux P (1992) The Mont Fleur scenarios: what will South Africa be like in the year 2002? Deeper News 7(1):1–26

Meadowlark Institute (2009) The Meadowlark Project Scenarios. http://www.meadowlarkproject.org/scenarios.asp. Cited10 Mar 2009

Mulligan M, Nadarajah Y (2008) Working on the sustainability of local communities with a "community-engaged" research methodology. Local Environ 13(2):81–94

Nadarajah Y, Mulligan M, Smith J et al (2009) Unexpected sources of hope: climate change, community and the future. Globalism Research Centre and Global Cities Institute, RMIT University. http://www.rmit.edu.au/browse;ID=90esk0wyyj33. Cited 23 Aug 2010

Nautilus Institute [NI] (2007) Open mind, open futures: how will Asia-Pacific communities respond to global insecurity? Global Scenarios Workshop report, Nautilus Institute, San Francisco

Ringland G (2002) Scenarios in public policy. Wiley, England

Robertson D, Wang QJ, Soste L et al (2007) Irrigation insights 8 – scenarios of the future: irrigation in the Goulburn Broken region. DPI, Future Farming Systems Research, Tatura

Rowe R, Thomas A (2008) Climate change adaptation: a framework for local action - Southern Grampians primary care partnership. Policy signpost no. 3. The McCaughey Centre, University of Melbourne, Melbourne

Scearce D, Fulton K (2004) What if? The art of scenario thinking for non-profits. Global Business Network. http://www.gbn.com/GBNDocumentDisplayServlet.srv?aid=32655&url=/UploadDocumentDisplayServlet.srv?id=34265. Cited 01 Aug 2008

Shell International (2003) Scenarios: an explorer's guide. http://www-static.shell.com/static/aboutshell/downloads/our_strategy/shell_global_scenarios/scenario_explorersguide.pdf. Cited 01 Aug 2008

Turnpenny J, O'Riordan T, Haxeltine A (2005) Developing regional and local scenarios for climate change mitigation and adaptation, part 2: scenario creation. Tyndall Centre Working paper no. 67, University of East Anglia, Norwich

UNFCCC Secretariat (2005) Compendium on methods and tools to evaluate impact of, and vulnerability and adaptation to climate change. UNFCCC and Stratus Consulting Inc.

United Nations Development Programme Global Environment Facility (2003) Developing socio-economic scenarios for use in vulnerability and adaptation assessments. National Communications Support Unit Handbook

Wang QJ, Robertson D, Soste L et al (2007) Irrigation Insights 9: regional scenario planning in practice: irrigation futures of the Goulburn Broken region, Dept of Primary Industries, Future Farming Systems Research, Tatura

Wannon Water (2007) Fact sheet: Wannon Water tackles climate change with 50 year plan Hamilton system and Glenthompson supply. Wannon Water, Victoria

Chapter 31
Coordinating Environmental Protection and Climate Change Adaptation Policy in Resource-Dependent Communities: A Case Study from the Tibetan Plateau

Julia A. Klein, Emily Yeh, Joseph Bump, Yonten Nyima, and Kelly Hopping

Abstract Resource-dependent communities are likely to be disproportionately affected by climate change. Yet, natural resource management policies continue to be developed and implemented without considering climate change adaptation. We highlight that this lack of coordination is potentially harmful to natural resources and resource-dependent communities with an example from the Tibetan Plateau, a region where climate is changing rapidly. Tibetan pastoralists inhabit rangelands that are the focus of recent development and management policies that promote fencing, sedentarization, individual rangeland use rights, and the elimination of grazing in some areas. These policies may have a negative effect on herders' ability to adapt to climate change. China's National Climate Change Programme lists controlling or eliminating grazing in some areas as key for adaptation to climate change, but experimental results indicate that grazing may buffer the rangelands from the negative effects of warming. These findings indicate that policies that support the well-developed strategies of resource-dependent communities for living in uncertain and variable environments can also enhance adaptation of these social and ecological systems to climate change. We conclude that management and environmental protection policies developed separately from climate change policy face increased failure potential and may decrease the ability of natural resources and the communities that depend upon them to successfully adapt to climate change.

J.A. Klein (✉) • J. Bump • K. Hopping
Department of Forest, Rangeland and Watershed Stewardship, College of Natural Resources,
Colorado State University, Campus Delivery 1472, Fort Collins, CO 80523–1472, USA
e-mail: jklein@cnr.colostate.edu; jklein@warnercnr.colostate.edu; joseph.bump@colostate.edu;
kelly.hopping@colostate.edu

E. Yeh • Y. Nyima
Department of Geography, University of Colorado, Boulder, CO, USA
e-mail: emily.yeh@colorado.edu; yy2161@gmail.com

J.D. Ford and L. Berrang-Ford (eds.), *Climate Change Adaptation in Developed Nations:*   423
From Theory to Practice, Advances in Global Change Research 42,
DOI 10.1007/978-94-007-0567-8_31, © Springer Science+Business Media B.V. 2011

Keywords Climate adaptation • Climate change • Ecosystem services • Grazing • Pastoralism • Rangeland policy • Social-ecological systems • Tibetan Plateau • Natural resource management • Resource-dependent communities • Property regimes • Environmental policy

Introduction

Climate change may have the greatest effects on resource-dependent communities, where livelihood activities are tightly coupled to the natural resource base. These communities tend to be economically and politically marginalized within their larger national context, even while their resources make significant contributions to the larger national economy. Rangelands – broadly defined here to include grasslands, shrublands, and tundra ecosystems – comprise almost half of the global land surface (Olson 1994). Rangeland systems provide many ecosystem services, i.e., benefits to people, ranging from valuable products to essential regulation of ecosystem health. For example, rangeland regions provide meat and dairy products, wool, oil and gas reserves, mineral resources, freshwater supplies, wildlife habitat, and opportunities for recreation and tourism. These regions also regulate climate, soil erosion, the quantity and quality of freshwater resources, and nutrient retention and cycling. Here, we focus on resource-dependent communities on the vast rangelands of the Tibetan Plateau region of China. Through this case study, we demonstrate that rangeland policy can influence the sensitivity of these ecosystems to climate change and potentially alter the capacity of the resource-dependent communities on the Tibetan Plateau to adapt.

China is not an Annex I party to the United Nations Framework Convention on Climate Change (UNFCCC). However, China's status in the international arena has been rapidly changing since the UNFCCC began 15 years ago. Today, China, the world's second largest economy, shares many characteristics with Annex I countries, including global economic clout, political power, regional and increasingly global leadership, and the highest annual greenhouse gas emissions in the world (Grumbine 2007). It also has large socioeconomic disparities between its urban, industrialized populations and its rural, resource-dependent populations. In the face of this large and growing inequality, wealthy, urban citizens of China arguably have more in common with citizens of wealthy, industrialized countries than with marginalized, resource-dependent Chinese citizens such as those who live on the Tibetan Plateau; similarly Tibetan herders have, in many respects, more in common with Saami reindeer herders than they do with many other Chinese citizens. Indeed, in considering questions of resource access and climate change adaptation, the analytically developed/developing or First/Third world divide can obscure more than it reveals (McCarthy 2005; Wainwright 2005).

Tibetan pastoralist communities have much in common with resource-dependent, indigenous communities of the industrialized countries of the Arctic, including the USA and Canada, as well as Norway, Sweden, Finland, and Russia, where rapid and significant climate change has occurred in combination with recent sociopolitical and environmental changes that increase vulnerability to climate change (Chapin et al. 2004; Ford et al. 2006, 2007; Tyler et al. 2007; Rees et al. 2008). The lesson this Tibetan Plateau case study provides is that policies that affect the status and use of natural resources will likely affect the capacity of those resources, and the communities who depend on them, to adapt to climate change. Therefore, environmental policies should be considered in the context of climate change and should complement climate adaptation policies. Moreover, policies that support resource-dependent communities' well-developed strategies for living in uncertain and variable environments will enhance adaptation of these socioecological systems to climate change.

The Tibetan Plateau

The Tibetan Plateau, situated at an average elevation of 4,500 m and covering 2.5 million km^2, is the largest and highest plateau in the world. Six of Asia's major rivers originate on the Tibetan Plateau, and approximately three billion people live on or downstream of the Plateau. The physical presence of the Tibetan Plateau influences the global climate system (Raymo and Ruddiman 1992; Kutzbach et al. 1993). Moreover, the timing and duration of the Tibetan Plateau snowpack regulates the timing and intensity of the Asian monsoon and thus has large regional climate effects (Wu and Qian 2003; Fasullo 2004; Zhao and Moore 2004; Zhang et al. 2004).

The physical environment of the Tibetan Plateau is highly variable both spatially and temporally. Approximately 85% of the Plateau landmass is above 3,000 m and 50% is above 4,500 m. However, the elevation range on the Plateau spans from less than 3,000 m to more than 8,000 m. Mean annual precipitation ranges from more than 700 mm to less than 50 mm, and mean annual temperatures range from more than 6°C to less than −10°C (Schaller 1998). Climatic conditions are also highly variable on seasonal, monthly, and daily timescales.

With the exception of its southeastern margins, the Tibetan Plateau is primarily a treeless landscape. The vegetation communities are largely determined by elevation and climate, although grazing and soils play a significant local role. Long et al. (2008) identified four major ecosystem types on the Tibetan Plateau: alpine meadow, alpine steppe, alpine desert, and alpine shrub. These vegetative communities combine to form the vast rangelands of the Tibetan Plateau, which is one of the most extensive grazing systems in the world (Miller 1990). These rangelands support Tibetan livestock and a diverse assemblage of wildlife (Schaller 1998).

Pastoralism has been the main form of subsistence for the Plateau's inhabitants for millennia, and animal husbandry is still the main source of livelihood over much of the Plateau (Miller 1999a) (Fig. 31.1). Today, grazing practices are semiresident, characterized by a winter area used as a home base and seasonal mobility to track favorable forage conditions (Long et al. 2008). Herders have a diverse mix of livestock, including sheep, goats, and yaks, with proportions varying

Fig. 31.1 Tibetan boy and girl riding a yak on the eastern region of the Tibetan Plateau

across the Plateau. Natural vegetation generally supports livestock year round, with forage availability and quality improving throughout summer, and herds primarily subsisting on senesced vegetation throughout winter. While livestock production is the base of the economy, other activities, such as the collection and trade of medicinal plants, are increasing in importance in some regions of the Plateau (Janes 1999; Law and Salick 2005; Winkler 2008).

Climate Change on the Tibetan Plateau

Several lines of evidence suggest that climate warming is already occurring on the Tibetan Plateau. This evidence includes ice core data (Thompson et al. 1993, 2000), long-term meteorological measurements (Liu and Chen 2000; Wu et al. 2007; Liu et al. 2008; You et al. 2008), glacial retreat (Yao et al. 2007), and local residents' observations of climate change on the eastern region of the Plateau (Byg and Salick 2009). Mean surface temperature has risen by up to 0.3°C every 10 years over the last 50 years (approximately three times the global warming rate), and future warming on the Plateau is predicted to be "well above the global mean" (Christensen et al. 2007). Eighty two percent of the Plateau's glaciers have retreated in the past half-century, and 10% of its permafrost has degraded in the past decade (Qui 2008). As these changes continue, or even accelerate, their effects will resonate far beyond the Plateau (Cyranoski 2005; Qui 2008).

Climate change is not only leading to warmer temperatures, but is also predicted to increase the frequency and intensity of extreme weather events. Precipitation extremes are predicted to increase, particularly in regions where positive trends in mean annual precipitation are expected (Meehl et al. 2007). The Tibetan Plateau is among the regions where increases in mean annual precipitation are predicted to occur (Christensen et al. 2007). Empirical evidence and models suggest it is very likely that extreme precipitation events will increase in the mid-latitudes, with greater increases in the frequency than in the magnitude of these events (Christensen et al. 2007). On the Tibetan Plateau, extreme weather events consist of snowstorms that cover the vegetation and prevent animals from accessing their primary food source. During the winter months, livestock tend to survive on intact, but senesced vegetation on the rangelands. Where supplementary feed is not available, livestock lose up to 30% of their body weight over winter (Miller 1998; Wu and Yan 2002). Therefore, large snowstorms that cover the vegetation for extended periods of time can cause high livestock mortality rates. Snowstorms are not a new phenomenon on the Tibetan Plateau. However, climate models predict an overall increase in winter precipitation on the Tibetan Plateau (Christensen et al. 2007), and data suggest that the frequency and intensity of snowstorms may be increasing. Folland and Karl (2001) conclude it is likely there has been a widespread increase in heavy and extreme precipitation events in the mid- and high latitudes of the Northern Hemisphere. Regionally, there is evidence for increasing spring snow accumulation over the past few decades on the Tibetan Plateau (Zhang et al. 2004; Niu et al. 2004). The observed changes in extremes are qualitatively consistent with changes in model simulations of future climate (Christensen et al. 2007).

Political-Economic Changes on the Tibetan Plateau

At the same time, dramatic political-economic changes have taken place on the Tibetan Plateau over the past 60 years, with the region's incorporation into the People's Republic of China. From the 1950s through the 1980s, pastures and livestock were collectivized and managed through communes (Goldstein and Beall 1990; Bauer 2008). Herd decollectivization to households took place in the 1980s, while pasture remained, at first, held in common. However, since China's economic reforms, the direction of policy has been toward greater marketization and privatization. Herders soon came to be seen as insufficiently market-oriented (Clarke 1987, 1998; Levine 1998, 1999). At the same time, policymakers began to be concerned about grassland degradation on the Tibetan Plateau, which, based on "tragedy of the commons" assumptions, they attributed to overgrazing and irrational management (Williams 2002; Harris 2008). Privatization, it was assumed, would be better both economically and environmentally.

This dual rationale led to the extension of the Household Responsibility System to pastoral areas, in which grassland use rights are leased to households for periods of 50 years. Implementation began in the eastern Tibetan Plateau in the mid-1990s, but is still ongoing in the Tibet Autonomous Region (TAR), where in some cases leasing has been to household groups or hamlets rather than households.

This grassland use rights policy was accompanied by further efforts at sedentarization as well as an extension of fencing. This included not only fencing of reserve pastures for the winter–spring period, but also in some places, fencing of boundaries between household groups and fencing of household winter pastures. It was believed that this suite of interventions would increase off-take rate, reduce overgrazing and thus degradation, and also encourage Tibetan herders to become more market-oriented commodity producers. However, there has been little evidence to date of improvement of range conditions. Instead, various problems have emerged with the implementation of fencing policies in different parts of the Plateau, including increased labor inputs in some areas, boundary conflicts, security concerns, problems for wildlife mobility, and greater difficulty for some herders in accessing water resources (Wu and Richard 1999; Yeh 2003; Yan et al. 2005; Yan and Wu 2005). While the overall benefits of the "enclosure movement" occurring on the Plateau thus remain in doubt, it is clear that the primary functional effect of the trend toward the fixing of particular pasture use rights to particular households, and the technology of fencing itself, has been to decrease mobility and move away from traditional systems of common property.

The assumption that traditional grazing patterns are environmentally destructive led to a new policy in 2003, tuimu huancao, or "converting pastures to grasslands." This calls for fencing off areas of purportedly degraded rangelands, some for a few months at a time, and others for 5–10 years, or even permanently. In its most dramatic form, found in the 150,000 km^2 Sanjiangyuan nature reserve in Qinghai Province, it is being implemented in conjunction with ecological migration and plans for the resettlement of 100,000 nomads to towns (Foggin 2008). In these cases, the policy represents a break from previous policy trends in calling not for a technological fix to purported degradation, but rather for a complete removal of grazing from the landscape (Yeh 2005; Harris 2008).

Tibetan Pastoral Management Strategies in a Variable Environment

Environmental, economic, and political risks have long shaped pastoral social organization and livestock management practices (McCabe 2004). Tibetan pastoralists, like their resource-dependent counterparts in other regions of the world, have a considerable body of practical knowledge about surviving in their uncertain environments, and historically practiced risk-reducing adaptive strategies (Goldstein and Beall 1990; Wu 1997; Miller 1998). Among the most important for the minimization of loss from extreme weather events or conditions are multispecies grazing, mobility (Miller 2000), reserving pasture, and networks of reciprocity. Multispecies grazing, practiced throughout most of the Plateau, maximizes use of rangeland resources because different species generally exhibit different forage and habitat preferences, and minimizes risk of livestock loss from disease or extreme weather events (Mace and Houston 1989; Mace 1990, 1993; Cincotta et al. 1992; Putman 1996; Tichit et al. 2004; Shrestha and Wegge 2008a,b). Consequently, management policies or economic incentives that favor livestock monocultures may increase herder vulnerability.

Studies in rangeland systems around the world have shown that livestock mobility is beneficial to the health of rangelands, while restriction often leads to rangeland degradation (Sneath 1998; Kerven and Alimaev 1998; Humphrey and Sneath 1999; McCabe 2004; Niamir-Fuller 2005; Bedunah et al. 2006, p. 127; Kerven et al. 2008). Mobility enables African pastoralists to opportunistically access rangeland resources when droughts occur (McCabe 2004; Niamir-Fuller 2005). Flexibility and mobility are among the most important assets European reindeer herders possess for adaptation to the challenges they face (Rees et al. 2008). Access to landscape diversity is a key contributor to resilience among resource-dependent communities in the northern regions of North America and Europe (Chapin et al. 2004). Furthermore, studies show much lower levels of rangeland degradation in Mongolia where mobile pastoralism is not restricted, compared to neighboring pastoral Russia and China, where sedentarization is promoted by the state (Sneath 1998). In Australia, agistment arrangements, where livestock are transferred between pastoral enterprises where forage shortages exist to those where forage excesses exist, foster livestock mobility and enhance pastoral success (McAllister et al. 2006). In Tibet, pastoralists report that it is constant trampling rather than grazing that is more important in driving deterioration of grassland conditions, a factor also reported as contributing to grassland degradation in Inner Asia (Humphrey and Sneath 1999); livestock movements can alleviate this stress by reducing concentrated trampling of the grasslands.

In Tibet, seasonal livestock migration and temporary migration are important strategies used by pastoralists to access superior resources and to manage natural disasters (Miller 1998, 1999a, b, 2000). Two of the most important environmental constraints that affect Tibetan herders' livelihoods are (1) severe snowstorms, especially in early winter or spring, that cover vegetation and prevent grazing, and (2) inadequate growing season soil moisture that results in insufficient vegetation. To cope with these constraints, Tibetan herders migrate individually or in groups to better vegetation conditions. Seasonal migration occurs annually, but the timing, distance, and duration vary. Temporary migration occurs in response to locally acute environmental conditions, for example, herders may move to places where they perceive there is less, or no, snow and better vegetation.

Our work suggests that rangeland policies on the Tibetan Plateau may be decreasing the capacity of that ecological and social system to adapt to climate change. The recent grazing removal policy may increase the sensitivity of Tibetan ecosystems to climate warming, which will require greater adaptation of the social and ecological system. Furthermore, the political-economic transformations documented above have changed herding and pasture management strategies that developed as means to survive in a variable and unpredictable environment.

Ecosystem Sensitivity to Climate Warming

In 1997, we initiated an experimental study to examine the separate and combined effects of warming and grazing in the northeastern Tibetan Plateau (Klein et al. 2004, 2007, 2008). This work was motivated in part by the need to investigate the controls of different factors on the structure and function of the Tibetan Plateau grassland ecosystem. We conducted this research at the Haibei research station, which is situated at latitude 37°37′N, longitude 101°12′E. We set up the experiment in a higher elevation, summer-grazed shrubland habitat and a lower elevation, winter-grazed meadow habitat. Within each habitat type, we fenced two 30 × 30 m areas within which we placed the experimental plots. Our treatments were simulated warming, grazing, and combined warming x grazing. We simulated warming using open top chambers (OTCs) (Fig. 31.2). The chambers mimic the greenhouse effect in that they allow transmission of shortwave radiation, but block and reradiate downwards the infrared radiation. OTCs are used by the International Tundra Experiment and are commonly employed to study the effects of climate warming on ecosystems (Marion et al. 1997; Arft et al. 1999). The OTCs consistently elevated growing season averaged mean daily air temperature at 10 cm above the soil surface by 1.0–2.0°C. We simulated the defoliation effects of grazing through selective clipping, which mimicked the grazing patterns of local livestock and had the same effects on measured plant properties (Klein et al. 2007, 2008). Additional details on the study site, the experimental design, and the microclimate effects of the treatments are documented in Klein et al. (2004, 2005, 2008).

We found that warming generally had negative effects on key vegetative properties. Warming in the absence of grazing caused a significant decrease in overall vegetative production and a decrease in overall plant diversity. These changes represent a decrease in the carrying capacity of the rangelands, a loss of potentially important species, and a potential loss in the ability of the system to respond to future perturbations. Warming also led to an expansion of the shrublands at the expense of meadow vegetation. Increasing shrubs on the landscape has implications for the system's energy balance, carbon storage, carrying capacity, and herd composition. We also examined how important ecosystem services responded to warming, specifically examining two: palatable livestock forage and medicinal plants. Palatable forages are plants that the animals prefer to graze, those that are nutritious and enhance livestock health and survival, and which thus form the basis of the entire animal husbandry system. Medicinal plants are utilized by pastoralists both for local health and as a source of income, as the domestic and international markets for Tibetan medicine are growing (Janes 1999; Law and Salick 2005). We found that warming also reduced the number of medicinal and palatable plant species, resulting in a potential warming-induced loss of ecosystem services on the Plateau.

Grazing buffered the system from the warming-induced losses of vegetative properties and ecosystem services. That is, when warming occurred in the presence

Fig. 31.2 Experimental warming plots (using open-top chambers) within the winter-grazed meadow site at the Haibei Research Station, northeastern Tibetan Plateau (latitude 37°37′N, longitude 101°12′E). Sheep are grazing in the background after a light April snowstorm. Curved, dark feature on the hill is a fence made from sod

as opposed to in the absence of grazing, there were fewer changes in vegetative characteristics and ecosystem services (Fig. 31.3). For example, by the fourth year of the experiment, warming in the absence of grazing significantly decreased total aboveground net primary production (ANPP) by 60 $g^{-2}year^{-1}$, while warming in the presence of grazing decreased total ANPP by 30 $g^{-2}year^{-1}$; furthermore, the amount of ANPP in the combined warming and grazing plots was not statistically different from that in the control plots (Klein et al. 2007). Not only did grazing mediate warming induced losses in total ANPP, it also mediated losses of overall plant diversity. For example, by the fourth year of the study, warming in the absence of grazing lead to a loss of nine species from the system; in contrast, warming in the presence of grazing had no effect on species richness (Klein et al. 2004). Thus, we found that grazing made the system less sensitive to warming. By dampening the negative impacts of warming, grazing reduced the amount of adaptation required to climate change. Current rangeland policies, which are removing grazing altogether from some regions of the Plateau, may increase the sensitivity of the system to climate warming and thus require greater adaptation to climate change in the region.

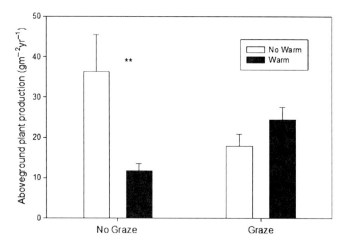

Fig. 31.3 Warming effects on the aboveground annual production of the medicinal plant *Gentiana straminea* in 2001. Warming significantly reduced the production of this medicinal plant in the non-grazed plots but had no significant effect on plant production in the grazed plots. The average production in warmed plots is represented by the *black bars*, while the average production in non-warmed plots is represented by the *white bars*. Nongrazed plots are on the *left*, while grazed plots are on the *right*. Bars represent means and standard errors. While this figure depicts results for a single species, this pattern was also observed with respect to total aboveground biomass and species richness (see Klein et al. 2004, 2007, 2008)

Adaptation to Snow Disasters

While phenomena such as drought and snow are climatic events, hunger and starvation are social ones (Sen 1981; Watts 1983; Watts and Bohle 1993). Thus, an analysis of vulnerability to snowstorms requires understanding not only physical conditions, but also political-economic processes. Miller (1998, 2000) suggests that as a result of mobility-based and other adaptive strategies, Tibetan herders did not historically experience large-scale famine following periodic drought and snowstorms. Our preliminary research shows that nevertheless, snowstorms did historically cause large-scale loss of livestock and in some cases, dramatic impoverishment. The snowstorms of 1997–1998, in which millions of livestock perished, were particularly devastating and were, in many places, the worst in living memory. The Chinese government called for emergency food relief for the herders, and in some areas such as Nagchu, implemented livestock restocking efforts for the hardest hit families.

The severity of the effects of the 1997–1998 snowstorms, including the need for emergency food aid, suggests the need to understand how Tibetan herders' vulnerability to snowstorms may be changing. As discussed above, climate models predict that extreme weather events, such as snowstorms, may become more severe and intense (Christensen et al. 2007). Furthermore, weaker livestock are more susceptible to perishing in snowstorms; thus, conditions that weaken livestock make herders more vulnerable to climate change. Our preliminary research in Nagchu, one of the hardest hit areas of the 1997–1998 storms, suggests that herders widely regard their livestock to have become weaker in recent years, due to deteriorating conditions of the rangelands. This in turn is likely to be driven in part by the effects of climate change, the mobility-reducing rangeland policies documented above, or interactions between the two (Williams 2002; Wu and Yan 2002; Yan et al. 2005; Yan and Wu 2005; Harris 2008). In addition, summer drought conditions preceding a severe snowstorm also weaken livestock, making them less likely to survive. Thus, the probable increase in frequency of all kinds of extreme weather events with climate change is also likely to act as a factor that increases herders' vulnerability.

In addition to livestock condition, the other key factor in determining livestock mortality in a severe snowstorm is ability to access foodstuff. Historically, temporary migration to pastures less covered with snow has been one of the only options available for coping with severe storms in the short term. However, the current trend toward hardened property boundaries and decreased mobility has the potential to exacerbate future vulnerability by reducing the ability to move in the short term. Decreased mobility may be offset by ongoing infrastructural improvements, such as the availability of telephones (information access) and better roads (to receive government aid and move livestock by truck), but the degree to which these can compensate for decreased flexibility remains to be seen. For example, herders may in the future have better information about good locations and even the ability to move their herds there through access to roads, but they may lack the rights or permission from other herders to graze their livestock on allocated pastures. While during the 1997–1998 events most herders were able to move and graze on other herders' pastures without being charged a fee, the ongoing reconfiguration of traditional social networks through economic reform and state development programs makes this an open question in future events. It also remains to be seen whether the government will remain flexible enough to follow the current policy, which calls for opening pastures, even those that have recently been declared permanently closed to grazing by tuimu huancao, to herders during snowstorms.

The other major option for reducing vulnerability to snowstorms would be to increase stored forage so that movement is unnecessary, even during a prolonged snowstorm. The increase in general income levels is leading to greater ability to store grain, and the government is making efforts toward hay production. However, the labor, skills, and capital inputs needed to produce adequate forage for such an event currently still fall far short of what would be required in many areas (Wu and Yan 2002). The government is also currently developing grain storage areas for distribution during snowstorms. What this indicates is an increasing trend toward relying on external aid during times of crisis, rather than traditional methods that herders could engage in themselves. In this sense, current rangeland policies may be fostering a system that is less internally resilient to extreme events that are expected to increase with climate change. Similarly, in a discussion of reindeer herders in Europe, Rees et al. (2008) assert that trucking reindeer to preferred pastures and providing artificial feed may increase herders' vulnerability to threats such as climate change, as reliance on these improvements increases herders' dependency on the state and decreases their flexibility to respond to these threats.

Conclusion

China's National Climate Change Programme suggests that its current rangeland policies constitute a "key area for adaptation to climate change" (PRC 2007). However, an examination of these current policies in their larger historical policy context suggests that they have been formulated and shaped by non-climate-related considerations, including a strong basis in tragedy of the commons assumptions about overgrazing and the benefits of private property, and the need to turn herders into commodity producers for development. Regardless of their origin, our experimental results indicate that current rangeland policies may do the opposite of what would be desirable from the perspective of adaptation to climate change. The grazing removal policy could increase the negative warming effects on the rangelands and require even greater adaptation. At the same time, the restriction on mobility through fencing and an emphasis on individual rangeland use rights limit the ability of herders to adapt to climate change.

The lesson to be learned from this case study is that policies that affect the status and use of natural resources will likely affect the capacity of those resources, and the communities who depend on them, to adapt to climate change. This is important for other resource-dependent communities such as those found in the Arctic, where mobility, resource flexibility, and strong social networks have (as in the Tibetan case) traditionally facilitated a mitigation of vulnerability, but where social, political, and biophysical changes have begun to undermine this adaptive capacity (Chapin et al. 2004; Tyler et al. 2007; Rees et al. 2008). In all cases, policymakers should take great care to develop environmental, economic, and climate change adaptation policies that are complementary. This will require significant institutional and bureaucratic coordination and cooperation, and a commitment from the top to ensure that all policies, and not only those that specifically target climate change, work toward, or at least do not contradict, the goal of climate change adaptation. Finally, policies that support the long-standing strategies developed by resource-dependent communities for living in highly variable environments will also enhance adaptation of these social and ecological systems to climate change.

References

Arft AM, Walker MD, Gurevitch J et al (1999) Responses of tundra plants to experimental warming: meta-analysis of the international tundra experiment. Ecol Monogr 69(4):491–511

Bauer K (2008) Land use, common property, and development among pastoralists in Central Tibet (1884–2004). Dissertation, Oxford University

Bedunah DJ, McArthur J, Durant E et al (comps) (2006) Rangelands of Central Asia: proceedings of the conference on transformations, issues, and future challenges, proceeding RMRS-P-39. U.S. Department of Agriculture, Forest Service, Rocky Mountain Research Station, Fort Collins

Byg A, Salick J (2009) Local perspectives on a global phenomenon: climate change in Eastern Tibetan villages. Global Environ Change 19:156–166

Chapin F, Peterson G, Berkes F et al (2004) Resilience and vulnerability of northern regions to social and environmental change. AMBIO J Hum Environ 33(6):344–349

Christensen JH, Hewitson B, Busuioc A et al (2007) Regional climate projections. In: Solomon S, Qin D, Manning M et al (eds) Climate change 2007: the physical science basis – contribution of Working Group I to the Fourth Assessment Report of the Intergovernmental Panel on Climate Change. Cambridge University Press, Cambridge

Cincotta RP, Zhang Y, Zhou X (1992) Transhumant alpine pastoralism in northeastern Qinghai Province: an evaluation of livestock population response during China's agrarian economic reform. Nomadic Peoples 30:3–25

Clarke GE (1987) China's reforms of Tibet and their effect on pastoralism, IDS discussion. Paper no. 237, Brighton

Clarke GE (1998) Development, society and environment in Tibet, and socio-economic change and the environment in a pastoral area of Lhasa municipality. In: Development, society, and environment in Tibet. Springer, New York

Cyranoski D (2005) The long-range forecast. Nature 438:275–276

Fasullo J (2004) A stratified diagnosis of the Indian monsoon-Eurasian snow cover relationship. J Clim 17(5):1110–1122

Foggin JM (2008) Depopulating the Tibetan grasslands: national policies and perspectives for the future of Tibetan herders in Qinghai Province, China. Mt Res Dev 28(1):26–31

Folland CK, Karl TR (2001) Observed climate variability and change. In: Houghton JT, Griggs YD, Noguer DJ et al (eds) Climate change 2001: the scientific basis – contribution of Working Group I to the Third Assessment Report of the Intergovernmental Panel on Climate Change. Cambridge University Press, Cambridge

Ford J, Smit B, Wandel J et al (2006) Vulnerability to climate change in Igloolik, Nunavut: what we can learn from the past and present. Polar Rec 42(2):127–138

Ford J, Pearce T, Smit B et al (2007) Reducing vulnerability to climate change in the Arctic: the case of the Nunavut Canada. Arctic 60(2):150–166

Goldstein MC, Beall CM (1990) Nomads of western Tibet: the survival of a way of life. University of California Press, Berkeley

Grumbine RE (2007) China's emergence and the prospects for global sustainability. BioScience 57(3):249–255

Harris R (2008) Wildlife conservation in China: preserving the habitat of China's wild west. M.E. Sharpe, Armonk

Humphrey C, Sneath D (1999) The end of nomadism? Pastoralism in the post-socialist economies of Central Asia. Duke University Press, Durham

Janes CR (1999) The health transition, global modernity and the crisis of traditional medicine: the Tibetan case. Soc Sci Med 48(12):1803–1820

Kerven C, Alimaev II (1998) Mobility and the market: economic and environmental impacts of privatization on pastoralists in Kazakhstan. Paper presented at the conference on strategic considerations on the development of Central Asia, Urumqi, 13–18 Sept 1998

Kerven C, Shenbaev K, Alimaev II et al (2008) Livestock mobility and degradation in Kazakhstan's semi-arid rangelands: scale of livestock mobility in Kazakstan. In: Behnke R (ed) The socio-economic causes and consequences of desertification in Central Asia. Springer, Dordrecht

Klein JA, Harte J, Zhao XQ (2004) Experimental warming causes large and rapid species loss, dampened by simulated grazing, on the Tibetan Plateau. Ecol Lett 7(12):1170–1179

Klein JA, Harte J, Zhao XQ (2005) Dynamic and complex microclimate responses to warming and grazing manipulations. Glob Chang Biol 11(9):440–1451

Klein JA, Harte J, Zhao XQ (2007) Independent and combined effects of experimental warming and grazing on forage production and nutrition: implications for rangeland quality on the Tibetan Plateau. Ecol Appl 17(2):541–557

Klein JA, Harte J, Zhao XQ (2008) Decline in medicinal and forage species with warming is mediated by plant traits on the Tibetan Plateau. Ecosystems 11(5):775–789

Kutzbach JE, Prell WL, Ruddiman WF (1993) Sensitivity of Eurasian climate to surface uplift of the Tibetan Plateau. J Geol 101(2):177–190

Law W, Salick J (2005) Human-induced dwarfing of Himalayan snow lotus, Saussurea laniceps (Asteraceae). Proc Natl Acad Sci USA 102(29):10218–10220

Levine NE (1998) From nomads to ranchers: managing pasture among ethnic Tibetans in Sichuan. In: Clarke G (ed) Development, society, and environment in Tibet. Springer, New York

Levine NE (1999) Cattle and the cash economy: responses to change among Tibetan pastoralists in Sichuan, China. Hum Organ 58(2):161–172

Liu X, Chen B (2000) Climatic warming in the Tibetan Plateau during recent decades. Int J Climatol 20(14):1729–1742

Liu W, Gui Q, Wang Y (2008) Temporal-spatial climate change in the last 35 years in Tibet and its geo-environmental consequences. Environ Geol 54:1747–1754

Long RJ, Ding LM, Shang ZH et al (2008) The yak grazing system on the Qinghai-Tibetan Plateau and its status. Rangel J 30:241–246

Mace R (1990) Pastoralists herd composition in unpredictable environments: a comparison of model predictions and data from camel-keeping groups. Agric Syst 33(1):1–11

Mace R (1993) Nomadic pastoralists adopt subsistence strategies that maximize long-term household viability. Behav Ecol Sociobiol 33(5):329–334

Mace R, Houston A (1989) Pastoralists strategies for survival in unpredictable environments: a model of herd composition that maximizes household viability. Agric Syst 31:185–204

Marion GM, Henry GHR, Freckman DW et al (1997) Open-top designs for manipulating field temperature in high-latitude ecosystems. Glob Chang Biol 3(Suppl-1):20–32

McAllister RRJ, Gordon IJ, Janssen MA et al (2006) Pastoralists' responses to variation of rangeland resources in time and space. Ecol Appl 16(2):572–583

McCabe T (2004) Cattle bring us to our enemies: Turkana ecology, politics and raiding in a disequilibrium system. University of Michigan Press, Ann Arbor

McCarthy J (2005) First world political ecology: directions and challenges. Environ Plann A 37(6):953–958

Meehl GA, Stocker TF, Collins WD et al (2007) Global climate projections. In: Solomon S, Qin D, Manning M et al (eds) Climate change 2007: the physical science basis – contribution of Working Group I to the Fourth Assessment Report of the Intergovernmental Panel on Climate Change. Cambridge University Press, Cambridge

Miller DJ (1990) Grasslands of the Tibetan Plateau. Rangel 12(3):159–163

Miller DJ (1998) Hard times on the plateau. Chinabrief 1:17–22

Miller DJ (1999a) Nomads of the Tibetan Plateau rangelands in Western China part one: pastoral history. Rangel 20(6):24–29

Miller DJ (1999b) Nomads of the Tibetan Plateau rangelands in Western China part two: pastoral production practices. Rangel 21(1):16–19

Miller DJ (2000) Tough times for Tibetan nomads in Western China: snowstorms, settling down, fences, and the demise of traditional nomadic pastoralism. Nomadic Peoples 4(1):83–109

Niamir-Fuller M (2005) Managing mobility in African rangelands. In: Collective action and property rights for sustainable rangeland management. Collective action and property rights research brief, pp 5–7

Niu T, Chen L, Zhou Z (2004) The characteristics of climate change over the Tibetan Plateau in the last 40 years and the detection of climatic jumps. Adv Atmos Sci 21(2):193–203

Olson JS (1994) Global ecosystem framework-definitions. USGS EROS Data Center Internal Report 39, Sioux Falls

People's Republic of China [PRC] (2007) China's national climate change programme. Prepared under the auspices of National Development and Reform Commission, People's Republic of China

Putman RJ (1996) Competition and resource partitioning in temperate ungulate assemblies. Chapman & Hall, New York

Qui J (2008) The third pole. Nature 454:393–396

Raymo ME, Ruddiman WR (1992) Tectonic forcing of late Cenozoic climate. Nature 359:117–122

Rees WG, Stammler FM, Danks FS et al (2008) Vulnerability of European reindeer husbandry to global change. Clim Change 87(1–2):199–217

Schaller GB (1998) Wildlife of the Tibetan Steppe. University of Chicago Press, Chicago

Sen A (1981) Poverty and famines: an essay on entitlement and deprivation. Oxford University Press, Oxford

Shrestha R, Wegge P (2008a) Habitat relationships between wild and domestic ungulates in Nepalese Trans-Himalaya. J Arid Environ 72(6):914–925

Shrestha R, Wegge P (2008b) Wild and domestic ungulates in Nepal Trans – Himalaya: competition or resource partitioning? Environ Conserv 35:1–12

Sneath D (1998) State policy and pasture degradation in Inner Asia. Science 281(5380):1147–1148

Thompson LG, Mosley-Thompson E, Davis ME et al (1993) Recent warming: ice core evidence from tropical ice cores with emphasis on Central Asia. Global Planet Change 7(1–3):145–156

Thompson LG, Yao T, Mosley-Thompson E et al (2000) A high-resolution millennial record of the South Asian Monsoon from Himalayan ice cores. Science 289(5486):1916–1919

Tichit M, Hubert B, Doyen L et al (2004) A viability model to assess the sustainability of mixed herds under climatic uncertainty. Anim Res 53(5):405–417

Tyler NJC, Turi JM, Sundset MA et al (2007) Saami reindeer pastoralism under climate change: applying a generalized framework for vulnerability studies to a sub-arctic social-ecological system. Global Environ Chang 17(2):191–206

Wainwright J (2005) The geographies of political ecology: after Edward Said. Environ Plann A 37(6):1033–1043

Watts M (1983) Silent violence: food, famine, and peasantry in Northern Nigeria. University of California Press, Berkeley

Watts M, Bohle H (1993) The space of vulnerability: the causal structure of hunger and famine. Prog Hum Geog 17:43–68

Williams DM (2002) Beyond great walls: environment, identity and development on the Chinese grasslands of Inner Mongolia. Stanford University Press, Stanford

Winkler D (2008) Yartsa Gunbu (Cordyceps sinensis) and the fungal commodification of Tibet's rural economy. Econ Bot 62(3):1–15

Wu N (1997) Indigenous knowledge and sustainable approaches for the maintenance of biodiversity in nomadic society – experiences from the Eastern Tibetan Plateau. Erde 128:67–80

Wu TW, Qian Z (2003) The Relation between the Tibetan winter snow and the Asian summer monsoon and rainfall: an observational investigation. J Clim 16(12):2038–2051

Wu N, Richard C (1999) The privatization process of rangeland and its impacts on pastoral dynamics in the Hindu-Kush Himalaya: the case of W. Sichuan, China, In: People and rangelands building the future. Proceedings of 6th international rangeland congress, Townsville, 19–23 July 1999

Wu N, Yan Z (2002) Climate variability and social vulnerability on the Tibetan Plateau: dilemmas on the road to pastoral reform. Erdkunde Band 56:2–14

Wu S, Yin Y, Zheng D et al (2007) Climatic trends over the Tibetan Plateau during 1971–2000. J Geogr Sci. doi:10.1007/s11442-007-0141-7

Yan ZL, Wu N (2005) Rangeland privatization and its impacts on the Zoige wetlands on the Eastern Tibetan Plateau. J Mt Sci 2(2):105–115

Yan ZL, Wu N, Yeshi D, Ru J (2005) A review of rangeland privatization and its implications in the Tibetan Plateau, China. Nomadic Peoples 9:31–51

Yao T, Pu J, Lu A et al (2007) Recent glacial retreat and its impacts on hydrological processes on the Tibetan plateau, China, and surrounding regions. Arct Antarct Alp Res 39(4):642–650

Yeh ET (2003) Tibetan range wars: spatial politics and authority on the grasslands of Amdo. Dev Change 34(3):499–523

Yeh ET (2005) Green governmentality and pastoralism in Western China: converting pastures to grasslands. Nomadic Peoples 9(1–2):9–29

You Q, Kang S, Aguilar E et al (2008) Changes in daily climate extremes in the eastern and central Tibetan Plateau during 1961–2005. J Geophys Res. doi:10.1029/2007JD009389

Zhang Y, Li T, Wang B (2004) Decadal change of the spring snow depth over the Tibetan Plateau: the associated circulation and influence on the East Asian summer monsoon. J Clim 17: 2780–2793

Zhao HX, Moore GWK (2004) On the relationship between Tibetan snow cover, the Tibetan plateau monsoon and the Indian summer monsoon. Geophys Res Lett 31:L14204

Chapter 32
Mainstreaming Climate Change Adaptation in Drinking Water Source Protection in Ontario: Challenges and Opportunities

Rob C. de Loë

Abstract Climate change will create numerous challenges for water managers. In Ontario, many parts of the province will experience decreased runoff and groundwater recharge, reduced lake levels, and more extreme precipitation events. Shift in patterns of precipitation also are expected, with more falling as rain rather than snow in winter, and with summers being drier. For water managers, the implications are serious. At the most basic level, long-term plans and investments that assume that future hydrological conditions in future will be the same as those of the past simply are not sensible. How can water managers at the local level, who already face considerable challenges in dealing with current climatic variability, adjust to this new reality? This chapter argues for adaptation through *mainstreaming*, in other words, building adaptation to climate change into existing decision-making and planning processes. Ongoing efforts to protect drinking water sources across Ontario currently afford the best opportunity to mainstream climate change into an existing water-related planning process. The chapter identifies challenges and opportunities for accomplishing this goal.

Keywords Source water protection • Mainstreaming • Adaptation • Ontario • Drinking water • Water management • Mainstreaming adaptation • Decision-making process • Climate variability

R.C. de Loë (✉)
Water Policy and Governance Group, Department of Environment and Resource Studies,
University of Waterloo, 200 University Avenue West, Waterloo, ON, N2L 3G1, Canada
e-mail: rdeloe@uwaterloo.ca

J.D. Ford and L. Berrang-Ford (eds.), *Climate Change Adaptation in Developed Nations:*
From Theory to Practice, Advances in Global Change Research 42,
DOI 10.1007/978-94-007-0567-8_32, © Springer Science+Business Media B.V. 2011

Introduction

Global climate change is transforming the hydrological cycle in new and unexpected ways. As a result, municipalities, water users, local water management agencies, and others will be required to implement adaptive responses (Bates et al. 2008). Simply put, decisions that currently are being made about land use and infrastructure should account for the ways in which future climate change will affect water resources. Considering the implications of climate change now, rather than in the future, will permit earlier and more effective adaptation. Unfortunately, water managers already are faced with formidable challenges and thus typically have little enthusiasm for tackling climate change (Kabat and van Schaik 2003). *Mainstreaming* climate change adaptation – in other words, integrating it with other decision-making and planning processes – is a way to address this problem (Kok et al. 2008; Smit and Wandel 2006; Swart and Raes 2007).

This chapter illustrates how climate change is being mainstreamed into an existing water-related planning process: source water protection (SWP) in the Province of Ontario, Canada. With 12.93 million inhabitants and an area of 1,068,580 km^2, Ontario is Canada's largest province by population and the second largest province by area. Most residents of the province live in urban areas, with approximately 5.5 million people residing in the Greater Toronto Area, an urban region that abuts Lake Ontario. Drinking water sources in the province are varied; most rural residents, and about 23% of the total population of the province, depend on groundwater (Simpson 2004). The remaining 77% of the population relies on surface water sources, including the Great Lakes, inland lakes, and rivers. The City of Toronto (population 2.48 million people) is entirely dependent on Lake Ontario for its drinking water supply, while communities such as Barrie and Guelph are served entirely by groundwater.

Threats to drinking water sources in Ontario are diverse and include spills from industrial activities, runoff from urban and agricultural land, contamination from septic systems and improperly maintained wells, and discharges from sewage treatment plants. Protecting drinking water sources from threats such as these has emerged as a priority in Ontario since the water supply of the Town of Walkerton was contaminated in 2000, leading to seven deaths and over 2,300 serious cases of illness (O'Connor 2002). Source water protection in Ontario currently takes place under the Clean Water Act (Ontario 2006) and its associated regulations. This statute was developed as part of the provincial government's response to the Walkerton tragedy. The focus of the Act is the protection of municipal drinking water sources at groundwater wells and surface water intakes.

Through the process of identifying threats, assessing vulnerability, and developing and implementing source protection plans, municipalities, conservation authorities (locally organized watershed management organizations), and other local stakeholders have an exceptional opportunity to mainstream climate change into water management and land use planning in Ontario. This chapter reviews anticipated impacts of climate change on water resources in Ontario, and then builds on that foundation to illustrate the challenges and opportunities of mainstreaming climate change into source protection planning.

Water Resources and Climate Change in Ontario

Ontario's climate has already changed during the past few decades. Measurable changes have been detected in variables such as air temperature and precipitation (CCME 2003). For example, mean annual temperature has increased by as much as 1.4°C since 1948 (Chiotti and Lavender 2008). At the same time, total precipitation in the Canadian portion of the Great Lakes–St. Lawrence Basin has increased between 1895 and 1995, with more precipitation falling as rain and less as snow. Impacts on hydrology from these kinds of changes are significant. For example, a trend towards an earlier spring freshet has been detected in southern Canada (Bruce et al. 2000).

These changes in temperature and precipitation reinforce the fact that climatic variability and change are normal. Indeed, water managers around the world have been adapting to changes in hydrology for millennia (Cech 2003). However, anticipated future climate changes are expected to occur much more rapidly in most parts of the world than has previously been experienced (Kabat and van Schaik 2003). Especially important for water managers is the fact that anticipated future changes in precipitation, evaporation, and runoff are expected to fall outside of the observed range of variability – meaning that past variability no longer will be a guidepost for future variability (Milly et al. 2008).

Recent projections suggest that Ontario's climate in future decades will be very different from the one experienced today (Bates et al. 2008). One area that has been studied extensively is the Great Lakes Basin, a region that is enormously important to Ontario's economy and society. The following changes are expected in this basin (Barnett et al. 2005; GLWQB 2003; Kling et al. 2003):

- Substantial temperature increases are expected in all seasons by the end of the century. By 2025–2035, projected spring and summer temperatures in the Great Lakes region are expected to be 1.5–2°C above current averages. Confidence around this projection is very strong.
- Projections for precipitation vary among studies. Some suggest slight increases in annual precipitation while others point to decreases. However, studies consistently point to a change in the seasonal distribution of precipitation, with more expected in winter in the form of rain and less in summer. Some studies suggest that winter precipitation could exceed summer precipitation by 50%, meaning that drier summers are likely in the region. Extremes, in the form of droughts and high-intensity rainfall events, also are expected to become more common.
- Evapotranspiration is expected to increase, although confidence in this projection is lower than for temperature projections. Uncertainty exists regarding the relationship between increased temperature and evaporation in studies of

future climates; some studies suggest that higher temperatures will increase evaporation, while others indicate that increased cloudiness associated with higher temperatures will decrease evaporation.

These changes are important because of their potential effect on the hydrological cycle. Broad-scale modeling of anticipated future climate changes on the hydrological cycle in Ontario points to a series of impacts that should be of great concern to water users and managers. These include significant changes to stream flows, lake levels, water quality, groundwater infiltration, and thus patterns of groundwater recharge and discharge. In the case of the Great Lakes Basin, studies have identified the following impacts (Bruce et al. 2000; GLWQB 2003; Kling et al. 2003; Lavender et al. 1998):

- Winter runoff is expected to increase, but total annual runoff is expected to decrease; summer and fall low flows are expected to be lower, and longer lasting.
- Groundwater recharge is expected to decrease due to reduced snow pack, and a greater frequency of droughts, and extreme precipitation events. Aquifers that are recharged rapidly from the surface, or which have short residence times, are expected to be most sensitive to changes in infiltration.
- Water temperature in rivers and streams is expected to rise as air temperatures increase and as summer baseflow (the portion of streamflow that comes from groundwater rather than runoff) is reduced.

Changes such as these will have profound impacts on humans and ecosystems. For example, decreased runoff during summer is likely to lead to reduced water quality, increased water treatment costs, and greater competition and conflict for reduced water supplies during drought periods. Warmer water temperatures, particularly in winter, could increase rates of survival and movement of water-borne pathogens, which could result in increased enteric disease. Water users dependent on groundwater for their supplies may expect increased costs because of a need to drill deeper wells. In rural areas, wells that are reliant on aquifers that are sensitive to reductions in infiltration are expected to dry up more often. Changes to wetland form and function are expected as groundwater discharge decreases, and stress on fish habitat is likely to increase with higher water temperature and lower flows.

Mainstreaming Climate Change into Source Protection

In the water field, adaptation has long been recognized as an essential response to climate change (Arnell and Delaney 2006; Bates et al. 2008; Bruce et al. 2000; de Loë et al. 2001; Frederick 1997). However, as noted earlier, water managers already are under considerable pressure as they attempt to deal with current problems (Kabat and van Schaik 2003). Hence, climate change simply is one of many emerging issues that decision makers must consider (Pittock 2003). Thus, as is the case in many other contexts (Kok et al. 2008; Smit and Wandel 2006; Swart and

Raes 2007), identifying opportunities to *mainstream* adaptation to climate change into existing planning and decision-making process is a sensible strategy to promote earlier and more effective adaptation. An opportunity currently exists in Ontario to integrate climate change into ongoing SWP activities.

The logic of source protection is simple: drinking water sources that are free from contamination need less treatment before being distributed, and are less likely to contain contaminants that conventional treatment processes cannot easily detect and remove. As the first line of defense in the multi-barrier approach to drinking water safety, source protection involves careful consideration of links between land use activities and water quality and quantity (CCME 2002). Source protection involves a wide range of activities that help to keep drinking water sources free from contamination. Examples pertinent to Ontario include: delineation of sensitive areas such as wellhead capture zones; bylaws (ordinances) to protect sensitive lands; outright acquisition of sensitive lands; and identification and monitoring of contaminant sources, wells, and septic systems (de Loë et al. 2005).

Prior to the Walkerton crisis in 2000, many Ontario communities already were using measures such as those described above to protect drinking water sources (de Loë et al. 2005). However, the extent to which SWP took place in a specific community depended very much on whether or not local leaders considered it a priority. Under the Clean Water Act and its regulations, source protection planning is now occurring on a systematic, province-wide basis. Locally organized Source Protection Committees (SPCs) are currently undertaking preliminary work that will provide the foundation for long-term planning. These committees are composed of representatives of municipalities, business and industry, agriculture, First Nations, and the public. Their work is supported by technical staff from provincial government agencies, and from the conservation authorities that comprise the Source Protection Regions in which the SPCs operate.

Detailed *assessment reports* are a key precursor for development of source protection plans. Two components of assessment reports that are especially pertinent in the context of this chapter include the *watershed characterization* and the *water budget*:

- Watershed characterizations are watershed-wide overviews of land uses and activities, water resources, and threats to drinking water safety.
- Water budgets describe and then model the relationships between physiography, surface water flows, groundwater levels and flows, climate variables, withdrawals of water, and other pertinent considerations.

Building on the foundation provided by the watershed characterizations and the water budgets, vulnerable areas will be delineated, and drinking water threats and issues will be identified and evaluated. A diverse range of tools (such as those identified above) will then be used by municipalities, conservation authorities, landowners, and others to protect source water in vulnerable areas.

As discussed in the previous section, climate change is expected to transform the hydrological cycle in Ontario, with implications for both water quantity and water quality. Thus, given the forward-looking perspective advanced by the Clean

Water Act, climate change clearly should be considered among other threats and issues in assessment reports. It also is important to emphasize that source protection planning is expected to be an ongoing process. Plans will be revised and updated over time, and the technical sophistication and accuracy of water budgets is expected to increase. Thus, addressing climate change specifically and directly in source protection planning is an important way to mainstream adaptation into a broad range of water and land management activities in Ontario. The extent to which this has been recognized in the institutions created by the provincial government has increased over time.

The Clean Water Act came into force in 2006, but preparatory work already was taking place in anticipation of future requirements. Draft guidance modules to support the creation of assessment reports were circulated by the Ontario Ministry of the Environment (OMOE) in 2005 to teams preparing to undertake SWP (OMOE 2005). In this lengthy and detailed document, climate change was addressed directly in only one place:

> Using historical climate trends and stream flow records, the team will evaluate future meteorological trends for the watershed. The team should consult Environment Canada to access appropriate climate change assumptions and existing predictive models. The team should provide a written description of the evaluation. (OMOE 2005, p. 17)

Evidently, at this early stage, it was not clear to the designers of the system how climate change should be addressed in source protection planning.

Recognizing the opportunity presented by source protection planning, Pollution Probe and the Canadian Water Resources Association commissioned a study that had the goal of identifying how climate change could be addressed systematically in the proposed process (de Loë and Berg 2006). Opportunities were identified specifically in relation to the preparation of watershed characterizations and water budgets. Detailed advice was provided on using Intergovernmental Panel on Climate Change (IPCC) scenarios, selecting Global Climate Models (GCMs), and choosing methods for downscaling GCM outputs for use in regional-scale models.

During 2006, the OMOE also appears to have given more thought to how climate change could be addressed in source protection planning. Draft guidance documents published in October of that year provided considerably more detail than earlier versions (OMOE 2006). To illustrate, in the 2006 document the OMOE acknowledged that historical climate and hydrological data can no longer be relied upon to predict future conditions; hence, teams charged with preparing assessment reports were urged to consider including climate change as a variable in watershed characterizations. Practical concerns regarding the incorporation of climate change into water budgets also were recognized in the document; thus, teams were asked to identify data and knowledge gaps and to recommend ways in which climate-related concerns can be addressed in future.

Ultimately, however, the OMOE decided that during the first cycle of source protection planning (which is ongoing as of 2009), the "impacts of future climate change will not be specifically evaluated" in assessment reports (OMOE 2006, p. 20). Instead, it was suggested that historical drought conditions should provide the basis for understanding how water budgets can be affected by climate change. This is the approach that is mandated in the final technical rules that were published in 2008 (OMOE 2008). These technical rules outlined the official requirements for preparation of assessment reports, and replaced the 2006 draft guidance modules. Under these rules, climate change is to be addressed as follows:

- Watershed characterizations must include an explanation of the "effects that projected changes in the climate over the following 25 years will have on the conclusions reached in the assessment report and a list of the information sources underlying those projected changes" (OMOE 2008, p. 10).
- Watershed characterizations also must include "conceptual" water budgets (descriptive overviews of hydrological processes and relationships rather than numerical models). In addition to other basic considerations (e.g., physiography, surface water bodies, groundwater systems, water takings), conceptual water budgets must describe the "climate of the area, including historical trends and existing projections related to changes in the climate of the area" (OMOE 2008, p. 15).
- For every sub-watershed in the source protection area, teams are required to prepare Tier One water budgets using geographic information systems, and, if warranted based on assessments of stress levels, more detailed Tier Two and Tier Three water budgets using computer-based three-dimensional groundwater flow models and continuous surface water flow models. Stress levels are determined using scenarios that are outlined in the technical rules. These scenarios include current climatic conditions, a 2-year drought, or a 10-year drought. The drought periods are based on the lowest mean annual precipitation in the study area during continuous 2- and 10-year periods, respectively.
- The three climate scenarios (current observed conditions and observed 2- and 10-year drought periods) are used again in the process of determining exposure to water quantity shortages in local areas.

The approach to incorporating climate change into SWP in Ontario that is mandated under the current technical rules (OMOE 2008) is a compromise. Requiring teams to evaluate stresses in their study areas in the context of two scenarios of reduced precipitation (the 2- and 10-year drought periods) recognizes climatic *variability*. However, anticipated changes to the climate in Ontario are expected to fall *outside* of the observed envelope of variability. Additionally, as noted earlier, climate change in Ontario is expected to affect not only the amount of precipitation, but also its timing. As a result, conditions during previous meteorological droughts may be a poor surrogate for anticipated climate changes.

The OMOE's decision regarding how climate change should be addressed in source protection planning appears to reflect a fundamentally pragmatic concern, namely, that integrating climate change into water budgets using IPCC scenarios and downscaling models is not feasible given the current technical capacity in Ontario. While several conservation authorities had previously demonstrated the necessary capacity to undertake this kind of modeling (de Loë and Berg 2006), the OMOE's decision was not unreasonable given well-documented variations in technical capacity among Ontario's 34 conservation authorities (Ivey et al. 2002), and a desire for source protection teams to produce assessment reports as quickly and effectively as possible. Nonetheless, there is cause for concern about the extent to which climate change can be addressed in future cycles of planning if it is not addressed more explicitly from the outset. Three concerns stand out.

First, the published technical rules (OMOE 2008) are the benchmark against which all assessment reports will be evaluated. Thus, there is little incentive for those preparing assessment reports to stray from this strictly defined path. Second, and following from this concern, there is little incentive for conservation authorities, who play key technical and leadership roles in the preparation of assessment reports, to build the technical capacity needed to more effectively integrate climate change into water budgets. Areas where capacity development is needed include use of the IPCC's scenarios, use of climate models, selection of methods for downscaling climate model output to regional and local scales, and incorporation of downscaled data into hydrological models. Third, there are technical considerations associated with integrating climate change into water budgets that should be addressed at the outset, and which may be more difficult to address retroactively (de Loë and Berg 2006). For instance, integrating climate change into hydrological models requires data that permit consideration of long-term climatic trends; calibration of models against observed data; and prediction of changes in hydrologic variables likely to be affected by climate change. Finally, in selecting hydrological models used in water budgets, suitability for predicting climate change impacts should be an important criterion (de Loë and Berg 2006).

Conclusion

Climate change is a pressing concern in Ontario. In the Great Lakes Basin, important changes in air temperature, and in the form and timing of precipitation, already have been observed. Projected climate warming is expected to produce many more changes like these, with significant implications for the hydrological cycle. Because of its impact on the hydrological cycle, climate change is a *local* problem that must be faced by conservation authorities, municipalities, provincial government officials, and water users. Decisions being made today regarding water management, land use planning, and infrastructure should account for the ways in which climate change will affect water resources. The challenge is to build climate change into day-to-day activities – in other words, to *mainstream* it in water management and land use planning.

Source protection planning in Ontario under the Clean Water Act provides an outstanding opportunity to mainstream climate change adaptation. Under the Clean Water Act, teams are currently developing assessment reports that will provide the foundation for creating source protection plans at the watershed scale. These assessment reports include systematic characterization of watersheds and detailed water budgets. Hence, they also can be a key resource for a host of other water-related decision-making processes where climate change is an important consideration. These include – but are by no means limited to – water allocation, municipal land use planning, and long-term water supply planning. Assessments reports that explicitly address climate change could be extremely useful in each of these areas. From this perspective, it is not hyperbole to suggest that source protection planning is the best opportunity in a generation to mainstream climate change into water management and land use planning in Ontario.

Opportunities to address climate change in source protection planning have been recognized. However, upon considering the scale of the challenge, the OMOE decided that in the first cycle of source protection planning, which currently is ongoing, climate change would only be recognized conceptually and indirectly. Thus, in the technical rules that were published in 2008, climate change has only been partially mainstreamed into the process of creating assessment reports. That it is addressed at all is a significant accomplishment given the rapid pace of source protection planning in Ontario. Furthermore, the door is open to giving climate change more consideration in future cycles of planning, for example, in revised water budgets. Nonetheless, there is cause for concern regarding the extent to which climate change will be addressed more effectively in future planning cycles. For instance, it cannot be assumed that the data needed to integrate climate change effectively into hydrological models are being collected now, or that the skills needed are being developed. An opportunity clearly exists to make climate change a central concern in a critical planning process that has significant ancillary benefits. The challenge will be to accomplish this goal given real constraints on the technical capacity of the people and organizations responsible.

Acknowledgments I would like to thank Aaron Berg, for his contribution to the technical report that provided some of the insights used in this chapter, Deb Lightman and Hugh Simpson, for their comments on a revised version, and the two anonymous reviewers who offered exceptionally useful and constructive advice regarding the first draft. I would also like to thank Pollution Probe for permission to use portions of the material contained in the report by de Loë and Berg (2006) in this chapter.

References

Arnell NW, Delaney EK (2006) Adapting to climate change: public water supply in England and Wales. Clim Change 78(2–4):227–255

Barnett TP, Adams JC, Lettenmaier DP (2005) Potential impacts of a warming climate on water availability in snow-dominated regions. Nature 438(17):303–309

Bates BC, Kundzewicz ZW, Wu S et al (2008) Climate change and water. Intergovernmental Panel on Climate Change, Geneva

Bruce J, Burton I, Martin H et al (2000) Water sector: vulnerability and adaptation to climate change. Prepared for the Climate Change Action Fund [unpublished]

Canadian Council of Ministers of the Environment [CCME] (2002) From source to tap: the multi-barrier approach to safe drinking water. Canadian Council of Ministers of the Environment, Winnipeg

CCME (2003) Climate, nature, people: indicators of Canada's changing climate. Canadian Council of Ministers of the Environment, Winnipeg

Cech TV (2003) Principles of water resources: history, development, management and policy, 2nd edn. Wiley, Hoboken

Chiotti Q, Lavender B (2008) Ontario. In: Lemmen DS, Warren FJ, Lacroix J et al (eds) From impacts to adaptation: Canada in a changing climate 2007. Government of Canada, Ottawa, Chapter 6

de Loë RC, Berg A (2006) Mainstreaming climate change in drinking water source protection planning in Ontario. Pollution Probe, Ottawa

de Loë R, Kreutzwiser R, Moraru L (2001) Adaptation options for the near term: climate change and the Canadian water sector. Global Environ Change 11(3):231–245

de Loë RC, Kreutzwiser RD, Neufeld D (2005) Local groundwater source protection in Ontario and the Provincial Water Protection Fund. Can Water Resour J 30(2):129–144

Frederick KD (1997) Adapting to climate impacts on the supply and demand for water. Clim Change 37(1):141–156

Great Lakes Water Quality Board (2003) Climate change and water quality in the Great Lakes Basin. International Joint Commission, Ottawa

Ivey JL, de Loë RC, Kreutzwiser RD (2002) Groundwater management by watershed agencies: an evaluation of the capacity of Ontario's conservation authorities. J Environ Manag 64(3):311–331

Kabat P, van Schaik H (2003) Climate changes the water rules: how water managers can cope with today's climate variability and tomorrow's climate change. Dialogue on Water and Climate, Delft

Kling GW, Hayhoe K, Johnson LB et al (2003) Confronting climate change in the Great Lakes region: impacts on our communities and ecosystems. UCS Publications, Cambridge, MA

Kok M, Metz B, Verhagen J et al (2008) Integrating development and climate policies: national and international benefits. Clim Policy 8(2):103–118

Lavender B, Smith JV, Koshida G et al (1998) Binational Great Lakes-St. Lawrence Basin climate change and hydrologic scenarios report. Environment Canada Environmental Adaptation Research Group, Downsview

Milly PCD, Betancourt J, Falkenmark M et al (2008) Stationarity is dead: whither water management. Science 319(5863):573–574

O'Connor DR (2002) Report of the Walkerton Inquiry: part one – the events of May 2000 and related issues. Queen's Printer for Ontario, Toronto

OMOE (2006) Assessment report: draft guidance modules, October 2006. Ontario Ministry of the Environment, Toronto

OMOE (2008) Technical rules: assessment report, Clean Water Act, 2006. Ontario Ministry of the Environment, Toronto

Ontario (2006) Clean Water Act. Statutes of Ontario, Chapter 22. Queen's Printer for Ontario, Toronto

Ontario Ministry of the Environment [OMOE] (2005) Assessment report: guidance modules. Source Water Implementation Group [draft 12 Dec 2005]

Pittock B (2003) Climate change: an Australian guide to the science and potential impacts. Australian Greenhouse Office, Australian Government, Canberra

Simpson H (2004) Promoting the management and protection of private water wells. J Toxicol Environ Health A 67(20–22):1679–1704

Smit B, Wandel J (2006) Adaptation, adaptive capacity and vulnerability. Glob Environ Chang Hum Policy Dimens 16(3):282–292

Swart R, Raes F (2007) Making integration of adaptation and mitigation work: mainstreaming into sustainable development policies? Clim Policy 7(4):288–303

Chapter 33
Opportunities and Barriers for Adaptation and Local Adaptation Planning in Canadian Rural and Resource-Based Communities

Robert A. McLeman, Michael Brklacich, Maureen Woodrow, Kelly Vodden, Patricia Gallaugher, and Renate Sander-Regier

Abstract This chapter describes various forces that influence the ability of decision-making and institutional structures in Canadian rural and resource-dependent communities to manage, plan for, and respond to future risks and uncertainties resulting from climate change. The context within which such communities make decisions related to capacity building is the outcome of historical development trajectories, interactions with higher levels of government, and macro-level economic structures and processes. The success of future capacity building and planning will be strongly influenced by such factors as improved coordination across different levels of government, the provision of locally geared information about environmental and climate change, economic diversification and the ability to adjust to and take advantage of rapidly changing demographic patterns in such communities.

R.A. McLeman (✉)
Department of Geography, University of Ottawa, Simard Hall 015, 60 University,
Ottawa, ON, K1N 6N5, Canada
e-mail: rmcleman@uottawa.ca

M. Brklacich
Department of Geography and Environmental Studies, Carleton University, Ottawa, ON, Canada
e-mail: michael_brklacich@carleton.ca

M. Woodrow
Telfer School of Management, University of Ottawa, Ottawa, ON, Canada
e-mail: woodrow@telfer.uottawa.ca

K. Vodden
Department of Geography, Memorial University of Newfoundland, St. John's, NL, Canada
e-mail: kvodden@mun.ca

P. Gallaugher
Department of Biological Sciences, Simon Fraser University, Burnaby, BC, Canada
e-mail: gallaugher@sfu.ca

R. Sander-Regier
Department of Geography, University of Ottawa, Ottawa, ON, Canada
e-mail: rsand071@uottawa.ca

J.D. Ford and L. Berrang-Ford (eds.), *Climate Change Adaptation in Developed Nations:*   449
From Theory to Practice, Advances in Global Change Research 42,
DOI 10.1007/978-94-007-0567-8_33, © Springer Science+Business Media B.V. 2011

Keywords Rural communities • Resource-based communities • Canada • Community-based research • Local governance • Adaptation planning • Capacity building • Social capital • Local adaptation • British Columbia • Newfoundland

Introduction

Residents and decision makers in rural and resource-dependent communities across Canada are increasingly aware that climatic variability and change may bring attendant risks to community livelihoods and well-being (Brklacich et al. 1997; Bryant et al. 2000; Reid et al. 2007). The work that has been done to date on climate change adaptation in Canada has often tended to focus on broad scale (i.e., national or regional) adaptation needs and on adaptation to the impacts of climate change on particular attributes of economic systems or sectors (e.g., Brklacich et al. 2007; Scott and McBoyle 2007; Spittlehouse and Stewart 2003). Research into community-level adaptation needs and implementation of responsive policies and planning is more recent and, of particular importance to this chapter, is not necessarily well known to local-level decision makers.

In this chapter, we describe and explore various forces that influence the ability of decision-making and institutional structures in Canadian rural and resource-dependent communities to manage, plan for and respond to future risks and uncertainties resulting from climate change. The examples are drawn from two participant communities in a multiyear comparative study of barriers to and opportunities for adaptation in selected Canadian communities where local livelihoods are tied strongly to land- or resource-based economic activities. The two participant communities described here – Alert Bay, British Columbia and Change Islands, Newfoundland (Fig. 33.1) – are both located on islands, but have very different historical trajectories, economies, demographic characteristics, and institutional arrangements. While some details are necessarily specific to the particular context of each community, the emphasis here is on describing broader insights that may be applicable to a range of rural and resource-based communities across Canada and in other developed nations.

Brief Description of Each Community, Climate-Related Stresses, and Institutional Context

Alert Bay, British Columbia

Alert Bay is a multicultural community of 1,300 people of First Nation and non-First Nation ancestry located on Cormorant Island, off the northeastern shore of much larger Vancouver Island. The relatively isolated island is home to the

Fig. 33.1 The two participant communities: Alert Bay, British Columbia and Change Islands, Newfoundland

people of the Village of Alert Bay, the 'Namgis First Nation, and a roughly 5-ha parcel of land set aside for use by people from outlying First Nations villages (called Whe-La-La-U). The development paths of the island's First Nation people, whose livelihoods have been strongly tied to traditional resource-based livelihood activities, and nonnative settlers, whose economy was also built on commercial resource extraction, are closely intertwined. Alert Bay has a relatively stable and large proportion of young people. Relations between the island's native and nonnative population are generally good, and there is ongoing cooperation between groups on a formal and informal basis. The historic Alert Bay Accord, a formally recognized infrastructure and economic development agreement, exemplifies the level of cooperation that assists the whole Island community in its progressive development.

Decades of overexploitation have led to a decline in the quality and quantity of two key resources: salmon and forests. The local economy has become increasingly diversified in recent years. Sales and services now represent the largest employment sector, and the 'Namgis First Nation government is the largest employer. An aggregate mine developed in 2006 as a joint venture between the 'Namgis First Nation and a private company could generate over $1 million annual revenue for the 'Namgis First Nation in coming years. New types of timber initiatives and small-scale run-of-the-river hydroelectric projects are also being undertaken. Tourists are being attracted to the area to watch the resident orca population, learn of the rich diversity of the area and, probably most importantly, experience the richness of

First Nation culture. Residents are concerned about the potential impacts of climate change on the region's forests and marine resources and have ongoing concerns about extreme weather events, storm surges, and the longer term implications of sea level change.

Cormorant Island has a complex institutional environment. The Village of Alert Bay, the 'Namgis First Nation Council, and the Whe-La-La-U Area Council all have distinct political and administrative responsibilities and areas, and the 'Namgis are in turn part of a larger tribal council representing four affiliated Kwakwaka'wakw Nations. All of their respective offices are located in Alert Bay. At least nine community and regional plans and four provincial and federal government-led plans related to land and resources are currently underway. The 'Namgis First Nation has a team of planning staff engaged in the treaty process, a sizable planning budget, and resources devoted to environmental monitoring. Ongoing efforts have been made to coordinate the planning activities and administrative tasks of the Village of Alert Bay and 'Namgis Councils.

Change Islands

Livelihoods in Change Islands off Newfoundland's Atlantic coast have historically been tied to commercial cod fishing. A century ago, its 1,000 residents formed a prosperous but isolated community. The population has since declined to approximately 300, as a result of ongoing structural changes in the fishing industry and a 1992 government-imposed moratorium on cod fishing following the collapse of North Atlantic cod stocks due to overfishing. Out-migration of employment-seeking young adults has left the community with an aging population.

In recent decades, winters have become milder and summers have become warmer, trends that residents believe may be affecting the composition of species found in inshore waters. Warmer water temperatures are bringing new species of fish and mammals to the area, while other species no longer come inshore in the numbers they once did. Fishing seasons prescribed by the federal Department of Fisheries and Oceans increasingly fall later than the times when the numbers of commonly fished species, such as lobster and crab, are at their highest in local waters. In addition to fishing, residents engage in land-based subsistence activities such as harvesting from woodlots, collecting wild berries, and maintaining vegetable gardens. Residents and local businesses are attempting to cater to and attract more tourists to decrease their dependency on fishing. Attempts to create small-scale manufacturing jobs locally have not proved successful so far. Despite ongoing efforts in economic diversification, the fish processing plant remains the largest single employer in the community.

Local municipal government is responsible primarily for maintaining roadways and water services within the municipal boundaries. Other critical services on which residents are dependent, such as ferry services, highways, health care, education, and fishing licenses, are controlled and managed by provincial or federal government ministries based elsewhere. Beyond economic development initiatives, little governance or planning occurs at the regional level. Community social networks and the social capital (Castle 2002) that has been built within them over many years are critical assets to ensuring the community's future ability to cope with and adapt to environmental and non-environmental stresses. This social capital is rapidly eroding in the face of population decline, which leaves fewer people available to participate in community activities.

Common Themes in the Formation of Community Vulnerability

In both communities, global environmental changes and societal changes interact to create exposure at the local level to particular environmental risks and to influence the capacity of the community to adapt to those risks. In Table 33.1, we summarize some critical themes that emerged during the project, with illustrations from the two island communities.

As might be expected from the contents of Table 33.1, it quickly emerged during this study that residents of these communities do not necessarily separate climate-related risks from the broader range of environmental stressors to which they currently adapt. Instead, variability in environmental conditions, including climatic variability, is typically seen as being a fact of life. There is no assumption that environmental conditions are ever entirely predictable or stable. Instead, residents tend to view vulnerability as emerging from the combined interaction of environmental and socioeconomic changes, with changes in climatic conditions and events being one element in such interactions.

The nature of these interactions is shaped by past histories and dynamic societal changes, which are experienced through consequent changes in governance structures, resource accessibility, local economic conditions, and demographic patterns, among others. Future trajectories, vulnerabilities, and adaptive capacity will be shaped and limited by past trajectories and become further modified by present institutional arrangements, such as resource management regulations prescribed by higher levels of government that are beyond the scope of the communities' direct influence.

Local adaptive capacity is strongly influenced by social networks and the social capital embedded within them. This social capital is sensitive to changes in the demographic conditions of the community, especially in terms of its age structure and cultural makeup, and is easily eroded by relying too often on the same individuals or groups to lead community initiatives. Related to this is a current recognition of the linkages between environmental stewardship and capacity building; this ethos differs significantly from past, exploitative approaches taken to resources, and sees the combined social and environmental well-being of the community for the long term as being preferable to short-term economic gains. That said, the ability to practice stewardship and the ways in which it can be done vary significantly between communities as a result of their different socioeconomic contexts. In Alert Bay,

Table 33.1 Key themes in vulnerability identified in Alert Bay and Change Islands

	Examples	
Theme	Alert Bay	Change Islands
Climate change is situated within broader environmental changes	Overexploitation has led to a decline in the quantity of salmon resources and the quantity and quality of forest resources	Intensive harvest has severely depleted fisheries resources
Historical trajectories shape future adaptation options	Current societal conditions occur in spite of the 'Namgis First Nation's long struggle to assert control over land and marine resources in face of exploitation by nonlocal interests	Fisheries collapse drives out-migration, in turn depleting social capital
Interactions across governance scales influence vulnerability	Size of salmon stocks and amount to be harvested influenced by Canadian, American fishing policies, and poor harvest strategies imposed by government	Fish plant operator is seeking outside "eco-friendly" certification to increase value of product
Demographic trends shape adaptive capacity	Formal and informal networks among Aboriginal and non-Aboriginal communities often work to mutual advantage	"Bridging" social capital needed to better integrate long-term residents and recently arrived groups
Stewardship ethos reduces vulnerability	"Fishery guardian" positions created to monitor local fisheries in absence of federal patrols	Individuals protect a rare breed of island horse and thereby create tourist attraction
Over-reliance on particular economic sectors increases vulnerability	Depletion of salmon and forest resources has driven communities to develop new partnerships, new activities, but these economic adjustment crisis have taken 20 years to develop	Tourism industry being pursued as means to reduce dependence on fish processing plant for jobs
Adaptive capacity is tied to social capital	Village of Alert Bay Advisory Planning Commission is volunteer-run	Volunteer committee found new operator for fish processing plant when former owner closed up

stewardship is carried out through formal structures such as the 'Namgis fishery guardian program which monitors local fisheries for conservation purposes. With its small population base, Change Islands sees stewardship practiced more on an individual level, such as a family that established a refuge to preserve the rare Newfoundland Pony, thereby creating a tourist attraction.

In both communities, reliance on a small number of economic activities is seen as a key driver of vulnerability to socioeconomic and environmental changes that must be overcome. The changing economy of Cormorant Island (Alert Bay) is still in recovery from the overexploitation of salmon stocks and local forests, a situation that is slowly being overcome through new developments in aggregate mining and small-scale hydroelectric generation and tourism. The community of Change Islands has yet to recover fully from the closure of the cod fishery and consequent large-scale out-migration of younger residents that shrank the local labor force and eroded social capital. While tourism to the area has been growing, it does not approach the scale of local economic benefits provided by the fish plant.

Advancing Adaptive Capacity in Rural and Resource-Dependent Communities

Despite the differences in the planning structures and environment between the study communities, residents identified four categories of shared challenges to increasing their capacity to plan for and adapt to future risks, climate change related or otherwise. Each of these is described in turn.

Inadequate Resources for Planning and Coordination with Higher Levels of Government

The first set of challenges is the creation and maintenance of a well-functioning mixture of formal and informal institutions at the local level, which is absolutely critical to ongoing community well-being and good governance. In turn, these local institutions must be able to interact efficiently with any number of other agencies in more senior levels of government to conduct short- and long-term planning. The efficiency of coordination among these various interacting levels of government can either support or impede a community's adaptive capacity. While higher levels of government tend to have the resources, expertise, and dedicated staff to administer and manage policies and programs and to continuously engage in forward planning, people engaged at the local community level must wear many hats. Often, those engaged in local planning are volunteers who must "learn the file" as they go along. This undermines the local community's potential to engage in long-term planning and to manage relations with higher levels of government. The need to engage volunteers in formal planning processes puts an extra draw on the community's social capital, thereby reducing its availability for other initiatives. Higher levels of government must recognize this fact and make additional professional resources available to local communities if they are to be partners in adaptation planning.

One-Size Regulations Do Not Fit All and Can Impede Capacity Building

The rules and regulations pertaining to any given activity or sector are formulated by higher levels of government to address a much wider range of situations than typically apply in any one local area. Complying with such directives can often have unintended negative impacts on local communities where specific problems that necessitate such regulations simply do not exist. For example, in Newfoundland regulations pertaining to tourism and hospitality – an economic sector critical to economic diversification and in Change Islands – are designed with year-round (typically urban) service establishments in mind. These regulations require seasonal operators (which describes most Change Islands' tourist businesses) to make modifications to premises, purchase expensive operating permits, and take other steps that can quickly make a seasonal operation unviable. Many other examples of this type suggest that one-size-fits-all regulations can retard community development and reduce the capacity of small communities to expand their economic base and secure their future livelihoods. While it may be argued that any system of institutional arrangements will have its share of inherent inefficiencies, at higher levels of government the costs of such inefficiencies are distributed across a wide geographic and population base. When it must be borne by rural communities with small populations and limited capital, the costs of institutional inefficiency are disproportionately higher.

Climate Change Information is Not Well Suited to Local Communities' Needs

A second common challenge to future adaptation planning is the unavailability of the types of information rural communities need about the potential impacts of climate change. Scientific information such as that generated from general circulation models (GCMs) is generally and most reliably expressed in terms of changes in average temperature conditions over large areas and, with a lesser degree of reliability, changes in precipitation regimes, again over large areas. The spatial scale of climate change information is typically too coarse to capture the local details rural community planners need, and the types of impact scenarios being generated and used by the climate science community are often not meaningful at the community level. Extreme events, such as the frequency and timing of hail or the potential for above-average accumulations of snow, are more pressing concerns for rural communities than changes in long-term average conditions, but climate science tends not to produce these specific analyses regularly or reliably. Furthermore, residents of rural communities lack the resources and expertise to translate climate change information, especially when expressed using scientific jargon, for use in a practical or quotidian sense at the local level.

How to Balance Economic Diversification While Maintaining Community Identity

These communities are socially, economically, and culturally rooted within their surrounding environments and resource bases. Residents often have a sophisticated knowledge of trends and patterns of change in those conditions and how to respond to such changes. At the same time, residents recognize future capacity building requires reducing their overdependence on small sets of resources and activities. In their view, diversification of the local economic base should not entail an abandonment of a community's traditional economic base, its accompanying sociocultural identity, and the social capital that has been generated by pursuing such livelihoods; rather, existing assets and activities should be leveraged to achieve diversification. For example, Alert Bay has used its First Nations heritage to develop its tourism industry, while Change Islands promotes its lack of development and isolated location as attractions for urban visitors. That said, tourism alone cannot provide a sufficient economic base to guarantee the long-term viability of either study community. Tourist numbers fluctuate considerably according to macroeconomic trends, and expanding and maintaining tourism infrastructure often requires access to financial capital beyond that available in rural communities. This in turn reveals additional needs that must be met to facilitate economic diversification, such as building multiple partnerships that would develop joint private–public sector ventures, promoting greater cooperation among neighboring communities in a given region, and facilitating engagement between formal and informal economic sectors. Such partnerships are seen as having the potential to expand the access of rural communities to capital, expertise, and information to build synergies and facilitate innovation.

Going Forward

Rural planning officials in both study communities, whether professional employees or volunteers, suggest they are at a crossroads. Both emphasize that more comprehensive formal planning will enhance their capacities to respond positively to future challenges and opportunities, climate-related or otherwise, if they have the necessary means to do so. They share a number of strong ideas on how to make adaptive planning both comprehensive and more inclusive. Local planners envisage supporting and coordinating a wide range of planning initiatives, including economic, social, and environmental planning. They would like to include all legitimate voices in the community, and particularly expand representation of youth, newcomers, and other groups who are often excluded from the planning process as it is currently conducted. Local planners would also like to improve coordination across sectors and governments. They would like to move away from less flexible plans that require firm predictions of the likelihood of future events, and shift toward contingency management of risks combined with deliberate building of capacity to adapt to risks known and unknown, one of which would be climate change and its impacts. But to what extent is this possible given the current situation? And how can the climate change research community help?

One place to start is to expand the interactions between climate change researchers and rural communities so that the former provide the latter with information consistent with spatial and temporal scales used in rural community planning. Rural residents are clearly interested in this initiative. Given the very nature of daily life in rural and resource-dependent communities, their residents are in many ways more appreciative of changes in environmental conditions and the risks associated with them than those who live in cities or suburbs. Residents are pragmatic in their attitudes toward climate change: if it is something that can be planned for, then it should be. However, they also see climate change as just one of many factors that must be considered in long-term rural community planning, and climate change-related risks must be weighed in relation to (and not in isolation from) other risks that will have to be managed over the long term.

This pragmatism is understandable. Increasingly fewer people in developed countries today live in rural and resource-dependent communities, and those who do are often falling behind urban residents in terms of the availability and accessibility of critical services and infrastructure. Any phenomenon that might threaten to further undermine the quality or viability of life in their communities, such as climate change, will be taken seriously. A critical question for the majority who do not live in rural or resource-dependent communities, yet who tend to be the ones that have the greatest influence over higher levels of government, is whether they want small, rural and resource-dependent communities to continue to exist and to thrive in the future. It will ultimately be urban residents who determine if the critical needs identified above – building linkages across levels of government, adapting constrictive institutional structures and regulations, and providing small communities the resources to act as full partners in planning – will be achieved.

Acknowledgments The community-based research described in this chapter received financial support from the Climate Change Impacts and Adaptation Program of Natural Resources Canada. The interest and enthusiasm of participants from each of the study communities made this research possible.

References

Brklacich M, McNabb D, Bryant C et al (1997) Adaptability of agriculture systems to global climatic change: a Renfrew County, Ontario, Canada pilot study. In: Ilbery B, Rickard T, Chiotti Q (eds) Agricultural restructuring and sustainability: a geographic perspective. CAB International, Wallingford

Brklacich M, Smit B, Wall E et al (2007) Impact-based approaches. In: Wall E, Smit B, Wandel J (eds) Farming in a changing climate: agricultural adaptation in Canada. UBC Press, Vancouver

Bryant C, Smit B, Brklacich M et al (2000) Adaptation in Canadian agriculture to climatic variability and change. Clim Change 45(1):181–201

Castle EN (2002) Social capital: an interdisciplinary concept. Rural Sociol 67(3):331–349

Reid S, Smit B, Caldwell W et al (2007) Vulnerability and adaptation to climate risks in Ontario agriculture. Mitig Adapt Strat Glob Change 12(4):609–637

Scott D, McBoyle G (2007) Climate change adaptation in the ski industry. Mitig Adapt Strat Glob Change 12(8):1411–1431

Spittlehouse DL, Stewart RB (2003) Adaptation to climate change in forest management. BC J Ecosyst Manag 4(1):1–11

Chapter 34
Anticipatory Adaptation in Marginalized Communities Within Developed Countries

Michelle Boyle and Hadi Dowlatabadi

Abstract The majority of anticipatory adaptation frameworks applied in developed countries tend to idealize the institutional and cultural readiness for their successful deployment. We explore the validity of these assumptions for Arctic Canada, where marginalized communities are experiencing extreme climate change, as well as contending with many other external and internal stresses. Collaborations with communities in Nunavut revealed that they lack the resources, institutional capacity, and expertise to employ long-term strategic planning processes and conventional analytical decision methods. More importantly, their priorities and cultural perspective are inconsistent with underlying Western theory and its implicit assumptions. In light of these challenges, we recommend that efforts to mainstream climate change adaptation rely on frameworks that can (1) respect community priorities and introduce resilience to climate change as one part of meeting other critical development goals and (2) accommodate key cultural differences in decision-making, values, and the use of information.

Keywords Anticipatory adaptation • Climate change • Marginalized communities • Nunavut • Canada • Arctic • Inuit • Adaptation frameworks • Resilience • Cultural differences • Traditional knowledge

M. Boyle
Institute for Resources, Environment and Sustainability, University of British Columbia, AERL, 422–2202 Main Mall, Vancouver, BC V6T 1Z4, Canada
e-mail: mboyle@ires.ubc.ca

H. Dowlatabadi (✉)
Institute of Resources, Environment and Sustainability and Liu Institute for Global Issues, University of British Columbia, AERL, 422–2202 Main Mall, Vancouver, BC, V6T 1Z4 Canada
e-mail: hadi.d@ubc.ca

J.D. Ford and L. Berrang-Ford (eds.), *Climate Change Adaptation in Developed Nations: From Theory to Practice*, Advances in Global Change Research 42, DOI 10.1007/978-94-007-0567-8_34, © Springer Science+Business Media B.V. 2011

Introduction

Established theories describing adaptation involve the introduction of a new stress, emerging awareness of its impacts, consideration of options for relief, and implementation of a chosen option (see Smithers and Smit 1997; Smit et al. 2000). Further refinement may classify responses according to whether they strengthen community resilience and robustness against similar future stresses (Fankhauser et al. 1999; Barnett 2001; Tompkins and Adger 2004; Allen 2006). A community's ability to achieve these responses has been used to characterize its adaptive capacity (Smit and Pilifosova 2001; Yohe and Tol 2002; Adger 2003). Various frameworks have identified human capital, information and technology, material resources and infrastructure, organization and social capital, political capital, wealth and financial capital, institutions and entitlement as the determinants of such capacity (Kestikalo 2008). We focus on some of these and some more fundamental prerequisites to successful adaptation in marginal communities.

Anticipatory adaptation takes a planned and strategic approach to responding to multiple stresses at a defined scale. It requires foreknowledge of climate change and how it is likely to manifest, but also must recognize that there are challenges and uncertainties associated with predicting impacts and adaptations through time. Therefore, the planning process itself must be iterative and allow for strategies to be adjusted as development goals and ecological changes coevolve. Structured processes and analytical decision aids have been developed to facilitate anticipatory adaptation (Dowlatabadi et al. 1994; Smith et al. 1996; and many others). Figure 34.1 illustrates one such process, referred to herein as "adaptive planning."

This chapter considers the application of adaptive planning in Inuit communities in the eastern arctic region of Canada (Nunavut), where climate change is one of many stressors. Early and significant climate changes are expected in the Arctic – including sea ice reduction, permafrost degradation, more hazardous weather, and alterations in biological systems – and impacts are feared to be disproportionately high due to reliance on climate-sensitive resources (ACIA 2005; IPCC 2007; Furgal and Prowse 2008). The fact that climate change may provide opportunity (e.g., easier access for resource exploitation) as well as adversity (e.g., amplifying pressures on traditional livelihoods and culture) highlights the need for measured and strategic responses that allow communities to capitalize on benefits and reduce risks.

Historically, Inuit groups relied directly on the land and survived with their exceptional skill in capturing and using scant resources efficiently. They adapted to seasonal and interannual variability in their environment by pursuing a nomadic lifestyle that followed natural rhythms and incorporated flexibility into harvesting patterns (Berkes and Jolly 2001; Bennett and Rowley 2004).

As recently as the 1950s and 1960s, the federal policy to centralize basic services drew Inuit off the land and into sedentary life. Ever since, arctic hamlets have depended largely on resource transfers from higher levels of government for their subsistence. In 1999, a vast area of land claimed by Inuit became an official

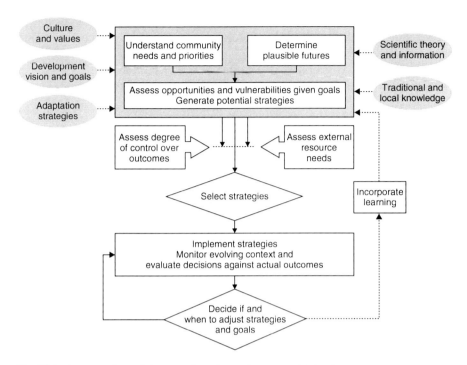

Fig. 34.1 A process for anticipatory adaptation at the community level

Canadian territory, now known as Nunavut, with a unique system of governance that inserts complementary governmental organizations to mind Inuit-specific interests in addition to general public administrative functions. Despite this achievement of self-determination and substantial improvements in many areas of community life, challenges related to health, housing, and education persist (Jenness 1961; Lotz 1970; Damas 2002; Berger 2006). Therefore, climate change responses will have to be part of a multi-stress adaptation strategy.

Adaptation to environmental and social change in this contemporary situation requires skills and knowledge that are novel to traditional Inuit culture, such as lobbying for resources, understanding the drivers and impacts of complex and distant global processes, and conducting abstract analyses of risks and trade-offs. It also requires decision-aiding approaches and institutional settings that are unfamiliar to traditional adaptive planning frameworks. In the absence of these, successful adaptive planning in Nunavut is unlikely. We suspect the existence of similar challenges in other marginalized communities.

The case of arctic communities is thus an opportunity to explore questions that must be considered in attempts to implement anticipatory adaptation in nonconventional settings. These questions include:

- *Priorities and agenda*: How are they defined? Where is climate change in the hierarchy of imperatives? Are communities empowered to act?

- *Financial resources*: Who controls these? How are funds allocated? Can climate change be reframed as representing opportunities for communities to access resources?
- *Information and knowledge*: Is there sufficient information about climate change, baseline conditions, and potential local impacts? How do communities use this information?
- *Culture and values*: How is one adaptation option chosen over another? Can there be deep cultural differences that render conventional adaptive planning inappropriate?
- *Mainstreaming adaptation*: Can existing institutions and planning processes adopt anticipatory adaptation? What support is needed?

Priorities and Agenda

Awareness of climate change is strong in Nunavut. For decades, researchers have been traveling to high latitudes to study the climate phenomenon. Local Inuit, embodying the oral history of generations, have offered their observations of climatic abnormalities to numerous studies (e.g., Fox 1998; Krupnik and Jolly 2002; Nickels et al. 2005; Ford 2006; Laidler 2007). Trusted Inuit leaders, such as Sheila Watt-Cloutier and Mary Simon, have even brought regional concerns of climate change and cultural impacts to international attention.

Climate change also figures prominently in the political rhetoric. Nunavut was the first of the provinces and territories in Canada to sign on to the federal climate change program and develop its own Climate Change Strategy in 2003 (George 2003). Conforming to international and national agendas, the policy's main objectives are to: control and reduce greenhouse gas emissions, identify and monitor climate change impacts, and to develop adaptation strategies (GN 2003).

These objectives are beginning to penetrate territorial policies and sector strategies, but attention to planning and adaptation at the community level in Nunavut is in its infancy. There are some modest initiatives to reduce risks to infrastructure and services within the capital city of Iqaluit. A workshop and conference have been held to share knowledge of impacts and adaptation options, and to identify planning needs. As well, an associated pilot project engaged volunteer consultants to develop community climate adaptation plans for two communities (see www.planningforclimatechange.ca, Arvai and Gregory 2007; Baksh and Render 2008).

Most Inuit perceive the greatest impacts from climate change are to harvesting activities; at present they are forced to adjust hunting patterns and equipment to accommodate seasonal and ice abnormalities (Ford et al. 2008). Elders feel particularly deeply the erosion of traditional knowledge, spiritual well-being and identity that occurs as the familiar "cultural landscape" morphs (Nelson 2003; see p. 6 of Ehrlich and Sian 2004 for definition).

In the larger picture, however, climate change ranks behind much more critical challenges facing communities. Sixty years of development pressures and
lagging investment in the needs of growing communities has resulted in dire
economic and social conditions. As Justice Berger, reporting on the fulfillment
of federal fiduciary responsibilities within the Nunavut Land Claim Agreement,
put it:

> Imagine the odds faced by a student attempting to do homework with 12 or 13 other people
> in the house ... [V]irtually every home has at least one resident smoker; oil heating may
> produce carbon monoxide and other pollutants. The fact that even one quarter of Inuit
> students graduate from high school is, under the circumstances, a testament to the tenacity
> of those students, their parents, and their communities. (Berger 2006, p. vi)

Our own efforts to understand local perspectives corroborate that, although
notable improvements have been made, the overriding priority remains meeting
basic needs (health, housing, and education/training). The imperative to expand
job opportunities for the burgeoning population of youth entering the workforce is
evident to all, even (if not especially) to elders who are preoccupied with ensuring
the well-being of future generations. Job opportunities are not to be achieved at
all costs, however; Inuit seek development solutions that allow them to maintain
a mixed economy and a distinct cultural identity (SEDS 2003; GN 2004). Thus,
it is not that climate change is unimportant to Nunavut communities, but that it is
eclipsed by more pressing concerns.

The tension between immediate needs and national climate change programs is
illustrated by a recent project to design energy-efficient public housing, through
a partnership between the Nunavut Housing Corporation, Infrastructure Canada,
and the Canadian Mortgage and Housing Corporation. A success by accounts, this
project constructed 70 units in its 2005 inaugural year (CMHC 2006). Yet, a report
in 2004 estimated that 3,300 units were needed now to alleviate critical shortages,
and 250 per year thereafter (ITK 2004). Only a large-scale deployment of these
new housing units could simultaneously address overcrowding and unemployment
(through construction contracts), as well as reduce energy use and greenhouse gas
emissions. Incremental actions also mean that, in order to receive the investment
they feel they deserve, each community must rely on being relatively worse off than
its fellows or more effective at lobbying to capture more of the allotted investment.
This situation promotes a moral hazard, rewarding the community most impacted
by failing to adapt to climate change.

Furthermore, the ability of communities to act on their own behalf is restricted.
In addition to the fact that many forces of change are outside local control (e.g.,
climate, long-range pollutants, globalization), hamlet governments are constrained
by insufficient jurisdictional authority. As a case in point, key local actors ranked
social and economic development action items as part of the Nunavut Economic
Development Strategy implementation. The exercise illustrated that community
control over priorities, whether they were sensitive to climate change or not, was
inversely related to their perceived importance (Boyle and Dowlatabadi 2005).

Moreover, several generations of top-down control and government intervention has undermined efforts to build the capacity of communities to manage their own affairs.

Financial Resources

Initiatives dedicated to climate change command a tiny proportion of the territorial budget. For example, the recently formed Nunavut Climate Centre, supported by one staff person, resides within a corner of the Department of Environment. The entire department receives only 2% of the governmental budget (GN 2008). If climate change is perceived as the key environmental stress, how can such small budgets adequately serve the needs of 26 communities?

Nor are the communities in a position to foot the bill, since they are fully dependent on external financial resource flows. Local governments in Nunavut (except Iqaluit) are not tax based and generate almost no revenue. There are few economic development opportunities, and most of these (e.g., fur products, tourism, Inuit art) are tied to international markets that can be fickle in demand. The most lucrative opportunity by far is mining, which historically has not left communities with enduring benefits (though government and industry efforts are working to improve this situation in the future).

Both cause and consequence of local economic conditions, nearly all essential public services and programs in communities are funded or delivered directly by the government or by Inuit organizations. As one example, over half of all dwellings in the territory are public housing units (OAG 2008). In general, the territory allocates resources among communities according to territorial funding cycles and centralized assessments of relative need. Nunavut itself relies on federal transfers, amounting to 92% of the territorial government's yearly budget, and a total of more than a billion dollars paid out to Inuit organizations in accordance with the Nunavut Land Claim Agreement (over a period of 15 years starting in 1999) (Government of Canada 1993; GN 2008).

Although communities could access a number of funding programs from all levels of government and other sources, there is, in effect, a perpetual shortage of resources to realize local initiatives. Criteria for eligibility and allocation of funds are frequently restrictive. Multiyear or seed funding is rare; usually community initiatives must demonstrate success and reapply each year for money to run programs and pay staff. Sometimes expected funds are reallocated to other programs or are discontinued. This system of piecemeal funding requires expertise in finding potential resources, writing proposals, and managing finances that may not be available in local organizations or even the community as a whole. The skills needed for quilting together the coordinated and consistent funding needed to implement long-term strategic plans are rarely present. Furthermore, ephemeral resources and programs do not constitute a stable platform from which to build capacity for long-term adaptation.

Information and Knowledge

Availability of information on climate, impacts, and adaptation options in Nunavut has increased dramatically in recent years. Qualitative studies on local observations of climate change are easily accessible and are most relevant to people in communities. This information documents past norms and benchmarks future climate change. When climate anomalies fall within known variability (established through oral history or scientific records), traditional knowledge provides insights about the possible impacts and how to respond. But estimating the impacts and effective responses to climate change beyond past experiences requires an understanding of local ecosystems and biogeochemistry beyond that currently at hand.

While model predictions of future climate are also available, they are rarely accessed or used in decision-making by local governments. Such predictions are reported at broad time and spatial scales and the uncertainties regarding if, when, and how impacts may manifest render the information too abstract for local planning purposes.

Unfortunately, socioeconomic and other determinants of vulnerability to climate change in Nunavut are not well characterized. This is both a matter of poor data coverage and characterization of key variables in terms that do not reflect their functional relevance locally. For example, the definition of a census "family" does not fit well with the Inuit notion of shared households and extended kin, and confounds the interpretation of related statistics. Data on land-based activities, which are pertinent to Inuit, are collected only at 10-year intervals and reported only at the territorial scale (see Boyle and Dowlatabadi 2006).

At the time of writing, the Nunavut Bureau of Statistics does not have the capacity to collect data, only to package and redistribute Nunavut-specific data collected by Statistics Canada. However, as noted above, these data often do not report on topics or indicators that are relevant to Nunavut communities. Several government departments collect socioeconomic data to varying degrees, but these are patchy, of varied quality, and often unavailable to the public for timely decision-making by communities.

While residents are well aware of socioeconomic conditions in their own communities and their relative severity in a qualitative sense, hamlet governments do not systematically gather data on residents. Periodic local level studies (e.g., academic research, project environmental assessments) are usually carried out for purposes that are not of importance to communities, and they are rarely repeated to examine trends. Yet, both governmental resource allocation mechanisms and long-term adaptive planning processes rely on availability of such information across Nunavut and over time.

Perhaps more important than the issues outlined above is Inuit mistrust of data and statistics. Scientific information represents a fundamentally different type of knowledge system than traditional knowledge. More damaging has been the lasting effect of real and perceived abuses in data collection and use by authorities and researchers in the past. In part because of these issues, many people prefer informal networks, the media, and trusted leaders as information sources.

Culture and Values

Conventional methods of decision-making and planning that are usually applied in developed countries grew out of a history of Western culture and science. As alluded to in the previous section, these methods embody a particular set of concepts and values that may not be well suited to Nunavut communities rooted in traditional Inuit culture.

Western and Inuit culture differ in their frameworks for understanding the world. Where Western science has pursued deconstruction and atomistic explanations, Inuit have holistic concepts and a highly contextual language for conveying their cosmology. For example, the Inuktitut word *sila* means weather; the spiritual force controlling the weather; outside; the great outdoors (the great beyond); it is also the root for the word for the universe (*silarjuaq*) and also is the root of the word for wisdom (*silatuniq*), which implies that the wise have taken something of the great outdoors into themselves (personal communication with J Bennett 9 June 2009). In contrast, Western science strives to differentiate weather from climate, emphasizing weather as an instantiation of the probability space described by climate. It is not surprising then to find that shared understanding of scientific concepts and methods used in decision-making and in translating scientific terminology meaningfully or consistently into aboriginal languages is challenging at best (Ellis 2005; Schuegraf and Fast 2005; Laidler 2006; Myers and Furgal 2006). Furgal et al. (2005), in attempting to communicate ideas related to contaminants in the arctic food chain, noted particular difficulties in communicating concepts such as "risk," "probabilities of risk," and "levels of safety." Productive discussions of priorities, values, and trade-offs are problematic within this cross-cultural context.

Assessing discounted benefits and costs and negotiating trade-offs is central to formal decision-making. However, these underlying rules and their implicit values may be too simplistic to be directly applicable for adaptive planning in Nunavut. For example, the number of caribou taken in a hunt depends on need (not want), season (hence lifecycle of caribou), respect for nature, and a host of other factors (Bennett and Rowley 2004). Echoes of this approach to maintaining livelihoods are transferred to modern economic decisions. In one exercise, when asked to choose between the set of opportunities and risks associated with two different development options – a pier to moor tourist cruise ships or a land-based hotel to host guests arriving by plane – community representatives chose "both." Their arguments were: if both are possible, why not have both? Why choose between facilities that create different opportunities? This example demonstrates our failure to state trade-offs in terms that are meaningful in Inuit setting and culture.

Anticipatory adaptation requires long time horizons. In livelihood decisions, the Inuit have exercised decisions over a harvest cycle by setting food caches for contingencies (Bennett and Rowley 2004). In kinship decisions, they have used multigenerational decision-making horizons to choose whom to marry, whose child to adopt, and so on. Thus, deliberated long-term decisions are not new to Inuit elders. However, translation of their traditional experiences into the domain of long-term planning for development and adaptation to climate change is still nascent.

Of course, Inuit values are not static; they have evolved and are evolving. Perhaps there will be greater comfort and competencies in employing formal long-term planning methods in future. However, given that Inuit culture has proven more resilient than predicted at the forefront of the development wave (Usher et al. 2003), traditional values are also likely to be integrated into the perspectives of future generations. The uncertainty in how this shifting baseline of Inuit values and aspirations will play out, however, exemplifies one of the most difficult challenges in the practice of anticipatory adaptation in general (Shepherd et al. 2006).

Mainstreaming Adaptation

Anticipatory adaptation processes are predicated on long-term strategic planning based on clear priorities and reliable resource flows, periodically modified by new information on social needs and environmental conditions (refer again to Fig. 34.1).

The chapter so far has revealed critical community needs, aggravated by climate change, and the ways in which lack of control over jurisdiction and resources forces hamlets to look to government and external sources to resolve many local concerns. Given, also, gaps in information, cultural differences, and planning mechanisms that mimic those in southern Canada, it is not at all surprising that the prerequisites for conventional strategic planning are not present in community planning processes (see Table 34.1 for examples). Investment in cultivating the specific skills and capacities necessary to implement adaptive planning will be constantly undermined until more basic challenges associated with planning are addressed.

The foregoing leads to the overall question of how best to tailor more broadly applied adaptation initiatives to meet the needs and context of marginalized communities. Adoption of a federal template will not be sufficient. In the case of Nunavut, nascent territorial government departments also need major injections of financial and human resources, expertise, and relevant data. Communication and coordination among departments needs to be expanded to the point of supporting multiagency, multi-sectoral planning.

In Nunavut, despite awareness of the importance of reducing risks and adapting to climate change, the current situation is at odds with the comprehensive and systematic planning and analysis that is required (Arvai and Gregory 2007; Baksh and Render 2008; Ford et al. 2007). Sustained support and efforts at effective capacity building by agencies dedicated to adaptation planning (e.g., Natural Resources Canada, Canadian Institute of Planners, Inuit Circumpolar Council, research institutes) are necessary for the time being.

Table 34.1 Challenges to implementing anticipatory adaptation into existing planning processes in Nunavut communities

Challenge	Examples
The need to constantly respond to crises and immediate concerns	• Keeping the community running smoothly, and the health and safety of residents takes priority.
	• Even basic services may not be easily available (e.g., electricians flown in for community repairs).
	• Events affect the whole community (e.g., flu epidemics, the passing of an elder).
High turnover rates and lack of institutional memory	• People in key planning positions change often. In our project, only one of five original contacts remained after 18 months (one position turned over twice).
	• Annual elections for Hamlet Council (for a 2-year position). Committees are reshuffled annually.
	• Average retention of employees for the Government of Nunavut is about 18 months.
Limited human resources	• Few people to carry out all the tasks required in running the community.
	• Skilled people become overwhelmed with participation in too many committees and initiatives.
	• Planning tasks are usually extraneous or marginal to full-time job responsibilities.
Use of information and analytical skills	• Preference for traditional knowledge and forms of decision-making.
	• Mistrust of data analysis and interpretation done by distant governments and external consultants/researchers.
	• Understanding/interpretation of key concepts (e.g., trade-off analysis, discounting, and risk) is not shared.
External planning structures and plans	• Most plans are tied to funding requirements from the government. Plans usually are written by external consultants; less than 50% of these plans are ever implemented.
	• The existing structure of governance is relatively new to communities. "Southern" planning models may be inconsistent with local culture and ways of decision-making.
Short planning horizons	• Budget allocations cover 1–3 years; community economic development plans usually look ahead 5 years. In comparison, mining companies plan 10–50 years ahead; the time horizon for climate change impacts stretches to 100 years.

Conclusion

Adaptation planning poses a scale challenge that is unfamiliar in climate mitigation planning. A global template for adaptation fails to reflect the context and multiple stresses confronting individual local communities. In addition, local governments do not have sufficient control over the agenda and resources to act upon their priorities, nor can agencies and community organizations rely on consistent funding and support. This situation forces reactive decision-making instead of long-term planning at the local scale. In Nunavut, the problem is exacerbated by at least two additional factors: (1) the human capacity and skills that are evolving to manage the territory are outpaced by emerging challenges and (2) long-term decision-making frameworks, and the science informing them, do not integrate a system of knowledge and values that can be shared by all parties.

However, using characteristic resourcefulness, Inuit and Nunavut as a whole are capturing new resources invested in climate change action to address long-standing priorities; cultural preservation and infrastructure improvements are prime examples. This approach reveals overall resilience, but also highlights the indispensable role that is currently played by higher levels of government and external entities that support communities with funding, human resources, and expertise.

Nonetheless, if adaptive planning is to succeed, significant investment will be necessary to build capacity and skills for planning in general, and for long-term strategic decision-making in particular. Changes to planning mechanisms and repair of jurisdictional fragmentation by the territorial government will be required to allow a more coordinated perspective at the local level. Nonetheless, long-term adaptive capacity for communities is contingent on their ability to regain control and become empowered to implement local initiatives and adaptation strategies so that they might realize their development goals.

Acknowledgments The authors are grateful for input from Dr. Susan Rowley, John Bennett, and anonymous reviewers. Editorial comments were provided by Eryn Kirkwood and Shane Roberts. This research was made possible through support from the Climate Decision Making Center (CDMC) located in the Department of Engineering and Public Policy. This Center has been created through a cooperative agreement between the National Science Foundation (SES-0345798) and Carnegie Mellon University. Research in Nunavut was supported by the Social Sciences and Humanities Research Council – Northern Research and Development Programme, Natural Resources Canada – Climate Change Impacts and Adaptations Programme, the Northern Scientific Training Programme, and the Oceans Management Research Network.

References

ACIA – Arctic Climate Impact Assessment (2005) Arctic climate impact assessment. Cambridge University Press, Cambridge

Adger WN (2003) Social capital, collective action, and adaptation to climate change. Econ Geogr 79(4):387–404

Allen K (2006) Community-based disaster preparedness and climate adaptation: local capacity-building in the Philippines. Disasters 30(1):81–101

Arvai J, Gregory R (2007) Final report: adaptation in Arctic communities. In: Nunavut climate change workshop, Government of Nunavut, Iqaluit, 6–8 Dec 2006

Baksh R, Render B (2008) Clyde River climate change adaptation action plan pilot project. Prepared for the Canadian Institute of Planners, March 2008

Barnett J (2001) Adapting to climate change in pacific island countries: the problem of uncertainty. World Dev 29(6):977–993

Bennett J, Rowley S (2004) Uqalurait: an oral history of Nunavut. McGill-Queen's University Press, Montreal

Berger T (2006) The Nunavut project: conciliator's final report. Nunavut Land Claims Agreement implementation contract negotiations for the second planning period 2003–2013

Berkes F, Jolly D (2001) Adapting to climate change: socio-ecological resilience in a Canadian western arctic community. Conserv Ecol 5(2), Article 18 [online]. www.consecol.org/vol5/iss2/art18. Cited 9 June 2009

Boyle M, Dowlatabadi H (2005) Sustainability planning in arctic resource communities. Presentation at Adapting to climate change in Canada 2005: understanding risks and building capacity, Natural Resources Canada, Montreal, 4–7 May 2005

Boyle M, Dowlatabadi H (2006) Socio-economic assessment and monitoring: a guide to collecting data for communities in Nunavut. Prepared for Nunavut Economic Developers Association

Canadian Mortgage and Housing Corporation [CMHC] (2006) Nunavut Housing Corporation 5-Plex: creating a template for affordable, energy-efficient housing in Canada's North. CMHC Housing Awards, Ottawa

Damas D (2002) Arctic migrants, Arctic villagers: the transformation of Inuit settlement in the Central Arctic. McGill-Queen's University Press, Montreal

Dowlatabadi H, Kandlikar M, Patwardhan A (1994) Exploring aggregate economic damage functions due to climate change. In: Proceedings of the Air and Waste Management Association international specialty conference: global climate change science, policy, and mitigation strategies, Phoenix, 5–8 April 1994

Ehrlich A, Sian S (2004) Cultural cumulative impact assessment in Canada's Far North. In: Proceedings of the 24th annual conference, International Association for Impact Assessment, Vancouver, 26–29 Apr 2004

Ellis S (2005) Meaningful consideration? A review of traditional knowledge in environmental decision making. Arctic 58(1):66–77

Fankhauser S, Smith J, Tol R (1999) Weathering climate change: some simple rules to guide adaptation decisions. Ecol Econ 30(1):67–78

Ford J (2006) Vulnerability to climate change in Arctic Canada (Nunavut). Dissertation, University of Guelph

Ford J, Pearce T, Smit B et al (2007) Reducing vulnerability to climate change in the Arctic: the case of Nunavut, Canada. Arctic 60(2):150–166

Ford J, Smit B, Wandel J et al (2008) Climate change in the Arctic: current and future vulnerability in two Inuit communities in Canada. Geogr J 174(1):45–62

Fox S (1998) Inuit knowledge of climate change. M.A. thesis, University of Waterloo

Furgal C, Prowse T (2008) Northern Canada. In: Lemmen D, Warren F, Lacroix J et al (eds) From impacts to adaptation: Canada in a changing climate 2007. Government of Canada, Ottawa

Furgal C, Powell S, Myers H (2005) Digesting messages about contaminants and country foods in the Canadian north: a review and recommendations for future research and action. Arctic 58(2):103–114

George J (2003) Nunavut signing on to Ottawa accord on climate change. Nunatsiaq News, 7 Nov 2003

GN (2004) Pinasuaqtavut: our commitment to building Nunavut's future. Legislative Assembly, Iqaluit

GN (2008) Main estimates 2008–9. Department of Finance, Iqaluit

Government of Canada (1993) Agreement between the Inuit of the Nunavut settlement area and her majesty the queen in right of Canada. The Minister of Indian Affairs and Northern Development and the Tungavik Federation of Nunavut, Ottawa

Government of Nunavut [GN] (2003) Nunavut climate change strategy. Government of Nunavut, Iqaluit

Intergovernmental Panel on Climate Change [IPCC] (2007) Climate change 2007: synthesis report – contribution of Working Group I, II and III to the Fourth Assessment Report of the Intergovernmental Panel on Climate Change. IPCC, Geneva

Inuit Tapiriit Kanatami [ITK] (2004) Backgrounder on Inuit and housing. Report for discussion at the Housing Sectoral Meeting, Ottawa, 24–25 Nov 2004

Jenness D (1961) The significance of communities and social capital in resource development in frontier regions. In: Proceedings of resources for tomorrow conference, vol 3. Department of Northern Affairs and Natural Resources Canada, Government of Canada, Ottawa

Kestikalo ECH (2008) Climate change and globalization in the Arctic: an integrated approach to vulnerability assessment. Earthscan, London

Krupnik I, Jolly D (eds) (2002) The Earth is faster now: Indigenous observations of arctic environment change. Arctic Research Consortium of the United States, Fairbanks

Laidler G (2006) Inuit and scientific perspectives on the relationship between sea ice and climate change: the ideal complement? Clim Change 78(2–4):407–444

Laidler G (2007) Ice, through Inuit eyes: characterizing the importance of sea ice processes, use, and change around three arctic communities. Dissertation, University of Toronto

Lotz J (1970) Northern realities. New Press, Toronto

Myers H, Furgal C (2006) Long-range transport of information: are arctic residents getting the message about contaminants? Arctic 59(1):47–60

Nelson O (2003) Climate change erodes Inuit knowledge, researchers say. Nunatsiaq News, 24 Jan 2003

Nickels S, Furgal C, Buell M et al (2005) Unikkaaqatigiit – putting the human face on climate change: perspectives from Inuit in Canada. Joint publication of Inuit Tapiriit Kanatami, Nasivvik Centre for Inuit Health and Changing Environments at Université Laval and the Ajunnginiq Centre at the National Aboriginal Health Organization, Ottawa

Office of the Auditor General [OAG] (2008) Report of the Auditor General of Canada to the Legislative Assembly of Nunavut: Nunavut Housing Corporation, Ottawa, May 2008

Schuegraf M, Fast H (2005) Sharing research findings in the North: experiences from the Western Arctic. Presentation at the oceans management research network national conference, Canada's oceans: research, management and the human dimension, Ottawa, 29 Sep–1 Oct 2005

Shepherd P, Tansey J, Dowlatabadi H (2006) Context matters: what shapes adaptation to water stress in the Okanagan? Clim Change 78(1):31–62

Sivummut Economic Development Strategy Group [SEDS] (2003) Nunavut economic development strategy: building a foundation for the future

Smit B, Pilifosova O (2001) Adaptation to climate change in the context of sustainable development and equity. In: McCarthy JJ, Canziani OF, Leary NA et al (eds) Climate change 2001: impacts, adaptation, and vulnerability – contribution of Working Group II to the Third Assessment Report to the Intergovernmental Panel on Climate Change. Cambridge University Press, Cambridge

Smit B, Burton I, Klein R et al (2000) An anatomy of adaptation to climate change and variability. Clim Change 45(1):223–251

Smith J, Ragland S, Pitts G (1996) A process for evaluating anticipatory adaptation measures for climate change. Water Air Soil Pollut 92(1–2):229–238

Smithers J, Smit B (1997) Human adaptation to climatic variability and change. Global Environ Change 7(2):129–146

Tompkins E, Adger WN (2004). Does adaptive management of natural resources enhance resilience to climate change?. Ecol Soc 9(2), Article 10 [online]. www.ecologyandsociety.org/vol9/iss2/art10. Cited 9 June 2009

Usher P, Duhaime G, Searles E (2003) The household as an economic unit in arctic aboriginal communities, and its measurement by means of a comprehensive survey. Soc Indic Res 61(2):175–202

Yohe G, Tol R (2002) Indicators for social and economic coping capacity – moving toward a working definition of adaptive capacity. Global Environ Change 12(1):25–40

Part VII
Future Directions

Chapter 35
Adaptation to Climate Change: Context, Status, and Prospects

Ian Burton

Abstract This chapter places the book in the context of the United Nations Framework Convention on Climate Change and describes how the developing countries were the leaders in recognizing the need for greater attention to adaptation. The developed countries are now beginning to move rapidly into adaptation and an explosion of interest is bringing to the fore a long list of unresolved questions about adaptation to climate change and its relation to development, and its appropriate management from the local to the national level and for different socioeconomic sectors and a wide variety of climate-related risks. This book is at the leading edge of a rapidly expanding literature with many questions remaining to be addressed.

Keywords Adaptation to climate change • Developed and developing countries • The evolution of adaptation • Prospects for adaptation

This Book

This book is the tip of an iceberg. It represents some of the first evidence in book form of the much larger volume of literature that is emerging and looks set to explode on the subject of adaptation to climate change in developed countries. But perhaps this iceberg metaphor is not so apt. Bodies of ice around the world are diminishing in size – ice sheets are getting thinner and glaciers are retreating. A better metaphor would be the tip of a volcano. There is beginning to be an eruption of hot literature, debate, policy innovations, and implementation related

I. Burton (✉)
Emeritus, Adaptation and Impacts Research Group (AIRG), Meteorological Service of Canada, Downsview, ON M3S 5T4, Canada

Emeritus, Institute for Environmental Studies, University of Toronto, Toronto, Canada
e-mail: Ian.Burton@ec.gc.ca

J.D. Ford and L. Berrang-Ford (eds.), *Climate Change Adaptation in Developed Nations:*   477
From Theory to Practice, Advances in Global Change Research 42,
DOI 10.1007/978-94-007-0567-8_35, © Springer Science+Business Media B.V. 2011

to adaptation in both developed and developing countries. The fact that this is now happening is exemplified in this book and in a similar edited volume on developing countries (Adger et al. 2009) and in other edited volumes (Leary et al. 2008a; Schipper and Burton 2009). Although the appearance of these books and the expansion of research is very welcome news, it is also the case that climate change has been on the international policy agenda since the Rio de Janeiro Conference of 1992 when the United Nations Framework Convention on Climate Change (UNFCCC, or "the Convention") was first opened for signature. Why has it taken nearly 20 years for adaptation to climate change to receive the attention from scientists, civil society, and policymakers that it so urgently needs? Why did it first become a matter of concern in developing countries? And now that it is emerging so strongly in developed countries, what is its current status, and more importantly what are its prospects? In short, what is the future of adaptation as an area of science and policy? Is there, or can there be an "adaptation science?" What sort of policy issue is adaptation? What departments of government should be involved? What kinds of professional expertise are required? How should adaptation be treated in schools and universities and how should curricula be structured? This book provides much evidence upon which we can begin to build some answers. This chapter draws upon the evidence in the book to suggest some ideas about the present status of adaptation and to explore future prospects. First, some background is provided in the form of recent history and context of adaptation. One final introductory point needs to be made. When in this chapter and elsewhere in the book the word "adaptation" is used by itself it can be taken to refer to adaptation to climate change, unless otherwise specified.

Recent Context: 1990–2005

Adaptation is a "loaded" word. That is to say, it comes with a lot of intellectual baggage. Some of this stems from its use in evolutionary (Darwinian) theory. The notion of the "survival of the fittest" was taken over by elite European social scientists and intellectuals and their political associates and applied to human societies in a way that could be used to help justify colonialism, imperialism, environmental determinism, and social and racial discrimination up to and including slavery, ethnic cleansing, and genocide. For this reason, in the more enlightened decades in the second half of the twentieth century, "adaptation" became a more or less unacceptable word – too loaded and with connotations that precluded its use without apology in the conventional social sciences.

It was surprising to many therefore when the negotiators of the UNFCCC chose the word "adaptation" to refer to all the many and varied responses to climate change other than the single "ultimate objective" – namely the control of greenhouse gas emissions in order to achieve "stabilization of greenhouse gas concentrations in the atmosphere at a level that would prevent dangerous anthropogenic interferences with the climate system" (UNFCCC 1992). The negotiators on the International

Negotiating Committee divided the potential responses to the climate change threat into a simple dichotomy – mitigation and adaptation. Mitigation – the control of greenhouse gas emissions and concentrations – was clearly the preferred option. Responding to climate risks by all and any other means was labeled as adaptation.

Why and how was this choice made? A large part of the explanation probably has to do with the recent history of international negotiations about two other issues – namely acid rain (precipitation) and ozone layer depletion. Both of these issues were seen primarily as pollution issues. The acid rain issue was finally put on a path to solution by agreement to curb the emissions of sulfur dioxide, both in Europe and in North America. Similarly, the depletion of the ozone layer was also put on a path to solution by a global agreement to curb and eventually eliminate the emission chlorofluorocarbons (CFCs and HCFCs) that were the known depleting agents. Thus, when these issues were closely followed by the emergence of yet another atmospheric issue (global warming or more correctly climate change), the negotiators and perhaps many of their scientific advisors assumed that the same sort of antipollution remedy should be applied and should be the "ultimate objective."

Observers who at an early stage pointed out the imbalance between mitigation and adaptation immediately had two issues to address. First, there was the often unspoken concern about the meaning of adaptation and its intellectual history and possible interpretation. Interestingly, the UNFCCC offers no definition of adaptation. Second, there was the idea that the proponents of adaptation were secretly trying to undermine the case for mitigation. "If we can adapt," went the supposed argument, "then why spend large amounts of money to restrict the use of fossil fuels and move to renewables?" For these and other reasons, adaptation struggled in the wings of the climate negotiations while efforts were made to reach agreement on targets and schedules for emission reductions, as exemplified in the Kyoto Protocol. Nevertheless the word "adaptation" will remain current if only because it is enshrined in the UNFCCC.

Developing Countries First

The Parties to the Convention most concerned about adaptation were the developing countries, especially those considered to be the most vulnerable among them. This includes the least developed (little capacity to adapt) and the small island developing states (in danger from sea level rise). These Parties were successful in persuading the developed countries to pledge some financial help. Article 4, Clause 4 of the Convention reads in part: "The developed country Parties . . . shall assist the developing country Parties that are particularly vulnerable to the adverse effects of climate change in meeting costs of adaptation to these adverse effects" (UNFCCC 1992).

Partly as a consequence of this promise, serious research on adaptation and the development of adaptation plans and programs began first in developing countries. At the outset, in the years following 1992, the developed countries continued to think that adaptation was an issue for the developing countries. It was assumed that climate change would take place relatively slowly and that the richer countries with greater adaptation capacity could take care of the problem in an incremental and responsive fashion as needed.

Developed Countries Waking Up

The initially complacent attitude of the developed countries is now largely reversed. As Moser writes (Chap. 3), we are now "entering the period of consequences." It is no longer acceptable to ignore the need for adaptation. There has been no single epiphany or magic moment, but among the many reasons for this transformation, four stand out as particularly persuasive. First, climate change has occurred at a faster rate than was anticipated at the time of the negotiations. In successive reports by the Intergovernmental Panel on Climate Change (IPCC), projections of the amount of warming have increased. Second, progress on the reduction of emissions of greenhouse gases has been extremely difficult to achieve in the international negotiations. Despite all the talk, emissions have continued to increase and atmospheric concentrations continue to rise. The pollution control or mitigation approach has not succeeded in the way that was initially hoped and intended. The virtual collapse of the negotiations at the 15th Conference of the Parties in Copenhagen in December 2009 was a final dramatic demonstration of the failure of mitigation. That is not to say that mitigation will never be achieved, but only that it will take much longer and be harder to accomplish, and that in the meantime, hopes that the increase in global mean surface temperature can be limited to 2°C are no longer being seriously entertained in the scientific community. Third, some impacts of climate change are now being observed and measured. This especially applies to impacts in natural ecosystems and the cryosphere – the retreat of glaciers, thinning of ice sheets, and the rapid reduction of the sea ice in the summer season in the Arctic Ocean. Fourth, there is growing evidence of the possible links between climate change and the magnitude of extreme weather events. While there is still much scientific uncertainty about the connection with specific events – the magnitude of tropical cyclones or the summer heat wave of 2003 in Western Europe, for example – these events are helping to change the view of climate change from a slowly creeping phenomenon to something that may have much shorter term and potentially catastrophic consequences.

The Cacophony of Adaptation Voices

In the early days of the adaptation debate when the concerns of the developing countries were dominant, a lot of emphasis was placed on the selection, design, and funding of specific adaptation projects. More recently, and especially as the developed countries are becoming more engaged, there is a move toward a wider and more encompassing view. Developed countries are beginning to design adaptation programs and strategies; and as the financial implications begin to emerge, a cacophony of voices can be heard, each emphasizing different aspects of the adaptation agenda, each according to their own needs and perceptions. These can be broadly grouped into the attributes of scale, risks, and sectors.

At the time of drafting and signing of the UNFCCC in 1992, it was assumed that adaptation would be a local, place-based activity. The mantra was "mitigation is global; adaptation is local." Many studies of vulnerability and adaptation in developing countries were directed to particular adaptation measures in particular places (Leary et al. 2008b), and this is reflected in a number of chapters in this volume (e.g., Willmott and Penny, Chap. 24). This perception has also evolved, and it is now widely accepted that place-based adaptation has to be assisted and guided by actions and policies at the national level and also at the sub-national (state or province) level, especially in large federal states such as the USA, Canada, and Australia (Report of the Expert Panel on Climate Change Adaptation 2009). This is reflected in most of the Chapters in Section I. Adaptation has also moved further into the international level beyond the question of financial help to developing countries. Concerns are growing about the international repercussions of the displacement of millions of people in coastal zones and impacts on international trade in climate-sensitive economic activities including agriculture. The question of scale is crucial in the emerging pattern of climate governance. What level of government is responsible for what kinds of adaptation? How will the costs of adaptation be allocated between the public and private sector and civil society, and the various levels of government, from local to regional, to national, to international? It will take time for these questions to be resolved and much more research and study is required. There are many voices advocating different positions and different interpretations around the questions of scale.

A second important dimension of adaptation is embodied in the question "Adaptation to what?" Increasingly, efforts to respond to this question are couched in terms of risks and hazards. Adaptation is required to respond to slowly mounting "chronic" risks such as sea level rise, ice sheet melting and glacier retreat, permafrost melt, and drought and desertification. Adaptation is also required to respond to more "acute" climate change-related risks such as floods, tropical cyclones, and other types of weather-related disaster risks. This is leading to a growing recognition of the needs and opportunities in bringing the fields of climate change adaptation and disaster risk management together in terms of research, practice, and policy. How this can best be achieved at different spatial scales is also a matter of much debate and contention.

A third salient dimension of adaptation pertains to sectors. As long as adaptation was thought of as a primarily place-based activity and dealing with some specific climate risk or impact then the broader question of sector-wide adaptation could be ignored or neglected. Now it is becoming recognized that no part of any economy or society or environment is immune to the effects of climate change. In 1992, the dominant impacts being considered were environmental or ecological – the effects on species and ecosystems. Then this broadened out to dimensions of a primarily social kind – health risks, effects on the poor and most vulnerable, on subsistence livelihoods, and the nonmarket parts of society. Now, the private economic sectors are coming strongly into the picture. This includes not only agriculture, forestry, and fisheries but also all parts of the built environment. The design codes and standards for public and private infrastructure (see Steenhof and Sparling, Chap. 17), as well as banking and commercial activities, may be increasingly disrupted and damaged by extreme events. One of the first to become acutely aware of the risks to themselves was the insurance industry (see Cook and Dowlatabadi, Chap. 18). Now, corporate board rooms across the world are hearing about plans to reduce exposure to climate change-related risks and to find better ways of exercising due diligence both in relation to shareholders and markets and clients.

The expression "cacophony of adaptation voices" perhaps now seems more justified in the light of this catalog of scales, risks, and sectors. And the opening volcano metaphor perhaps seems more apt. How then can progress be made on adaptation? This book lays out some of the topics that need to be addressed. Can they be addressed altogether or will this pattern of fragmented chaos remain indefinitely?

Prospects

It would be a brave and foolhardy person who would pretend to have a clear and integrated vision of the future of adaptation to climate change. There remains much uncertainty about how much adaptation will be required, where and by whom. It is also unclear how rapidly the projected climate change impacts will occur, and at what magnitude.

Nevertheless, some things can be said without much fear of contradiction. Clearly, ways will have to be found of managing adaptation at all scales from local to global. All countries and all sizes of communities from villages to mega cities must be involved. Similarly, adaptation to climate change must be incorporated into risk management and disaster risk reduction across a wide range of hazards. It is also important to factor adaptation into economic and sustainable development and environmental protection in rich countries, the rapidly emerging new economies, as well as the least developed.

More knowledge of adaptation measures, strategies, and policies is required. Can and should this be developed into a science of adaptation? Professional fields such as agronomy, water resources management, hydrology, engineering, public health, social science, planning, economics, and many others are necessarily involved. Is it possible or desirable to develop a community or practice or a professional expertise across so many fields to respond to the shared problem of climate change? Clearly many issues remain unresolved and many decisions remain to be made.

Adaptation Selection

An important difference between adaptation in a Darwinian sense and the adaptation process that humanity now faces is that natural selection took care of itself. It happened and worked on an unknowing and unaware diversity of species including *Homo sapiens*. Darwin was very knowledgeable about selective breeding. He knew that the enormous range of types of domesticated dogs, for example, had been produced by the exercise of human choice on wild populations. Natural selection was the opposite of plant and animal breeding or selection. We are now entering the period of consequences and this involves cultural selection – or more specifically, the conscious selection of adaptations. The chapters in this book give a glimpse of the wide range of choices and decisions involved. Can humanity develop a collective sense of judgment and responsibility to enable the world to start to develop in a climate-resilient and climate-sensitive way? Or will competitiveness and self-interest prevent the successful transformation that we face? In seeking to adapt to climate change, we face an existential choice as great as any before in the long history and prehistory of humanity. This is not a trivial book.

References

Adger WN, Lorenzoni I, O'Brien KL (eds) (2009) Adapting to climate change: thresholds, values, governance. Cambridge University Press, Cambridge

Leary N, Adejuwon J, Barros V et al (eds) (2008a) Climate change and adaptation. Earthscan, London

Leary N, Conde C, Kulkarni J et al (2008b) Climate change and vulnerability. Earthscan, London

Report of the Expert Panel on Climate Change Adaptation (2009) Adapting to climate change in Ontario: towards the design and implementation of a strategy and action plan. Ministry of the Environment, Queen's Printer, Toronto

Schipper L, Burton I (eds) (2009) The Earthscan reader on adaptation to climate change. Earthscan, London

UNFCCC (1992) The United Nations Framework Convention on Climate Change. www.unfccc.int. Cited 15 Sept 2010

Index